CAMBRIDGE TRACTS IN MATHEMATICS

General Editors

---

194 Mathematics of Two-Dimensional Turbulence

CAMBRIDGE TRACTS IN MATHEMATICS

A complete list of books in the series can be found at www.cambridge.org/mathematics.

Recent titles include the following:

165. Projective Differential Geometry Old and New. By V. Ovsienko and S. Tabachnikov
166. The Lévy Laplacian. By M. N. Feller
167. Poincaré Duality Algebras, Macaulay's Dual Systems, and Steenrod Operations. By D. Meyer and L. Smith
168. The Cube-A Window to Convex and Discrete Geometry. By C. Zong
169. Quantum Stochastic Processes and Noncommutative Geometry. By K. B. Sinha and D. Goswami
170. Polynomials and Vanishing Cycles. By M. Tibăr
171. Orbifolds and Stringy Topology. By A. Adem, J. Leida, and Y. Ruan
172. Rigid Cohomology. By B. Le Stum
173. Enumeration of Finite Groups. By S. R. Blackburn, P. M. Neumann, and G. Venkataraman
174. Forcing Idealized. By J. Zapletal
175. The Large Sieve and its Applications. By E. Kowalski
176. The Monster Group and Majorana Involutions. By A. A. Ivanov
177. A Higher-Dimensional Sieve Method. By H. G. Diamond, H. Halberstam, and W. F. Galway
178. Analysis in Positive Characteristic. By A. N. Kochubei
179. Dynamics of Linear Operators. By F. Bayart and É. Matheron
180. Synthetic Geometry of Manifolds. By A. Kock
181. Totally Positive Matrices. By A. Pinkus
182. Nonlinear Markov Processes and Kinetic Equations. By V. N. Kolokoltsov
183. Period Domains over Finite and p-adic Fields. By J.-F. Dat, S. Orlik, and M. Rapoport
184. Algebraic Theories. By J. Adámek, J. Rosický, and E. M. Vitale
185. Rigidity in Higher Rank Abelian Group Actions I: Introduction and Cocycle Problem. By A. Katok and V. Niţică
186. Dimensions, Embeddings, and Attractors. By J. C. Robinson
187. Convexity: An Analytic Viewpoint. By B. Simon
188. Modern Approaches to the Invariant Subspace Problem. By I. Chalendar and J. R. Partington
189. Nonlinear Perron Frobenius Theory. By B. Lemmens and R. Nussbaum
190. Jordan Structures in Geometry and Analysis. By C.-H. Chu
191. Malliavin Calculus for Lévy Processes and Infinite-Dimensional Brownian Motion. By H. Osswald
192. Normal Approximations with Malliavin Calculus. By I. Nourdin and G. Peccati
193. Distribution Modulo One and Diophantine Approximation. By Y. Bugeaud
194. Mathematics of Two-Dimensional Turbulence. By S. Kuksin and A. Shirikyan
195. A Universal Construction for R-free Groups. By I. Chiswell and T. Müller
196. The Theory of Hardy's Z-Function. By A. Ivić
197. Induced Representations of Locally Compact Groups. By E. Kaniuth and K. F. Taylor
198. Topics in Critical Point Theory. By K. Perera and M. Schechter
199. Combinatorics of Minuscule Representations. By R. M. Green

Mathematics of Two-Dimensional Turbulence

SERGEI KUKSIN
Ecole Polytechnique, Palaiseau

ARMEN SHIRIKYAN
Université de Cergy-Pontoise

CAMBRIDGE
UNIVERSITY PRESS

CAMBRIDGE UNIVERSITY PRESS
Cambridge, New York, Melbourne, Madrid, Cape Town,
Singapore, São Paulo, Delhi, Mexico City

Cambridge University Press
The Edinburgh Building, Cambridge CB2 8RU, UK

Published in the United States of America by Cambridge University Press, New York

www.cambridge.org
Information on this title: www.cambridge.org/9781107022829

First published 2012

Printed and bound in the United Kingdom by the MPG Books Group

A catalogue record for this publication is available from the British Library

Library of Congress Cataloguing in Publication data
Kuksin, Sergej B., 1955–
Mathematics of two-dimensional turbulence / Sergei Kuksin, Armen Shirikyan.
p. cm. – (Cambridge tracts in mathematics ; 194)
ISBN 978-1-107-02282-9 (hardback)
1. Hydrodynamics – Statistical methods. 2. Turbulence – Mathematics.
I. Shirikyan, Armen. II. Title.
QA911.K85 2012
532′.052701519 – dc23 2012024345

ISBN 978-1-107-02282-9 Hardback

To

Yulia, Nikita, Masha

and

Anna, Rafael, Gabriel

Contents

Preface

Equations and forces

Two-dimensional (2D) statistical hydrodynamics studies statistical properties of the velocity field $u(t,x)$ of a (imaginary) two-dimensional fluid satisfying the stochastic 2D Navier–Stokes equations

$$\begin{aligned} &\dot{u}(t,x) + \langle u, \nabla\rangle u - \nu\Delta u + \nabla p = f(t,x), \qquad \operatorname{div} u = 0, \\ &u = (u^1, u^2), \qquad x = (x_1, x_2). \end{aligned} \tag{1}$$

Here $\nu > 0$ is the kinematic viscosity, $u = u(t,x)$ is the velocity of the fluid, $p = p(t,x)$ is the pressure, and f is the density of an external force applied to the fluid. The space variable x belongs to a two-dimensional domain, which in this book is supposed to be bounded. Suitable boundary conditions are assumed. For example, one may consider the case when the domain is a rectangle $(0,a) \times (0,b)$, where a and b are positive numbers, and the equations are supplemented with periodic boundary conditions; that is to say, the space variable x belongs to the torus $\mathbb{R}^2/(a\mathbb{Z} \oplus b\mathbb{Z})$ (in the case of periodic boundary conditions we will assume that space-mean values of the force f and the solution u vanish). Equations (1) are *stochastic* in the sense that the initial condition $u_0 = u(0,x)$, or the force f, or both of them, are random, i.e., depend on a random parameter. So the solutions u are random vector fields. The task is to study various characteristics of u averaged in ensemble, or to study their properties which hold for most values of the random parameter. In this book, we assume that the force is random and refer the reader to [FMRT01] for a mathematical treatment of the Navier–Stokes equations with zero (or deterministic) force and random initial data.

The Reynolds number R of a random velocity field $u(t,x)$ is defined as

$$R = \frac{\langle\text{characteristic scale for } x\rangle \cdot \bigl(\mathbb{E}E(u)\bigr)^{1/2}}{\nu},$$

where $E(u) = \frac{1}{2}\int |u(x)|^{1/2}dx$ is the kinetic energy of the fluid and $\mathbb{E}$ denotes the average in ensemble. Since the forces we consider are smooth, then the solutions u of (1) are regular in x and their space-scale is of order one. So $R \sim \nu^{-1}(\mathbb{E}E(u))^{1/2}$. A velocity field u is called *turbulent* if $R \gg 1$. Turbulent solutions for (1) are of prime interest.

If the motion of a "physical" three-dimensional fluid is parallel to the (x_1, x_2)-plane and its velocity depends only on (x_1, x_2), i.e., $u = u(t, x_1, x_2)$ and $u^3 = 0$, then $(u^1, u^2)(t, x_1, x_2)$ satisfy (1). Such flows are called *two-dimensional*. Turbulent flows of real fluids are never two-dimensional (i.e., two-dimensional flows are never observed in experiments with high Reynolds number). Still, the 2D equations (1) and the 2D turbulence which they describe are now intensively studied by mathematicians, physicists, and engineers since, firstly, they appear in physics outside the realm of hydrodynamics (e.g., they describe flows of 2D films, see Figure 1 on p. xv), secondly, they provide a model[1] for the 3D Navier–Stokes equations and 3D turbulence, and, thirdly, the 3D statistical hydrodynamics in thin domains is approximately two-dimensional; see Section 6.1 of this book. Accordingly, two-dimensional statistical hydrodynamics is important for meteorology to model intermediate-scale flows in atmosphere (see Figures 6.1 and 6.2 in Chapter 6).

Statistical properties of the random force f are very important. It is natural and traditional to assume that

(a) the random field $f(t, x)$ is smooth in x, and
(b) it is stationary in t with rapidly decaying correlations.
If the space domain is unbounded, we should also assume that
(c) the space correlations of f decay rapidly.

However, (c) is not relevant for this book since we only consider flows in bounded domains.

In mathematics, the point of view[2] that turbulence in dimensions 2 and 3 should be described by the Navier–Stokes equations with a random force satisfying (a)–(c) goes back to A. N. Kolmogorov; see in [VF88]. Also see that book for some results on stochastic Navier–Stokes equations in the whole space $\mathbb{R}^d$, $d = 2$ or 3, with a random force satisfying (a)–(c).

We consider three classes of random forces:

Kick forces. These are random fields of the form

$$f(t, x) = h(x) + \sum_{k \in \mathbb{Z}} \delta(t - \tau_k)\eta_k(x), \tag{2}$$

[1] This model is not perfect since it is well known that the Navier–Stokes equations in dimensions 2 and 3 are very different. Still, it may be the best available now. Another popular model for the 3D Navier–Stokes system is the Burgers equation; see the review [BK07] by Bec and Khanin. For the stochastic 1D Burgers equation, see [Bor12].

[2] which is not at all a unique insight on turbulence!

where h is a smooth deterministic function, $\tau_k = k\tau$ with some $\tau > 0$, and $\{\eta_k\}$ are independent identically distributed random vector functions, which we assume to be divergence-free. For $t \in (\tau_{k-1}, \tau_k)$ (i.e., between two consecutive kicks) a solution $u(t, x)$ for (1), (2) satisfies the deterministic equations $(1)_{f=h}$, and at the time τ_k, when the kth kick $\eta_k(x)$ comes, it has an instant increment equal to that kick; see Section 2.3. The kick forces are singular in t and are not stationary in t, but statistically periodic (the difference between the two notions is not great if the time t is much longer than the period τ between the kicks). An advantage of this class of random forces is that the kicks η_k may have any statistics.

White in time forces. These are random fields of the form

$$f(t, x) = h(x) + \frac{d}{dt}\zeta(t, x), \tag{3}$$

where h is as above and $\zeta(t) = \zeta(t, \cdot)$ is a Wiener process in the space of smooth divergence-free vector functions. Such random fields are stationary and singular in t. A disadvantage is that they must be Gaussian; see Section 2.4.

Compound Poisson processes. These are kick forces (2) for which the periods $\tau_k - \tau_{k-1}$ between kicks are independent exponentially distributed random variables.

A big technical advantage of these three classes of random forces is that the corresponding solution $u(t, x)$, regarded as a random process $u(t, \cdot) =: u(t)$ in the space of vector fields, is a Markov process. At the time of writing it is not clear how to extend the results of this book to arbitrary smooth random forces f satisfying (a) and (b).

What is in this book?

We are concerned with basic problems and questions, interesting for physicists and engineers working in the theory of turbulence. Accordingly Chapters 3–5 (which form the main part of this book) end with sections where we explain the physical relevance of the obtained results. These sections also provide brief summaries of the corresponding chapters.

In Chapters 3 and 4, our main goal is to justify, for the 2D case, the statistical properties of a fluid's velocity field $u(t, x)$ which physicists assume in their work. We refer the reader to the books [Bat82; Fri95; Gal02], written in a sufficiently rigorous way and where the underlying assumptions are formulated in a clear manner.[3] The first postulate in the physical theory of turbulence is that

[3] Apart from a few pages at the end, the book [Bat82] is about 3D flows. But all discussions and most of the results may be literally translated to the 2D case.

statistical properties of a turbulent flow $u(t, x)$ converge, as time goes to infinity, to a statistical equilibrium independent of the initial data. Mathematically speaking, this means that a process $u(t, \cdot)$, defined by Eq. (1) in the space of vector fields, has a unique stationary measure, and every solution converges to this measure in distribution. That is, the law of the random field $x \mapsto u(t, x)$ (which is a time-dependent measure in a function space) converges, when $t \to \infty$, to the measure in question. Random processes possessing this property of "short-range memory" are said to be *mixing*.

In Chapter 3, we study the problem of convergence to a statistical equilibrium for Markov processes corresponding to equations with the three classes of random forces as above. We prove abstract theorems which establish the exponential mixing for certain classes of Markov processes. Next we show that these theorems apply to Eq. (1) if a random force f satisfies certain mild non-degeneracy assumptions. This establishes the convergence to a unique statistical equilibrium and proves that it is exponentially fast.

If the viscosity ν and the force f continuously depend on a parameter in such a way that the former stays positive and the latter stays non-degenerate, then the stationary measure continuously depends on this parameter. For any fixed initial data $u(0)$ the law of the corresponding solution $u(t)$ continuously depends on the parameter as well. In Section 4.3, we show that this continuity is uniform in time $t \geq 0$. That is, in two space dimensions the statistical hydrodynamics is stable, no matter how big the Reynolds number, whereas the "usual" hydrodynamics of large Reynolds numbers is very unstable.

The mixing has a number of important consequences, well-known in physics, but taken for granted there. Namely, consider any observable quantity $F(u)$, such as the first or second component of the velocity field $u = (u^1, u^2)$, or the energy $E = \frac{1}{2}\int |u|^2 dx$, or the enstrophy $\frac{1}{2}\int (\operatorname{curl} u)^2 dx$. Then $F(t) = F(u(t, \cdot))$ is an ergodic process. That is, *its time average converges to the ensemble average with respect to the stationary measure.* We show that the difference between the two mean values (in time and in ensemble) decays as $T^{-\gamma}$, where $\gamma < 1/2$ and T is the time of averaging; see Section 4.1.1. Next, if the ensemble average for an observable $F(u)$ vanishes, then the process $F(t)$ satisfies the central limit theorem: the law of the random variable

$$\frac{1}{\sqrt{T}} \int_0^T F(t)\, dt$$

converges, as $T \to \infty$, to a normal distribution $N(0, \sigma)$. For non-trivial observables F, the dispersion σ is strictly positive. In particular, for large T the random variables $T^{-1/2} \int_0^T u^j(t, x) dt$, $j = 1, 2$, are almost Gaussian. Physicists say that *on large time-scales a turbulent velocity field is approximately Gaussian.* These and some other related results are proved in Chapter 4.

In Chapter 5 we study velocity fields $u(t, x)$, corresponding to solutions of (1) with a force (2) or (3) where $h = 0$, when the viscosity ν is small and the Reynolds number is large. There we only discuss stationary measures and stationary-in-time solutions u_ν (i.e., solutions $u_\nu(t, x)$ such that the law $\mathcal{D}(u_\nu(t))$ for each t equals the stationary measure). First we observe that for a limit of order one to exist as $\nu \to 0$, the force f should be proportional to $\sqrt{\nu}$; see Section 5.2.4. So the equations read as

$$\dot{u}(t, x) + \langle u, \nabla \rangle u - \nu \Delta u + \nabla p = \sqrt{\nu}\, f(t, x), \quad \operatorname{div} u = 0,$$

where f is the force (2) or (3) with $h \equiv 0$. This is in sharp contrast with the 3D theory, where it is believed that a limit of order one exists for the original scaling (1), without the additional factor $\sqrt{\nu}$ on the right-hand side.[4] In that chapter we restrict ourselves to the case when the space domain is the square torus $\mathbb{T}^2 = \mathbb{R}^2/2\pi\mathbb{Z}^2$. The results remain true for the non-square tori $\mathbb{R}^2/(a\mathbb{Z} \oplus b\mathbb{Z})$, but the argument does not apply to the equations in a bounded domain with the Dirichlet boundary condition.

Denote by μ_ν the unique stationary measure. We show that the set of measures $\{\mu_\nu, 0 < \nu \leq 1\}$ is tight (i.e., relatively compact) and that any limit point $\mu_0 = \lim_{\nu_j \to 0} \mu_{\nu_j}$ is a non-trivial invariant measure for the Euler system

$$\dot{u}(t, x) + \langle u, \nabla \rangle u + \nabla p = 0, \quad \operatorname{div} u = 0.$$

It is supported by the set of divergence-free vector fields from the Sobolev space H^2 of order two. This result agrees well with the popular belief that *the Euler equation is "responsible" for 2D turbulence*. We do not know if a limiting measure μ_0 is unique, i.e., if $\mu_0 = \lim_{\nu \to 0} \mu_\nu$. But we know that the measures μ_ν satisfy, uniformly in $\nu > 0$, infinitely many algebraical relations, called the *balance relations*. These relations depend only on two scalar characteristics of the force f. This indicates some *universality features of 2D turbulence*. Such universality is another physical belief. In Section 5.1.3, we use the balance relations to prove that for any t and x the random variables $u_\nu(t, x)$ and $\operatorname{curl} u_\nu(t, x)$ have finite exponential moments uniformly in $\nu \geq 0$. In Section 5.2, we study further properties of the limiting measures μ_0. In particular, we establish that any μ_0 has no atoms and that its support is an infinite-dimensional set.

The results of Chapter 5 provide a foundation of the mathematical theory of space-periodic 2D turbulence. In Section 5.3, we discuss the relation of these results with the existing heuristic theory of 2D turbulence, originated by Batchelor and Kraichnan.

4 Note that for the small-viscosity Burgers equation the right scaling of the force is also trivial, i.e., without any additional factor; see [BK07; Bor12].

The difference between 2D turbulence and real physical 3D turbulence is very great. In Chapter 6, we discuss a few rigorous results on 3D turbulence, related to the 2D theory presented in the preceding sections. Namely, in Section 6.1 we discuss (without proof) the convergence of the statistical characteristics of a flow in a thin 3D layer, corresponding to the 3D Navier–Stokes system with a random kick force, to those of a 2D flow in the limiting 2D surface. In contrast with similar deterministic results, the convergence holds uniformly in time. So a class of *anisotropic 3D turbulent flows may be approximated by 2D flows* like those which we consider in our book. Section 6.2 contains a discussion of results due to Da Prato, Debussche, and Odasso, and Flandoli and Romito, showing that weak solutions of the stochastic 3D Navier–Stokes system perturbed by a white-in-time random force (which a priori are non-unique) may be arranged as a Markov process. This process is mixing if the force is rough as a function of the space variable. Finally, in Section 6.3, we invoke the methods of control theory to study further properties of stationary measures for Eqs. (1), (3).

Other equations

The abstract theorems from Chapters 3 and 4 and the methods developed there to study the solutions of Eq. (1) apply to many other stochastic equations. For instance, one can consider the stochastic complex Ginzburg–Landau equation with a conservative nonlinearity,

$$\dot{u} + i\Delta u - i|u|^{2m}u = \Delta u - u + f(t,x), \tag{4}$$

where $x \in \mathbb{T}^d$, $d \leq 3$. If $d = 1$ or 2, then $m \geq 0$, while if $d = 3$, then one can take, say, $m \in [0, 1]$. Such equations describe *optical turbulence*. If f is a bounded kick force, then direct analogues of the theorems in Chapters 3 and 4 remain true for (4) with the same proof.

However, if the force f is white in time, then the methods of Chapters 3 and 4 apply only to Eq. (4) with $m = 1$ if $d = 1$ and $m < 1$ if $d \geq 2$ (while the equation defines a good Markov process for any m as above). That is, for some deep reason, the arguments developed to treat the stochastic Navier–Stokes equations (1) with white-in-time forces apply only to PDEs with conservative nonlinearities of degree ≤ 3;[5] see Section 3.5.5.

Readers of this book

The book is aimed at mathematicians and physicists with some background in PDEs and in stochastic methods. Standard university courses on these subjects are sufficient since the book is provided with preliminaries on function

[5] But the method applies to Eq. (4) with $m > 1$ if we add a strong nonlinear damping $-|u|^{2m'}u$, $m' \geq m$, on the right-hand side.

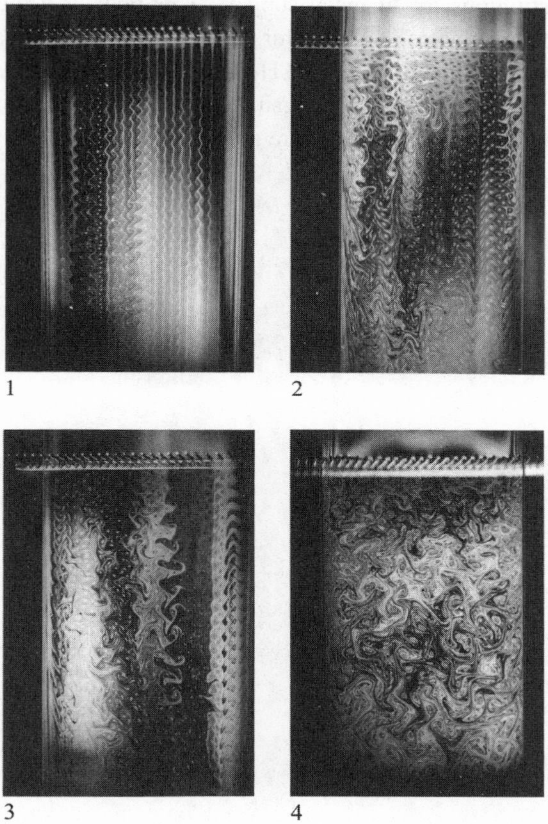

Figure 1 The onset of 2D turbulence. Panels 1–4 represent the down-motion of a soap film, punctured by a comb at the top. The Reynolds number is increasing from panel to panel. This is 2D turbulent motion described by the 2D Navier–Stokes system (1). Reprinted with permission from [Rut96]. Copyright 1996, American Institute of Physics.

spaces (Section 1.1), on the 2D Navier–Stokes equations (Chapter 2), and on stochastics (Sections 1.2 and 1.3). There the reader will find all the needed non-standard results.

Acknowledgements

Our interest in the stochastic 2D Navier–Stokes equations (1) and the problem of 2D turbulence appeared during research on the qualitative theory of randomly forced nonlinear PDEs which we undertook at Heriot–Watt University in Edinburgh during the years 1999–2002. The second author was a post-doc, supported by two EPSRC grants and by the university's funds. We are grateful to the two institutions for financial support. This book is a much extended version of the lecture notes [Kuk06a] for a course which the first author taught at the Mathematical Department of ETH-Zürich during the winter term of the year 2004/05. He thanks the Forschungsinstitut at ETH for their hospitality and for help in the preparation of the lecture notes.

1
Preliminaries

1.1 Function spaces

1.1.1 Functions of the space variables

Let Q be a domain in $\mathbb{R}^d$ (i.e., a connected open subset of $\mathbb{R}^d$) or the torus $\mathbb{T}^d = \mathbb{R}^d/2\pi\mathbb{Z}^d$. We shall say that a domain Q is *Lipschitz* if its boundary ∂Q is locally Lipschitz.[1] We shall need *Lebesgue* and *Sobolev* spaces on Q and some embedding and interpolation theorems.

Lebesgue spaces

We denote by $L^p(Q;\mathbb{R}^n)$, $1 \le p \le \infty$, the usual Lebesgue space of vector-valued functions and abbreviate $L^p(Q;\mathbb{R}) = L^p(Q)$. We write $\langle\cdot,\cdot\rangle$ for the L_2 scalar product and $|\cdot|_p$ for the standard norm in $L^p(Q;\mathbb{R}^n)$.

Sobolev spaces

We denote by $C_0^\infty(Q;\mathbb{R}^n)$ the space of infinitely smooth functions $\varphi : Q \to \mathbb{R}^n$ with compact support. Let u and v be two locally integrable scalar functions on Q and let $\alpha = (\alpha_1, \dots, \alpha_d)$ be a multi-index. We say that v *is the* α^{th} *weak partial derivative of* u if

$$\int_Q u D^\alpha \varphi \, dx = (-1)^{|\alpha|} \int_Q v\varphi \, dx \quad \text{for all } \varphi \in C_0^\infty(Q;\mathbb{R}),$$

where $|\alpha| := \alpha_1 + \cdots + \alpha_d$ and $D^\alpha = \partial_1^{\alpha_1} \cdots \partial_d^{\alpha_d}$. In this case, we write $D^\alpha u = v$.

Let $m \ge 0$ be an integer. The space $H^m(Q,\mathbb{R}^n)$ consists of all locally integrable functions $u : Q \to \mathbb{R}^n$ such that the derivative $D^\alpha u$ exists in the weak

[1] This means that ∂Q can be represented locally as the graph of a Lipschitz function.

sense for each multi-index α with $|\alpha| \le m$ and belongs to $L^2(Q;\mathbb{R}^n)$. We write $H^m(Q;\mathbb{R}) = H^m(Q)$ and define the norm in $H^m(Q;\mathbb{R}^n)$ as

$$\|u\|_m := \Big(\sum_{|\alpha|\le m} |D^\alpha u|_2^2\Big)^{1/2}.$$

In the case $Q = \mathbb{T}^d$, it is easy to define $H^m(\mathbb{T}^d;\mathbb{R}^n)$ for all $m \in \mathbb{R}$. To this end, let us expand a function $u \in L^2(\mathbb{T}^d,\mathbb{R}^n)$ in a Fourier series:

$$u(x) = \sum_{s\in\mathbb{Z}^d} u_s e^{isx}.$$

Define the following norm, which is equivalent to $\|\cdot\|_m$ for non-negative integers m:

$$\|u\|_m = \Big(\sum_{s\in\mathbb{Z}^d} \big(1+|s|^2\big)^m |u_s|^2\Big)^{1/2}. \tag{1.1}$$

The space $H^m(\mathbb{T}^d;\mathbb{R}^n)$ is defined as the closure of $C^\infty(\mathbb{T}^d,\mathbb{R}^n)$ with respect to the norm $\|\cdot\|_m$. It is easy to see that if $m \ge 0$ is an integer, then the two definitions of $H^m(\mathbb{T}^d;\mathbb{R}^n)$ give the same function space. The following result is a simple consequence of the definition of $\|\cdot\|_m$.

Lemma 1.1.1 *For any $m \in \mathbb{R}$ and any multi-index α, the linear map D^α is continuous from $H^m(\mathbb{T}^d;\mathbb{R}^n)$ to $H^{m-|\alpha|}(\mathbb{T}^d;\mathbb{R}^n)$. Accordingly, the Laplace operator $\Delta : H^m(\mathbb{T}^d;\mathbb{R}^n) \to H^{m-2}(\mathbb{T}^d;\mathbb{R}^n)$ is continuous. Similar assertions are true for any open domain $Q \subset \mathbb{R}^d$ and any integer $m \ge 0$.*

Now let $u \in H^m(\mathbb{T}^d;\mathbb{R}^n)$ be a function with zero mean value, that is,

$$\langle u\rangle := (2\pi)^{-d}\int_{\mathbb{T}^d} u(x)\,dx = 0, \tag{1.2}$$

where the integral is understood in the sense of the theory of distributions if $m < 0$. In this case, the first Fourier coefficient of u is zero, $u_0 = 0$, and therefore the norm

$$\|u\|_m = \Big(\sum_{s\ne 0} |s|^{2m} |u_s|^2\Big)^{1/2}$$

is equivalent to (1.1) on the space

$$\dot H^m(\mathbb{T}^d;\mathbb{R}^n) = \{u \in H^m(\mathbb{T}^d;\mathbb{R}^n) : \langle u\rangle = 0\}.$$

In particular, $\|u\|_1^2 = |\nabla u|_2$ is a norm on $\dot H^1(\mathbb{T}^d;\mathbb{R}^n)$.

Finally, let us define the *Sobolev space* $H^m(Q;\mathbb{R}^n)$ in a bounded Lipschitz domain $Q \subset \mathbb{R}^d$ for an arbitrary $m \ge 0$. Namely, without loss of generality, we can assume that $Q \subset \mathbb{T}^d$. We shall say that a function $u \in L^2(Q,\mathbb{R}^n)$ belongs to $H^m(Q;\mathbb{R}^n)$ if there is a function $\tilde u \in H^m(\mathbb{T}^d;\mathbb{R}^n)$ whose restriction to Q coincides with u. In this case, we define $\|u\|_m$ as the infimum of $\|\tilde u\|_m$ over all possible extensions $\tilde u \in H^m(\mathbb{T}^d;\mathbb{R}^n)$ for u.

Property 1.1.2 *Sobolev embeddings*. Let Q be either a Lipschitz domain in $\mathbb{R}^d$ or the torus $\mathbb{T}^d$.

1. If $m \le \frac{d}{2}$ and $2 \le q \le \frac{2d}{d-2m}$, $q < \infty$, then
$$H^m(Q;\mathbb{R}^n) \subset L^q(Q;\mathbb{R}^n). \tag{1.3}$$
2. If $m \ge \frac{d}{2} + \alpha$ with $0 < \alpha < 1$, then
$$H^m(Q;\mathbb{R}^n) \subset C_b^\alpha(Q;\mathbb{R}^n), \tag{1.4}$$
where $C_b^\alpha(Q)$ denotes the space of functions that are bounded and Hölder continuous with exponent α. In particular, if $m > \frac{d}{2}$, then $H^m(Q;\mathbb{R}^n)$ is continuously embedded into the space $C_b(Q;\mathbb{R}^n)$ of bounded continuous functions.
3. If Q is bounded, then we have the compact embedding
$$H^{m_1}(Q;\mathbb{R}^n) \Subset H^{m_2}(Q;\mathbb{R}^n) \quad \text{for } m_1 > m_2. \tag{1.5}$$
It follows that embeddings (1.3) and (1.4) are compact for $q < \frac{2d}{d-2m}$ and $m > \frac{d}{2} + \alpha$, respectively.

Property 1.1.3 *Duality*. The spaces $H^m(\mathbb{T}^d;\mathbb{R}^n)$ and $H^{-m}(\mathbb{T}^d;\mathbb{R}^n)$ are dual with respect to the L^2-scalar product $\langle\cdot,\cdot\rangle$. That is,
$$\|u\|_m = \sup_v |\langle u, v\rangle| \quad \text{for any } u \in C^\infty(\mathbb{T}^d;\mathbb{R}^n), \tag{1.6}$$
where the supremum is taken over all $v \in C^\infty(\mathbb{T}^d;\mathbb{R}^n)$ such that $\|v\|_{-m} \le 1$. Relation (1.6) implies that the scalar product in L^2 extends to a continuous bilinear map from $H^m(\mathbb{T}^d;\mathbb{R}^n) \times H^{-m}(\mathbb{T}^d;\mathbb{R}^n)$ to $\mathbb{R}$.

Property 1.1.4 *Interpolation inequality*. Let $Q \subset \mathbb{R}^d$ be a Lipschitz domain, let $a < b$ be non-negative integers, and let $0 \le \theta \le 1$ be a constant. Then
$$\|u\|_{\theta a+(1-\theta)b} \le \|u\|_a^\theta \|u\|_b^{1-\theta} \quad \text{for any } u \in H^b(Q;\mathbb{R}^n). \tag{1.7}$$
In the case of the torus, inequality (1.7) holds for any real numbers $a < b$ and any $\theta \in [0,1]$.

Proof for the case of a torus We have
$$\begin{aligned}\|u\|^2_{\theta a+(1-\theta)b} &= \sum_{s\in\mathbb{Z}^d} \left(1+|s|^2\right)^{\theta a+(1-\theta)b} |u_s|^2\\ &= \sum_{s\in\mathbb{Z}^d} \Big(\left(1+|s|^2\right)^{\theta a}|u_s|^{2\theta}\Big)\Big(\left(1+|s|^2\right)^{(1-\theta)b}|u_s|^{2(1-\theta)}\Big)\\ &\le \Big(\sum_{s\in\mathbb{Z}^d}(1+|s|^2)^a|u_s|^2\Big)^\theta \Big(\sum_{s\in\mathbb{Z}^d}(1+|s|^2)^b|u_s|^2\Big)^{1-\theta},\end{aligned}$$
where we used Hölder's inequality in the last step. □

Example 1.1.5 Let Q be either a Lipschitz domain in $\mathbb{R}^2$ or the torus $\mathbb{T}^2$. Then the Sobolev embedding (1.3), with $m = 1/2$ and $q = 4$, and the interpolation inequality (1.6) with $a = 0$, $b = 1$, and $\theta = \frac{1}{2}$, imply that

$$|u|_4 \le C_1 \|u\|_{1/2} \le C_2 \sqrt{|u|_2 \|u\|_1} \quad \text{for any } u \in H^1(Q;\mathbb{R}^n). \tag{1.8}$$

This is *Ladyzhenskaya's inequality*.

A proof of Properties 1.1.2–1.1.4 can be found in [BIN79; Ste70; Tay97].

1.1.2 Functions of space and time variables

Solutions of the equations mentioned in the introduction are functions depending on the time t and the space variables x. We fix any $T > 0$ and view a solution $u(t, x)$ with $0 \le t \le T$ as a map

$$[0, T] \longrightarrow \text{“space of functions of } x\text{”}, \qquad t \mapsto u(t, \cdot).$$

Let us introduce the corresponding functional spaces.

For a Banach space X, we denote by $C(0, T; X)$ the space of continuous functions $u : [0, T] \to X$ and endow it with the norm

$$\|u\|_{C(0,T;X)} = \sup_{0\le t\le T} \|u(t)\|_X,$$

where $\|\cdot\|_X$ stands for the norm in X. We denote by $S(0, T; X)$ the space of functions of the form

$$u(t) = \sum_{k=1}^{N} u_k \mathbb{I}_{\Gamma_k}(t),$$

where $N \ge 1$ is an integer depending on the function, $u_k \in X$ are some vectors, Γ_k are Borel-measurable subsets of $[0, T]$ (see Section 1.2.1), and $\mathbb{I}_\Gamma$ stands for the indicator function of Γ. If X is separable, then for $p \in [1, \infty]$ define $L^p(0, T; X)$ as the completion of the space $S(0, T; X)$ with respect to the norm

$$\|u\|_{L^p(0,T;X)} = \begin{cases} \left(\int_0^T \|u(t)\|_X^p dt\right)^{1/p} & \text{for } 1 \le p < \infty, \\ \operatorname*{ess\,sup}_{0\le t\le T} \|u(t)\|_X & \text{for } p = \infty. \end{cases}$$

Note that, in view of Fubini's theorem, we have

$$L^p\big(0, T; L^p(Q;\mathbb{R}^n)\big) = L^p\big((0, T) \times Q;\mathbb{R}^n\big) \quad \text{for } p < \infty.$$

A more detailed discussion of these spaces can be found in [Lio69; Yos95].

We shall also need the space of continuous functions on an interval with range in a metric space. Namely, let $J \subset \mathbb{R}$ be a closed interval and let X be a Polish space, that is, a complete separable metric space with a distance dist_X.

We denote by $C(J;X)$ the space of continuous functions from J to X. When J is bounded, $C(J;X)$ is a Polish space with respect to the distance

$$\|u-v\|_{C(J;X)}=\max_{t\in J}\operatorname{dist}_X\bigl(u(t),v(t)\bigr).$$

In the case of an unbounded interval J, we endow $C(J;X)$ with the metric

$$\operatorname{dist}(u,v)=\sum_{k=1}^{\infty}2^{-k}\frac{\|u-v\|_{C(J_k;X)}}{1+\|u-v\|_{C(J_k;X)}},\tag{1.9}$$

where $J_k=J\cap[-k,k]$. Note that, for a sequence $\{u_j\}\subset C(J,X)$, we have $\operatorname{dist}(u_j,u)\to 0$ as $j\to\infty$ if and only if $\|u_j-u\|_{C(J_k;X)}$ for each k. That is, (1.9) is the *metric of uniform convergence on bounded intervals*. When $J=\mathbb{Z}$ (or J is a countable subset of $\mathbb{Z}$), formula (1.9) may be used to define a distance on X^J. This distance corresponds to the *Tikhonov topology* on X^J.

Exercise 1.1.6 Prove that if $J\subset\mathbb{R}$ is an unbounded closed interval, then $C(J;X)$ is a Polish space. Prove also that if X is a separable Banach space, then $C(J;X)$ is a separable Fréchet space.

1.2 Basic facts from measure theory

In this section, we first recall the concept of a σ-algebra, together with some related definitions, and formulate without proof three standard results on the passage to the limit under Lebesgue's integral. We next discuss various metrics on the space of probability measures on a Polish space and establish some results on (maximal) couplings of measures.

1.2.1 σ-algebras and measures

Let Ω be an arbitrary set and let $\mathcal{F}$ be a family of subsets of Ω. Recall that $\mathcal{F}$ is called a *σ-algebra* if it contains the sets $\varnothing$ and Ω, and is invariant under taking the complement and countable union of its elements. Any pair $(\Omega,\mathcal{F})$ is called a *measurable space*. If $(\Omega_i,\mathcal{F}_i)$, $i=1,2$, are measurable spaces, then a mapping $f:\Omega_1\to\Omega_2$ is said to be *measurable* if $f^{-1}(\Gamma)\in\mathcal{F}_1$ for any $\Gamma\in\mathcal{F}_2$. If μ is a (positive) measure on $(\Omega_1,\mathcal{F}_1)$, then its *image* under f is the measure $f_*(\mu)$ on $(\Omega_2,\mathcal{F}_2)$ defined by $f_*(\mu)(\Gamma)=\mu(f^{-1}(\Gamma))$ for any $\Gamma\in\mathcal{F}_2$. Note that f_* is a linear mapping on the space of positive measures:

$$f_*(c_1\mu_1+c_2\mu_2)=c_1f_*(\mu_1)+c_2f_*(\mu_2)\quad\text{for any } c_1,c_2\geq 0.$$

The *product* of two measurable spaces $(\Omega_i,\mathcal{F}_i)$, $i=1,2$, is defined as the set $\Omega_1\times\Omega_2$ endowed with the minimal σ-algebra $\mathcal{F}_1\otimes\mathcal{F}_2$ generated by subsets of the form $\Gamma_1\times\Gamma_2$ with $\Gamma_i\in\mathcal{F}_i$. The product of finitely or countably many σ-algebras is defined in a similar way.

Given a probability measure μ on a measurable space $(\Omega,\mathcal{F})$, we denote by $\mathcal{N}_\mu$ the family of subsets $A\subset\Omega$ such that $A\subset B$ for some $B\in\mathcal{F}$ with

$\mu(B) = 0$. A σ-algebra $\mathcal{F}$ is said to be *complete* with respect to a measure μ if it contains all sets from $\mathcal{N}_\mu$. The *completion of $\mathcal{F}$ with respect to μ* is defined as the minimal σ-algebra generated by $\mathcal{F} \cup \mathcal{N}_\mu$ and is denoted by $\mathcal{F}_\mu$. This is the minimal complete σ-algebra which contains $\mathcal{F}$. A subset $\Gamma \subset \Omega$ is said to be *universally measurable* if it belongs to $\mathcal{F}_\mu$ for any probability measure μ on $(\Omega, \mathcal{F})$. If μ is a measure on $(\Omega_1, \mathcal{F}_1)$, $\mathcal{F}_1$ is complete with respect to μ, and a map $f : \Omega_1 \to \Omega_2$ is a μ-almost sure limit of a sequence of measurable maps, then f is measurable.

Now let X be a Polish space, that is, a complete separable metric space. We denote by dist_X the metric on X. The *Borel σ-algebra* $\mathcal{B} = \mathcal{B}(X)$ is defined as the minimal σ-algebra containing all open subsets of X. The pair $(X, \mathcal{B}(X))$ is called a *measurable Polish space*. If X_1 and X_2 are Polish spaces, then a map $f : X_1 \to X_2$ is said to be *measurable* if $f^{-1}(\Gamma) \in \mathcal{B}(X_1)$ for any $\Gamma \in \mathcal{B}(X_2)$. In particular, a function $f : X \to \mathbb{R}$ is called *measurable* if it is measurable with respect to the Borel σ-algebras on X and $\mathbb{R}$. An important property of Polish spaces is that any probability measure on it is *regular*. Namely, *Ulam's theorem* says that, for any probability measure μ on a Polish space X and any $\varepsilon > 0$, there is a compact set $K \subset X$ such that $\mu(K) \geq 1 - \varepsilon$. A proof of this result can be found in [Dud02] (see theorem 7.1.4).

Recall that, for any probability measure $\mathbb{P}$ on a measurable space $(\Omega, \mathcal{F})$, the triple $(\Omega, \mathcal{F}, \mathbb{P})$ is called a probability space. A probability space $(\Omega, \mathcal{F}, \mathbb{P})$ is said to be *complete* if $\mathcal{F}_\mathbb{P} = \mathcal{F}$. We shall often consider a probability space together with a family $\{\mathcal{F}_t \subset \mathcal{F}\}$ of σ-algebras that depend on a parameter t varying either in $\mathbb{R}_+$ or in $\mathbb{Z}_+$. In this case, we shall always assume that $\mathcal{F}_t$ is non-decreasing with respect to t. The quadruple $(\Omega, \mathcal{F}, \mathcal{F}_t, \mathbb{P})$ is called a *filtered probability space*. We shall say that $(\Omega, \mathcal{F}, \mathcal{F}_t, \mathbb{P})$ satisfies the *usual hypotheses* if $(\Omega, \mathcal{F}, \mathbb{P})$ is complete and $\mathcal{F}_t$ contains all $\mathbb{P}$-null sets of $\mathcal{F}$.

If X is a Polish space, then an X-valued *random variable* is a measurable map ξ from a probability space $(\Omega, \mathcal{F}, \mathbb{P})$ into X. The *law* or the *distribution* of ξ is defined as the image of $\mathbb{P}$ under ξ and is denoted by $\mathcal{D}(\xi)$, i.e., $\mathcal{D}(\xi) = \xi_*(\mathbb{P})$. If we need to emphasise that the distribution of a random variable is considered with respect to a probability measure μ, then we write $\mathcal{D}_\mu(\xi)$. An X-valued *random process* is defined as a collection of a probability space $(\Omega, \mathcal{F}, \mathbb{P})$ and a family of X-valued random variables $\{\xi_t\}$ on Ω (where t varies in $\mathbb{R}_+$ or $\mathbb{Z}_+$). If the underlying probability space is equipped with a filtration $\mathcal{F}_t$, then we shall say that the process ξ_t is *adapted to* $\mathcal{F}_t$ if ξ_t is $\mathcal{F}_t$-measurable for any $t \geq 0$. Finally, a random process ξ_t defined on a filtered probability space $(\Omega, \mathcal{F}, \mathcal{F}_t, \mathbb{P})$ is said to be *progressively measurable* if for any $t \geq 0$ the map $(s, \omega) \mapsto \xi_s(\omega)$ from $[0, t] \times \Omega$ to X is measurable. It is clear that if t varies in $\mathbb{Z}_+$, then these two concepts coincide.

1.2.2 Convergence of integrals

In what follows, we shall systematically use well-known results on the passage to the limit under Lebesgue's integrals. For the reader's convenience, we state them here without proof, referring the reader to section 4.3 of [Dud02].

Let $(\Omega, \mathcal{F})$ be a measurable space, let μ be an arbitrary σ-finite measure on it (so $\mu(\Omega) \leq \infty$), and let $f_n : \Omega \to \mathbb{C}$ be a sequence of integrable functions. The following result, called *Lebesgue's theorem on dominated convergence*, gives a sufficient condition for the convergence of the integrals of f_n to that of the limit function.

Theorem 1.2.1 *Assume that $\{f_n\}_{n\geq 1}$ is a sequence of functions that converge μ-almost surely and satisfy the inequality*

$$|f_n(\omega)| \leq g(\omega) \quad \textit{for } \mu\textit{-almost every } \omega \in \Omega, \tag{1.10}$$

where $g : \Omega \to \mathbb{R}_+$ is a μ-integrable function. Then

$$\lim_{n\to\infty} \int_\Omega f_n d\mu = \int_\Omega \left(\lim_{n\to\infty} f_n\right) d\mu. \tag{1.11}$$

In the case when the functions f_n are real-valued and form a monotone sequence, the bound (1.10) can be replaced by a weaker condition, which a posteriori turns out to be equivalent to the former. Namely, we have the following *monotone convergence theorem*.

Theorem 1.2.2 *Let $f_n : \Omega \to \mathbb{R}$ be a non-decreasing* (*or non-increasing*) *sequence that converges μ-almost surely and satisfies the condition*

$$\sup_{n\geq 1} \left| \int_\Omega f_n d\mu \right| < \infty.$$

Then relation (1.11) *holds.*

Finally, the following result, called *Fatou's lemma*, is useful when estimating the integral of the limit for a sequence of non-negative functions.

Theorem 1.2.3 *Let $f_n : \Omega \to \mathbb{R}_+$ be an arbitrary sequence of μ-integrable functions. Then*

$$\int_\Omega \left(\liminf_{n\to\infty} f_n\right) d\mu \leq \liminf_{n\to\infty} \int_\Omega f_n d\mu.$$

In particular, the three theorems above apply if Ω is the set $\mathbb{N}$ of non-negative integers with the counting measure. In this case, they describe passage to the limit for sums of infinite series.

1.2.3 Metrics on the space of probabilities and convergence of measures

In what follows, we denote by X a Polish space with a metric d_X. Define $C_b(X)$ as the space of bounded continuous functions $f : X \to \mathbb{R}$ endowed with the norm

$$\|f\|_\infty = \sup_{u\in X} |f(u)|,$$

and denote by $L_b(X)$ the space of bounded Lipschitz functions on X. That is, of functions $f \in C_b(X)$ for which

$$\mathrm{Lip}(f) := \sup_{u_1,u_2\in X} \frac{|f(u_1) - f(u_2)|}{\mathrm{dist}_X(u_1, u_2)} < \infty.$$

The space $L_b(X)$ is endowed with the norm

$$\|f\|_L = \|f\|_\infty + \mathrm{Lip}(f).$$

Note that $C_b(X)$ and $L_b(X)$ are Banach spaces with respect to the corresponding norms. The following exercise summarises some further properties of these spaces.

Exercise 1.2.4 Let X be a Polish space.

(i) Prove that $C_b(X)$ is separable if and only if X is compact.
(ii) Prove that $L_b(X)$ is not separable for the space $X = [0, 1]$ with the usual metric.

Hint: To prove that $C_b(X)$ is separable for a compact metric space X, use the existence of a finite ε-net and a partition of unity on X. To show that if X is not compact, then $C_b(X)$ is not separable, use the existence of a sequence $\{x_k\} \subset X$ such that $\mathrm{dist}_X(x_k, x_m) \geq \varepsilon > 0$. Finally, to prove (ii), construct a continuum $\{\varphi_\alpha\} \subset L^\infty(X)$ such that the distance between any two functions is equal to 1, and use the integrals of φ_α.

Let us denote by $\mathcal{P}(X)$ the set of probability measures on $(X, \mathcal{B}(X))$ and by $\mathcal{P}_1(X)$ the subset of those measures $\mu \in \mathcal{P}(X)$ for which

$$\mathfrak{m}_1(\mu) := \int_X \mathrm{dist}_X(u, u_0)\, \mu(du) < \infty, \tag{1.12}$$

where $u_0 \in X$ is an arbitrary point. The triangle inequality implies that the class $\mathcal{P}_1(X)$ does not depend on the choice of u_0. We shall need the following three metrics.

Total variation distance:

$$\|\mu_1 - \mu_2\|_{\mathrm{var}} := \frac{1}{2} \sup_{\substack{f \in C_b(X)\\ \|f\|_\infty \leq 1}} \big|(f, \mu_1) - (f, \mu_2)\big|, \quad \mu_1, \mu_2 \in \mathcal{P}(X). \tag{1.13}$$

This is the distance induced on $\mathcal{P}(X)$ by its embedding into the space dual to $C_b(X)$. It can be extended to probability measures on an *arbitrary* measurable space; see Remark 1.2.8 below.

Dual-Lipschitz distance:

$$\|\mu_1-\mu_2\|_L^* := \sup_{\substack{f\in L_b(X)\\ \|f\|_L\le 1}} \big|(f,\mu_1)-(f,\mu_2)\big|, \quad \mu_1,\mu_2\in\mathcal{P}(X). \tag{1.14}$$

This is the distance induced on $\mathcal{P}(X)$ by its embedding into the space dual to $L_b(X)$.

Kantorovich distance:

$$\|\mu_1-\mu_2\|_K := \sup_{\substack{f\in L_b(X)\\ \mathrm{Lip}(f)\le 1}} \big|(f,\mu_1)-(f,\mu_2)\big|, \quad \mu_1,\mu_2\in\mathcal{P}_1(X). \tag{1.15}$$

Exercise 1.2.5 Show that the symmetric functions (1.13)–(1.15) define metrics on the sets $\mathcal{P}(X)$ and $\mathcal{P}_1(X)$. *Hint:* The only non-trivial point is that if the measures μ_1 and μ_2 satisfy the relation $\|\mu_1-\mu_2\|_L^*=0$, then $\mu_1=\mu_2$. This can be done with the help of the monotone class technique; see Corollary A.1.3 in the appendix.

An immediate consequence of definitions (1.13)–(1.15) and the inequalities $\|f\|_\infty\le\|f\|_L$ and $\mathrm{Lip}(f)\le\|f\|_L$ is that

$$\|\mu_1-\mu_2\|_L^* \le 2\|\mu_1-\mu_2\|_{\mathrm{var}} \quad \text{for } \mu_1,\mu_2\in\mathcal{P}(X), \tag{1.16}$$

$$\|\mu_1-\mu_2\|_L^* \le \|\mu_1-\mu_2\|_K \quad \text{for } \mu_1,\mu_2\in\mathcal{P}_1(X). \tag{1.17}$$

Furthermore, if the space X is bounded, that is, there is an element $u_0\in X$ and a constant $d_0>0$ such that

$$\mathrm{dist}_X(u,u_0)\le d_0 \quad \text{for all } u\in X,$$

then, for any function $f\in C_b(X)$ vanishing at $u_0\in X$, we have

$$\|f\|_\infty\le d_0\,\mathrm{Lip}(f),$$

where the right-hand side may be infinite. It follows that in this case

$$\|\mu_1-\mu_2\|_K\le 2d_0\|\mu_1-\mu_2\|_{\mathrm{var}} \quad \text{for } \mu_1,\mu_2\in\mathcal{P}_1(X).$$

It turns out that the distance $\|\cdot\|_L^*$ is equivalent to the one obtained by replacing $L_b(X)$ in (1.14) with the space of bounded Hölder-continuous functions. Namely, for $\gamma\in(0,1)$ we denote by $C_b^\gamma(X)$ the space of continuous functions

$f : X \to \mathbb{R}$ such that

$$|f|_\gamma := \|f\|_\infty + \sup_{0<\mathrm{dist}_X(u,v)\le 1} \frac{|f(u)-f(v)|}{\mathrm{dist}_X(u,v)^\gamma} < \infty.$$

Let us set

$$\|\mu_1-\mu_2\|_\gamma^* := \sup_{\substack{f\in C_b^\gamma(X)\\ |f|_\gamma\le 1}} \big|(f,\mu_1)-(f,\mu_2)\big|, \quad \mu_1,\mu_2\in\mathcal{P}(X). \tag{1.18}$$

Proposition 1.2.6 *For any $\gamma\in(0,1)$ and $\mu_1,\mu_2\in\mathcal{P}(X)$, we have*

$$\|\mu_1-\mu_2\|_L^* \le \|\mu_1-\mu_2\|_\gamma^* \le 5\big(\|\mu_1-\mu_2\|_L^*\big)^{\frac{1}{2-\gamma}}.$$

Proof The lower bound of the inequality is obvious, and therefore we shall confine ourselves to the proof of the upper bound. For any continuous function $f : X \to \mathbb{R}$, we define an approximation for it by the relation

$$f_\varepsilon(u) = \inf_{v\in X}\big(\varepsilon^{-1}d(u,v)+f(v)\big), \quad u\in X, \tag{1.19}$$

where $\varepsilon>0$ is an arbitrary constant. It is a matter of direct verification to show that if $f\in C_b^\gamma(X)$ and $\|f\|_\gamma\le 1$, then

$$\|f_\varepsilon\|_L \le 1+\varepsilon^{-1}, \qquad 0\le f(u)-f_\varepsilon(u)\le \varepsilon^{\frac{1}{1-\gamma}} \quad \text{for } u\in X. \tag{1.20}$$

We now fix $\delta>0$ and find a function $f\in C_b^\gamma(X)$ with $\|f\|_\gamma\le 1$ such that

$$\|\mu_1-\mu_2\|_\gamma^* \le |(f,\mu_1)-(f,\mu_2)|+\delta. \tag{1.21}$$

It follows from (1.20) that, for any $\varepsilon>0$, we have

$$\begin{aligned}|(f,\mu_1)-(f,\mu_2)| &\le |(f_\varepsilon-f,\mu_1)|+|(f_\varepsilon-f,\mu_2)|+|(f_\varepsilon,\mu_1)-(f_\varepsilon,\mu_2)|\\ &\le 2\varepsilon^{\frac{1}{1-\gamma}}+\big(1+\varepsilon^{-1}\big)\|\mu_1-\mu_2\|_L^*.\end{aligned}$$

Choosing $\varepsilon=(\|\mu_1-\mu_2\|_L^*)^{\frac{1-\gamma}{2-\gamma}}$ and noting that $\|\mu_1-\mu_2\|_L^*\le 2$, we get

$$|(f,\mu_1)-(f,\mu_2)| \le 5\big(\|\mu_1-\mu_2\|_L^*\big)^{\frac{1}{2-\gamma}}.$$

Combining this with (1.21) and recalling that $\delta>0$ was arbitrary, we arrive at the required assertion. □

The following proposition gives an alternative description of the total variation distance and provides some formulas for calculating it.

Proposition 1.2.7 *For any $\mu_1,\mu_2\in\mathcal{P}(X)$, we have*

$$\|\mu_1-\mu_2\|_{\mathrm{var}} = \sup_{\Gamma\in\mathcal{B}(X)} |\mu_1(\Gamma)-\mu_2(\Gamma)|. \tag{1.22}$$

Furthermore, if μ_1 and μ_2 are absolutely continuous with respect to a given measure $m\in\mathcal{P}(X)$, then

$$\|\mu_1-\mu_2\|_{\mathrm{var}} = \frac{1}{2}\int_X \big|\rho_1(u)-\rho_2(u)\big|\,dm = 1-\int_X(\rho_1\wedge\rho_2)(u)\,dm, \tag{1.23}$$

where $\rho_i(u)$ is the density of μ_i with respect to m.

Remark 1.2.8 Let us note that a measure $m \in \mathcal{P}(X)$ with respect to which μ_1 and μ_2 are absolutely continuous always exists. For instance, we can take $m = \frac{1}{2}(\mu_1 + \mu_2)$. Furthermore, relation (1.22) enables one to extend the definition of the *total variation distance* to an arbitrary measurable space.

Proof of Proposition 1.2.7

Step 1. Let us denote by $\|\mu_1 - \mu_2\|'_{\text{var}}$ the right-hand side of (1.22) and show that relation (1.23) is true for it. Setting $\rho = \rho_1 \wedge \rho_2$ and integrating the obvious relation

$$\tfrac{1}{2}\,|\rho_1 - \rho_2| = \tfrac{1}{2}\,(\rho_1 + \rho_2) - \rho$$

over X with respect to m, we obtain the second equality in (1.23).

We now show that

$$\|\mu_1 - \mu_2\|'_{\text{var}} \le 1 - \int_X \rho(u)\,dm. \tag{1.24}$$

Let us define the set $Y = \{u \in X : \rho_1(u) > \rho_2(u)\}$. Since $\rho = \rho_2$ on Y, we have

$$\begin{aligned}\mu_1(\Gamma) - \mu_2(\Gamma) &= \int_\Gamma (\rho_1 - \rho_2)\,dm \le \int_{\Gamma\cap Y} (\rho_1 - \rho_2)\,dm\\ &= \int_{\Gamma\cap Y} (\rho_1 - \rho)\,dm \le \int_X (\rho_1 - \rho)\,dm = 1 - \int_X \rho(u)\,dm\end{aligned}$$

for any $\Gamma \in \mathcal{B}(X)$. By symmetry, this inequality implies (1.24).

To prove the converse inequality, we note that $\rho = \rho_1$ on Y^c and $\rho = \rho_2$ on Y. It follows that

$$\begin{aligned}\mu_1(Y) - \mu_2(Y) &= \int_Y (\rho_1 - \rho_2)\,dm\\ &= \Big(\int_Y \rho_1 dm + \int_{Y^c} \rho\,dm\Big) - \Big(\int_Y \rho_2 dm + \int_{Y^c} \rho\,dm\Big)\\ &= \Big(\int_Y \rho_1 dm + \int_{Y^c} \rho_1 dm\Big) - \Big(\int_Y \rho\,dm + \int_{Y^c} \rho\,dm\Big)\\ &= 1 - \int_X \rho\,dm.\end{aligned}$$

This completes the proof of (1.23) for $\|\mu_1 - \mu_2\|'_{\text{var}}$.

Step 2. We now prove (1.22). Using Step 1, for any $f \in C_b(X)$ with $\|f\|_\infty \le 1$, we derive

$$\big|(f, \mu_1) - (f, \mu_2)\big| \le \int_X \big|f(u)\big(\rho_1(u) - \rho_2(u)\big)\big|\,dm \le 2\,\|\mu_1 - \mu_2\|'_{\text{var}},$$

which implies that

$$\|\mu_1 - \mu_2\|_{\text{var}} \le \|\mu_1 - \mu_2\|'_{\text{var}}.$$

To establish the converse inequality, let us consider a function $f(u)$ that is equal to 1 on Y and to -1 on Y^c. We have

$$\begin{aligned}(f,\mu_1)-(f,\mu_2)&=\int_X f(u)\bigl(\rho_1(u)-\rho_2(u)\bigr)\,dm\\&=\int_X\bigl|\rho_1(u)-\rho_2(u)\bigr|\,dm=2\,\|\mu_1-\mu_2\|'_{\text{var}},\end{aligned}\tag{1.25}$$

where we used the first relation in (1.23). To complete the proof of (1.22), let us choose a sequence $f_n\in C_b(X)$ such that (see Exercise 1.2.12 below)

$$\|f_n\|_\infty\le 1\quad\text{for all } n\ge 1,$$
$$f_n(u)\to f(u)\quad\text{as } n\to\infty \text{ for } m\text{-a.e. } u\in X.$$

It is easy to see that the difference $(f_n,\mu_1)-(f_n,\mu_2)$ tends to the left-hand side of (1.25) as $n\to\infty$. This completes the proof of the proposition. □

Exercise 1.2.9 Analysing the proof of relation (1.23) show that it remains valid for any positive Borel measure m (that is, we no longer require that $m(X)=1$) with respect to which μ_1 and μ_2 are absolutely continuous.

Exercise 1.2.10 Let X be a Polish space and let $L^\infty(X)$ be the space of bounded measurable functions endowed with the norm $\|\cdot\|_\infty$. Prove that for any measures $\mu_1,\mu_2\in\mathcal{P}(X)$ we have (cf. (1.13))

$$\|\mu_1-\mu_2\|_{\text{var}}:=\frac{1}{2}\sup_{\substack{f\in L^\infty(X)\\ \|f\|_\infty\le 1}}\bigl|(f,\mu_1)-(f,\mu_2)\bigr|.\tag{1.26}$$

Two measures $\mu,\nu\in\mathcal{P}(X)$ are said to be *mutually singular* if there is a Borel subset $A\subset X$ such that $\mu(A)=1$ and $\nu(A)=0$. Relation (1.23) of Proposition 1.2.7 implies the following result.

Corollary 1.2.11 *Two measures $\mu,\nu\in\mathcal{P}(X)$ are mutually singular if and only if $\|\mu-\nu\|_{\text{var}}=1$.*

Exercise 1.2.12 Let X be a Polish space and let $m\in\mathcal{P}(X)$. Show that for any bounded measurable function $f\colon X\to\mathbb{R}$ there is a sequence of continuous functions that are uniformly bounded by $\|f\|_\infty$ and converge to f for m-almost all $u\in X$. *Hint:* It suffices to prove that f can be approximated by continuous functions in the space $L^1(X,m)$. To do this, show that any bounded measurable function can be approximated (in the sense of uniform convergence) by finite linear combinations of indicator functions and that the indicator function of any measurable set can be approximated (in the sense of convergence in $L^1(X,m)$) by bounded continuous functions. For the latter property, one could use Ulam's theorem on interior regularity of Borel measures; see Section 1.2.1.

Exercise 1.2.13

(i) Let $(\Omega_i, \mathcal{F}_i)$, $i = 1, 2$, be two measurable spaces and let $f : \Omega_1 \to \Omega_2$ be a measurable mapping. Prove that, for any measures μ_1, μ_2 on $(\Omega_1, \mathcal{F}_1)$,

$$\|f_*(\mu_1) - f_*(\mu_2)\|_{\text{var}} \le \|\mu_1 - \mu_2\|_{\text{var}};$$

see Remark 1.2.8 and relation (1.22) for the definition of the total variation distance in the case of measurable spaces.

(ii) Let X be a Polish space and let ξ_1, ξ_2 be two X-valued random variables defined on a probability space $(\Omega, \mathcal{F}, \mathbb{P})$ such that $\xi_1 = \xi_2 \circ \Phi$ almost surely, where $\Phi : \Omega \to \Omega$ is a measurable transformation. Show that

$$\|\mathcal{D}(\xi_1) - \mathcal{D}(\xi_2)\|_{\text{var}} \le \|\mathbb{P} - \Phi_*(\mathbb{P})\|_{\text{var}}.$$

Hint: Use (i) with $f = \xi_2$, $\mu_2 = \mathbb{P}$, and $\mu_1 = \Phi_*(\mathbb{P})$.

We shall say that a sequence $\{\mu_k\} \subset \mathcal{P}(X)$ *converges weakly to* $\mu \in \mathcal{P}(X)$ if

$$(f, \mu_k) \to (f, \mu) \quad \text{as } k \to \infty \tag{1.27}$$

for any $f \in C_b(X)$. In this case, we write $\mu_k \to \mu$. Recall that a family $\{\mu_\alpha, \alpha \in \mathcal{A}\} \subset \mathcal{P}(X)$ is said to be *tight* in X if for any $\varepsilon > 0$ there is a compact set $K_\varepsilon \subset X$ such that $\mu_\alpha(K_\varepsilon) \ge 1 - \varepsilon$ for any $\alpha \in \mathcal{A}$. In what follows, we shall need the following well-known result called *Prokhorov's theorem*, which gives a necessary and sufficient condition for the compactness of a family of measures in the weak topology. Its proof can be found in [Dud02, theorem 11.5.4].

Theorem 1.2.14 *Let X be a Polish space and let $\{\mu_\alpha, \alpha \in \mathcal{A}\}$ be a family of probability measures on X. Then $\{\mu_\alpha\}$ is relatively compact in the weak topology of $\mathcal{P}(X)$ if and only if it is tight.*

The following result of fundamental importance shows, in particular, that the weak convergence of measures is equivalent to the convergence in the dual-Lipschitz distance.

Theorem 1.2.15

(i) *The set $\mathcal{P}(X)$ endowed with the total variation distance is a complete metric space. Furthermore, a sequence $\{\mu_k\} \subset \mathcal{P}(X)$ converges to a measure μ in this space if and only if* (1.27) *holds uniformly in $f \in C_b(X)$ with $\|f\|_\infty \le 1$. In particular, $\mathcal{P}(X)$ is naturally embedded in the dual space of $C_b(X)$ as its closed subspace.*

(ii) *The set $\mathcal{P}(X)$ endowed with the dual-Lipschitz distance is a complete metric space. Furthermore, a sequence $\{\mu_k\} \subset \mathcal{P}(X)$ converges to a measure μ in this space if and only if either $\{\mu_k\}$ converges weakly to μ or* (1.27) *holds for any $f \in L_b(X)$.*

Proof Assertion (i) follows easily from the definition and basic properties of the metric $\|\cdot\|_{\mathrm{var}}$. Indeed, if $\{\mu_k\} \subset \mathcal{P}(X)$ is a Cauchy sequence, then by (1.22) for each Borel set $\Gamma \subset X$ there exists a limit $\lim_{k\to\infty} \mu_k(\Gamma) =: \mu(\Gamma)$. If we show that μ is a probability measure, then it will be the limit of $\{\mu_k\}$ for the total variation distance.

It is obvious that the mapping $\Gamma \mapsto \mu(\Gamma)$ is additive and that $\mu(\Gamma) = 1$. To complete the proof of (i), it remains to establish the σ-additivity of μ. To this end, it suffices to verify that, for any decreasing sequence $\{\Gamma_n\}$ of Borel subsets in X, we have $\mu(\cap_n \Gamma_n) = \lim_{n\to\infty} \mu(\Gamma_n)$. However, this relation follows from (1.22) since each μ_k is a measure. The second claim of (i) is an immediate consequence of (1.13).

We now prove assertion (ii). To simplify the presentation, we carry out the proof for compact metric spaces X, referring the reader to [Dud02, chapter 11] for the general case.

We first show that if (1.27) holds for any $f \in L_b(X)$, then $\mu_k \to \mu$ in the dual-Lipschitz distance. Indeed, using the compactness of X, it is easy to show that for any $\varepsilon > 0$ the ball $\{f \in L_b(X) : \|f\|_L \le 1\}$ has a finite ε-net in $C_b(X)$ that consists of functions of $L_b(X)$. Therefore convergence (1.27) for any $f \in L_b(X)$ implies the convergence of $\{\mu_k\}$ in the space $\mathcal{P}(X)$ endowed with the dual-Lipschitz distance.

We now prove that if $\mu_k \to \mu$ in the dual-Lipschitz distance, then (1.27) holds for any function $f \in C_b(X)$. Indeed, without loss of generality, we can assume that $f \ge 0$. The compactness of X implies that the function $f_\varepsilon \in L_b(X)$ defined by (1.19) converges to f in the norm $\|\cdot\|_\infty$ as $\varepsilon \to 0$. Furthermore,

$$|(f,\mu_k)-(f,\mu)| \le |(f,\mu_k)-(f_\varepsilon,\mu_k)|+|(f_\varepsilon,\mu_k)-(f_\varepsilon,\mu)|+|(f_\varepsilon,\mu)-(f,\mu)|.$$

The first and third terms on the right-hand side of this inequality go to zero as $\varepsilon \to 0$ uniformly in k, while the second can be made arbitrarily small by choosing a sufficiently large k.

Finally, let us prove that the space $\mathcal{P}(X)$ endowed with the dual-Lipschitz distance is complete. Let $\{\mu_k\} \subset \mathcal{P}(X)$ be a Cauchy sequence. Since X is a compact space, by Prokhorov's theorem we can find a subsequence $\{\mu_{k_j}\}$ that converges weakly to a measure $\mu \in \mathcal{P}(X)$, that is, (1.27) holds for any $f \in C_b(X)$. If we show that the limiting measure μ does not depend on the subsequence, we can conclude that the entire sequence converges to μ weakly and, hence, in the space $\mathcal{P}(X)$ as well.

Since $\{\mu_k\}$ is a Cauchy sequence, for any $f \in L_b(X)$ we have

$$\lim_{k\to\infty} (f,\mu_k) = \lim_{k_j\to\infty} (f,\mu_{k_j}) = (f,\mu).$$

Thus, the limiting measure is uniquely defined. This completes the proof of the theorem in the case of compact metric spaces. □

Let us emphasise that if $\mu_k \to \mu$ weakly in $\mathcal{P}(X)$, then, in general, it is *not* true that

$$\mu_k(\Gamma) \to \mu(\Gamma) \quad \text{as } k \to \infty \tag{1.28}$$

for any $\Gamma \in \mathcal{B}(X)$. However, the well-known *portmanteau theorem* claims that $\mu_k \to \mu$ if and only if one of the following conditions is satisfied:

$$\liminf_{k\to\infty} \mu_k(G) \geq \mu(G) \quad \text{for any open set } G \subset X, \tag{1.29}$$

$$\liminf_{k\to\infty} \mu_k(F) \leq \mu(F) \quad \text{for any closed set } F \subset X. \tag{1.30}$$

It is also equivalent to convergence (1.28) for any Borel subset $\Gamma \subset X$ such that $\mu(\partial\Gamma) = 0$, where $\partial\Gamma$ stands for the boundary of Γ. We refer to theorem 11.1.1 in the book [Dud02] for a proof of these results.

A simple, but important consequence of the portmanteau theorem is the following description of the weak convergence of measures on the real line. Given a probability measure $\mu \in \mathcal{P}(\mathbb{R})$, we denote by $F_\mu(x)$ its *distribution function*, defined by $F_\mu(x) = \mu((-\infty, x])$ for $x \in \mathbb{R}$. Note that the distribution function of a measure is always non-decreasing and right-continuous at any point, and if the measure has no atoms, then its distribution function is continuous.

Lemma 1.2.16 *Let $\{\mu_n\} \subset \mathcal{P}(\mathbb{R})$ be a sequence. Then $\{\mu_n\}$ converges weakly to a measure $\mu \in \mathcal{P}(\mathbb{R})$ if and only if*

$$F_{\mu_n}(x) \to F_\mu(x) \quad \textit{as } n \to \infty,$$

where $x \in \mathbb{R}$ is an arbitrary point of continuity for F_μ.

Another remarkable property of weak convergence is that, under some assumptions, one can pass to the limit under the integrals, even if the integrand is not continuous. To formulate the corresponding result, recall that a continuous mapping $\pi : X \to X$ acting in a Polish space X is called a *projection* if $\pi \circ \pi = \pi$.

Lemma 1.2.17 *Let X be a Polish space, let $\pi_n : X \to X$, $n \geq 1$, be continuous projections, and let $\{\mu_k\} \subset \mathcal{P}(X)$ be a sequence converging weakly to a measure $\mu \in \mathcal{P}(X)$. Assume that $f : X \to \mathbb{R} \cup \{+\infty\}$ is a Borel functional such that $\{f \circ \pi_n\}$ is a sequence of bounded continuous functions converging to f pointwise and*

$$(f \circ \pi_n, \mu_k) \leq C \quad \textit{for any } k, n \geq 1. \tag{1.31}$$

Then $(f, \mu) \leq C$, provided that either $f \geq 0$ or $\{f \circ \pi_n\}$ is non-decreasing.

Note that if $\{f \circ \pi_n\}$ is non-decreasing, then inequality (1.31) will be satisfied if we assume that $(f, \mu_k) \leq C$ for any $k \geq 1$.

Proof We can pass to the limit in inequality (1.31) as $k \to \infty$. This results in $(f \circ \pi_n, \mu) \leq C$. Now the required assertion follows from Fatou's lemma in the first case and from the monotone convergence theorem in the second case. □

We complete this subsection by two exercises establishing some further properties of the space of probability measures and of weak convergence.

Exercise 1.2.18

(i) Let X be a Polish space. Prove that the complete metric space $(\mathcal{P}(X), \|\cdot\|_L^*)$ is separable.
(ii) Prove that the space $(\mathcal{P}(\mathbb{R}), \|\cdot\|_{\mathrm{var}})$ is not separable.

Exercise 1.2.19 Let X be a Polish space and let $\zeta_m^n, \zeta_m, \zeta^n, \zeta$ be some X-valued random variables such that

$$\mathcal{D}(\zeta_m^n) \to \mathcal{D}(\zeta^n) \quad \text{as } m \to \infty \text{ for any } n \geq 1,$$
$$\sup_{m\geq 1} \mathbb{E}\left(\mathrm{dist}_X(\zeta_m, \zeta_m^n) + \mathrm{dist}_X(\zeta^n, \zeta)\right) \to 0 \quad \text{as } n \to \infty.$$

Show that $\mathcal{D}(\zeta_m) \to \mathcal{D}(\zeta)$ as $m \to \infty$.

1.2.4 Couplings and maximal couplings of probability measures

Definition 1.2.20 Let $\mu_1, \mu_2 \in \mathcal{P}(X)$. A pair of random variables (ξ_1, ξ_2) defined on the same probability space is called a *coupling for* (μ_1, μ_2) if

$$\mathcal{D}(\xi_j) = \mu_j \quad \text{for } j = 1, 2. \tag{1.32}$$

The law $\mathcal{D}(\xi_1, \xi_2) =: \mu$ is a measure on the product space $X \times X$. If we denote by π_1 and π_2 the projections of $X \times X$ to the first and second factor, respectively, then

$$(\pi_1)_*\mu = \mu_1, \quad (\pi_2)_*\mu = \mu_2. \tag{1.33}$$

The other way round, if μ is a measure on $X \times X$ satisfying conditions (1.33), then the random variables $\xi_1 = \pi_1$ and $\xi_2 = \pi_2$ defined on the probability space $(X \times X, \mathcal{B}(X \times X), \mu)$ satisfy (1.32). So a measure μ on $X \times X$ satisfying (1.33) is an alternative definition of the coupling. In this form, the coupling was systematically used by L. Kantorovich starting from the late 1930s, e.g., see [KA82].

In what follows, an important role is played by the *maximal coupling* of measures. Let (ξ_1, ξ_2) be a coupling for (μ_1, μ_2). For any $\Gamma \in \mathcal{B}(X)$, we have

$$\begin{aligned}\mu_1(\Gamma) - \mu_2(\Gamma) &= \mathbb{E}\big(\mathbb{I}_\Gamma(\xi_1) - \mathbb{I}_\Gamma(\xi_2)\big) \\ &= \mathbb{E}\big(\mathbb{I}_{\{\xi_1 \neq \xi_2\}}\big(\mathbb{I}_\Gamma(\xi_1) - \mathbb{I}_\Gamma(\xi_2)\big)\big) \leq \mathbb{P}\{\xi_1 \neq \xi_2\}.\end{aligned}$$

Therefore,

$$\mathbb{P}\{\xi_1 \neq \xi_2\} \geq \|\mu_1 - \mu_2\|_{\mathrm{var}}.$$

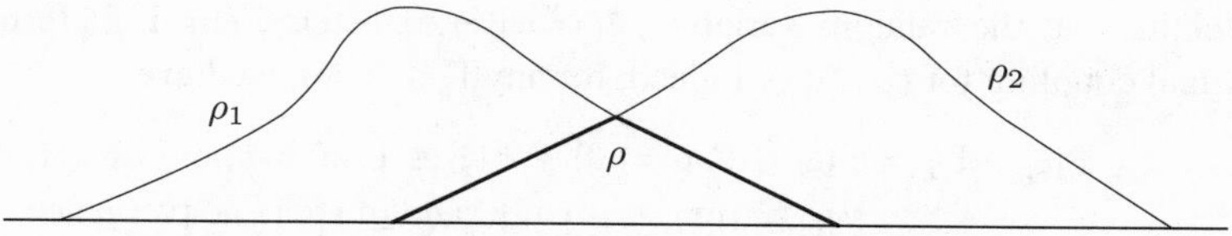

Figure 1.1 The densities ρ_1, ρ_2, and ρ marked by thin and thick graphs.

Definition 1.2.21 A coupling (ξ_1, ξ_2) is said to be *maximal* if

$$\mathbb{P}\{\xi_1 \neq \xi_2\} = \|\mu_1 - \mu_2\|_{\text{var}},$$

and the random variables ξ_1 and ξ_2 conditioned on the event $N = \{\xi_1 \neq \xi_2\}$ are independent. The latter condition means that, for any $\Gamma_1, \Gamma_2 \in \mathcal{B}(X)$, we have[2]

$$\mathbb{P}\{\xi_1 \in \Gamma_1, \xi_2 \in \Gamma_2 \mid N\} = \mathbb{P}\{\xi_1 \in \Gamma_1 \mid N\}\,\mathbb{P}\{\xi_1 \in \Gamma_1 \mid N\}.$$

Exercise 1.2.22 Let $\mu_1, \mu_2 \in \mathcal{P}(X)$ be such that $\|\mu_1 - \mu_2\|_{\text{var}} = 1$. Show that a coupling (ξ_1, ξ_2) for (μ_1, μ_2) is maximal if and only if ξ_1 and ξ_2 are independent.

Exercise 1.2.23 Let (ξ_1, ξ_2) be any pair of random variables that are independent on the event $N = \{\xi_1 \neq \xi_2\}$. Show that

$$\mathbb{P}\{\xi_1 \in \Gamma, \xi_2 \in \Gamma\} \geq \mathbb{P}\{\xi_1 \in \Gamma\}\,\mathbb{P}\{\xi_2 \in \Gamma\} \quad \text{for any } \Gamma \in \mathcal{B}(X). \tag{1.34}$$

The following result is often referred to as the *coupling lemma* or *Dobrushin's lemma*. It makes an effective tool to study the total variation distance between measures.

Lemma 1.2.24 *For any two measures $\mu_1, \mu_2 \in \mathcal{P}(X)$ there exists a maximal coupling (ξ_1, ξ_2).*

Proof Let us set $\delta := \|\mu_1 - \mu_2\|_{\text{var}}$. If $\delta = 1$, then, by Exercise 1.2.22, any pair (ξ_1, ξ_2) of independent random variables with $\mathcal{D}(\xi_i) = \mu_i$, $i = 1, 2$, is a maximal coupling for (μ_1, μ_2). If $\delta = 0$, then $\mu_1 = \mu_2$, and for any random variable ξ with distribution μ_1 the pair (ξ, ξ) is a maximal coupling. Hence, we can assume that $0 < \delta < 1$.

Let $m = \frac{1}{2}(\mu_1 + \mu_2)$ and let (see Figure 1.1)

$$\rho_i = \frac{d\mu_i}{dm}, \quad \rho = \rho_1 \wedge \rho_2, \quad \hat{\rho}_i = \delta^{-1}(\rho_i - \rho). \tag{1.35}$$

Direct verification shows that the measures $\hat{\mu}_i = \hat{\rho}_i dm$ and $\mu = (1 - \delta)^{-1}\rho\, dm$ are probabilities on X. Let ζ_1, ζ_2, ζ, and α be independent random variables defined on the same probability space such that

$$\mathcal{D}(\zeta_i) = \hat{\mu}_i, \quad \mathcal{D}(\zeta) = \mu, \quad \mathbb{P}\{\alpha = 0\} = \delta, \quad \mathbb{P}\{\alpha = 1\} = 1 - \delta. \tag{1.36}$$

[2] In the case $\mathbb{P}(N) = 0$, this condition should be omitted.

We claim that the random variables $\xi_i = \alpha\zeta + (1-\alpha)\zeta_i$, $i = 1, 2$, form a maximal coupling for (μ_1, μ_2). Indeed, for any $\Gamma \in \mathcal{B}(X)$, we have

$$\begin{aligned}\mathbb{P}\{\xi_i \in \Gamma\} &= \mathbb{P}\{\xi_i \in \Gamma, \alpha = 0\} + \mathbb{P}\{\xi_i \in \Gamma, \alpha = 1\}\\ &= \mathbb{P}\{\alpha = 0\}\mathbb{P}\{\zeta_i \in \Gamma\} + \mathbb{P}\{\alpha = 1\}\mathbb{P}\{\zeta \in \Gamma\}\\ &= \delta \int_\Gamma \hat{\rho}_i(u)\,dm + \int_\Gamma \rho(u)\,dm = \mu_i(\Gamma), \end{aligned} \tag{1.37}$$

where we used the independence of $(\zeta_1, \zeta_2, \zeta, \alpha)$ and the relation $\rho_i = \rho + \delta\hat{\rho}_i$. Furthermore,

$$\begin{aligned}\mathbb{P}\{\xi_1 \neq \xi_2\} &= \mathbb{P}\{\xi_1 \neq \xi_2, \alpha = 0\} + \mathbb{P}\{\xi_1 \neq \xi_2, \alpha = 1\}\\ &= \mathbb{P}\{\alpha = 0\}\mathbb{P}\{\zeta_1 \neq \zeta_2\} = \delta,\end{aligned}$$

where we used again the independence of $(\zeta_1, \zeta_2, \zeta, \alpha)$ and also the relation

$$\mathbb{P}\{\zeta_1 = \zeta_2\} = \delta^{-2} \iint_{\{u_1 = u_2\}} \hat{\rho}_1(u_1)\hat{\rho}_2(u_2)\, m(du_1) m(du_2) = 0,$$

which follows from the identity $\hat{\rho}_1(u)\hat{\rho}_2(u) \equiv 0$. A similar argument shows that the random variables ξ_1 and ξ_2 conditioned on $\{\xi_1 \neq \xi_2\}$ are independent. This completes the proof of Lemma 1.2.24. □

Relation (1.37), Corollary 1.2.11, and Exercise 1.2.22 imply the following alternative version[3] of the coupling lemma.

Corollary 1.2.25 *Any two measures $\mu_1, \mu_2 \in \mathcal{P}(X)$ admit a representation*

$$\mu_j = (1-\delta)\mu + \delta\nu_j, \quad j = 1, 2,$$

where $\delta = \|\mu_1 - \mu_2\|_{\mathrm{var}}$, $\mu, \nu_1, \nu_2 \in \mathcal{P}(X)$, and the measures ν_1 and ν_2 are mutually singular.

We shall call $(1-\delta)\mu$ the *minimum* of μ_1 and μ_2 and denote it by $\mu_1 \wedge \mu_2$. Another important corollary is the following result on the conditional law of the random variables that form a maximal coupling. Recall that the *conditional law* $\mathcal{D}(\xi \mid N)$ of an X-valued random variable ξ given an event N of non-zero probability is defined by the relation

$$\mathcal{D}(\xi \mid N)(\Gamma) = \frac{\mathbb{P}(\{\xi \in \Gamma\} \cap N)}{\mathbb{P}(N)}, \quad \Gamma \in \mathcal{B}(X).$$

Lemma 1.2.26 *Let (ξ_1, ξ_2) be a maximal coupling for a pair of measures $\mu_1, \mu_2 \in \mathcal{P}(X)$ such that $\mu_1 \wedge \mu_2(X) > 0$. Then $\mathbb{P}\{\xi_1 = \xi_2\} > 0$, and we have*

$$\mathcal{D}(\xi_1 \mid \{\xi_1 = \xi_2\}) = \mathcal{D}(\xi_2 \mid \{\xi_1 = \xi_2\}) = \frac{\mu_1 \wedge \mu_2}{\mu_1 \wedge \mu_2(X)}. \tag{1.38}$$

[3] In this form, the coupling lemma was intensively used by Dobrushin; e.g., see [Dob74].

Proof The case in which $\xi_1 = \xi_2$ almost surely is trivial, and we assume that $\xi_1 \neq \xi_2$ with positive probability. Proposition 1.2.7 implies that

$$\mathbb{P}\{\xi_1 \neq \xi_2\} = \|\mu_1 - \mu_2\|_{\text{var}} = 1 - \mu_1 \wedge \mu_2(X) < 1, \tag{1.39}$$

whence we conclude that $\mathbb{P}\{\xi_1 = \xi_2\} > 0$, and the conditional laws in (1.38) are well defined. Let us set $\mu = \mathcal{D}(\xi_1 \mid \{\xi_1 = \xi_2\})$ and $\tilde{\mu}_i = \mathcal{D}(\xi_i \mid \{\xi_1 \neq \xi_2\})$. Then

$$\mu_i = \mathbb{P}\{\xi_1 = \xi_2\}\mu + \mathbb{P}\{\xi_1 \neq \xi_2\}\tilde{\mu}_i, \quad i = 1, 2.$$

It follows that

$$\mu_1 \wedge \mu_2 = \mathbb{P}\{\xi_1 = \xi_2\}\mu + \mathbb{P}\{\xi_1 \neq \xi_2\}\tilde{\mu}_1 \wedge \tilde{\mu}_2.$$

Combining this with (1.39), we see that $\tilde{\mu}_1 \wedge \tilde{\mu}_2 = 0$, and the above relation immediately gives (1.38). □

In what follows, we deal with pairs of measures depending on a parameter, and we shall need a maximal coupling for them that depends on the parameter in a measurable manner. More precisely, let Z be a Polish space endowed with its Borel σ-algebra and let $\{\mu(z, du), z \in Z\}$ be a family of probability measures on X. We shall say that $\mu(z, du)$ *is a random probability measure on* X if for any Borel set $\Gamma \subset X$ the function $z \mapsto \mu(z, \Gamma)$ is measurable from Z to $\mathbb{R}$.

Exercise 1.2.27 Prove that $\{\mu(z, du), z \in Z\}$ is a random probability measure if and only if the mapping $z \mapsto \mu(z, \cdot)$ is measurable from Z to the space $\mathcal{P}(X)$ endowed with the Borel σ-algebra corresponding to the dual-Lipschitz metric.

Theorem 1.2.28 *Let X and Z be Polish spaces, and let $\{\mu_i(z, du), z \in Z\}$, $i = 1, 2$, be two random probability measures on X. Then there is a probability space $(\Omega, \mathcal{F}, \mathbb{P})$ and two measurable functions $\xi_i(z, \omega) : Z \times \Omega \to X$, $i = 1, 2$, such that $(\xi_1(z, \cdot), \xi_2(z, \cdot))$ is a maximal coupling for $(\mu_1(z, du), \mu_2(z, du))$ for any $z \in Z$.*

Proof In view of Theorem A.2.2 in the appendix, any Polish space is a standard measurable space. Since the objects considered in the theorem are invariant with respect to measurable isomorphisms, we can assume from the very beginning that X coincides with one of the spaces described in Definition A.2.1. To be precise, we shall assume that X is the interval $[0, 1]$ endowed with its Borel σ-algebra.

We shall repeat the scheme used in the proof of Lemma 1.2.24, controlling the dependence of the resulting random variables on the parameter z. To this end, we first note that if $\mu(z, du)$ is a random probability measure on X, then there is a probability space $(\Omega, \mathcal{F}, \mathbb{P})$ and a measurable function $\xi(z, \omega) : Z \times \Omega \to X$ such that the law of $\xi(z, \cdot)$ coincides with $\mu(z, du)$ for any $z \in Z$. For instance, we can take for the probability space the interval $[0,1]$, endowed with its Borel σ-algebra and the Lebesgue measure, and define ξ by the relation

$$\xi(z,\omega)=\min\{t\in[0,1]:F(z,t)\geq\omega\},$$

where $F(z,t)=\mu(z,(-\infty,t])$ is the probability distribution function for the measure μ.

We now introduce the family

$$\nu(z,du)=\mu_1(z,du)\wedge\mu_2(z,du),$$

where $\nu_1\wedge\nu_2$ denotes the minimum of two measures $\nu_1,\nu_2\in\mathcal{P}(X)$, and define the following families of probability measures (cf. (1.35)):

$$\hat{\mu}_i(z,du)=\begin{cases}\delta(z)^{-1}\big(\mu_i(z,du)-\nu(z,du)\big) & \text{for } \delta(z)\neq 0,\\ \mu_1(z,du) & \text{otherwise,}\end{cases}$$

$$\mu(z,du)=\begin{cases}\big(1-\delta(z)\big)^{-1}\nu(z,du) & \text{for } \delta(z)\neq 1,\\ \lambda & \text{otherwise,}\end{cases}$$

where $\delta(z)=\|\mu_1(z,\cdot)-\mu_2(z,\cdot)\|_{\mathrm{var}}$, and $\lambda\in\mathcal{P}(X)$ is any fixed measure. It can be shown that δ is a measurable function of $z\in Z$, and $\hat{\mu}_i(z,\cdot)$ and $\mu(z,\cdot)$ are random probability measures on X; see Exercise 1.2.29 below.

What has been said implies that there is a probability space $(\Omega,\mathcal{F},\mathbb{P})$ and measurable functions ζ_1,ζ_2,ζ, and α of the variable $(z,\omega)\in Z\times\Omega$ such that, for any $z\in Z$, the random variables $\zeta_i(z,\cdot)$, $\zeta(z,\cdot)$, and $\alpha(z,\cdot)$ are independent, and

$$\mathcal{D}(\zeta_i(z,\cdot))=\hat{\mu}_i(z,du),\qquad \mathcal{D}(\zeta(z,\cdot))=\mu(z,du),$$

$$\mathbb{P}\{\alpha(z,\cdot)=0\}=\delta(z),\qquad \mathbb{P}\{\alpha(z,\cdot)=1\}=1-\delta(z);$$

cf. (1.36). The proof of the theorem can now be completed by a literal repetition of the argument used to establish Lemma 1.2.24. □

Exercise 1.2.29 Show that $\delta(z)$ is a measurable function of $z\in Z$, and $\hat{\mu}_i(z,\cdot)$ and $\mu(z,\cdot)$ are random probability measures on X. *Hint:* Use a parameter version of the Radon–Nikodym theorem (e.g., see [Nov05]): if $\lambda(z,du)$ and $\nu(z,du)$ are random probability measures on X such that $\lambda(z,du)$ is absolutely continuous with respect to $\nu(z,du)$ for any $z\in Z$, then the Radon–Nikodym derivative $\frac{d\lambda(z,\cdot)}{d\nu(z,\cdot)}$ can be chosen to be a measurable function of $(z,u)\in Z\times X$.

The existence result of the following exercise is useful when one constructs a coupling for solutions of stochastic PDEs; see [Mat02b; Oda08] and Section 3.5.2.

Exercise 1.2.30 Let X and Y be Polish spaces.

(i) Show that for any pair of measures $\mu_1,\mu_2\in\mathcal{P}(X)$ and any measurable mapping $f:X\to Y$ there is a coupling (ξ_1,ξ_2) for (μ_1,μ_2) such that $(f(\xi_1),f(\xi_2))$ is a maximal coupling for $(f_*(\mu_1),f_*(\mu_2))$.

(ii) Let Z be a Polish space and let $f : X \times Z \to Y$ be a measurable mapping. Show that for any random probability measures $\mu_1(z, du)$ and $\mu_2(z, du)$ on X there is a probability space $(\Omega, \mathcal{F}, \mathbb{P})$ and X-valued measurable functions $\xi_i(z, \omega)$, $i = 1, 2$, defined on $Z \times \Omega$ such that, for any $z \in Z$, the pair $(f(z, \xi_1), f(z, \xi_2))$ is a coupling for $\big(f_*(z, \mu_1(z, \cdot)), f_*(z, \mu_2(z, \cdot))\big)$, where $f_*(z, \mu_i(z, \cdot))$ stands for the image of $\mu_i(z, \cdot)$ under the mapping $u \mapsto f(z, u)$.

1.2.5 Kantorovich functionals

Let F be a measurable symmetric function on $X \times X$ such that

$$F(u_1, u_2) \geq \mathrm{dist}_X(u_1, u_2) \quad \text{for all } u_1, u_2 \in X. \tag{1.40}$$

We define the *Kantorovich functional* corresponding to F as the following function $\mathcal{K}_F$ on $\mathcal{P}(X) \times \mathcal{P}(X)$:

$$\mathcal{K}_F(\mu_1, \mu_2) = \inf\{\mathbb{E}\, F(\xi_1, \xi_2)\}, \tag{1.41}$$

where the infimum is taken over all couplings (ξ_1, ξ_2) for (μ_1, μ_2). The function F is called the *(Kantorovich) density* of the functional $\mathcal{K}_F$.

Lemma 1.2.31 *For any $\mu_1, \mu_2 \in \mathcal{P}(X)$, we have*[4]

$$\|\mu_1 - \mu_2\|_L^* \leq \mathcal{K}_F(\mu_1, \mu_2). \tag{1.42}$$

Proof Let (ξ_1, ξ_2) be a coupling for (μ_1, μ_2). Then, for any $g \in L_b(X)$ with $\|g\|_L \leq 1$, we have

$$(g, \mu_1 - \mu_2) = \mathbb{E}\,\big(g(\xi_1) - g(\xi_2)\big) \leq \mathbb{E}\, \mathrm{dist}_X(\xi_1, \xi_2) \leq \mathbb{E}\, F(\xi_1, \xi_2).$$

Taking first the supremum in g and then the infimum with respect to all couplings (ξ_1, ξ_2), we obtain (1.42). □

In the case when the function F coincides with dist_X, we have the following result, which is the celebrated *Kantorovich–Rubinstein theorem.*

Theorem 1.2.32 *For any probability measures $\mu_1, \mu_2 \in \mathcal{P}(X)$, we have*

$$\|\mu_1 - \mu_2\|_K = \inf\{\mathbb{E}\, \mathrm{dist}_X(\xi_1, \xi_2)\} = \mathcal{K}_{\mathrm{dist}_X}(\mu_1, \mu_2), \tag{1.43}$$

where the infimum is taken over all couplings (ξ_1, ξ_2) for (μ_1, μ_2). Moreover, the infimum is attained at some coupling (ξ_1, ξ_2).

In the case when the space X is endowed with the discrete topology (that is, $\mathrm{dist}_X(u_1, u_2) = 1$ for any $u_1 \neq u_2$), Theorem 1.2.32 is a straightforward consequence of the existence of a maximal coupling of measures (see Lemma 1.2.24).

[4] The right-hand side of inequality (1.42) may be infinite.

For the general case, we refer the reader to [KA82, section VIII.4] or [Dud02, theorem 11.8.2].

In his celebrated research on the mass-transfer problem (for which he was given a Nobel prize in economics), L. Kantorovich interpreted $\mathbb{E}\,\mathrm{dist}_X(\xi_1, \xi_2)$ as the work needed to transport mass points $\xi_1(\omega)$ to $\xi_2(\omega)$, and used relation (1.43) to estimate the work via the distance between the measures μ_1 and μ_2. We shall use that equality the other way round, that is, as a tool to estimate this distance. Accordingly, inequality (1.42) will be sufficient for our purposes.

1.3 Markov processes and random dynamical systems

1.3.1 Markov processes

Let X be a Polish space. A *Markov family of random processes in* X (or simply a *Markov process*) is defined as a collection of the following objects (e.g., see [KS91, section 2.5]):

- a measurable space $(\Omega, \mathcal{F})$ with a filtration $\{\mathcal{F}_t, t \in \mathcal{T}_+\}$;
- a family of probability measures $\{\mathbb{P}_v, v \in X\}$ on $(\Omega, \mathcal{F})$ such that the mapping $v \mapsto \mathbb{P}_v(A)$ is universally measurable[5] for any $A \in \mathcal{F}$;
- an X-valued random process $\{u_t, t \in \mathcal{T}_+\}$ adapted to the filtration $\mathcal{F}_t$ and satisfying the conditions below for any $v \in X$, $\Gamma \in \mathcal{B}(X)$, and $t, s \in \mathcal{T}_+$:

$$\mathbb{P}_v\{u_0 = v\} = 1, \tag{1.44}$$

$$\mathbb{P}_v\{u_{t+s} \in \Gamma \mid \mathcal{F}_s\} = P_t(u_s, \Gamma) \quad \text{for } \mathbb{P}_v\text{-almost every } \omega \in \Omega. \tag{1.45}$$

Here P_t stands for the *transition function* of the family defined as the law of u_t under the probability measure $\mathbb{P}_v$:

$$P_t(v, \Gamma) := \mathbb{P}_v\{u_t \in \Gamma\}, \qquad v \in X, \qquad \Gamma \in \mathcal{B}(X). \tag{1.46}$$

Relation (1.45) is called the *Markov property*, and the above Markov family is denoted by $(u_t, \mathbb{P}_v)$. If $\mathcal{T}_+ = \mathbb{Z}_+$, then the object we have just defined is also called a *family of Markov chains* (or a *discrete-time Markov process* or simply a *Markov chain*).

Given $\lambda \in \mathcal{P}(X)$ and a Markov process $(u_t, \mathbb{P}_v)$, we define the probability measure

$$\mathbb{P}_\lambda(\Gamma) = \int_X \mathbb{P}_v(\Gamma)\lambda(dv), \qquad \Gamma \in \mathcal{F}, \tag{1.47}$$

and denote by $\mathbb{E}_\lambda$ the corresponding mean value. The following exercise gives a simple generalisation of the Markov property.

[5] See Section A.3 for the definition of universal measurability.

Exercise 1.3.1 Show that if $(u_t, \mathbb{P}_v)$ is a Markov process, then for $f \in L^\infty(X)$ and $\lambda \in \mathcal{P}(X)$, we have

$$\mathbb{E}_\lambda\{f(u_{t+s}) \,|\, \mathcal{F}_s\} = \mathbb{E}_{u_s} f(u_t) \quad \mathbb{P}_\lambda\text{-almost surely.} \tag{1.48}$$

More generally, for any $m \geq 1$, any $0 < t_1 < \cdots < t_m$, and any bounded measurable function $f : X \times \cdots \times X \to \mathbb{R}$, we have

$$\mathbb{E}_\lambda\{f(u_{t_1+s}, \ldots, u_{t_m+s}) \,|\, \mathcal{F}_s\} = \mathbb{E}_{u_s} f(u_{t_1}, \ldots, u_{t_m}) \quad \mathbb{P}_\lambda\text{-almost surely.} \tag{1.49}$$

We now establish the so-called *Kolmogorov–Chapman relation*, which will imply, in particular, that every Markov process generates an evolution in the space of probability measures.

Lemma 1.3.2 *For any $t, s \in \mathcal{T}_+$, $v \in X$, and $\Gamma \in \mathcal{B}(X)$, the transition function P_t satisfies the relation*

$$P_{t+s}(v, \Gamma) = \int_X P_t(v, dz) P_s(z, \Gamma). \tag{1.50}$$

Proof In view of (1.45), we have

$$\begin{aligned} P_{t+s}(v, \Gamma) &= \mathbb{E}_v\, \mathbb{I}_\Gamma(u_{t+s}) = \mathbb{E}_v \left\{ \mathbb{E}_v \left(\mathbb{I}_\Gamma(u_{t+s}) \,|\, \mathcal{F}_t \right) \right\} \\ &= \mathbb{E}_v\, P_s(u_t, \Gamma) = \int_X P_t(v, dz) P_s(z, \Gamma), \end{aligned}$$

where we used the fact that the law of u_t under $\mathbb{P}_v$ coincides with $P_t(v, \cdot)$. □

To each Markov process there correspond two families of linear operators acting in the spaces of bounded measurable functions $L^\infty(X)$ and of probability measures $\mathcal{P}(X)$. They are called *Markov semigroups* and are defined in terms of the transition function by the following relations:

$$\mathfrak{P}_t : L^\infty(X) \to L^\infty(X), \qquad \mathfrak{P}_t f(v) = \int_X P_t(v, dz) f(z),$$

$$\mathfrak{P}_t^* : \mathcal{P}(X) \to \mathcal{P}(X), \qquad \mathfrak{P}_t^* \mu(\Gamma) = \int_X P_t(v, \Gamma) \mu(dv),$$

where $t \in \mathcal{T}_+$. It is straightforward to check that $\mathfrak{P}_t f \in L^\infty(X)$ for $f \in L^\infty(X)$ and $\mathfrak{P}_t^* \mu \in \mathcal{P}(X)$ for $\mu \in \mathcal{P}(X)$. The following exercise justifies the term *semigroup* and establishes some simple properties.

Exercise 1.3.3 Show that the families $\mathfrak{P}_t$ and $\mathfrak{P}_t^*$ form semigroups, that is, $\mathfrak{P}_0 = \mathrm{Id}$ and $\mathfrak{P}_{t+s} = \mathfrak{P}_s \circ \mathfrak{P}_t$, and similarly for $\mathfrak{P}_t^*$. Show also that they satisfy the following duality relation:

$$(\mathfrak{P}_t f, \mu) = (f, \mathfrak{P}_t^* \mu) \quad \text{for any } f \in L^\infty(X),\ \mu \in \mathcal{P}(X).$$

Hint: Use the Kolmogorov–Chapman relation (1.50) for the semigroup property and Fubini's theorem for the duality relation.

Exercise 1.3.4 Let $(u_t, \mathbb{P}_v)$ be a Markov process in X, let $\lambda \in \mathcal{P}(X)$, and let $\mathbb{P}_\lambda$ be the measure defined by (1.47). Prove that the $\mathbb{P}_\lambda$-law of u_t coincides with $\mathfrak{P}_t^*\lambda$. Use this property to show that if $f \in L^\infty(X)$, then

$$\mathbb{E}_\lambda f(u_t) = \int_X \mathfrak{P}_t f(v)\,\lambda(dv). \tag{1.51}$$

Definition 1.3.5 A measure $\mu \in \mathcal{P}(X)$ is said to be *stationary* for $(u_t, \mathbb{P}_v)$ if $\mathfrak{P}_t^*\mu = \mu$ for all $t \geq 0$.

Example 1.3.6 Let H be a separable Banach space, let $S : H \to H$ be a continuous mapping, and let $\{\eta_k, k \geq 1\}$ be a sequence of i.i.d. random variables in H defined on a complete probability space $(\Omega, \mathcal{F}, \mathbb{P})$. We fix $v \in H$ and consider a sequence $\{v_k, k \geq 0\}$ defined by the rule

$$v_0 = v, \qquad v_k = S(v_{k-1}) + \eta_k, \quad k \geq 1. \tag{1.52}$$

This system defines a discrete-time Markov process in H. Indeed, let us denote by $(\widetilde{\Omega}, \widetilde{\mathcal{F}})$ the product of the measurable spaces $(H, \mathcal{B}(H))$ and $(\Omega, \mathcal{F})$ (that is, $\widetilde{\Omega} = H \times \Omega$ and $\widetilde{\mathcal{F}} = \mathcal{B}(H) \otimes \mathcal{F}$) and provide it with the filtration $\{\widetilde{\mathcal{F}}_k, k \geq 0\}$, where $\widetilde{\mathcal{F}}_k = \mathcal{B}(H) \otimes \mathcal{F}_k$ and $\mathcal{F}_k$ is the σ-algebra generated by $\eta_1, \dots, \eta_k$ (so that $\mathcal{F}_0$ is the trivial σ-algebra). For any $v \in H$, we take $\mathbb{P}_v = \delta_v \otimes \mathbb{P}$ and define a process $\{u_k, k \geq 0\}$ as follows: for $\widetilde{\omega} = (v, \omega)$ we set $u_k^{\widetilde{\omega}} = v_k^\omega$, where $\{v_k\}$ is given by (1.52). Denote by $P_k(v, \cdot)$ the law of v_k, where $k \geq 0$ and $v \in H$. It is easy to see that the objects introduced above satisfy (1.44) and (1.45).

Now let v be an H-valued random variable independent of $\{\eta_k\}$ with a law μ. Then, by Exercise 1.3.4, we have $\mathcal{D}_\mu(u_k) = \mathfrak{P}_k^*\mu$ for $k \geq 0$. In particular, if μ is a stationary measure for the Markov process $(u_k, \mathbb{P}_v)$, then $\mathcal{D}_\mu(u_k) = \mu$ for all $k \geq 0$. In what follows, a stationary measure for $(u_k, \mathbb{P}_v)$ is called a *stationary measure* for Eq. (1.52) and the sequence $\{v_k\}$ defined by (1.52) is called a *stationary solution*.

An important feature of Markov processes is the *strong Markov property*. It says, roughly speaking, that relation (1.45) remains valid if s is replaced by a random time σ, provided that it satisfies an addition property of the "independence of the future". To formulate the corresponding result, we first introduce the concept of a *stopping time*, formalising that property.

Definition 1.3.7 A random variable $\tau : \Omega \to \mathcal{T}_+ \cup \{\infty\}$ is called a *stopping time* for a filtration $\{\mathcal{F}_t, t \in \mathcal{T}_+\}$ if the event $\{\tau \leq t\}$ belongs to $\mathcal{F}_t$ for any $t \in \mathcal{T}_+$.

Exercise 1.3.8

(i) Show that if τ and σ are stopping times, then the random variables $\tau+\sigma$, $\tau\wedge\sigma$, and $\tau\vee\sigma$ are also stopping times.
(ii) Suppose that the function $t\mapsto u_t$ is continuous from $\mathbb{R}_+$ to X. Show that for any closed subset $A\subset X$, the random variable[6]

$$\tau(A)=\min\{t\in\mathcal{T}_+ : u_t\in A\}$$

is a stopping time. *Hint:* See exercise 2.7 in [KS91].

For any stopping time τ, we shall denote by $\mathcal{F}_\tau$ the σ-algebra of the events $\Gamma\in\mathcal{F}$ such that

$$\Gamma\cap\{\tau\le t\}\in\mathcal{F}_t \quad \text{for any } t\in\mathcal{T}_+.$$

Exercise 1.3.9 Show that any stopping time τ is $\mathcal{F}_\tau$-measurable.

In what follows, we shall always assume the Markov processes we deal with satisfy the following additional hypotheses:

Feller property. For any $f\in C_b(X)$ and $t\ge 0$, we have $\mathfrak{P}_t f\in C_b(X)$.

Time continuity. The trajectories $u_t(\omega)$, $\omega\in\Omega$, are continuous in time.[7]

Theorem 1.3.10 *Let $(u_t,\mathbb{P}_u)$ be a Markov process, let $f: X\to\mathbb{R}$ be a bounded measurable function, and let τ be a stopping time. Then, for any almost surely finite $\mathcal{F}_\tau$-measurable random variable $\sigma:\Omega\to\mathcal{T}_+$ and any measure λ on $(X,\mathcal{B}(X))$, we have*

$$\mathbb{E}_\lambda\big(\mathbb{I}_{\{\tau<\infty\}}f(u_{\tau+\sigma})\,|\,\mathcal{F}_\tau\big)=\mathbb{I}_{\{\tau<\infty\}}(\mathfrak{P}_\sigma f)(u_\tau) \quad \mathbb{P}_\lambda\text{-almost surely.} \tag{1.53}$$

Relation (1.53) is called the *strong Markov property* of the Markov family $(u_t,\mathbb{P}_u)$.

Proof of Theorem 1.3.10 We confine ourselves to the case in which $\mathcal{T}_+=\mathbb{Z}_+$. The proof in the case of continuous-time Markov processes can be carried out with the help of the approximation of τ by stopping times with range in an increasing sequence of discrete subsets of $\mathbb{R}_+$; e.g., see section III.3 in [RY99].

Step 1. It suffices to show that

$$\mathbb{E}_\lambda\big(\mathbb{I}_{\{\tau<\infty\}}f(u_{\tau+m})\,|\,\mathcal{F}_\tau\big)=\mathbb{I}_{\{\tau<\infty\}}(\mathfrak{P}_m f)(u_\tau), \tag{1.54}$$

[6] As usual, the minimum over an empty set is equal to $+\infty$.
[7] This condition is imposed only if $\mathcal{T}_+=\mathbb{R}_+$.

where $m \geq 0$ is an arbitrary integer, and the equality holds $\mathbb{P}_\lambda$-almost surely. Indeed, if (1.54) is established, then we can write

$$\begin{aligned}\mathbb{E}_\lambda\bigl\{\mathbb{I}_{\{\tau<\infty\}} f(u_{\tau+\sigma}) \,|\, \mathcal{F}_\tau\bigr\} &= \sum_{m=0}^{\infty} \mathbb{E}_\lambda\bigl\{\mathbb{I}_{\{\tau<\infty\}}\mathbb{I}_{\{\sigma=m\}} f(u_{\tau+m}) \,|\, \mathcal{F}_\tau\bigr\}\\ &= \mathbb{I}_{\{\tau<\infty\}} \sum_{m=0}^{\infty} \mathbb{I}_{\{\sigma=m\}}(\mathfrak{P}_m f)(u_\tau)\\ &= \mathbb{I}_{\{\tau<\infty\}}(\mathfrak{P}_\sigma f)(u_\tau).\end{aligned}$$

Step 2. We now prove (1.54). To this end, it suffices to show that

$$\mathbb{E}_\lambda\bigl(\mathbb{I}_{\Gamma\cap\{\tau<\infty\}} f(u_{\tau+m})\bigr) = \mathbb{E}_\lambda\bigl(\mathbb{I}_{\Gamma\cap\{\tau<\infty\}}(\mathfrak{P}_m f)(u_\tau)\bigr), \tag{1.55}$$

where $\Gamma \in \mathcal{F}_\tau$ is an arbitrary subset. Let us note that

$$\mathbb{I}_{\Gamma\cap\{\tau<\infty\}} = \sum_{n=0}^{\infty} \mathbb{I}_{\Gamma\cap\{\tau=n\}}. \tag{1.56}$$

Since $\Gamma_n := \Gamma \cap \{\tau = n\}$ belongs to $\mathcal{F}_n$, the Markov property (1.48) implies that

$$\begin{aligned}\mathbb{E}_\lambda\bigl(\mathbb{I}_{\Gamma_n} f(u_{\tau+m})\bigr) &= \mathbb{E}_\lambda\bigl(\mathbb{I}_{\Gamma_n} f(u_{n+m})\bigr)\\ &= \mathbb{E}_\lambda\bigl(\mathbb{I}_{\Gamma_n}\mathbb{E}\,\{f(u_{n+m}) \,|\, \mathcal{F}_n\}\bigr)\\ &= \mathbb{E}_\lambda\bigl(\mathbb{I}_{\Gamma_n}(\mathfrak{P}_m f)(u_n)\bigr).\end{aligned}$$

Combining this relation with (1.56), we obtain (1.55). □

Corollary 1.3.11 *Let τ be a stopping time such that $\tau \leq t$ almost surely. Then, for any $\Gamma \in \mathcal{B}(X)$ and $u \in X$, we have*

$$P_t(u, \Gamma) = \mathbb{E}_u\, P_{t-\tau}(u_\tau, \Gamma). \tag{1.57}$$

More generally, if $f : X \to \mathbb{R}$ is a bounded measurable function, then

$$\mathfrak{P}_t f(u) = \mathbb{E}_u\,(\mathfrak{P}_{t-\tau} f)(u_\tau) \quad \textit{for any } u \in X. \tag{1.58}$$

Proof To prove (1.57), let us set $\sigma = t - \tau$ and apply (1.53) to the function $f = \mathbb{I}_\Gamma$. This results in the relation

$$\begin{aligned}P_t(u, \Gamma) &= \mathbb{E}_u \mathbb{I}_\Gamma(u_t) = \mathbb{E}_u\, \mathbb{E}_u\bigl(\mathbb{I}_\Gamma(u_{\tau+\sigma}) \,|\, \mathcal{F}_\tau\bigr)\\ &= \mathbb{E}_u(\mathfrak{P}_\sigma \mathbb{I}_\Gamma)(u_\tau) = \mathbb{E}_u P_\sigma(u_\tau, \Gamma).\end{aligned}$$

Relation (1.58) is a straightforward consequence of (1.57). □

In conclusion, we discuss yet another version of the strong Markov property. Recall that $C(\mathcal{T}_+, X)$ stands for the space of continuous functions[8] from $\mathcal{T}_+$ to X with the topology of uniform convergence on bounded subsets of $\mathcal{T}_+$; cf. (1.9). In view of Exercise 1.1.6, $C(\mathcal{T}_+, X)$ is a Polish space, and we endow it with its Borel σ-algebra.

Exercise 1.3.12 Let $f : C(\mathcal{T}_+, X) \to \mathbb{R}$ be a measurable function that is either bounded or non-negative. Show that, for any $u \in X$ and any stopping time τ, we have

$$\mathbb{E}_u \left(\mathbb{I}_{\{\tau<\infty\}} f(u_{\tau+\cdot}) \mid \mathcal{F}_\tau\right) = \mathbb{I}_{\{\tau<\infty\}} \left(\mathbb{E}_v f(u_\cdot)\right)\Big|_{v=u_\tau} \quad \mathbb{P}_u\text{-almost surely.} \tag{1.59}$$

1.3.2 Random dynamical systems

In Example 1.3.6, we constructed a Markov chain starting from a dynamical system depending on a random parameter. That construction can be extended to more general situations, enabling one to associate a Markov process with some PDE subject to random perturbations. A key point here is the concept of a random dynamical system, which is introduced below. The relation between random dynamical systems and Markov processes is described in the next subsection.

Let $(\Omega, \mathcal{F}, \mathbb{P})$ be a complete probability space, let $\mathcal{T}$ be either $\mathbb{R}$ or $\mathbb{Z}$, and let $\boldsymbol{\theta} = \{\theta_t : \Omega \to \Omega,\ t \in \mathcal{T}\}$ be a group of measurable mappings. We shall say that $\boldsymbol{\theta}$ is *measure-preserving* if $(\theta_t)_*\mathbb{P} = \mathbb{P}$ for all $t \in \mathcal{T}$; that is,

$$\mathbb{P}\big(\theta_t(\Gamma)\big) = \mathbb{P}(\Gamma) \quad \text{for any } t \in \mathcal{T}, \Gamma \in \mathcal{F}. \tag{1.60}$$

Example 1.3.13 Let U be a Polish space and let μ be a probability measure on U. Define a probability space $(\Omega, \mathcal{F}, \mathbb{P})$ by the following rules.

- Ω is the set of functions $\omega : \mathcal{T} \to U$; we shall write ω_t for the value of ω at the point t.
- $\mathcal{F}$ is the minimal σ-algebra generated by the cylindrical sets

$$\Gamma = \big\{\omega \in \Omega : \omega_{t_i} \in \Gamma_i \text{ for } i = 1, \dots, n\big\}, \tag{1.61}$$

  where $t_1, \dots, t_n \in \mathcal{T}$ and $\Gamma_1, \dots, \Gamma_n \in \mathcal{B}(U)$.
- $\mathbb{P}$ is the unique probability measure on $(\Omega, \mathcal{F})$ such that

$$\mathbb{P}(\Gamma) = \prod_{i=1}^{n} \mu(\Gamma_i)$$

  for any cylindrical set of the form (1.61).

[8] In the case $\mathcal{T}_+ = \mathbb{Z}_+$, we obtain the space of all functions from $\mathbb{Z}_+$ to X.

With a slight abuse of notation, we shall denote by the same symbol the completion of the probability space defined above.

Let $\boldsymbol{\theta} = \{\theta_t : \Omega \to \Omega\}$ be the family of shifts on Ω:

$$(\theta_t\omega)_s = \omega_{t+s} \quad \text{for } s \in \mathcal{T}.$$

Then $\boldsymbol{\theta}$ is a group of measure-preserving transformations on Ω. We refer the reader to the book [Str93] for more details on this construction.

Below in this section, given a probability space $(\Omega, \mathcal{F}, \mathbb{P})$, we shall always assume that a group of measure-preserving transformations $\boldsymbol{\theta} = \{\theta_t, t \in \mathcal{T}\}$ acts on it. Let $\mathcal{T}_+ = \{t \in \mathcal{T} : t \geq 0\}$ and let X be a Polish space endowed with its Borel σ-algebra $\mathcal{B}(X)$. Consider a family $\boldsymbol{\Phi} = \{\varphi_t : \Omega \times X \to X, t \in \mathcal{T}_+\}$ of measurable mappings.

Definition 1.3.14 We shall say that $\boldsymbol{\Phi}$ is a (continuous) *random dynamical system* (or an *RDS*) *in X over $\boldsymbol{\theta}$* if the following properties are satisfied.

(i) *Continuity*. For any $\omega \in \Omega$ and $t \in \mathcal{T}_+$, the mapping $\varphi_t^\omega : X \to X$ is continuous.

(ii) *Cocycle property*. For any $t, s \in \mathcal{T}_+$ and $\omega \in \Omega$, we have

$$\varphi_0^\omega = \mathrm{Id}_X, \quad \varphi_{t+s}^\omega = \varphi_t^{\theta_s\omega} \circ \varphi_s^\omega. \tag{1.62}$$

If $\mathcal{T} = \mathbb{R}$, we assume in addition that the trajectories $\varphi_t^\omega(u)$ are continuous in $t \in \mathcal{T}_+$ for any $\omega \in \Omega$ and $u \in X$. If $\mathcal{T} = \mathbb{Z}$, we shall sometimes say $\boldsymbol{\Phi}$ is a *discrete-time RDS*, to emphasise the difference between the two cases.

In what follows, we often omit ω from the notation and write $\varphi_t u$ instead of $\varphi_t^\omega(u)$. The following example gives a canonical way for constructing an RDS associated with a random perturbation of a deterministic mapping.

Example 1.3.15 In the setting of Example 1.3.6, let us construct a discrete-time RDS in H whose trajectories have the same distribution as $\{v_k, k \in \mathbb{Z}_+\}$.

Let $(\Omega, \mathcal{F}, \mathbb{P})$ be the complete probability space defined in Example 1.3.13, where $\mathcal{T} = \mathbb{Z}$, $U = H$, and μ is the law of η_k. Let $\boldsymbol{\theta} = \{\theta_k : \Omega \to \Omega\}$ be the corresponding group of shifts on Ω. We now define measurable mappings $\varphi_k : \Omega \times H \to H$ by the formulas

$$\varphi_1^\omega u = S(u) + \omega_1, \quad \varphi_k^\omega u = S(\varphi_{k-1}^\omega u) + \omega_k, \quad k \geq 1, \tag{1.63}$$

where $\omega = (\omega_j, j \in \mathbb{Z})$. It is easy to verify that $\boldsymbol{\Phi} = \{\varphi_k, k \in \mathbb{Z}_+\}$ is an RDS over $\boldsymbol{\theta}$. Moreover, for any $u \in H$ the distribution of the corresponding trajectory $\{u_k\}$ under the law $\mathbb{P}$ coincides with that of the sequence $\{v_k\}$ defined by (1.52) with $v = u$.

An important class of RDS is formed by those possessing an additional property of independence of the past and the future. To each such RDS there corresponds a Markov process. We now turn to a description of those RDS and a construction of the associated Markov processes.

1.3.3 Markov RDS

As before, let $(\Omega, \mathcal{F}, \mathbb{P})$ be a complete probability space on which acts a group $\boldsymbol{\theta} = \{\theta_t, t \in \mathcal{T}\}$ of measure-preserving transformations, let X be a Polish space, and let $\boldsymbol{\Phi} = \{\varphi_t^\omega, t \in \mathcal{T}_+\}$ be an RDS in X over $\boldsymbol{\theta}$. For any $p, q \in \mathcal{T}$ with $p < q$, we denote by $\mathcal{F}_{[p,q]} \subset \mathcal{F}$ the sub-σ-algebra generated by the subsets of $\mathcal{F}$ of zero measure and the X-valued random variables $\varphi_t^{\theta_s\omega} u$, where

$$u \in X, \quad t, s \in \mathcal{T}, \quad p \le s < q, \quad 0 < t \le q - s. \tag{1.64}$$

The definition implies that, for any $t, s \in \mathcal{T}$ and $u \in X$, the random variable $\varphi_t^{\theta_s\omega} u$ is $\mathcal{F}_{[s,s+t]}$-measurable. We also define the σ-algebras

$$\begin{aligned} \mathcal{F}_{[-\infty,q]} &= \sigma(\mathcal{F}_{[p,q]} : p \in \mathcal{T}, p < q), \\ \mathcal{F}_{[p,+\infty]} &= \sigma(\mathcal{F}_{[p,q]} : q \in \mathcal{T}, q > p), \\ \mathcal{F}_{[-\infty,+\infty]} &= \sigma(\mathcal{F}_{[p,q]} : p, q \in \mathcal{T}, p < q). \end{aligned}$$

Exercise 1.3.16 Describe the σ-algebra $\mathcal{F}_{[p,q]}$ for the RDS of Example 1.3.15. *Hint:* $\mathcal{F}_{[p,q]}$ coincides with the σ-algebra of the sets of the form

$$\{(\omega_j, j \in \mathbb{Z}) : (\omega_{p+1}, \dots, \omega_q) \in \Gamma\}, \qquad \Gamma \in \underbrace{\mathcal{B}(H) \otimes \cdots \otimes \mathcal{B}(H)}_{q-p \text{ times}}. \tag{1.65}$$

generated by ω_k, $p + 1 \le k \le q$, and the subsets of zero measure.

Exercise 1.3.17 Show that for any finite or infinite numbers $p < q$, we have

$$\theta_t^{-1}(\mathcal{F}_{[p,q]}) = \mathcal{F}_{[p+t,q+t]}, \quad t \in \mathcal{T}.$$

We now introduce an important concept of Markov RDS. Let us set

$$\mathcal{F}^- = \mathcal{F}_{[-\infty,0]}, \quad \mathcal{F}^+ = \mathcal{F}_{[0,+\infty]}.$$

The σ-algebras $\mathcal{F}^-$ and $\mathcal{F}^+$ are called the *past* and the *future* of $\boldsymbol{\Phi}$, respectively.

Definition 1.3.18 The RDS $\boldsymbol{\Phi}$ is said to be *Markov* if its past and future are independent.

Exercise 1.3.19 Show that the discrete-time RDS defined in Example 1.3.15 is Markov.

Exercise 1.3.20 Let $\mathcal{F}_t^- = \mathcal{F}_{[-\infty,t]}$ and $\mathcal{F}_t^+ = \mathcal{F}_{[t,+\infty]}$. Show that, in the case of Markov RDS, the σ-algebras $\mathcal{F}_t^-$ and $\mathcal{F}_t^+$ are independent for any $t \in \mathcal{T}$.

In what follows, we assume that the RDS $\boldsymbol{\Phi}$ under study is Markov. We write $\mathcal{F}_t$ instead of $\mathcal{F}_t^-$ and call $\{\mathcal{F}_t, t \in \mathcal{T}\}$ the *filtration generated by* $\boldsymbol{\Phi}$. Note that the very definition of $\mathcal{F}_t$ implies that the filtered probability space $(\Omega, \mathcal{F}, \mathcal{F}_t, \mathbb{P})$ satisfies the usual hypotheses.

An important property of Markov RDS is that the distribution of the random variable $\varphi_t u$, where u is an X-valued random variable, depends only on the distribution of the initial state u, provided that the latter is $\mathcal{F}_0$-measurable. In other words, a Markov RDS defines an evolution in the space of probability measures on the phase space X. To prove this assertion, we shall need the Markov property; cf. (1.45).

Proposition 1.3.21 *Let $f : X \to \mathbb{R}$ be a bounded measurable function. Then, for any $\mathcal{F}_0$-measurable random variable $u : \Omega \to X$ and any $s, t \in \mathcal{T}_+$, we have*

$$\mathbb{E}\big\{f(\varphi_{s+t}u) \,|\, \mathcal{F}_s\big\} = \big\{\mathbb{E}\, f(\varphi_t v)\big\}\big|_{v=\varphi_s^\omega u} \quad \mathbb{P}\text{-almost surely.} \tag{1.66}$$

Proof Let us consider a function $g : \Omega \times X \to \mathbb{R}$ defined as

$$g(\omega, v) = f(\varphi_t^{\theta_s\omega} v), \quad \omega \in \Omega, \quad v \in X.$$

The definition of the σ-algebras $\mathcal{F}_{[p,q]}$ implies that, for any $v \in X$, the function $g(\omega, v)$ is $\mathcal{F}_s^+$-measurable and that $\varphi_s^\omega u$ is $\mathcal{F}_s$-measurable. Since $\mathcal{F}_s$ and $\mathcal{F}_s^+$ are independent, it follows that (for instance, see problem 9 in section 10.1 of [Dud02])

$$\mathbb{E}\{g(\omega, \varphi_s^\omega u) \,|\, \mathcal{F}_s\} = \big\{\mathbb{E}\, g(\omega, v)\big\}\big|_{v=\varphi_s^\omega u} \quad \text{almost surely.} \tag{1.67}$$

On the other hand, in view of the cocycle property, we have

$$g(\omega, \varphi_s^\omega u) = f(\varphi_t^{\theta_s\omega} \circ \varphi_s^\omega u) = f(\varphi_{s+t}^\omega u). \tag{1.68}$$

Combining (1.67) and (1.68), we obtain (1.66). □

We now introduce the *transition function* for $\boldsymbol{\Phi}$; cf. (1.46). For any $v \in X$ and $t \in \mathcal{T}_+$, we denote by $P_t(v, \cdot)$ the law of $\varphi_t v$:

$$P_t(v, \Gamma) = \mathbb{P}\{\varphi_t v \in \Gamma\}, \quad v \in X, \quad \Gamma \in \mathcal{B}(X). \tag{1.69}$$

The following result is the analogue for Markov RDS of the property described in Exercise 1.3.4.

Corollary 1.3.22 *Let $u : \Omega \to X$ be an $\mathcal{F}_0$-measurable random variable and let $\mu = \mathcal{D}(u)$. Then the distribution μ_t of $\varphi_t^\omega u$ is given by the formula*

$$\mu_t(\Gamma) = \int_X P_t(v, \Gamma)\mu(dv). \tag{1.70}$$

In particular, the measure μ_t depends only on μ (and not on the random variable u).

Proof In view of (1.66) with $f = \mathbb{I}_\Gamma$, for any $\Gamma \in \mathcal{B}(X)$, we have

$$\mathbb{E}\,\mathbb{I}_\Gamma(\varphi_t u) = \mathbb{E}\left\{\mathbb{E}\left(\mathbb{I}_\Gamma(\varphi_t u) \mid \mathcal{F}_0\right)\right\} = \mathbb{E}\left\{\left(\mathbb{E}\,\mathbb{I}_\Gamma(\varphi_t v)\right)\Big|_{v=u}\right\}.$$

It remains to note that $\mathbb{E}\,\mathbb{I}_\Gamma(\varphi_t v) = \mathbb{P}\{\varphi_t v \in \Gamma\} = P_t(v, \Gamma)$. □

Proposition 1.3.21 implies that the transition function (1.69) satisfies the Kolmogorov–Chapman relation. So it is the transition function of some Markov process; see section III.1 in [RY99]. Using the Markov RDS $\boldsymbol{\Phi}$, it is possible to construct a Markov process with this transition function in a canonical way. We now present this construction.

Let us denote by Ω' the product space $X \times \Omega$ endowed with the product σ-algebra $\mathcal{B}(X) \otimes \mathcal{F}$, and introduce the filtration $\mathcal{F}_t' = \mathcal{B}(X) \otimes \mathcal{F}_t$. Let us define the process $u_t(\omega') = \varphi_t^\omega u$, where $\omega' = (u, \omega) \in \Omega'$, and the family of probability measures $\mathbb{P}_u = \delta_u \otimes \mathbb{P}$, where $\delta_u \in \mathcal{P}(X)$ is the Dirac measure concentrated at u. It is straightforward to see that $(u_t, \mathbb{P}_u)$ is a Markov family whose transition function coincides with (1.69). We denote by $\mathfrak{P}_t$ and $\mathfrak{P}_t^*$ the corresponding Markov semigroups. The continuity of $\varphi_t u$ with respect to u implies that $(u_t, \mathbb{P}_u)$ possesses the Feller property; cf. Exercise 1.3.23 below. In what follows, we shall often drop the prime from the notation and write $\omega, \Omega, \mathcal{F}, \mathcal{F}_t$ instead of $\omega', \Omega', \mathcal{F}', \mathcal{F}_t'$. This will not lead to confusion.

The following exercise implies, in particular, that the strong Markov property is true for Markov processes corresponding to continuous Markov RDS. In particular, relations (1.57)–(1.59) hold for them.

Exercise 1.3.23 Show that if $\boldsymbol{\Phi}$ is a continuous Markov RDS, then the corresponding Markov process possesses the Feller property. Combine this fact with Theorem 1.3.10 to conclude that the strong Markov property holds. *Hint:* Use Lebesgue's theorem on dominated convergence.

We conclude this subsection by a simple exercise describing the Markov semigoups in the case of deterministic dynamical systems.

Exercise 1.3.24 Let $\boldsymbol{\Phi} = \{\varphi_k^\omega : X \to X\}$ be an RDS such that

$$\varphi_1^\omega u = \psi(u) \quad \text{for any } \omega \in \Omega, u \in X,$$

where $\psi : X \to X$ is a continuous operator. Describe the Markov semigroups $\mathfrak{P}_k$ and $\mathfrak{P}_k^*$ in terms of ψ.

1.3.4 Invariant and stationary measures

To every RDS $\boldsymbol{\Phi} = \{\varphi_t^\omega, t \in \mathcal{T}_+\}$ there corresponds a semigroup of measurable mappings in an extended phase space. Namely, we consider the product space $(\Omega \times X, \mathcal{F} \otimes \mathcal{B}(X))$ and define a family of mappings $\boldsymbol{\Theta} = \{\Theta_t, t \in \mathcal{T}_+\}$ by the relation

$$\Theta_t(\omega, u) = \big(\theta_t\omega, \varphi_t^\omega u\big), \quad (\omega, u) \in \Omega \times X.$$

Exercise 1.3.25 Show that the family $\boldsymbol{\Theta}$ is a semigroup of measurable transformations.

We now introduce the concept of an invariant measure for $\boldsymbol{\Phi}$. Let $\mathcal{P}(\Omega \times X, \mathbb{P})$ be the set of probability measures on $(\Omega \times X, \mathcal{F} \times \mathcal{B}(X))$ whose projection to Ω coincides with $\mathbb{P}$.

Definition 1.3.26 A measure $\mathfrak{M} \in \mathcal{P}(\Omega \times X, \mathbb{P})$ is said to be *invariant* for $\boldsymbol{\Phi}$ if $(\Theta_t)_*\mathfrak{M} = \mathfrak{M}$ for any $t \in \mathcal{T}_+$. In other words, $\boldsymbol{\Theta}$ is a semigroup of measure-preserving transformations of the space $(\Omega \times X, \mathcal{F} \otimes \mathcal{B}(X), \mathfrak{M})$ (cf. (1.60)).

In what follows, we shall need an alternative description of invariant measures for $\boldsymbol{\Phi}$. Recall that a family $\{\mu_\omega, \omega \in \Omega\} \subset \mathcal{P}(X)$ is called a *random probability measure* if for any $\Gamma \in \mathcal{B}(X)$ the function $\omega \mapsto \mu_\omega(\Gamma)$ from Ω to $\mathbb{R}$ is $(\mathcal{F}, \mathcal{B}(\mathbb{R}))$-measurable. It is well known that any measure $\mathfrak{M} \in \mathcal{P}(\Omega \times X, \mathbb{P})$ admits a *disintegration*

$$\mathfrak{M}(d\omega, du) = \mu_\omega(du)\,\mathbb{P}(d\omega), \tag{1.71}$$

where $\{\mu_\omega\}$ is a random probability measure on X. Relation (1.71) means that if $f : \Omega \times X \to \mathbb{R}$ is a bounded measurable function, then

$$\iint_{\Omega\times X} f(\omega, u)\mathfrak{M}(d\omega, du) = \int_\Omega \left\{\int_X f(\omega, u)\mu_\omega(du)\right\}\mathbb{P}(d\omega).$$

Moreover, representation (1.71) is unique up to a set of zero measure in the sense that if $\{\mu'_\omega, \omega \in \Omega\}$ is another random probability measure for which (1.71) holds, then $\mu_\omega = \mu'_\omega$ for almost every $\omega \in \Omega$. We refer the reader to section 10.2 of [Dud02] for the proof of these results. Note that if $\mathfrak{M}$ is represented in the form (1.71), then the projection of $\mathfrak{M}$ to X coincides with the mean value of μ_ω. That is,

$$(\Pi_X)_*\,\mathfrak{M} = \mathbb{E}\,\mu_\cdot := \int_\Omega \mu_\omega \mathbb{P}(d\omega).$$

Proposition 1.3.27 *A measure $\mathfrak{M} \in \mathcal{P}(\Omega \times X, \mathbb{P})$ is invariant for $\boldsymbol{\Phi}$ if and only if for any $t \in \mathcal{T}_+$ its disintegration $\{\mu_\omega\}$ satisfies the relation*

$$(\varphi_t^\omega)_* \mu_\omega = \mu_{\theta_t\omega} \quad \textit{for almost all } \omega \in \Omega. \tag{1.72}$$

Proof Suppose that $\mathfrak{M}$ is an invariant measure for $\boldsymbol{\Phi}$. Then for any bounded measurable functions $f : X \to \mathbb{R}$ and $g : \Omega \to \mathbb{R}$ we have

$$\int_\Omega \int_X f(\varphi_t^\omega u) g(\theta_t\omega)\, \mu_\omega(du)\mathbb{P}(d\omega) = \int_\Omega \int_X f(u) g(\omega)\, \mu_\omega(du)\mathbb{P}(d\omega). \tag{1.73}$$

Replacing $g(\omega)$ by $g(\theta_{-t}\omega)$ and using the invariance of $\mathbb{P}$ with respect to $\boldsymbol{\theta}$, we see that

$$\int_\Omega \biggl\{ \int_X f(\varphi_t^\omega u)\, \mu_\omega(du) \biggr\} g(\omega)\mathbb{P}(d\omega) = \int_\Omega \biggl\{ \int_X f(u)\, \mu_{\theta_t\omega}(du) \biggr\} g(\omega)\, \mathbb{P}(d\omega).$$

Since this relation is true for any function g, we conclude that

$$\int_X f(\varphi_t^\omega u)\, \mu_\omega(du) = \int_X f(u)\, \mu_{\theta_t\omega}(du) \quad \text{for almost all } \omega \in \Omega. \tag{1.74}$$

Now note that the Borel σ-algebra on a Polish space is countably generated (see Corollary A.1.3 in the Appendix). Therefore (1.74) implies that (1.72) holds for any $t \in \mathcal{T}_+$.

Conversely, if (1.72) holds, then reversing the above arguments, we arrive at (1.73) for any bounded measurable functions f and g. Application of the monotone class technique (see Theorem A.1.1 in the Appendix) shows that $\mathfrak{M}$ is invariant for $\boldsymbol{\Phi}$. □

We conclude this subsection with the concept of a stationary measure for a Markov RDS. Recall that the Markov semigroups for such an RDS were introduced in the foregoing subsection.

Definition 1.3.28 A measure $\mu \in \mathcal{P}(X)$ is said to be *stationary* for an RDS $\boldsymbol{\Phi}$ if it is stationary for the corresponding Markov process, i.e., $\mathfrak{P}_t^* \mu = \mu$ for any $t \in \mathcal{T}_+$.

Note that, in contrast to invariant measures, stationary measures are defined only for Markov RDS.

Exercise 1.3.29 Let $\mathcal{F}_t$ be the filtration generated by $\boldsymbol{\Phi}$ and let $u : \Omega \to X$ be an $\mathcal{F}_0$-measurable random variable such that $\mathcal{D}(u) = \mu$. Show that if μ is a stationary measure for $\boldsymbol{\Phi}$, then

$$\mathcal{D}(\varphi_t u) = \mu \quad \text{for all } t \in \mathcal{T}_+.$$

Let $\mathcal{A} \subset X$ be a closed subset. We shall say that $\mathcal{A}$ is *absorbing* for the RDS $\boldsymbol{\Phi}$ if for any $u \in X$ there is an almost surely finite random time $T_u \in \mathcal{T}_+$ such that, with probability 1, we have

$$\varphi_t^\omega u \in \mathcal{A} \quad \text{for any } t \geq T_u.$$

Lemma 1.3.30 *If $\mathcal{A}$ is an absorbing set for the RDS $\boldsymbol{\Phi}$ and μ is a stationary measure for $\boldsymbol{\Phi}$, then* $\operatorname{supp} \mu \subset \mathcal{A}$.

Proof We need to show that if a function $f \in C_b(X)$ vanishes on $\mathcal{A}$, then $(f, \mu) = 0$. To this end, note that

$$(f, \mu) = (f, \mathfrak{P}_t^* \mu) = (\mathfrak{P}_t f, \mu) = \int_H \mathfrak{P}_t f(u)\, \mu(du).$$

Since $\mathcal{A}$ is an absorbing set, we have $f(\varphi_t u) = 0$ for $t \geq T_u$, whence, by Lebesgue's theorem on dominated convergence, we conclude that

$$\mathfrak{P}_t f(u) = \mathbb{E}\, f(\varphi_t u) \to 0 \quad \text{as } t \to \infty.$$

Hence, $(f, \mu) = 0$ for any $f \in C_b(X)$ vanishing on $\mathcal{A}$. This completes the proof of the lemma. □

We conclude this chapter with a result showing that the global stability of a Markov RDS implies the uniqueness of a stationary measure. For simplicity, we confine ourselves to the case of a separable Banach space X with norm $\|\cdot\|$.

Proposition 1.3.31 *Let $\boldsymbol{\Phi} = \{\varphi_t, t \in \mathcal{T}_+\}$ be a Markov RDS in X such that*

$$\|\varphi_t^\omega u - \varphi_t^\omega v\| \leq \psi_{u,v} \alpha(t) \|u - v\|, \quad t \in \mathcal{T}_+, \quad \omega \in \Omega, \quad u, v \in X, \quad (1.75)$$

where $R > 0$ is arbitrary, $\psi_{u,v}$ is an almost surely finite random variable measurable with respect to its arguments, and $\alpha(t)$ is a function going to zero as $t \to +\infty$. Then $\boldsymbol{\Phi}$ has at most one stationary measure. Moreover, if a stationary measure μ exists, then for any $R \geq 1$, $u \in X$, and $t \in \mathcal{T}_+$, we have

$$\|P_t(u, \cdot) - \mu\|_L^* \leq 2\mu\big(B_X(R)^c\big) + \int_{B_X(R)} \mathbb{E}\Big(2 \wedge \big(\psi_{u,v}\alpha(t)\|u - v\|\big)\Big)\mu(dv). \quad (1.76)$$

Proof Let $\mu, \nu \in \mathcal{P}(X)$ be two stationary distributions. Then for any bounded measurable function $f : X \to \mathbb{R}$ we have

$$(f, \mu) - (f, \nu) = (\mathfrak{P}_t f, \mu) - (\mathfrak{P}_t f, \nu) = \iint_{X \times X} \mathbb{E}\big(f(\varphi_t u) - f(\varphi_t v)\big)\mu(du)\nu(dv).$$

Assuming that $f \in L_b(X)$ with $\|f\|_L \le 1$ and using inequality (1.76), for any $R > 0$ we derive

$$|(f, \mu) - (f, \nu)| \le 2\mu\big(B_X^c(R)\big) + 2\nu\big(B_X^c(R)\big) + 2 \iint_{X \times X} \mathbb{E}\big(1 \wedge R\psi_{u,v}\alpha(t)\big)\mu(du)\mu(dv).$$

Passing to the limit as $t \to +\infty$ and then $R \to +\infty$, we see that $(f, \mu) = (f, \nu)$ for any $f \in L_b(X)$. This implies that $\mu = \nu$.

We now prove (1.76). To this end, we essentially repeat the above argument. Namely, take an arbitrary function $f \in C_b(X)$ and write

$$(f, P_t(u, \cdot)) - (f, \mu) = (\mathfrak{P}_t f, \delta_u) - (\mathfrak{P}_t f, \mu) = \int_X \mathbb{E}\left(f(\varphi_t u) - f(\varphi_t v)\right) \mu(dv).$$

Using (1.75) to bound the right-hand side of this relation and taking the supremum over all $f \in L_b(X)$ with $\|f\|_L \le 1$, we arrive at the required result. □

Notes and comments

The basic facts about Sobolev spaces used in this monograph are very well known, and their proof is widely available in the literature; see the books of Lions and Magenes [LM72] and Taylor [Tay97]. A comprehensive treatment of the theory of Sobolev spaces can be found, for instance, in the books [Ada75; Maz85] by Adams and Maz'ja. Classes of measurable functions with range in a Banach or Hilbert space and the concept of Bochner's integral on an abstract measurable space are studied in chapter V of [Yos95], and a rich source of their applications in the theory of nonlinear PDEs is the book [Lio69].

An excellent book about measure theory on metric spaces is the one by Dudley [Dud02]. A lot of useful information can also be found in Bogachev's two-volume book [Bog07]. The idea of coupling goes back to Doeblin [Doe38; Doe40], but its systematic application in the theory of stochastic processes started much later; see the papers [Har55; Pit74] by Harris and Pitman. Lemma 1.2.24 was often used by Dobrushin [Dob68; Dob74] in his celebrated work on Gibbs systems.

The general theory of random dynamical systems and Markov processes is treated in many textbooks and monographs; e.g., see the books of Revuz [Rev84], Kifer [Kif86], Karatzas and Shreve [KS91], and L. Arnold [Arn98]. Our presentation of this subject essentially follows one of these references.

2

Two-dimensional Navier–Stokes equations

In this chapter we present some well-known results on the 2D Navier–Stokes equations. We begin with the deterministic case. Galerkin approximations are used to prove the existence, uniqueness, and regularity of solutions. Some important estimates due to Foiaş and Prodi are also established. We next consider the Navier–Stokes system perturbed by a random external force and prove the well-posedness of the Cauchy problem and some a priori estimates for solutions. Combining these results with the classical Bogolyubov–Krylov argument, it is then shown that the randomly forced Navier–Stokes equations have at least one stationary measure.

2.1 Cauchy problem for the deterministic system

2.1.1 Equations and boundary conditions

The 2D Navier–Stokes (NS) equations have the form

$$\begin{cases} \dot{u} + \langle u, \nabla \rangle u - \nu \Delta u + \nabla p = f(t, x), \\ \operatorname{div} u = 0. \end{cases} \tag{2.1}$$

Here $u = (u^1, u^2)$ and p are the unknown velocity field and pressure, $\nu > 0$ is the kinematic viscosity, f is the density of an external force, and $\langle u, \nabla \rangle$ stands for the differential operator $u^1 \partial_1 + u^2 \partial_2$. Equations (2.1) are considered either in a bounded domain $Q \subset \mathbb{R}^2$ with a smooth boundary ∂Q or on the standard torus $\mathbb{T}^2 = \mathbb{R}^2 / 2\pi \mathbb{Z}^2$. The latter means that the space variable $x = (x_1, x_2)$ belongs to $\mathbb{R}^2$, and the functions u, p, and f are assumed to be 2π-periodic with respect to x_i, $i = 1, 2$. All the results and proofs remain valid without any change for non-standard tori $\mathbb{T}^2 / a(\mathbb{Z} \oplus b\mathbb{Z})$, where $a, b > 0$. In the case of a bounded domain, Eqs. (2.1) are supplemented with the no-slip boundary condition:

$$u\big|_{\partial Q} = 0. \tag{2.2}$$

The Cauchy problem for the Navier–Stokes system consists in finding a pair of functions (u, p) of appropriate regularity that satisfy (2.1) and (2.2) in a sense to be specified, as well as the initial condition

$$u(0, x) = u_0(x), \tag{2.3}$$

where u_0 is a given function. In the case of the torus, the boundary condition (2.2) should be omitted.

In what follows, to simplify the presentation, the results of this chapter are usually proved for the torus. If the corresponding proof in the case of a bounded domain is different, or the result is not valid, we mention this explicitly.

2.1.2 Leray decomposition

Let $H^m(\mathbb{T}^2; \mathbb{R}^2)$ be the space of vector fields on $\mathbb{T}^2$ whose components belong to the Sobolev space of order m and let

$$H^m_\sigma = \{u \in H^m(\mathbb{T}^2; \mathbb{R}^2) : \operatorname{div} u = 0 \text{ on } \mathbb{T}^2\},$$

where the divergence is taken in the sense of distributions. It is clear that H^m_σ is a closed subspace of $H^m(\mathbb{T}^2; \mathbb{R}^2)$.

The following result can be easily established with the help of Fourier expansions of square-integrable functions.

Exercise 2.1.1 Show that H^m_σ coincides with the closure in $H^m(\mathbb{T}^2; \mathbb{R}^2)$ of the space

$$\mathcal{V} = \{u \in C^\infty(\mathbb{T}^2; \mathbb{R}^2) : \operatorname{div} u = 0 \text{ on } \mathbb{T}^2\}.$$

Let us recall that $\dot{H}^m = \dot{H}^m(\mathbb{T}^2)$ stands for the space of functions in $H^m(\mathbb{T}^2)$ with zero mean value (see (1.2)), and let ∇H^{m+1} be the space of functions $u \in H^m(\mathbb{T}^2; \mathbb{R}^2)$ that are representable in the form $u = \nabla p$ for some $p \in H^{m+1}$. Since $\|\nabla p\|_k$ is a norm on $\dot{H}^{m+1}$, we see that ∇H^{m+1} is a closed subspace in $H^m(\mathbb{T}^2; \mathbb{R}^2)$. The following result due to Helmholtz has been a common tool in the theory of Navier–Stokes equations since the work of Leray.

Theorem 2.1.2 *For any $m \in \mathbb{R}$, the space $H^m(\mathbb{T}^2; \mathbb{R}^2)$ admits the direct decomposition*

$$H^m(\mathbb{T}^2; \mathbb{R}^2) = H^m_\sigma \dotplus \nabla H^{m+1}. \tag{2.4}$$

Moreover, the sum is orthogonal for $m = 0$, and decompositions (2.4) *corresponding to two different values m_1 and m_2 give the same representation for any function belonging to the intersection $H^{m_1}(\mathbb{T}^2; \mathbb{R}^2) \cap H^{m_2}(\mathbb{T}^2; \mathbb{R}^2)$.*

Proof We first note that the intersection of H_σ^m and ∇H^{m+1} consists of the zero function. Let us expand a function $u \in H^m(\mathbb{T}^2;\mathbb{R}^2)$ in the Fourier series:

$$u(x) = \sum_{s\in\mathbb{Z}^2} u_s e^{isx}, \quad u_s \in \mathbb{C}^2.$$

Denoting $s^\perp = (-s_2, s_1)$, we write

$$u(x) = \sum_{s\in\mathbb{Z}^2} \frac{\langle u_s, s^\perp\rangle s^\perp}{|s|^2} e^{isx} + \sum_{s\in\mathbb{Z}^2} \frac{\langle u_s, s\rangle s}{|s|^2} e^{isx}. \tag{2.5}$$

To prove (2.4), it remains to note that the first sum on the right-hand side is an element of H_σ^m, while the second is the gradient of the function

$$-i\sum_{s\in\mathbb{Z}^2} \frac{\langle u_s, s\rangle}{|s|^2} e^{isx},$$

which belongs to H^{m+1}. The fact that (2.4) is an orthogonal sum for $m = 0$ is easy to show with the help of integration by parts, and compatibility of decompositions (2.4) corresponding to different values of m follows from (2.5). □

In what follows, we denote by $\Pi : H^m(\mathbb{T}^2;\mathbb{R}^2) \to H_\sigma^m$ the projection associated with decomposition (2.4). It is called the *Leray projection*. Even though the definition of Π depends on m, Theorem 2.1.2 implies that the function Πu depends only on u, and not on the space $H^m(\mathbb{T}^2;\mathbb{R}^2)$ in which the operator Π is considered. Therefore we do not indicate the dependence of the projection operators on m.

An analogue of the Leray–Helmholtz decomposition (2.4) is true for any bounded Lipschitz domain. Let us formulate the corresponding result in the case $m = 0$. We set

$$L_\sigma^2(Q;\mathbb{R}^2) = \{u \in L^2(Q;\mathbb{R}^2) : \operatorname{div} u = 0 \text{ in } Q,\ \langle u, \boldsymbol{n}\rangle = 0 \text{ on } \partial Q\},$$

where $\boldsymbol{n}$ denotes the outward unit normal to ∂Q. Furthermore, define ∇H^1 as the space of functions $u \in L^2(Q;\mathbb{R}^2)$ that are representable in the form $u = \nabla p$ for some $p \in H^1(Q)$. A proof of the following theorem can be found in [Tem79, chapter 1].

Theorem 2.1.3 *The subspaces L_σ^2 and ∇H^1 are closed in $L^2(Q;\mathbb{R}^2)$, and we have the orthogonal decomposition*

$$L^2(Q;\mathbb{R}^2) = L_\sigma^2 \oplus \nabla H^1. \tag{2.6}$$

As before, we denote by $\Pi : L^2(Q;\mathbb{R}^2) \to L_\sigma^2$ the Leray projection. It can be shown that Π is continuous in $H^1(Q,\mathbb{R}^2)$ and admits a continuous extension to an operator from $H^{-1}(Q,\mathbb{R}^2)$ to the dual space of $L_\sigma^2 \cap H_0^1(Q,\mathbb{R}^2)$; see section I.1 in [Tem79]. We refer the reader to chapter 17 of the book [Tay97] for a detailed presentation of the corresponding results in the case when Q is a compact manifold.

2.1.3 Properties of some multilinear maps

In this subsection, we establish some elementary estimates of linear and nonlinear functionals relevant to the Navier–Stokes equations. We shall need two auxiliary lemmas.

Lemma 2.1.4 *Let $X_1, \dots, X_r$ and Y be Banach spaces and let F be an r-linear map from the direct product $X_1 \times \cdots \times X_r$ to Y such that*

$$\|F(u_1, \dots, u_r)\|_Y \le C\|u_1\|_{X_1} \dots \|u_r\|_{X_r} \quad \text{for all } u_j \in X_j,\ j = 1, \dots, r. \tag{2.7}$$

Then F is continuous. Moreover, if $V_i \subset X_i$, $i = 1, \dots, r$, are dense vector spaces, F is defined for $u_j \in V_j$, and inequality (2.7) holds for $u_j \in V_j$, then F extends to an r-linear continuous map from $X_1 \times \cdots \times X_r$ to Y.

The proof of this lemma is straightforward. To formulate the second result, we introduce some notation. For any sufficiently regular vector fields u and v on the torus $\mathbb{T}^2$, we set

$$Lu = -\Pi\Delta u, \quad B(u, v) = \Pi\big(\langle u, \nabla\rangle v\big). \tag{2.8}$$

For the periodic boundary conditions, Lu equals $-\Delta u$, but this is not the case for bounded domains and Dirichlet boundary conditions. It is easy to check with the help of a Fourier expansion that L is a self-adjoint operator in L^2_σ with domain $D(L) = H^2_\sigma$. It is called the *Stokes operator*.

Define the space

$$\mathcal{H} = \{u \in L^2(0, T; H^1_\sigma) : \dot{u} \in L^2(0, T; H^{-1}_\sigma)\}$$

and endow it with the norm

$$\|u\|_{\mathcal{H}} = \left(\int_0^T \big(\|u(t)\|_1^2 + \|\dot{u}(t)\|_{-1}^2\big)\, dt\right)^{1/2},$$

where the time derivative is understood in the sense of distributions. Note that $\mathcal{H}$ is a Hilbert space. A proof of the following result can be found in [LM72] (see theorem 3.1 in chapter 1) and [Lio69] (see theorem 5.1 in chapter 1).

Lemma 2.1.5 *The space $\mathcal{H}$ is continuously embedded in $C(0, T; L^2_\sigma)$ and compactly embedded in $L^2(0, T; H^m_\sigma)$ for any $m < 1$. In particular, $u(t)$ is a well-defined vector in L^2_σ for any $u \in \mathcal{H}$ and $t \in [0, T]$.*

We now turn to some estimates that will be needed in the sequel. The first result essentially is an integration-by-parts formula.

Proposition 2.1.6 *For any $u, v \in \mathcal{H}$ and $0 \le t \le T$, we have*

$$\int_0^T \langle Lu(t),\ v(t)\rangle\, dt = \int_0^T \langle \nabla u(t),\ \nabla v(t)\rangle\, dt\,, \tag{2.9}$$

$$\int_0^t \langle \dot u, u\rangle\, ds = \frac{1}{2}\big(|u(t)|_2^2 - |u(0)|_2^2\big)\,. \tag{2.10}$$

Proof In view of Lemma 2.1.4, it suffices to establish (2.9) and (2.10) for smooth functions u and v. In this case, both relations are obvious. □

Let us note that the function $t \mapsto \langle \dot u, u\rangle$ belongs to $L_1(0, T)$ for any $u \in \mathcal{H}$. It follows from (2.10) that $|u(t)|^2$ is an absolutely continuous function, and

$$\frac{d}{dt}\, |u(t)|^2 = 2\langle \dot u, u\rangle \quad \text{for any } u \in \mathcal{H}\,.$$

Proposition 2.1.7 *For any divergence-free smooth vector fields u, v, and w, we have*

$$\langle B(u, v), v\rangle = 0, \tag{2.11}$$

$$\langle B(u, v), w\rangle = -\langle B(u, w), v\rangle, \tag{2.12}$$

$$|\langle B(u, v), w\rangle| \le C\|u\|_{1/2}\|v\|_{1/2}\|w\|_1, \tag{2.13}$$

$$\|B(u, v)\|_{-1} \le C\|u\|_{1/2}\,\|v\|_{1/2}\,, \tag{2.14}$$

$$\|B(u, v)\|_{-3} \le C\,|u|_2\,|v|_2\,, \tag{2.15}$$

where $C > 0$ is a constant not depending on the functions.

Proof Integrating by parts, we derive

$$\langle B(u, v), v\rangle = \sum_{j,l=1}^{2} \int_{\mathbb{T}^2} u^j(\partial_j v^l)\, v^l\, dx = \sum_{j,l=1}^{2} \frac{1}{2} \int_{\mathbb{T}^2} u^j\, \partial_j |v|^2\, dx$$

$$= -\frac{1}{2}\int_{\mathbb{T}^2} (\operatorname{div} u)|u|^2 dx = 0\,.$$

This proves (2.11). To establish (2.12), it suffices to use (2.11) with v replaced by $v + w$.

Let us prove (2.13). Applying consecutively relation (2.12), the Hölder inequality, and the continuous embedding $H^{1/2} \subset L^4$ (see Property 1.1.2), we obtain

$$|\langle B(u, v), w\rangle| = |\langle B(u, w), v\rangle| \le C_1 \int_{\mathbb{T}^2} |u|\ |\nabla w|\ |v|\, dx$$

$$\le C_2\,|\nabla w|_2\,|u|_4\,|v|_4 \le C_3\|w\|_1\,\|u\|_{1/2}\,\|v\|_{1/2}\,. \tag{2.16}$$

Inequality (2.14) follows from (2.13) by duality. Finally, to establish (2.15), it suffices to estimate the integral in (2.16) by the product $|\nabla w|_\infty\,|u|_2\,|v|_2$ and apply the Sobolev embedding $H^2 \subset L^\infty$. □

Combining Lemma 2.1.4 and Propositions 2.1.6 and 2.1.9, we obtain the following result.

Corollary 2.1.8 *The operator B defined initially on smooth vector fields extends to a continuous bilinear map from $H_\sigma^{1/2} \times H_\sigma^{1/2}$ to H_σ^{-1} and from $L^2_\sigma \times L^2_\sigma$ to H_σ^{-3}.*

Consider now a trilinear map b defined on smooth non-autonomous vector fields on $\mathbb{T}^2$ by the relation

$$b(u, v, w) = \big\langle B\big(u(t,\cdot), v(t,\cdot)\big), w(t,\cdot)\big\rangle.$$

Proposition 2.1.9 *The map $b(u, v, w)$ is continuous from $\mathcal{H} \times \mathcal{H} \times L^2(0, T; H^1_\sigma)$ to $L_1(0, T)$, and*

$$b(u, v, v) = 0 \quad \textit{for any } u, v \in \mathcal{H}, \tag{2.17}$$

where the equality holds in the space $L_1(0, T)$. In particular, for any $u, v \in \mathcal{H}$ the function $B(u, v)$ belongs to $L^2(0, T; H_\sigma^{-1})$.

Proof We first note that $\mathcal{H} \subset L^4(0, T; H^{1/2})$. Indeed, since $\|u\|_{1/2}^4 \le |u|_2^2\, \|u\|_1^2$, we have

$$\|u\|^4_{L^4(0,T;H^{1/2})} = \int_0^T \|u(t)\|_{1/2}^4\, dt \le \Big(\sup_{0\le t\le T} |u(t)|_2^2\Big) \int_0^T \|u(t)\|_1^2\, dt \le \|u\|_{\mathcal{H}}^4 .$$

Hence, using (2.14), for smooth functions $u, v \in \mathcal{H}$ and $w \in L^2(0, T; H^1_\sigma)$, we obtain

$$\begin{aligned}
&\int_0^T |b(u, v, w)|\, ds \\
&\le C \int_0^T \|w\|_1\, \|u\|_{1/2}\, \|v\|_{1/2}\, ds \\
&\le \Big(\int_0^T \|w\|_1^2\, ds\Big)^{1/2} \Big(\int_0^T \|u\|_{1/2}^4\, ds\Big)^{1/4} \Big(\int_0^T \|v\|_{1/2}^4\, ds\Big)^{1/4} \\
&\le C\, \|w\|_{L^2(0,T;H^1)} \|u\|_{\mathcal{H}} \|v\|_{\mathcal{H}} .
\end{aligned} \tag{2.18}$$

Lemma 2.1.4 now implies the continuity of b. By duality, it follows from inequality (2.18) that $B(u, v)$ belongs to $L^2(0, T; H_\sigma^{-1})$. Finally, relation (2.17), which is true for smooth functions in view of (2.11), extends to the general case by continuity. □

2.1.4 Reduction to an abstract evolution equation

We now introduce the concept of a solution for the Navier–Stokes system. Let us fix a constant $T > 0$ and functions $f \in L^2(0, T; H^{-1})$ and $u_0 \in L^2_\sigma(\mathbb{T}^2, \mathbb{R}^2)$. We shall denote by $\mathcal{D}'$ the space of $\mathbb{R}^2$-valued distributions on $(0, T) \times \mathbb{T}^2$.

Definition 2.1.10 A pair of functions (u, p) is called a *solution of Eq.* (2.1) on $(0, T) \times \mathbb{T}^2$ if $u \in \mathcal{H}$, $p \in L^2(0, T; L^2)$, and (2.1) holds in $\mathcal{D}'$. If, in addition, relation (2.3) is also satisfied, then (u, p) is called a *solution of the Cauchy problem* (2.1), (2.3).

Let us note that (2.1) is not a system of evolution equations in the sense that it does not contain the time derivative of the unknown function p. However, it is possible to exclude the pressure from the problem in question and to obtain a nonlocal nonlinear PDE which can be regarded as an evolution equation in a Hilbert space.

To this end, let us formally apply the Leray projection Π to the first equation in (2.1). Since $\Pi(\nabla p) = 0$, $u(t) \in L^2_\sigma$ for any t, and $\Pi \dot{u} = \dot{u}$, using notation (2.8) we obtain

$$\dot{u} + \nu L u + B(u) = \Pi f(t), \tag{2.19}$$

where $B(u) = B(u, u)$. Let us introduce the concept of a solution for (2.19).

Definition 2.1.11 A function $u \in \mathcal{H}$ is called a *solution of Eq.* (2.19) on $(0, T) \times \mathbb{T}^2$ if (2.19) holds in $\mathcal{D}'$. If, in addition, the initial condition (2.3) is also satisfied, then u is called a *solution of the Cauchy problem* (2.19), (2.3).

Note that the functions Δu and $\langle u, \nabla \rangle u$ are elements of $L^2(0, T; H^{-1})$, and therefore their Leray projections are well-defined and belong to $L^2(0, T; H^{-1}_\sigma)$. Thus, all the terms in (2.2) belong to $\mathcal{D}'$, and Definition 2.1.11 makes sense. The following result shows that the Cauchy problems for (2.1) and (2.19) are equivalent.

Theorem 2.1.12 *Let a pair* $(u, p) \in \mathcal{H} \times L^2(0, T; L^2)$ *be a solution of* (2.1) *on* $(0, T) \times \mathbb{T}^2$. *Then* $u \in \mathcal{H}$ *is a solution of* (2.19). *Conversely, if* $u \in \mathcal{H}$ *is a solution of* (2.19), *then there is a* $p \in L^2(0, T; L^2)$ *such that* (u, p) *is a solution of* (2.1).

Proof Let (u, p) be a solution for (2.1). Then integrating the first equation in (2.1) with respect to time, we obtain

$$u(t) = u_0 + \int_0^t \big(\nu \Delta u - \langle u, \nabla \rangle u - \nabla p + f\big)\, ds, \quad 0 \le t \le T, \tag{2.20}$$

where the equality holds in H^{-1}. Applying to (2.20) the Leray projection Π, which is continuous in H^{-1}, and using the fact that Π commutes with the integration in time, we obtain

$$u(t) = u_0 + \int_0^t \big(-\nu L u - B(u) + \Pi f\big)\, ds, \quad 0 \le t \le T. \tag{2.21}$$

This implies that (2.19) holds in $\mathcal{D}'$. Conversely, let $u \in \mathcal{H}$ be a solution of (2.19). Then $\dot{u} \in L^2(0, T; H_\sigma^{-1})$. Consider the equation

$$\nabla p(t) = g(t) := -\dot{u} + \nu \Delta u - \langle u, \nabla \rangle u + f, \tag{2.22}$$

whose right-hand side g belongs to $L^2(0, T; H^{-1})$. If we show that

$$g(t) \in \nabla L^2 \quad \text{for almost all } t \in [0, T], \tag{2.23}$$

then we can resolve (2.23) with respect to p and find a solution $p \in L^2(0, T; L^2)$. In this case, the construction will imply that (u, p) satisfies (2.1) in the sense of distributions.

To prove (2.23), note that u satisfies (2.21). It follows that

$$\Pi\biggl(\int_0^t g(s)\, ds\biggr) = \int_0^t \Pi g(s)\, ds = 0 \quad \text{for } 0 \le t \le T,$$

whence we conclude that $\Pi g(t) = 0$ for almost all $t \in [0, T]$. This implies the required relation (2.23). □

An analogue of Theorem 2.1.12 is true for the Navier–Stokes system in a bounded domain. However, its proof is more complicated, and we refer the reader to the book [Tem79] for the details.

2.1.5 Existence and uniqueness of solution

In this subsection, we prove that problem (2.19), (2.3) possesses a unique solution $u \in \mathcal{H}$ for any functions $u_0 \in L_\sigma^2$ and $f \in L^2(0, T; H^{-1})$. In view of Theorem 2.1.12, this will imply that problem (2.1), (2.3) is also well posed in an appropriate functional space.

Theorem 2.1.13 *Let $u_0 \in L_\sigma^2$ and $f \in L_2(0, T; H^{-1})$. Then problem* (2.19), (2.3) *has a unique solution $u \in \mathcal{H}$, which satisfies the inequality*

$$|u(t)|_2^2 + \nu \int_0^t |\nabla u(s)|_2^2\, ds \le |u_0|_2^2 + \nu^{-1} \int_0^t \|f(s)\|_{-1}^2\, ds, \quad 0 \le t \le T. \tag{2.24}$$

Proof To simplify the presentation, we shall assume that $\nu = 1$.

Uniqueness. Let $u_1, u_2 \in \mathcal{H}$ be two solutions equal to u_0 at $t = 0$. Then the difference $u = u_2 - u_1$ satisfies the equation

$$\dot{u} + Lu + B(u_2, u) + B(u, u_1) = 0. \tag{2.25}$$

Multiplying this relation in H by $u(s)$ and integrating in time, we derive

$$\int_0^t \bigl(\langle \dot{u}, u\rangle + \langle Lu, u\rangle + \langle B(u, u_1), u\rangle\bigr)\, ds = 0. \tag{2.26}$$

Since $\langle Lu, u\rangle = |\nabla u|_2^2$ and $u(0) = 0$, combining (2.26) with (2.10), (2.13) and using the second inequality in (1.8), we obtain

$$\frac{1}{2}|u(t)|_2^2 + \int_0^t |\nabla u(s)|_2^2 ds = -\int_0^t \langle B(u, u_1), u\rangle ds$$
$$\leq C_1 \int_0^t |\nabla u_1|_2 |\nabla u|_2 |u|_2\, ds$$
$$\leq \frac{1}{2}\int_0^t |\nabla u(s)|_2^2\, ds + C_2 \int_0^t |\nabla u_1|_2^2\, |u|_2^2\, ds.$$

It follows that

$$|u(t)|^2 + \int_0^t |\nabla u(s)|_2^2 ds \leq 2C_2 \int_0^t |\nabla u_1(s)|_2^2\, |u(s)|_2^2 ds.$$

Since $t \mapsto |\nabla u(t)|_2^2$ is an integrable function, Gronwall's lemma implies that $u \equiv 0$.

Existence. The proof of existence is based on a well-known general approach called the *Galerkin method*. Roughly speaking, we construct solutions for finite-dimensional approximations of the equation and then show that they have a limit point which satisfies the original equation. The proof is divided into three steps.

Step 1: A priori estimate. Let $u(t, x)$ be a *smooth* solution of (2.19). We consider the function $u \mapsto \frac{1}{2}|u|_2^2$ and calculate its derivative along trajectories of (2.19):

$$\frac{1}{2}\frac{d}{dt}|u(t)|_2^2 = \langle u(t), \dot{u}(t)\rangle = \langle u, -Lu - B(u) + \Pi f(t)\rangle$$
$$= -|\nabla u|_2^2 + \langle u, f\rangle \leq -|\nabla u|_2^2 + |\nabla u|_2\, \|f\|_{-1}$$
$$\leq -\frac{1}{2}|\nabla u|_2^2 + \frac{1}{2}\|f\|_{-1}^2,$$

where we used (2.11). Integrating in time, we obtain inequality (2.24) with $\nu = 1$. It follows that

$$\|u\|_{L^\infty(0,T;H)} + \|u\|_{L^2(0,T;H^1)} \leq |u_0|_2 + \|f\|_{L_2(0,T;H^{-1})} =: C_1(u_0, f). \quad (2.27)$$

Furthermore, using (2.14) and the interpolation inequality, we obtain

$$\|B(u)\|_{-1} \leq C\, \|u\|_{1/2}^2 \leq C\, |u|\, |\nabla u|_2,$$

whence it follows that

$$\|B(u)\|_{L^2(0,T;H^{-1})} \leq C_2(u_0, f).$$

Expressing $\dot{u}$ in terms of u from (2.19) and recalling (2.27), we get

$$\|\dot{u}\|_{L^2(0,T:H^{-1})} \leq C_3(u_0, f).$$

We have thus shown that

$$\|u\|_{\mathcal{H}} \le C(u_0, f). \tag{2.28}$$

Step 2: Galerkin approximations. Let us introduce the standard trigonometric basis in L^2_σ. Denote by Z^2_0 the set of non-zero integer vectors $s = (s_1, s_2)$. For $s \in \mathbb{Z}^2_0$, let

$$e_s = \begin{cases} c_s s^\perp \sin\langle s, x\rangle, & s \in \mathbb{Z}^2_+, \\ c_s s^\perp \cos\langle s, x\rangle, & s \in \mathbb{Z}^2_-, \end{cases} \tag{2.29}$$

where $c_s = (\sqrt{2}\pi|s|)^{-1}$, $s^\perp = (-s_2, s_1)$, $\mathbb{Z}^2_+$ stands for the set of vectors $s \in \mathbb{Z}^2_0$ such that either $s_1 > 0$ or $s_1 = 0$ and $s_2 > 0$, and $\mathbb{Z}^2_-$ is the complement of $\mathbb{Z}^2_+$ in $\mathbb{Z}^2_0$. Given an integer $N > 0$, we write

$$H_{(N)} = \operatorname{span}\big(\{e_s, |s| \le N\} \cup \{(1,0), (0,1)\}\big) \tag{2.30}$$

and denote by

$$\mathsf{P}_N : L^2(\mathbb{T}^2; \mathbb{R}^2) \to L^2(\mathbb{T}^2; \mathbb{R}^2)$$

the orthogonal projection onto $H_{(N)}$. It is clear that

$$H_{(N)} \subset C^\infty \cap L^2_\sigma, \quad \dim H_{(N)} < \infty,$$

and that the operator L maps the subspace $H_{(N)}$ into itself. Let us apply P_N to (2.19):

$$\mathsf{P}_N \dot{u} + \mathsf{P}_N L u + \mathsf{P}_N B(u) = \mathsf{P}_N f.$$

A curve $u(t) \in H_{(N)}$ satisfies this relation if and only if

$$\dot{u} + Lu + \mathsf{P}_N B(u) = \mathsf{P}_N f. \tag{2.31}$$

This is an ODE in $H_{(N)}$ defined by a smooth vector field, with a square-integrable right-hand side. Let us supplement it with the initial condition

$$u(0) = \mathsf{P}_N u_0. \tag{2.32}$$

Problem (2.31), (2.32) has a unique solution $u = u_N$ defined on a time interval $[0, T_N)$, where either $T_N = T$, or $T_N < T$, and $u_N(t)$ blows up as $t \to T_N^-$. We claim that $T_N = T$ and

$$u_N \in C(0, T; H_{(N)}), \quad \dot{u}_N \in L^2(0, T; H_{(N)}). \tag{2.33}$$

To prove these claims, we shall derive an a priori estimate for u_N.

Let us calculate the derivative of the functional $\frac{1}{2}|u|^2$ along the trajectories of (2.31). We have

$$\begin{aligned} \frac{1}{2}\frac{d}{dt}|u(t)|^2 &= \langle u, \dot{u}\rangle = \langle u, -Lu - \mathsf{P}_N B(u) + \mathsf{P}_N f\rangle \\ &= \langle u, -Lu - B(u) + f(t)\rangle. \end{aligned}$$

Repeating the argument used in Step 1, we see that $u = u_N$ satisfy inequalities (2.27) and (2.28) with $T = T_N$ uniformly in $N \geq 1$. It follows that $T_N = T$ for all N. Thus, the Galerkin approximations $\{u_N\}$ form a bounded sequence in $\mathcal{H}$.

Step 3: Passage to the limit. We now wish to pass to the limit in Eq. (2.31) as $N \to \infty$. Since $\mathcal{H}$ is a Hilbert space, and a closed ball in a Hilbert space is weakly compact, there is a sequence $N_j \to \infty$ such that $\{u_{N_j}\}$ converges weakly to an element $u \in \mathcal{H}$ as $j \to \infty$. Since the linear operators

$$\mathcal{H} \to L_2(0, T; H_\sigma^{-1}),\ u \mapsto \dot{u},$$
$$\mathcal{H} \to L_2(0, T; H_\sigma^{-1}),\ u \mapsto Lu,$$

are continuous, we have

$$\dot{u}_{N_j} \rightharpoonup \dot{u} \quad \text{in } L_2(0, T; H_\sigma^{-1}), \tag{2.34}$$
$$L u_{N_j} \rightharpoonup Lu \quad \text{in } L_2(0, T; H_\sigma^{-1}). \tag{2.35}$$

Consider now the nonlinear term $\{B(u_{N_j})\}$. By Lemma 2.1.5,

$$u_{N_j} \to u \quad \text{in } L^2(0, T; H_\sigma^{1/2}). \tag{2.36}$$

Noting that, by (2.14), the bilinear form $u \mapsto B(u)$ is continuous from the space $L^2(0, T; H_\sigma^{1/2})$ to $L^1(0, T; H_\sigma^{-1})$, we see that

$$B(u_{N_j}) \to B(u) \quad \text{in } L^1(0, T; H_\sigma^{-1}). \tag{2.37}$$

Finally, due to (2.36),

$$u_{N_j}(0) \to u(0) \quad \text{in } L_\sigma^2. \tag{2.38}$$

Let us take any integer $m \geq 1$ and apply the projection P_m to Eq. (2.31) with $N = N_j \geq m$. Since $\mathsf{P}_m \circ P_{N_j} = \mathsf{P}_m$, we obtain

$$\mathsf{P}_m \dot{u}_{N_j} + \mathsf{P}_m L u_{N_j} + \mathsf{P}_m B(u_{N_j}) = \mathsf{P}_m f.$$

Letting $N_j \to \infty$ and using (2.34)–(2.37), we derive

$$\mathsf{P}_m \dot{u} + \mathsf{P}_m Lu + \mathsf{P}_m B(u) = \mathsf{P}_m f,$$

where the equality holds in the space $L^1(0, T; H_\sigma^{-1})$. Since m is arbitrary, we conclude that

$$\dot{u} + Lu + B(u) = f.$$

Furthermore, due to (2.38),

$$u(0) = \lim_{j \to \infty} u_{N_j}(0) = \lim_{j \to \infty} P_{N_j} u_0 = u_0.$$

We have thus proved the existence of a solution. It remains to establish inequality (2.24).

To this end, note that inequality (2.24) remains valid for $u = u_N$. Since $\{u_{N_j}\}$ converges weakly to $u \in \mathcal{H}$, we see that

$$\|u\|_{L^2(0,t;H^1)} \le \liminf_{j\to\infty} \|u_{N_j}\|_{L^2(0,t;H^1)} \quad \text{for any } t \le T.$$

Combining this with convergence (2.36), we conclude that one can pass to the limit in (2.24) with $u = u_{N_j}$ as $j \to \infty$. This results in inequality (2.24) for the constructed solution $u(t, x)$. The proof of Theorem 2.1.13 is complete. □

Exercise 2.1.14 Show that any solution $u(t, x)$ of the Navier–Stokes system (2.19) satisfies the relation

$$|u(t)|_2^2 + 2\nu \int_0^t |\nabla u(s)|_2^2 ds = |u(0)|_2^2 + 2 \int_0^t \langle f(s), u(s)\rangle ds, \qquad (2.39)$$

which is called the *energy balance*. *Hint:* Relation (2.39) can be obtained formally by taking the scalar product of Eq. (2.19) with u and integrating in time. To justify this calculation, use relation (2.10).

Remark 2.1.15 Our proof shows that any sequence of Galerkin approximations u_{N_j} contains a subsequence that converges to a solution of the Navier–Stokes system. By the uniqueness, the limiting function must coincide with the solution $u(t, x)$ constructed in Theorem 2.1.13. Hence, the whole sequence $\{u_N\}$ converges to u weakly in $\mathcal{H}$ and, by Lemma 2.1.5, strongly in $L^2(0, T; H_\sigma^r)$ with $r < 1$:

$$u_N \rightharpoonup u \quad \text{in} \quad \mathcal{H}, \qquad u_N \to u \quad \text{in} \quad L^2(0, T; H_\sigma^r). \qquad (2.40)$$

Theorem 2.1.13 on the existence and uniqueness of a solution is true for 2D Navier–Stokes equations in a bounded domain with the Dirichlet boundary condition, and the proof remains essentially the same; see chapter III in [Tem79].

We conclude this subsection with a remark on solutions with zero mean value. Recall that the mean value $\langle u\rangle$ of a function $u \in H^m(\mathbb{T}^2)$ with $m \ge 0$ is defined in (1.2). In the case of an arbitrary $m \in \mathbb{R}$, it can be defined as the zero-order coefficient in the Fourier expansion of u. Let $u(t, x)$ be a solution of (2.19). Integrating Eq. (2.19) in time and taking the mean value of both sides, we obtain

$$\langle u(t)\rangle + \int_0^t \langle -\Delta u + B(u)\rangle\, ds = \langle u(0)\rangle + \int_0^t \langle f(s)\rangle\, ds.$$

Now note that the integrand on the left-hand side vanishes, and therefore

$$\langle u(t)\rangle = \langle u(0)\rangle + \int_0^t \langle f(s)\rangle\, ds.$$

So if the mean value of $f(t)$ is zero for almost all $t \in [0, T]$, then $\langle u(t)\rangle$ is constant. In what follows, we shall study problem (2.19), (2.3) for the case when the mean values of $f(t)$ and u_0 vanish. What has been said implies that in this case $\langle u(t)\rangle = 0$ for all $t \in [0, T]$.

Let us introduce the function spaces

$$H = \{u \in L^2_\sigma(\mathbb{T}^2, \mathbb{R}^2) : \langle u\rangle = 0\}, \quad V = H^1 \cap H, \quad V^k = H^k \cap H \tag{2.41}$$

and denote by V^* the dual space of V with respect to the scalar product in L^2. Note that V^* can be regarded as the quotient space H^{-1}/W, where W is the space of functionals $f \in H^{-1}$ vanishing on ∇H^2. It follows from Poincaré's inequality that the usual Sobolev norm on V^m is equivalent to $\langle L^m u, u\rangle^{1/2}$. Slightly abusing the notation, we sometimes denote the latter norm by $\|\cdot\|_m$. Finally, for a bounded domain Q, we write $H = L^2_\sigma$ and $V = H^1_0(Q; \mathbb{R}^2) \cap L^2_\sigma$ and denote by V^* the dual of V.

When dealing with stochastic Navier–Stokes equations, we shall very often apply Itô's formula to processes that take values in one space and possess stochastic differentials in a larger space. In this context, the concept of a *Gelfand triple* will play an important role; see Section A.6. Note that the spaces $V \subset H \subset V^*$ form a Gelfand triple, as do the spaces $V^{k+1} \subset V^k \subset V^{k-1}$ for any integer $k \geq 1$.

2.1.6 Regularity of solutions

The Navier–Stokes system is a parabolic-type equation, and in the 2D case it has good smoothing properties. The proof of this fact is particularly simple for the problem on the torus due to the following lemma.

Lemma 2.1.16 *If $u \in H \cap H^2$, then $\langle B(u), \Delta u\rangle = 0$.*

The proof of this result is based on the existence of a *stream function* for divergence-free vector fields.

Exercise 2.1.17 Show that for any vector field $u \in H$ on the 2D torus there is a function $\psi \in H^1(\mathbb{T}^2)$ such that $u = \operatorname{curl}\psi := (-\partial_2\psi, \partial_1\psi)$. Moreover, the function ψ is unique up to an additive constant, and if $u \in H^k(\mathbb{T}^2, \mathbb{R}^2)$, then $\psi \in H^{k+1}(\mathbb{T}^2)$. *Hint:* Use the Fourier expansion.

Proof of Lemma 2.1.16 By Exercise 2.1.17, there is a function $\psi \in H^3(\mathbb{T}^2)$ such that $u = \operatorname{curl}\psi$. Furthermore, it follows from Theorem 2.1.2 that for any $u \in H \cap H^2$ there is $p \in H^1(\mathbb{T}^2)$ such that $B(u) = \langle u, \nabla\rangle u - \nabla p$. Thus, setting $\operatorname{curl} u = \partial_1 u^2 - \partial_2 u^1$ and integrating by parts, we obtain

$$\begin{aligned}\langle B(u), \Delta u\rangle &= \int_{\mathbb{T}^2} \big(\langle u, \nabla\rangle u - \nabla p\big)\operatorname{curl}(\Delta\psi)\,dx \\ &= \int_{\mathbb{T}^2} \langle u, \nabla\rangle(\operatorname{curl} u)\,\Delta\psi\,dx, \end{aligned}\tag{2.42}$$

where we used the relations

$$\operatorname{curl}(\langle u, \nabla\rangle u) = \langle u, \nabla\rangle(\operatorname{curl} u), \quad \operatorname{curl}(\nabla p) = 0.$$

Now note that $\operatorname{curl} u = \operatorname{curl}(\operatorname{curl} \psi) = \Delta\psi$. Substituting this into (2.42), integrating by parts, and using that $\operatorname{div} u = 0$, we arrive at the required result. □

We can now state a first regularisation result for 2D Navier–Stokes equations. It says, roughly speaking, that the solution immediately becomes more regular than the initial function.

Theorem 2.1.18 *Let $u_0 \in H$ and $f \in L^2(0, T; H)$. Then the solution $u(t, x)$ of problem* (2.19), (2.3) *belongs to the space $C(t_0, T; H^1) \cap L^2(t_0, T; H^2)$ for any $t_0 > 0$. Moreover, for $0 \le t \le T$, we have*

$$t\,\|u(t)\|_1^2 + \nu \int_0^t s\,\|u(s)\|_2^2 ds \le |u_0|_2^2 + \nu^{-1} \int_0^t s\,|f(s)|_2^2 ds + \int_0^t \|f(s)\|_{-1}^2\, ds. \tag{2.43}$$

Proof We shall confine ourselves to a formal derivation of the a priori estimate (2.43) for smooth solutions. Its justification can be carried out with the help of the Galerkin approximation; cf. proof of (2.24). Once (2.43) is established, we conclude immediately that $u \in L^\infty(t_0, T; H^1) \cap L^2(t_0, T; H^2)$ for any $t_0 > 0$. Combining this with Eq. (2.19), we see that $\dot{u} \in L^2(t_0, T; H)$, whence it follows that $u \in C(t_0, T; H^1)$.

To prove (2.43), consider the functional $\varphi(t, u) = t\,\langle Lu, u\rangle$. Differentiating it with respect to time and using Eq. (2.19) and Lemma 2.1.16, we derive

$$\begin{aligned}\partial_t \varphi(t, u) &= \|u\|_1^2 + 2t\langle Lu, \dot{u}\rangle\\ &= \|u\|_1^2 + 2t\langle Lu, -\nu Lu - B(u) + \Pi f\rangle\\ &= \|u\|_1^2 - 2\nu t\,|Lu|_2^2 + 2t\langle Lu, f\rangle.\end{aligned}$$

Using the inequality $2t\langle Lu, f\rangle \le \nu t\,|Lu|_2^2 + \nu^{-1} t\,|f|_2^2$, we obtain

$$\partial_t\left(t\,\|u\|_1^2\right) \le \|u\|_1^2 - \nu t\,\|u\|_2^2 + \nu^{-1} t\,|f|_2^2 .$$

Integrating from 0 to t, we get

$$t\,\|u(t)\|_1^2 + \nu \int_0^t s\,\|u(s)\|_2^2\, ds \le \nu^{-1} \int_0^t s\,|f(s)|_2^2\, ds + \int_0^t \|u(s)\|_1^2\, ds .$$

Using (2.24), we see that u satisfies (2.43). □

An analogue of Theorem 2.1.18 is true for the Navier–Stokes system in a bounded domain. However, in this case the proof becomes more complicated, because Lemma 2.1.16 is no longer true, and the term $\langle B(u), Lu\rangle$ does not vanish. The corresponding argument is similar to that used below to prove the higher regularity of solutions. For simplicity, we assume again that $\nu = 1$.

Theorem 2.1.19 *Let $u_0 \in H$, and let $f \in L^2(0, T; H^{m-1})$ for some integer $m \ge 2$. Then a solution $u(t, x)$ of problem* (2.19), (2.3) *belongs to the space*

$C(t_0, T; H^m) \cap L^2(t_0, T; H^{m+1})$ for any $t_0 > 0$. Moreover, there is a constant $C_m > 0$ such that the following inequality holds for $0 \le t \le T$:

$$t^m \|u(t)\|_m^2 + \int_0^t s^m \|u\|_{m+1}^2 \, ds \le \int_0^t s^m \|f(s)\|_{m-1}^2 ds + C_m\big(|u_0|_2^2 + |u_0|_2^{4m+2} + \|f\|_{L^2(0,T;H)}^2 + \|f\|_{L^2(0,T;H)}^{4m+2}\big). \quad (2.44)$$

Proof As in the proof of the previous theorem, we shall confine ourselves to the formal derivation of (2.44). We argue by induction. Note that, in view of Theorem 2.1.18, inequality (2.44) holds for $m = 1$. We now assume that $m \ge 2$ and set $\varphi_m(t, u) = t^m \|u\|_m^2 = t^m \langle L^m u, u\rangle$. Then, for a smooth solution u, we have

$$\begin{aligned}\partial_t \varphi_m(t, u) &= m\, t^{m-1} \|u\|_m^2 + 2t^m \langle L^m u, \dot u\rangle \\ &= m\, t^{m-1} \|u\|_m^2 - 2\nu t^m \|u\|_{m+1}^2 - 2t^m \langle L^m u, B(u) - f\rangle\rangle. \quad (2.45)\end{aligned}$$

To estimate the nonlinear term, we need the following proposition, whose proof if given at the end of this subsection.

Lemma 2.1.20 *For any integer $m \ge 2$ there is a constant $C_m > 0$ such that*

$$|\langle L^m u, B(u)\rangle| \le C_m \|u\|_{m+1}^{\frac{4m-1}{2m}} \|u\|_1^{\frac{m+1}{2m}} |u|_2^{1/2} \quad \text{for } u \in H \cap H^{m+1}. \quad (2.46)$$

Substituting (2.46) into (2.45) and carrying out some simple transformations, we derive

$$\partial_t\big(t^m \|u\|_m^2\big) + t^m \|u\|_{m+1}^2 \le m\, t^{m-1} \|u\|_m^2 + C_1 t^m \|u\|_1^{2(m+1)} |u|_2^{2m} + t^m \|f\|_{m-1}^2. \quad (2.47)$$

By Theorems 2.1.13 and 2.1.18 with $\nu = 1$, we have

$$t\|u\|_1^2 |u|_2^2 \le C\left(|u_0|_2^2 + \|f\|_{L^2(0,T,H)}^2\right)^2.$$

Substituting this into (2.47) and integrating in time, we obtain

$$\begin{aligned} t^m \|u\|_m^2 + \int_0^t s^m \|u(s)\|_{m+1}^2 ds &\le m \int_0^t s^{m-1} \|u(s)\|_m^2 ds \\ &\quad + \int_0^t s^m \|f\|_{m-1}^2 \, ds + C_2\big(|u_0|_2^2 \\ &\quad + \|f\|_{L^2(0,T,H)}^2\big)^{2m} \int_0^t \|u(s)\|_1^2 ds. \quad (2.48)\end{aligned}$$

Using now inequality (2.24) and the induction hypothesis to estimate the first and third integrals on the right-hand side of (2.48), we arrive at the required estimate (2.44). □

Proof of Lemma 2.1.20 We first note that

$$\langle B(u), L^m u\rangle = \sum_{|\alpha|=m} C_\alpha \langle D^\alpha B(u), D^\alpha u\rangle, \quad (2.49)$$

where C_α are some constants. Each term under the sum can be written as

$$\langle D^\alpha B(u), D^\alpha u\rangle = \sum_{\beta\le\alpha} \binom{\alpha}{\beta} \langle B(D^{\alpha-\beta}u, D^\beta u), D^\alpha u\rangle.$$

Since $\langle B(u, D^\alpha u), D^\alpha u\rangle = 0$, the term with $\beta = \alpha$ vanishes, and using (2.13), we get

$$\begin{aligned} |\langle D^\alpha B(u), D^\alpha u\rangle| &\le C \sum_{\beta<\alpha} \|D^{\alpha-\beta}u\|_{1/2}\|D^\beta u\|_1\|D^\alpha u\|_{1/2} \\ &\le C \sum_{\beta<\alpha} \|u\|_{1/2+m-|\beta|}\, \|u\|_{1+|\beta|}\|u\|_{m+\frac{1}{2}}, \end{aligned} \tag{2.50}$$

where $m = |\alpha|$ and $0 \le |\beta| < m$. Note that the numbers $\frac{1}{2} + m - |\beta|$, $1 + |\beta|$ and $m + \frac{1}{2}$ lie between 1 and $m + \frac{1}{2}$. By the interpolation inequality (1.7), for any $a \in [1, m + \frac{1}{2}]$, we have

$$\|u\|_a \le |u|_2^{1-\frac{a}{m+1}}\, \|u\|_{m+1}^{\frac{a}{m+1}} := X, \qquad \|u\|_a \le \|u\|_1^{1-\frac{a-1}{m}}\, \|u\|_{m+1}^{\frac{a-1}{m}} := Y.$$

Take any term in the sum in the right-hand side of (2.50). Estimating each of its factors by $X^\theta Y^{1-\theta}$ with a suitable θ, we obtain

$$|\langle D^\alpha B(u), D^\alpha u\rangle| \le C\, \|u\|_{m+1}^{\frac{4m-1}{2m}}\, \|u\|_1^{\frac{m+1}{2m}}\, |u|_2^{1/2}.$$

Substituting this into (2.49), we arrive at (2.46). □

Theorem 2.1.19 implies, in particular, that if the right-hand side is infinitely smooth, then so is the solution of (2.19) for $t > 0$. It can also be shown that the space–time analyticity of f implies a similar property for the solution. We refer the reader to [FT89; DG95] for an exact formulation and proof of the corresponding result.

An analogue of Theorem 2.1.19 is true for bounded domains with smooth boundary. In this case, to prove the regularity of u we have to assume that the right-hand side is smooth both in space and time, and the initial function u_0 satisfies some compatibility condition. These results are not needed for studying ergodic properties of the Navier–Stokes flow, and therefore we do not give any further details, referring the reader to [Tay97] for a systematic study of regularity of solutions to nonlinear PDEs.

2.1.7 Navier–Stokes process

Let us consider the Navier–Stokes system (2.19) in which $f \in L^2_{\mathrm{loc}}(\mathbb{R}_+, H^{-1})$ and $\nu = 1$. By Theorem 2.1.13, for any $\tau \in \mathbb{R}$ and any $u_0 \in H$, Eq. (2.19) has a unique solution $u \in C(\mathbb{R}_\tau, H) \cap L^2_{\mathrm{loc}}(\mathbb{R}_\tau, V)$ such that $u(\tau) = u_0$, where $\mathbb{R}_\tau = [\tau, \infty)$. We introduce the resolving operator $S_{t,\tau} : H \to H$ by the relation

$S_{t,\tau}(u_0) = u(t)$. It is straightforward to verify that the family $\{S_{t,\tau}, t \geq \tau \geq 0\}$ forms a *process*:

$$S_{\tau,\tau} = \mathrm{Id}_H, \qquad S_{t,\tau} = S_{t,s} \circ S_{s,\tau} \quad \text{for any } t \geq s \geq \tau \geq 0,$$

where Id_H stands for the identity mapping in H. We shall call $S_{t,\tau}$ the *Navier–Stokes process* or, for short, the *NS process*. Let us set $S_t = S_{t,0}$. The following proposition establishes a dissipativity property of the NS process.

Proposition 2.1.21

(i) *Let* $u_0 \in H$ *and* $f \in L^2_{\mathrm{loc}}(\mathbb{R}_+, H^{-1})$. *Then*

$$|S_t(u_0)|_2^2 \leq e^{-\alpha_1 t}|u_0|_2^2 + \int_0^t e^{-\alpha_1(t-s)}\|f(s)\|_{-1}^2 ds, \tag{2.51}$$

where $\alpha_1 > 0$ *stands for the first positive eigenvalue*[1] *of the Laplacian.*

(ii) *Let* $u_0 \in V$ *and* $f \in L^2_{\mathrm{loc}}(\mathbb{R}_+, H)$. *Then*

$$\|S_t(u_0)\|_1^2 \leq e^{-\alpha_1 t}\|u_0\|_1^2 + \int_0^t e^{-\alpha_1(t-s)}|f(s)|_2^2 ds. \tag{2.52}$$

Proof In view of (2.24) with $\nu = 1$, we have

$$|S_t(u_0)|_2^2 + \int_0^t |\nabla S_s(u_0)|_2^2 ds \leq |u_0|_2^2 + \int_0^t \|f(s)\|_{-1}^2\, ds.$$

Using Poincaré's inequality $|\nabla u|_2^2 \geq \alpha_1 |u|_2^2$ to minorise the integrand on the left-hand side and applying the Gronwall inequality, we arrive at (2.51). The proof of (2.52) can be carried out by a similar argument, using Lemma 2.1.16; cf. the proof of Theorem 2.1.18. □

Exercise 2.1.22 Prove inequality (2.52).

Exercise 2.1.23 Show that any solution of the Navier–Stokes system satisfies the inequality

$$|u(0)|_2^2 \leq C\,(t^{-1} \vee 1) \int_0^t \left(\|S_s(u(0))\|_1^2 + \|f(s)\|_{-1}^2\right) ds, \tag{2.53}$$

where $C > 0$ is a constant not depending on t and u_0. *Hint:* Use (2.51) to estimate from above the left-hand side of (2.39).

Note that inequality (2.51) remains true for the Navier–Stokes system in a bounded domain, and the proof is literally the same as in the case of a torus. On the other hand, inequality (2.52) is no longer valid, because Lemma 2.1.16 does not hold in general. Still, the NS process is bounded from H to V, and

[1] In the case of the standard torus $\mathbb{T}^2 = \mathbb{R}^2/2\pi\mathbb{Z}^2$, we have $\alpha_1 = 1$.

some explicit bounds for $\|S_t(u)\|_1$ can be obtained by repeating the arguments used in the proof of Theorem 2.1.18.

Corollary 2.1.24 *Suppose that $f \equiv 0$. Then the function $u = 0$ is the only fixed point of the NS semigroup, and it is globally exponentially stable in H. Moreover, in the case of a torus, it is also globally exponentially stable in V.*

We now turn to the continuity properties of the NS process. We shall write $S_t(u_0, f)$ to indicate the dependence of the resolving operator on the right-hand side of the NS equation. The following result shows that S_t is uniformly Lipschitz continuous on bounded subsets of H.

Proposition 2.1.25

(i) *There exists a constant $C > 0$ such that, for any functions $u_{01}, u_{02} \in H$ and $f_1, f_2 \in L^2_{\text{loc}}(\mathbb{R}_+, H^{-1})$, we have*

$$\begin{aligned}\big|S_t(u_{01}, f_1) - S_t(u_{02}, f_2)\big|_2^2 &\le |u_{01} - u_{02}|_2^2 \exp\left(C\int_0^t \|S_r(u_{01}, f_1)\|_1^2 dr\right)\\ &\quad + \int_0^t \exp\left(C\int_s^t \|S_r(u_{01}, f_1)\|_1^2 dr\right)\|f_1(s)\\ &\quad - f_2(s)\|_{-1}^2 ds. \qquad (2.54)\end{aligned}$$

(ii) *There exists a constant $C > 0$ such that, for $0 < t \le 1$ and any functions $u_{01}, u_{02} \in H$ and $f_1, f_2 \in L^2_{\text{loc}}(\mathbb{R}_+, H)$, we have*

$$\begin{aligned}&\big\| S_t(u_{01}, f_1) - S_t(u_{02}, f_2)\big\|_1^2\\ &\quad \le C\int_0^t |f_1 - f_2|_2^2 ds + A(t)\, t^{-3}\left(|u_{01} - u_{02}|_2^2 + \int_0^t \|f_1 - f_2\|_{-1}^2 ds\right), \qquad (2.55)\end{aligned}$$

where we set

$$A(t) = \exp\left(C\int_0^t \left(\|S_r(u_{01}, f_1)\|_1^2 + \|S_r(u_{02}, f_2)\|_1^2 + |f_1|_2^2 + |f_2|_2^2\right) ds\right).$$

The proof of this result is rather standard and uses a well-known idea. Namely, one should take the difference between the equations corresponding to two solutions, multiply (in the sense of the scalar product in H) the resulting relation by an appropriate function, estimate the nonlinear terms with the help of Sobolev embedding and interpolation inequalities, and finally apply Gronwall's lemma. The realisation of this scheme is somewhat technical, and to keep the presentation on an elementary level, we postpone the proof of Proposition 2.1.25 until Section 2.6. The reader not interested in those technical details may safely skip it.

An analogue of Proposition 2.1.25 holds for the Navier–Stokes system in a bounded domain. Namely, in that case, inequality (2.54) and its proof remain true without any changes. As for (2.55), its proof used inequality (2.43), which does not hold for a general bounded domain. One can show, however, that the quantity $\|S_t(u_{01}, f_1) - S_t(u_{02}, f_2)\|_{1/2}^2$ is bounded by the right-hand side of (2.55) in which t^{-3} is replaced by $t^{-5/2}$.

Exercise 2.1.26 Prove the above claims for the Navier–Stokes system in a bounded domain with the Dirichlet boundary condition.

The following exercise establishes the continuity of the resolving operator for the Navier–Stokes system in different norms. This will be important in Section 4.3.

Exercise 2.1.27

(i) Prove that, for any functions $u_0 \in H$, $h \in L^2(0, T; H)$, and $g \in L^\infty(0, T; V)$, Eq. (2.19) with the right-hand side $f = h + \partial_t g$ has a unique solution $u \in \mathcal{X}_T := C(0, T; H) \cap L^2(0, T; V)$ issued from u_0.

(ii) Let $u_1, u_2 \in \mathcal{X}_T$ be two solutions of Eq. (2.19) with $f = \partial_t g_i$, $i = 1, 2$, where $g_i \in L^\infty(0, T; V)$. Then for any $R > 0$ there is a constant $C_R > 0$ such that if $|u_i(0)|_2 \le R$ and $\|g_i\|_{L^\infty(0,T;V)} \le R$, then

$$\|u_1 - u_2\|_{\mathcal{X}_T} \le C_R\big(|u_1(0) - u_2(0)|_2 + \|g_1 - g_2\|_{L^\infty(0,T;V)}\big).$$

2.1.8 Foiaş–Prodi estimates

The Foiaş–Prodi inequalities enable one to establish the exponential convergence of two solutions for the Navier–Stokes system, on condition that one has good control over their low Fourier modes. We shall need two versions of these estimates that correspond to discrete- and continuous-time perturbations.

Let us fix an integer $N \ge 1$ and denote by $H_{(N)}$ the vector span[2] of the basis functions e_s with $|s| \le N$; see (2.29). We write P_N for the orthogonal projection in L^2 onto $H_{(N)}$. Along with the Navier–Stokes system (2.19), let us consider the equation

$$\dot{u} + Lu + B(u) + \lambda \mathsf{P}_N\big(u - u'(t)\big) = \Pi f(t), \tag{2.56}$$

where $u'(t)$ is a given function and $\lambda > 0$ is a large parameter. Equation (2.56) is a Navier–Stokes-type system, and the arguments in the proof of Theorem 2.1.13 can be used to show that the Cauchy problem for (2.56) has a unique solution $u \in C(\mathbb{R}_+, H) \cap L^2_{\mathrm{loc}}(\mathbb{R}_+, V)$ for any $u' \in L^2_{\mathrm{loc}}(\mathbb{R}_+, H)$.

[2] This space is slightly different from the one defined by (2.30), because we exclude the constant functions (1, 0) and (0, 1).

Theorem 2.1.28 *Let $f \in L^2_{\mathrm{loc}}(\mathbb{R}_+, H^{-1})$ and let $u' \in C(\mathbb{R}_+, H) \cap L^2_{\mathrm{loc}}(\mathbb{R}_+, V)$ be a solution of the Navier–Stokes system* (2.19) *such that*

$$\int_s^t \|u'(r)\|_1^2 dr \le \rho + K(t-s) \quad \text{for } s \le t \le s+T, \tag{2.57}$$

where s, T, ρ, and K are non-negative constants. Then for any $M > 0$ there are constants $N \ge 1$ and $\lambda > 0$ depending on M and ρ such that for any solution $u \in C(\mathbb{R}_+, H) \cap L^2_{\mathrm{loc}}(\mathbb{R}_+, V)$ of (2.56) *we have*

$$|u(t) - u'(t)|_2 \le e^{-M(t-s)+C\rho} |u(s) - u'(s)|_2 \quad \text{for } s \le t \le s+T, \tag{2.58}$$

where $C > 0$ is an absolute constant.

Proof We confine ourselves to the formal derivation of (2.58). The difference $w = u - u'$ satisfies the equation

$$\dot{w} + Lw + B(u) - B(u') + \lambda \mathsf{P}_N w = 0.$$

Taking the scalar product in L^2 of this equation with $2w$ and using (2.11) and (2.13), we derive

$$\begin{aligned} \partial_t |w|_2^2 + 2|\nabla w|_2^2 + 2\lambda\, |\mathsf{P}_N w|_2^2 &\le C_1 \|w\|_1 |w|_2 \|u'\|_1 \\ &\le |\nabla w|_2^2 + C_2 \|u'\|_1^2 |w|_2^2. \end{aligned} \tag{2.59}$$

Now note that, by Poincaré's inequality, we have

$$|\nabla w|_2^2 \ge N^2 \, |(I - \mathsf{P}_N) w|_2^2.$$

Substituting this into (2.59), we obtain

$$\partial_t |w|_2^2 + \bigl(\lambda_1 - C_2 \|u'\|_1^2\bigr) \, |w|_2^2 \le 0,$$

where $\lambda_1 = \min\{N^2, 2\lambda\}$. Application of the Gronwall inequality results in

$$\begin{aligned} |w(t)|_2^2 &\le |w(s)|_2^2 \exp\biggl(-\lambda_1 (t-s) + C_2 \int_s^t \|u'(r)\|_1^2 dr\biggr) \\ &\le |w(s)|_2^2 \exp\bigl(C_2 \rho - (\lambda_1 - C_2 K)(t-s)\bigr). \end{aligned}$$

Choosing λ and N so large that $\lambda_1 \ge C_2 K + 2M$, we arrive at (2.58) with $C = C_2/2$. □

Remark 2.1.29 The above proof does not use the fact that the space variables belong to the torus, and the result remains true for a bounded domain. In this case, P_N denotes the orthogonal projection in L^2 to the subspace spanned by the first N eigenfunctions of the *Stokes operator* $L = -\Pi\Delta$. Furthermore, the regularity in time of the function f was not used either. In particular, if the right-hand side has the form $f = h + \partial_t \zeta$, where $h \in L^2_{\mathrm{loc}}(\mathbb{R}_+, H)$ and $\zeta \in C(\mathbb{R}_+, H)$, then the conclusion of Theorem 2.1.28 remains valid; cf. Exercise 2.1.27 for the existence of a solution. This observation will be important for the case in which the right-hand side is the sum of a deterministic function and the time derivative of a Brownian motion.

We now turn to the discrete version of the Foiaş–Prodi estimate. To simplify the presentation, we shall assume that the right-hand side f is independent of time and belongs to the space H. In this case, the resolving operator $S_{t,\tau}$ of the Navier–Stokes equation (2.19) depends only on the difference $t-\tau$, and we shall write $S_{t-\tau}$. Let us fix a constant $T>0$ and consider four sequences $u_k, u_k', \zeta_k, \zeta_k' \in H$ satisfying the relations

$$u_k = S(u_{k-1}) + \zeta_k, \quad u_k' = S(u_{k-1}') + \zeta_k', \quad k \geq 1, \tag{2.60}$$

where $S = S_T$.

Theorem 2.1.30 *Suppose that*

$$\mathsf{P}_N u_k = \mathsf{P}_N u_k', \quad (I - \mathsf{P}_N)\zeta_k = (I - \mathsf{P}_N)\zeta_k' \quad \textit{for } l+1 \leq k \leq m, \tag{2.61}$$

where $m > l \geq 0$ are some integers. Then there is an absolute constant $C>0$ such that

$$|u_k - u_k'|_2 \leq (CN^{-1})^{k-l} \exp\Big(C(k-l)\big(\langle \|\boldsymbol{u}\|_1^2\rangle_l^k + \langle \|\boldsymbol{u}'\|_1^2\rangle_l^k + 1\big)\Big)|u_l - u_l'|_2, \tag{2.62}$$

where $l \leq k \leq m$, and for a sequence $\{v_j\} \subset H$ we set

$$\langle \|\boldsymbol{v}\|_1^2\rangle_l^k = \frac{1}{k-l}\sum_{j=l}^{k-1}\int_0^T \|S_t(v_j)\|_1^2 dt.$$

Proof Let us set $\Delta_k = |u_k - u_k'|_2$. It follows from (2.55), (2.61), and the Poincaré inequality that

$$\begin{aligned}\Delta_k &= |(I-\mathsf{P}_N)(u_k - u_k')|_2 = \big|(I-\mathsf{P}_N)(S(u_{k-1} - S(u_{k-1}'))\big|_2 \\ &\leq C_1 N^{-1}\big\|S(u_{k-1} - S(u_{k-1}')\big\|_1 \\ &\leq C_2 N^{-1}\Delta_{k-1}\exp\left(C_2\int_0^T \big(\|S_t(u_{k-1})\|_1^2 + \|S_t(u_{k-1}')\|_1^2 + |f|_2^2\big)\,dt\right).\end{aligned}$$

Iteration of this inequality results in (2.62). □

Remark 2.1.31 As in the case of continuous-time perturbations, the conclusion of Theorem 2.1.30 remains true for a bounded domain. The only difference is that the constant N^{-1} in (2.62) should be replaced by α_N^{-1}, where α_j stands for the jth eigenvalue of the Stokes operator L.

2.1.9 Some hydrodynamical terminology

As we mentioned in the Introduction, the behaviour of an incompressible fluid occupying a domain $Q \subset \mathbb{R}^d$ (or the d-dimensional torus $\mathbb{T}^d$), with $d = 2$ or 3, is described by the $d + 1$ equations (2.1), supplemented by suitable boundary conditions if necessary. In those equations, $u(t, x)$, $p(t, x)$, and ν denote, respectively, the velocity, pressure, and the kinematic viscosity of the fluid. The quantity $\frac{1}{2}|u(t)|_2^2 = \frac{1}{2}\int_Q |u(t, x)|^2 dx$ is called the *energy of the fluid* (at time t).

Let us assume that $f = 0$ and that the fluid satisfies the no-slip condition, i.e., $u = 0$ on ∂Q. Multiplying the first equation in (2.1) by $u(t)$ in L_2, we obtain

$$\frac{1}{2}\frac{d}{dt}|u(t)|_2^2 + \nu \int_Q |\nabla u|^2 dx = 0.$$

Accordingly, the quantity $\varepsilon = \nu \int_Q |\nabla u|^2 dx$ is called the *rate of dissipation of energy*.

The *Reynolds number* of the flow is defined as

$$R = \frac{\langle \text{characteristic scale for } x\rangle \cdot \langle \text{characteristic scale for } u\rangle}{\nu}.$$

The terms in the numerator are ambiguous, and for the purposes of this book, we understand them as follows:

$$\langle \text{characteristic scale for } x\rangle = \text{diameter of } Q \text{ (or the period of the torus)},$$

$$\langle \text{characteristic scale for } u\rangle = \frac{1}{\sqrt{2}}|u|_2 =: \sqrt{E}.$$

If u depends on a random parameter, then we modify this definition:

$$\langle \text{characteristic scale for } u\rangle = \left(\mathbb{E}\, E\right)^{1/2}.$$

In the 2D case, we denote by $v = \operatorname{curl} u = \partial_1 u^2 - \partial_2 u^1$ the *vorticity* of the flow, and call $\Omega = \frac{1}{2}\int v^2 dx$ the *enstrophy*.

Exercise 2.1.32 Show that $\Omega = \frac{1}{2}\int_Q |\nabla u|^2 dx$.

Applying the curl operator to the first equation in (2.1) and using the relation $\operatorname{curl}(\langle u, \nabla\rangle u) = \langle u, \nabla\rangle v$, we derive

$$\dot{v} - \nu\Delta v + \langle u, \nabla\rangle v = \operatorname{curl} f. \tag{2.63}$$

Let us assume that $f = 0$ and consider the case of a torus. Multiplying (2.63) by v in L_2, we obtain

$$\frac{1}{2}\frac{d}{dt}|v|_2^2 + \nu \int_{\mathbb{T}^2} |\nabla v|^2\, dx = 0.$$

We deduce that $\nu \int_{\mathbb{T}^2} |\nabla v|^2 dx$ is the *rate of dissipation of enstrophy*.

Exercise 2.1.33 Show that, in the case of a torus, we have $\int_{\mathbb{T}^2} |\nabla v|^2 dx = |\Delta u|_2^2$.

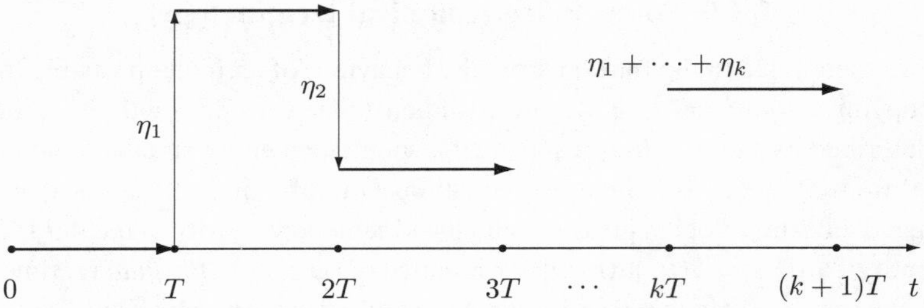

Figure 2.1 Time integral of a random kick force.

2.2 Stochastic Navier–Stokes equations

We now turn to the Navier–Stokes system with a random right-hand side. More precisely, we shall consider Eqs. (2.1) in which f has the form

$$f(t,x)=h(x)+\eta(t,x), \tag{2.64}$$

where h is a deterministic function and η is a stochastic process. For the latter, we study the following three cases that give rise to a Markov process in phase space: a random kick force, spatially regular white noise, and the time derivative of a compound Poisson process. Each of these three types of forces has its own advantages: the first and third ones are more realistic in the sense that they allow non-Gaussian perturbations, while the second can be considered as a limiting model for a large class of stationary processes whose time integral does not necessarily satisfy the condition of independence of increments; cf. Theorem 2.2.2 below and théorème 4.3 in [Rio00].

Let us describe the above-mentioned models in more detail. We shall say that a stochastic process η is a *random kick force on* $\mathbb{T}^2$ if it has the form (see Figure 2.1)

$$\eta(t,x)=\sum_{k=1}^{\infty}\eta_k(x)\delta(t-kT), \tag{2.65}$$

where $T>0$ is a constant and $\{\eta_k\}$ is a sequence of i.i.d. random variables in H.

We shall say that a stochastic process η is a *spatially regular white noise* if

$$\eta(t,x)=\frac{\partial}{\partial t}\zeta(t,x),\quad \zeta(t,x)=\sum_{j=1}^{\infty}b_j\beta_j(t)e_j(x),\quad t\geq 0, \tag{2.66}$$

where $\{e_j\}$ is an orthonormal basis in H, $b_j\geq 0$ are some constants such that

$$\mathfrak{B}:=\sum_{j=1}^{\infty}b_j^2<\infty, \tag{2.67}$$

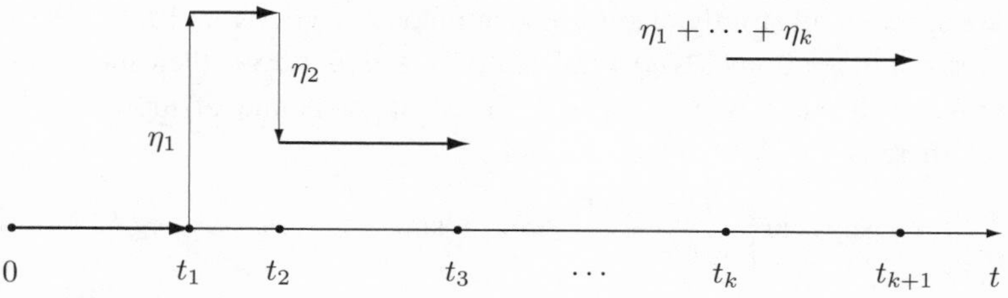

Figure 2.2 Compound Poisson process.

and $\{\beta_j\}$ is a sequence of independent standard Brownian motions. The following result is a simple consequence of Doob's moment inequality (A.56) and Levy's equivalence theorem[3] (see theorem 9.7.1 in [Dud02]).

Exercise 2.2.1 Show that for any $T > 0$ the series in (2.66) converges almost surely in the space $C(0, T; H)$. *Hint:* One can apply the argument used in the proof of Theorem 2.2.2 below to prove the convergence of the series in probability; see the solution of Exercise 2.4.17.

Finally, we shall say that η is the *time derivative of a compound Poisson process* if (see Figure 2.2)

$$\eta(t, x) = \sum_{k=1}^{\infty} \eta_k(x)\delta(t - t_k). \tag{2.68}$$

Here $\{\eta_k\}$ are i.i.d. random variables in H, $\{t_k\}$ is a sequence independent of $\{\eta_k\}$ such that[4] $\tau_k = t_k - t_{k-1}$ are i.i.d. random variables with an exponential distribution; see example 1.3.10 in [App04].

We shall mostly study the first two types of noise, confining ourselves to a brief discussion of the results for the third class.

The concept of a solution for the Navier–Stokes system with random kicks and spatially regular white noise is defined in the next two subsections. Here we recall a result due to Donsker which shows the *universality of white noise*. Namely, we shall prove that an appropriately normalised high-frequency random kick force is close to a spatially regular white noise.

Let $t_0 > 0$ be a constant, let X be a separable Banach space, and let $D(0, t_0; X)$ be the space of functions that are right-continuous and have left-hand limits at any point of the interval $[0, t_0]$. It is a well-known fact that the space $D(0, t_0; X)$ is complete with respect to the modified Skorokhod metric, which is defined in the following way (see [Bil99, section 12]). Let Λ

[3] In [Dud02], this result is proved for real-valued random variables. However, analysing the proof, it is easily seen that if a series of independent random variables with range in a separable Banach space converges in probability, then it converges also almost surely.

[4] By definition, we set $t_0 = 0$.

be the space of all strictly increasing continuous functions $\lambda : [0, t_0] \to [0, t_0]$ such that $\lambda(0) = 0$ and $\lambda(t_0) = t_0$. If $\zeta_1, \zeta_2 \in D(0, t_0; X)$, then the modified Skorokhod distance is defined as the least upper bound of those $\delta > 0$ for which there is $\lambda \in \Lambda$ such that

$$\sup_{0 \le s < t \le t_0} \ln \left| \frac{\lambda(t) - \lambda(s)}{t - s} \right| \le \delta, \quad \sup_{0 \le t \le t_0} \|\zeta_1 - \zeta_2 \circ \lambda\|_X \le \delta.$$

In the case $X = \mathbb{R}$, we shall write $D(0, t_0)$ instead of $D(0, t_0; \mathbb{R})$.

Let us consider a family of random kick forces depending on a parameter $\varepsilon > 0$ and defined by

$$\eta_\varepsilon(t, x) = \sqrt{\varepsilon} \sum_{k=1}^{\infty} \eta_k(x) \delta(t - k\varepsilon). \tag{2.69}$$

Assume that $\{\eta_k\}$ is a sequence of i.i.d. random variables of the form

$$\eta_k(x) = \sum_{j=1}^{\infty} b_j \xi_{jk} e_j(x), \tag{2.70}$$

where $\{e_j\}$ and $\{b_j\}$ are the same as in (2.66), and $\{\xi_{jk}\}$ is a family of independent scalar random variables whose laws do not depend on k and satisfy the relations

$$\mathbb{E}\,\xi_{jk} = 0, \quad \mathbb{E}\,\xi_{jk}^2 = 1.$$

Let us define the time integrals of processes (2.69) and (2.66):

$$\zeta_\varepsilon(t, x) = \sqrt{\varepsilon} \sum_{k=1}^{\infty} \eta_k(x) \theta(t - k\varepsilon) = \sqrt{\varepsilon} \sum_{j=1}^{\infty} b_j e_j(x) \sum_{k=1}^{\infty} \xi_{jk} \theta(t - k\varepsilon),$$

$$\zeta(t, x) = \sum_{j=1}^{\infty} b_j \beta_j(t) e_j(x). \tag{2.71}$$

Here θ is the Heaviside function, that is, $\theta(t) = 1$ for $t \ge 0$ and $\theta(t) = 0$ for $t < 0$. It is clear that, for any $t_0 > 0$, almost every trajectory of ζ_ε and ζ belongs to $D(0, t_0; H)$.

Theorem 2.2.2 *For any $t_0 > 0$, the family $\mathcal{D}(\zeta_\varepsilon)$ converges to $\mathcal{D}(\zeta)$ weakly in the space $\mathcal{P}(D(0, t_0; H))$.*

Proof Let us write ζ_ε as

$$\zeta_\varepsilon(t, x) = \sum_{j=1}^{\infty} b_j \beta_j^\varepsilon(t) e_j(x), \quad \beta_j^\varepsilon(t) = \sqrt{\varepsilon} \sum_{k=1}^{\infty} \xi_{jk} \theta(t - k\varepsilon).$$

Donsker's theorem implies that, for any $t_0 > 0$,

$$\mathcal{D}(\beta_j^\varepsilon) \to \mathcal{D}(\beta_j) \quad \text{in } \mathcal{P}\big(D(0, t_0)\big) \text{ as } \varepsilon \to 0^+;$$

see section 14 of [Bil99]. It follows that, for each $n \geq 1$,

$$\mathcal{D}(\zeta_\varepsilon^n) \to \mathcal{D}(\zeta^n) \quad \text{in } \mathcal{P}\big(D(0, t_0; H)\big) \text{ as } \varepsilon \to 0^+, \tag{2.72}$$

where we set

$$\zeta_\varepsilon^n(t, x) = \sum_{j=1}^{n} b_j \beta_j^\varepsilon(t) e_j(x),$$

and $\zeta^n(t, x)$ is defined in a similar way. In view of Doob's moment inequality (A.56), we have

$$\mathbb{E} \sup_{0 \leq t \leq t_0} |\zeta^n(t) - \zeta(t)|_2^2 \leq 4\, \mathbb{E}\, |\zeta^n(t_0) - \zeta(t_0)|_2^2 \leq 4 \sum_{j=n+1}^{\infty} b_j^2 \to 0 \tag{2.73}$$

as $n \to \infty$, and a similar estimate holds for ζ_ε^n and ζ_ε. Combining (2.72) and (2.73) and using Exercise 1.2.19, we arrive at the required result. □

In Chapter 4, we shall need another version of Theorem 2.2.2 that concerns convergence to ζ in the weak topology of the space of probability measures on $C(0, t_0; H)$. Namely, let us denote by $\tilde{\zeta}_\varepsilon$ a continuous process that coincides with ζ_ε at the points $t_k^\varepsilon = k\varepsilon$ and is linear between any two consecutive points of that form. In this case, it is easy to check that

$$\sup_{t_{k-1}^\varepsilon \leq t \leq t_k^\varepsilon} |\zeta_\varepsilon(t) - \tilde{\zeta}_\varepsilon(t)|_2 \leq \sqrt{\varepsilon}\, |\eta_k|_2, \quad k \geq 1,$$

whence it follows that

$$\sup_{0 \leq t \leq t_0} |\zeta_\varepsilon(t) - \tilde{\zeta}_\varepsilon(t)|_2 \leq \sqrt{\varepsilon} \max\big\{|\eta_k|_2,\ 1 \leq k \leq [t_0/\varepsilon] + 1\big\}. \tag{2.74}$$

Exercise 2.2.3 Assuming that $\mathbb{E}|\eta_1|_2^q < \infty$ for some $q > 2$, show that

$$\mathbb{P}\Big\{\sup_{0 \leq t \leq t_0} |\zeta_\varepsilon(t) - \tilde{\zeta}_\varepsilon(t)|_2 > \delta\Big\} \to 0 \quad \text{as } \varepsilon \to 0, \tag{2.75}$$

where $t_0 > 0$ and $\delta > 0$ are arbitrary constants.

Combining this result with Theorem 2.2.2, it is not difficult to show that $\mathcal{D}(\tilde{\zeta}_\varepsilon)$ converges to $\mathcal{D}(\zeta)$ weakly in the space $\mathcal{P}(D(0, t_0; H))$. The following exercise shows that the convergence holds in fact for a stronger topology without any additional assumption on η_k.

Exercise 2.2.4 Prove that, for any $t_0 > 0$, the family $\mathcal{D}(\tilde{\zeta}_\varepsilon)$ converges to $\mathcal{D}(\zeta)$ weakly in the space $\mathcal{P}(C(0, t_0; H))$. *Hint:* Repeat the scheme used in the proof of Theorem 2.2.2, applying Donsker's theorem given in section 8 of [Bil99].

2.3 Navier–Stokes equations perturbed by a random kick force

This section is devoted to a systematic study of the Navier–Stokes system (2.1) with right-hand side given by (2.64), (2.65). We shall always assume that $h \in H$ is a deterministic function and $\{\eta_k\}$ is a sequence of H-valued i.i.d. random variables defined on a complete probability space $(\Omega, \mathcal{F}, \mathbb{P})$. After projecting to the space H, the problem takes the form (cf. (2.19))

$$\dot{u} + \nu L u + B(u) = h + \sum_{k=1}^{\infty} \eta_k \delta(t - kT). \tag{2.76}$$

As before, we shall consider the case of a torus and make remarks concerning the problem in a bounded domain. Accordingly, the problem will be studied in the functional spaces (2.41).

2.3.1 Existence and uniqueness of solution

We begin with a definition of the concept of a solution for (2.76). Let us denote by $S_t : H \to H$ the resolving operator for Eq. (2.19) with $f(t, x) = h(x)$:

$$\dot{u} + \nu L u + B(u) = h. \tag{2.77}$$

In other words, S_t takes u_0 to $u(t)$, where $u \in C(\mathbb{R}_+, H) \cap L^2_{\text{loc}}(\mathbb{R}_+, V)$ stands for the solution of (2.77) satisfying $u(0) = u_0$.

A *filtration for* (2.76) is defined as any non-decreasing family of σ-algebras $\{\mathcal{G}_t, t \geq 0\}$ of the space $(\Omega, \mathcal{F})$ such that $\mathcal{G}_t = \mathcal{G}_{(k-1)T}$ for $(k-1)T \leq t < kT$ and η_k is $\mathcal{G}_{kT}$-measurable and independent of $\mathcal{G}_{(k-1)T}$ for any integer $k \geq 1$. Recall that a random process $v(t)$, $t \geq 0$, is said to be *adapted to* $\mathcal{G}_t$ if the random variable $v(t)$ is $\mathcal{G}_t$-measurable for any $t \geq 0$. Let us define the interval $J_k = [(k-1)T, kT)$ and the space

$$\mathcal{H}(J_k) = \{u \in L^2(J_k, V) : \dot{u} \in L^2(J_k, V^*)\},$$

where the space V is defined in (2.41) and V^* denotes its dual. In view of Lemma 2.1.5, every element of $\mathcal{H}(J_k)$ extends to a continuous curve $\bar{J}_k \to H$.

Definition 2.3.1 A random process $u(t), t \geq 0$, is called a *solution of Eq.* (2.76) if it is adapted to the filtration $\mathcal{G}_t$ for (2.76), and almost every trajectory of u satisfies the following conditions for any integer $k \geq 1$:

- the restriction of u to J_k belongs to $\mathcal{H}(J_k)$ and satisfies (2.77) (in particular, $u : J_k \to H$ is a continuous curve which has a limit at the right endpoint of J_k);
- we have the relation

$$u(t_k^+) - u(t_k^-) = \eta_k, \tag{2.78}$$

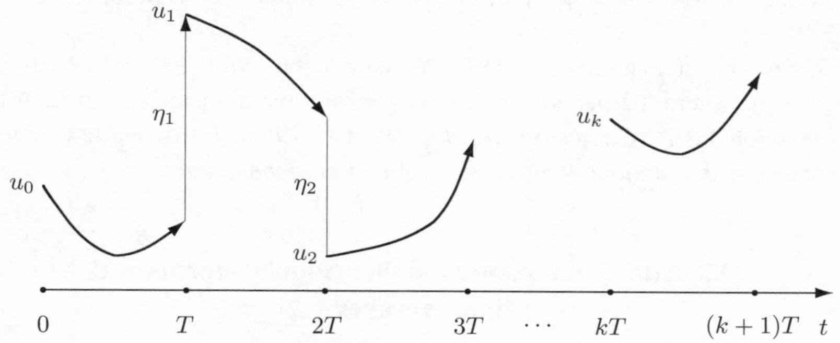

Figure 2.3 Evolution defined by Eq. (2.76).

where $u(t_k^-)$ and $u(t_k^+)$ denote the left- and right-hand limits of u at the point $t_k := kT$.

That is, on the interval (t_{k-1}, t_k) the function u is a solution of the "free" Navier–Stokes system (see (2.77)), and at any point t_k it has an instantaneous increment equal to the kth kick η_k (see Figure 2.3). We shall always normalise the solutions of (2.76) by the condition of right-continuity at the points t_k, $k \geq 1$. That is, denoting $u_k = u(kT)$, $k = 0, 1, \ldots$, we can write

$$u_k = S_T(u_{k-1}) + \eta_k, \quad k \geq 1, \tag{2.79}$$

and for $t = kT + \tau$ with $0 \leq \tau < T$, we have $u(t) = S_\tau(u_k)$.

The following theorem establishes the existence and uniqueness of a solution for the Cauchy problem for (2.76). Its proof is a straightforward consequence of Theorem 2.1.13 and is left to the reader as an exercise.

Theorem 2.3.2 *Let u_0 be an H-valued $\mathcal{G}_0$-measurable random variable. Then Eq.* (2.76) *has a solution satisfying the initial condition* (2.3) *for all $\omega \in \Omega$. Moreover, the solution is unique in the sense that if $\tilde{u}$ is another random process with the above properties, then*

$$\mathbb{P}\{u(t) = \tilde{u}(t) \textit{ for all } t \geq 0\} = 1.$$

2.3.2 Markov chain and RDS

Theorem 2.3.2 enables one to define a discrete-time process and an RDS in H which are associated with solutions of the Navier–Stokes system, evaluated at times kT, $k \in \mathbb{Z}_+$. Indeed, the sequence $\{u_k, k \geq 0\}$, where $u_k = u(kT)$, satisfies Eq. (2.79). By Example 1.3.6, the latter defines a Markov family of random processes[5] on the extended probability space $H \times \Omega$, and by Example 1.3.15,

[5] Equation (2.79) itself defines a Markov chain (starting from u_0) on the original probability space.

it defines an RDS $\boldsymbol{\Phi} = \{\varphi_k^\omega\}$ in H, which is Markov in view of Exercise 1.3.19.

Recall that a measure $\mu \in \mathcal{P}(H)$ is said to be *stationary* for (2.79) if $\mathfrak{P}_1^*\mu = \mu$, where $\mathfrak{P}_k^*$ denotes the Markov semigroup associated with (2.79); see Section 1.3.1. A trajectory $\{u_k, k \geq 0\}$ of (2.79) such that the law of u_k coincides with a stationary measure is called a *stationary solution*.

2.3.3 Additional results: higher Sobolev norms and time averages

Our next goal is to establish some estimates for solutions of (2.76). We begin with moments of solutions, evaluated at deterministic or random times proportional to T. To simplify the presentation, we assume that $\nu = 1$; all the results remain valid for any $\nu > 0$ if we suitably adjust the constants.

Proposition 2.3.3

(i) *In addition to the hypotheses of Theorem 2.3.2, assume that*

$$\mathbb{E}\,|u_0|_2^m < \infty, \qquad K_m := \mathbb{E}\,|\eta_1|_2^m < \infty \tag{2.80}$$

for an integer $m \geq 1$. Then there are positive constants $C_m = C_m(T)$ and $q < 1$ such that

$$\mathbb{E}\,|u(kT)|_2^m \leq q^{kT}\,\mathbb{E}\,|u_0|_2^m + C_m(K_m + 1), \quad k \geq 1. \tag{2.81}$$

(ii) *If $K_{2m} < \infty$ and σ is a stopping time with range in $T\mathbb{Z}_+ \cup \{\infty\}$ satisfying the condition $E := \sum_l \bigl(\mathbb{P}\{\sigma = lT\}\bigr)^{1/2} < \infty$, then we have*

$$\mathbb{E}\left(\mathbb{I}_{\{\sigma<\infty\}}|u(\sigma)|_2^m\right) \leq \mathbb{E}\left(q^\sigma|u_0|_2^m\right) + C_m E\bigl(K_{2m}^{1/2} + 1\bigr), \tag{2.82}$$

where we set $q^\infty = 0$.

Proof Recall that $u_k = u(kT)$. Inequality (2.51) implies that

$$|S_T(v)|_2 \leq e^{-\alpha_1 T/2}|v|_2 + C_1|h|_2 \quad \text{for } v \in H.$$

Furthermore, it follows from (2.79) that

$$|u_k|_2 = |S_T(u_{k-1}) + \eta_k|_2 \leq |S_T(u_{k-1})|_2 + |\eta_k|_2.$$

Combining the last two relations, we obtain

$$|u_k|_2^m \leq q^T|u_{k-1}|_2^m + C_2\bigl(|\eta_k|_2^m + |h|_2^m\bigr), \tag{2.83}$$

where $q < 1$ is a constant depending only on the first eigenvalue α_1, and $C_2 > 0$ depends on m. Iterating this inequality, we get

$$|u_k|_2^m \le q^{kT}|u_0|_2^m + C_2\sum_{l=1}^{k} q^{(k-l)T}\big(|\eta_l|_2^m + |h|_2^m\big). \tag{2.84}$$

Taking the mean value, we arrive at (2.81).

We now prove (2.82). Taking $k = (\sigma/T)\wedge n$ in (2.84), where $n \ge 1$ is an integer, and multiplying the resulting relation by $\mathbb{I}_{\{\sigma<\infty\}}$, we get

$$\begin{aligned}\mathbb{I}_{\{\sigma<\infty\}}|u_{(\sigma/T)\wedge n}|_2^m &\le \mathbb{I}_{\{\sigma<\infty\}}q^{\sigma\wedge(nT)}|u_0|_2^m \\ &\quad + C_2\mathbb{I}_{\{1\le\sigma<\infty\}}\sum_{l=1}^{(\sigma/T)\wedge n} q^{\sigma\wedge(nT)-lT}\big(|\eta_l|_2^m + |h|_2^m\big) \\ &\le q^{\sigma\wedge(nT)}|u_0|_2^m + C_2\sum_{s=1}^{\infty}\sum_{l=1}^{s\wedge n}\big(q^T\big)^{s\wedge n-l}\mathbb{I}_{\{\sigma=sT\}}\big(|\eta_l|_2^m + |h|_2^m\big).\end{aligned}$$

Taking the expectation and using Fatou's lemma to pass to the limit as $n\to\infty$ in the left-hand side of the above inequality, we derive

$$\mathbb{E}\left(\mathbb{I}_{\{\sigma<\infty\}}|u_{\sigma/T}|^m\right) \le \mathbb{E}\left(q^\sigma|u_0|_2^m\right) + C_2\mathcal{E}_m, \tag{2.85}$$

where

$$\mathcal{E}_m = \lim_{n\to\infty}\sum_{s=1}^{\infty}\sum_{l=1}^{s\wedge n}\big(q^T\big)^{s\wedge n-l}\mathbb{E}\big(\mathbb{I}_{\{\sigma=sT\}}\big(|\eta_l|_2^m + |h|_2^m\big)\big).$$

Applying the Cauchy–Schwarz inequality, we get

$$\begin{aligned}\mathcal{E}_m &\le \lim_{n\to\infty}\sum_{s=1}^{\infty}\sum_{l=1}^{s\wedge n}\big(q^T\big)^{s\wedge n-l}\big(\mathbb{P}\{\sigma=sT\}\big)^{\frac12}\big((\mathbb{E}\,|\eta_l|_2^{2m})^{\frac12} + |h|_2^m\big) \\ &= \big(K_{2m}^{\frac12}+1\big)\lim_{n\to\infty}\sum_{s=1}^{\infty}\sum_{r=0}^{s\wedge n-1} q^{Tr}\big(\mathbb{P}\{\sigma=sT\}\big)^{\frac12} \\ &\le \big(K_{2m}^{\frac12}+1\big)\sum_{s=1}^{\infty}\sum_{r=0}^{s-1} q^{Tr}\big(\mathbb{P}\{\sigma=sT\}\big)^{\frac12}.\end{aligned}$$

Substituting this inequality into (2.85), we obtain (2.82). □

Proposition 2.3.3 implies that if all moments of the initial function u_0 and those of the random perturbations η_k are finite, then so are moments of the solution. If we replace this hypothesis by the stronger condition of finiteness of a second exponential moment, then we can obtain some bounds of a similar quantity for solutions. More precisely, we have the following result.

Proposition 2.3.4 *In addition to the hypotheses of Theorem 2.3.2, assume that for some $\varkappa_0 > 0$ we have*

$$\mathbb{E}\exp\big(\varkappa_0|u_0|_2^2\big) < \infty, \quad Q := \mathbb{E}\exp\big(\varkappa_0|\eta_1|_2^2\big) < \infty. \tag{2.86}$$

Then there are positive constants $\varkappa$ and $q < 1$ such that

$$\mathbb{E}\exp\big(\varkappa\,|u(kT)|_2^2\big) \le C\left(\mathbb{E}\exp\big(\varkappa\,|u_0|_2^2\big)\right)^{q^{kT}}, \quad k \ge 0, \tag{2.87}$$

where we set $C = \big(Q\exp(\varkappa_0|h|_2^2)\big)^\gamma$ and $\gamma = (1-q^T)^{-1}$.

Proof It follows from inequality (2.83) with $m = 2$ that

$$\begin{aligned}\mathbb{E}\exp\big(\varkappa|u_k|_2^2\big) &\le \mathbb{E}\exp\big(\varkappa q^T|u_{k-1}|_2^2 + C_2\varkappa(|\eta_k|_2^2 + |h|_2^2)\big)\\ &= \mathbb{E}\exp\big(\varkappa q^T|u_{k-1}|_2^2\big)\,\mathbb{E}\exp\big(C_2\varkappa|\eta_k|_2^2\big)\exp(C_2\varkappa|h|_2^2),\end{aligned}$$

where we used the independence of η_k and u_{k-1}. Choosing $\varkappa = \varkappa_0/C_2$ and applying Hölder's inequality, we derive

$$\mathbb{E}\exp\big(\varkappa|u_k|_2^2\big) \le Q\,\exp(\varkappa_0|h|_2^2)\big(\mathbb{E}\exp\big(\varkappa|u_{k-1}|_2^2\big)\big)^{q^T}.$$

Iterating this inequality, we obtain (2.87). □

We now turn to some estimates for the mean value of higher Sobolev norms. Since the corresponding proofs are similar to those of Propositions 2.3.3 and 2.3.4, we shall confine ourselves to formulating the results.

Proposition 2.3.5

(i) *In addition to the hypotheses of Theorem 2.3.2, assume that*

$$\mathbb{E}\,|u_0|_2^m < \infty, \quad \mathbb{E}\,\|\eta_1\|_s^m < \infty \quad \textit{for all } m \ge 1, \tag{2.88}$$

where $s \ge 1$ is an integer. Then there are integers $l_{sm} \ge m$ and a constant C_{sm} such that

$$\mathbb{E}\,\|u(kT)\|_s^m \le C_{sm}\big(\mathbb{E}\,|u((k-1)T)|_2^{l_{sm}} + \mathbb{E}\,\|\eta_1\|_s^m + 1\big), \quad k \ge 1. \tag{2.89}$$

(ii) *In addition to the hypotheses of Theorem 2.3.2, assume that*

$$\mathbb{E}\exp\big(\varkappa_0|u_0|_2^2\big) < \infty, \quad \mathbb{E}\exp\big(\varkappa_0\|\eta_1\|_s^2\big) < \infty, \tag{2.90}$$

where $s \ge 1$ is an integer. Then there are positive constants p_s, $\varkappa_s$, and C_s such that

$$\mathbb{E}\exp\big(\varkappa_s\|u(kT)\|_s^{p_s}\big) \le C_s\mathbb{E}\exp\big(\varkappa\,|u((k-1)T)|_2^2\big)\,\mathbb{E}\exp\big(\varkappa_0\|\eta_1\|_s^2\big), \tag{2.91}$$

where $k \ge 1$ is an arbitrary integer and $\varkappa$ is the constant in Proposition 2.3.4(ii).

Let us note that, in view of Propositions 2.3.3 and 2.3.4, the right-hand sides of (2.89) and (2.91) are finite. Thus, to ensure that the moments of higher Sobolev norms of solutions are finite, it suffices to assume that so are the moments of the L^2-norm of the initial function. More precisely, we have the following result on uniform bounds for the mean value of higher Sobolev norms of solutions.

Corollary 2.3.6 *Under the hypotheses of parts (i) or (ii) of Proposition 2.3.5, we have respectively the following estimates for the solutions of Eq.* (2.76) *such that* $u(0) = u_0$:

$$\mathbb{E}\,\|u(kT)\|_s^m \le C_{sm}(u_0), \quad k \ge 1, \tag{2.92}$$

$$\mathbb{E}\exp\bigl(\varkappa_s\|u(kT)\|_s^{p_s}\bigr) \le C_{sm}(u_0), \quad k \ge 1, \tag{2.93}$$

where C_{sm} and C_s are some constants depending on the initial condition and the right-hand side of (2.76).

Exercise 2.3.7 Prove Proposition 2.3.5 and Corollary 2.3.6. *Hint:* Use Theorem 2.1.19 to establish (2.89) and (2.91); a proof can be found in the appendix of [KS01b].

We now derive some estimates for the stopping time at which the time average of solutions becomes large. Given a function $u(t)$ with range in H^1 and integers $l < k$, we define

$$\langle\|u\|_1^2\rangle_l^k = \frac{1}{k-l}\int_{lT}^{kT}\|u(t)\|_1^2 dt.$$

Let us fix a constant $M > 0$ and consider the stopping time

$$\tau(M) = T\min\{k \ge 1 : \langle\|u\|_1^2\rangle_0^{k+1} > M\}, \tag{2.94}$$

where we set $\tau(M) = +\infty$ if the condition in the brackets is never satisfied.

Proposition 2.3.8 *Assume that the conditions of Proposition 2.3.4 are fulfilled. Then there are positive constants $\delta = \delta(\varkappa_0, T)$ and $C = C(T, h)$ such that*

$$\mathbb{P}\{\tau(M) = kT\} \le R\,(CQ)^k e^{-\delta M(k+1)} \quad \textit{for } k \ge 1, \tag{2.95}$$

where $R = \mathbb{E}\,e^{\varkappa_0|u_0|_2^2}$. Furthermore, if $\delta M > \ln(R + CQ)$, then

$$\mathbb{P}\{\tau(M) = +\infty\} \ge \frac{R}{R+CQ} > 0. \tag{2.96}$$

Proof

Step 1. We first show that, for any integer $k \ge 1$,

$$|u(t_k)|_2^2 + \int_0^{t_k}\|u(t)\|_1^2 \le 2|u_0|_2^2 + C_1\biggl(\sum_{l=1}^k|\eta_l|_2^2 + t_k\|h\|_{-1}^2\biggr), \tag{2.97}$$

where $t_k = kT$, and C_1 is a constant depending on T. Indeed, on any interval $J_k = [t_{k-1}, t_k]$ the solution $u(t,x)$ satisfies the deterministic Navier–Stokes equations (2.77) with $\nu = 1$. Thus, by (2.24), we have

$$|u(t_l^-)|_2^2 + \int_{t_{l-1}}^{t_l} \|u(t)\|_1^2 dt \le |u(t_{l-1})|_2^2 + T\|h\|_{-1}^2.$$

Furthermore, relation (2.78) implies that

$$\begin{aligned} |u(t_l)|_2^2 &= |u(t_l^-)|_2^2 + |\eta_l|_2^2 + 2\langle \eta_l, u(t_l^-)\rangle \\ &\le (1+\mu)|u(t_l^-)|_2^2 + (1+\mu^{-1})|\eta_l|_2^2, \end{aligned}$$

where the constant $\mu > 0$ will be chosen later. Combining these two inequalities, we derive

$$(1+\mu)^{-1}|u(t_l)|_2^2 + \int_{t_{l-1}}^{t_l} \|u(t)\|_1^2 dt \le |u(t_{l-1})|_2^2 + T\|h\|_{-1}^2 + \mu^{-1}|\eta_l|_2^2.$$

Taking the sum over $l = 1, \dots, k$, we obtain

$$\begin{aligned} (1+\mu)^{-1}|u(t_k)|_2^2 + \int_0^{t_k} \|u(t)\|_1^2 dt &\le |u_0|_2^2 + \frac{\mu}{1+\mu}\sum_{l=1}^{k-1} |u(t_l)|_2^2 + t_k\|h\|_{-1}^2 \\ &\quad + \mu^{-1}\sum_{l=1}^{k} |\eta_l|_2^2. \end{aligned}$$

It follows from inequality (2.53), applied to the intervals J_l, $1 \le l \le k-1$, that

$$\sum_{l=1}^{k-1} |u(t_l)|_2^2 \le C_2 \int_0^{t_k} \|u(t)\|_1^2 dt + C_2 t_k \|h\|_{-1}^2,$$

where $C_2 > 0$ is a constant depending on T. Choosing the constant $\mu \in (0,1)$ so small that $2\mu C_2 \le 1+\mu$, we derive (2.97) from these two inequalities.

Step 2. Let us set

$$I_k = (k+1)\langle \|u\|_1^2 \rangle_0^{k+1}.$$

Inequality (2.97) implies that

$$I_k \le 2|u_0|_2^2 + C_3\biggl(\sum_{l=1}^{k} |\eta_l|_2^2 + t_{k+1}\|h\|_{-1}^2\biggr).$$

Let the constant $\delta > 0$ be so small that $(C_3 \vee 2)\delta \le \varkappa_0$. Then it follows from this inequality and (2.86) that

$$\mathbb{E}\, e^{\delta I_k} \le R\,(CQ)^k.$$

Chebyshev's inequality now implies that

$$\mathbb{P}\{I_k > M(k+1)\} \le R\,(C\,Q)^k e^{-\delta M(k+1)}.$$

It remains to note that $\{\tau(M) = k\} \subset \{I_k > M(k+1)\}$.

Step 3. It remains to establish (2.96). Note that if $\delta M > \ln(R + C\,Q)$, then in view of (2.95) we have

$$\mathbb{P}\{\tau(M) < \infty\} = \sum_{k=1}^{\infty} \mathbb{P}\{\tau(M) = k\} \le \sum_{k=1}^{\infty} R\,(C\,Q)^k e^{-\delta M(k+1)} \le \frac{C\,Q}{R + C\,Q}.$$

This inequality readily implies the required result. □

2.4 Navier–Stokes equations perturbed by spatially regular white noise

2.4.1 Existence and uniqueness of solution, and Markov process

Let us turn to the equation

$$\dot{u} + \nu L u + B(u) = h + \eta, \quad \eta = \frac{\partial \zeta}{\partial t}, \quad \zeta(t, x) = \sum_{j=1}^{\infty} b_j \beta_j(t) e_j(x), \quad (2.98)$$

where $h \in H$ is a deterministic function, and b_j, β_j, and e_j are the same as in (2.66). In particular, we assume that

$$\mathfrak{B} = \sum_{j=1}^{\infty} b_j^2 < \infty.$$

This ensures that $\zeta(t, \cdot)$ is a continuous process with range in H. Since for almost every $\omega \in \Omega$ the curve $\zeta^{\omega}(t)$ is only H-continuous, i.e., has law smoothness, we cannot apply deterministic tools to construct solutions of (2.98); see Exercise 2.1.27. Using instead a stochastic approach, we shall show that, for a.e. ω and for any initial function $u(0) = u_0 \in H$, Eq. (2.98) has a unique solution $u(t, x)$. Moreover, solutions can be written as $u = U(u_0, \zeta^{\omega}(\cdot))$, where U is a Borel mapping of its arguments.

Let us start with the concept of the solution for (2.98). We shall always assume that the Brownian motions $\{\beta_j\}$ are defined on a complete probability space $(\Omega, \mathcal{F}, \mathbb{P})$ with a filtration $\mathcal{G}_t$, $t \ge 0$, and that the σ-algebras $\mathcal{G}_t$ are completed with respect to $(\mathcal{F}, \mathbb{P})$, that is, $\mathcal{G}_t$ contains all the $\mathbb{P}$-null sets $A \in \mathcal{F}$. In this case, we shall say that the filtered probability space $(\Omega, \mathcal{F}, \mathcal{G}_t, \mathbb{P})$ satisfies the *usual hypotheses*.

Definition 2.4.1 An H-valued random process $u(t)$, $t \ge 0$, is called a *solution for* (2.98) if:

(a) The process $u(t)$ is adapted to the filtration $\mathcal{G}_t$, and its almost every trajectory belongs to the space[6]

$$\mathcal{X} := C(\mathbb{R}_+; H) \cap L^2_{\text{loc}}(\mathbb{R}_+, V).$$

(b) Equation (2.98) holds in the sense that, with probability 1,

$$u(t) + \int_0^t \big(\nu L u + B(u)\big)\, ds = u(0) + th + \zeta(t), \quad t \geq 0, \tag{2.99}$$

where the equality holds in the space H^{-1}.

It is easy to see that, with probability 1, the left- and right-hand sides of (2.99) belong to the space $C(\mathbb{R}_+, H^{-1})$, so relation (2.99) makes sense. The construction of a solution for (2.98) given below is based on a reduction to an equation with random coefficients. Namely, let us consider the stochastic Stokes equation

$$\dot{z} + \nu L z = \eta(t, x) \quad t \geq 0. \tag{2.100}$$

The concept of a solution for (2.100) is similar to that for (2.98), and we do not repeat its definition. The following result uses the existence of an analytic semigroup $e^{-\nu t L}$ generated by the Stokes operator, and the concept of a stochastic convolution. We refer the reader to chapter 1 of [Hen81] and section 5.1 of [DZ92] for the corresponding definitions and results.[7]

Proposition 2.4.2 *Equation* (2.100) *has a solution $z(\cdot) \in \mathcal{X}$ satisfying the initial condition $z(0) = 0$. Moreover, this solution is unique in the sense that if $\tilde{z}(t)$ is another solution, then $z \equiv \tilde{z}$ almost surely. Finally, the solution z can be represented as the stochastic convolution*

$$z(t) = \int_0^t e^{-\nu(t-s)L} d\zeta(s), \tag{2.101}$$

and for almost every ω we have $z^\omega = Z(\zeta^\omega)$, where $Z : C(\mathbb{R}_+; H) \to C(\mathbb{R}_+; V^*)$ is the continuous linear mapping defined by the right-hand side of (2.102).

Proof The uniqueness of a solution can be established by the same argument as in the deterministic case (see the proof of Theorem 2.1.13), and therefore we shall confine ourselves to the proof of its existence and to representation (2.101).

[6] We recall that $C(\mathbb{R}_+; X)$ is a Fréchet space endowed with distance (1.9).

[7] The stochastic convolution is encountered only in this chapter. The reader not interested in that concept may regard the right-hand side of (2.102) as the definition of the stochastic convolution (the proof of Proposition 2.4.2 uses only representation (2.102)). Alternatively, the convolution $z(t)$ can be defined by duality:

$$\langle \xi, z(t) \rangle = \int_0^t \langle e^{-\nu(t-s)L}\xi, d\zeta(s) \rangle = \sum_{j=1}^\infty b_j \int_0^t \xi_j(s) d\beta_j(s), \quad \xi \in H,$$

where $\xi_j(s)$ is the jth component of the vector $e^{-\nu(t-s)L}\xi$.

Integrating by parts, we rewrite the stochastic integral (2.101) in the Wiener form (cf. section 2.1 in [McK69])

$$\int_0^t e^{-\nu(t-s)L} d\zeta(s) = \zeta(t) - \nu \int_0^t Le^{-\nu(t-s)L}\zeta(s)\,ds. \tag{2.102}$$

In view of Exercise 2.2.1, we have $\zeta \in C(0, T; H)$ almost surely. Since the operator $Le^{-\nu tL}$ is continuous from H to V^* with a norm bounded by $C(\nu t)^{-1/2}$, it follows from (2.102) that $z(t)$ is a random process in V^* that has continuous trajectories and is adapted to the filtration $\mathcal{G}_t$. The required assertions will be established if we prove that, with probability 1, the trajectories of z belong to the space $\mathcal{X}$ introduced in Definition 2.4.1 and satisfy the relation

$$z(t) + \nu \int_0^t Lz(s)\,ds = \zeta(t), \quad t \geq 0. \tag{2.103}$$

The proof of this fact is rather simple in the case when the orthonormal basis $\{e_j\}$ entering the definition of ζ (see (2.98)) consists of the eigenfunctions of the Stokes operator L. To simplify the presentation, we shall consider only that particular case, leaving the general situation to the reader as an exercise.

Let us consider a sequence of processes z_n defined by

$$z_n(t) = \zeta^n(t) - \nu \int_0^t Le^{-\nu(t-s)L}\zeta^n(s)\,ds,$$

where $\zeta^n(t) = \sum_{j=1}^n b_j \beta_j(t) e_j$. Since the subspace $H_n = \operatorname{span}\{e_1, \dots, e_n\}$ is invariant under L and $\|Le^{-cL}\|_{\mathcal{L}(H,V^*)} \leq c^{-1}$, we see that z_n is an H_n-valued process with continuous trajectories such that

$$\sup_{0 \leq t \leq T} \mathbb{E}\, \|z_n(t) - z(t)\|_{V^*}^2 \to 0 \quad \text{as } n \to \infty, \tag{2.104}$$

where $T > 0$ is an arbitrary constant. Moreover, it is easy to see that

$$z_n(t) + \nu \int_0^t Lz_n(s)\,ds = \zeta^n(t), \quad t \geq 0. \tag{2.105}$$

We claim that there is a subsequence z_{n_k} converging almost surely to a limit $\bar{z}$ in the space $\mathcal{X}_T = C(0, T; H) \cap L^2(0, T; V)$. Indeed, for any integers $m < n$ the H_n-valued process $z_{mn} = z_n - z_m$ satisfies the stochastic differential equation

$$\dot{z}_{mn} + \nu L z_{mn} = \sum_{j=m+1}^n b_j \dot{\beta}_j(t) e_j(x).$$

Applying Itô's formula to $|z_{mn}(t)|_2^2$, we derive

$$|z_{mn}(t)|_2^2 = \int_0^t \bigl(-2\nu\, |\nabla z_{mn}(s)|_2^2 + F_{mn}\bigr)\,ds + 2 \sum_{j=m+1}^n b_j \int_0^t \langle z_{mn}, e_j\rangle d\beta_j, \tag{2.106}$$

where $F_{mn} = \sum_{j=m+1}^{n} b_j^2$. Taking the mean value in (2.106), we see that

$$\sup_{0\le t\le T} \mathbb{E}\,|z_{mn}(t)|_2^2 + 2\nu\,\mathbb{E}\int_0^T \|z_{mn}(s)\|_1^2 dt \le C_4 T F_{mn}. \tag{2.107}$$

Furthermore, application of Doob's moment inequality (A.56) to the stochastic integral in (2.106) results in (cf. the derivation of (2.73))

$$\mathbb{E}\sup_{0\le t\le T} |z_{mn}(t)|_2^2 \le F_{mn}T + C_6\,\mathbb{E}\int_0^T \sum_{j=1}^{n} b_j^2 (z_{mn}(t), e_j)^2 dt. \tag{2.108}$$

Now note that $F_{mn} \to 0$ as $m, n \to \infty$. Combining this with (2.107) and (2.108), we conclude that

$$\mathbb{E}\sup_{0\le t\le T} |z_{mn}(t)|_2^2 + \mathbb{E}\int_0^T \|z_{mn}(s)\|_1^2 dt \to 0 \quad \text{as } m, n \to \infty. \tag{2.109}$$

Thus, the sequence $\{z_n\}$ converges to a limit $\bar z$ in the space $L^2(\Omega, \mathcal{X}_T)$. It follows that a subsequence $\{z_{n_k}\}$ converges almost surely to $\bar z$ in $\mathcal{X}_T$. Since

$$z_{n_k}(t) + \nu\int_0^t L z_{n_k}(s)\,ds = \zeta^{n_k}(t), \quad 0 \le t \le T,$$

where the equality holds in H^{-1} for almost all $\omega \in \Omega$, then passing to the limit as $k \to \infty$, we arrive at relation (2.103) with $z = \bar z$. Thus, the function $\bar z$ is a solution of (2.100) vanishing at $t = 0$. The fact that $\bar z$ can be written in the form (2.101) can easily be established by passing to the limit in a similar relation for z_{n_k}. □

Remark 2.4.3 Denote

$$H_\zeta = \overline{\operatorname{span}\{e_j : b_j \ne 0\}} \subset H, \tag{2.110}$$

and denote by m_ζ the law of ζ. This is a measure on $C(\mathbb{R}_+; H)$ whose support equals $C(\mathbb{R}_+; H_\zeta)$. We write $\mathcal{F}_\zeta$ for the m_ζ-completion of the Borel σ-algebra on $C(\mathbb{R}_+; H)$. The process $z_n(t)$ constructed in the proof has the form $z_n = Z_n(\zeta)$, where Z_n is a continuous operator from $C(\mathbb{R}_+; H)$ to $\mathcal{X}$, and our proof shows that m_ζ-almost surely the sequence $\{Z_n\}$ converges to Z. In particular, it follows that

$$Z : \big(C(\mathbb{R}_+; H), \mathcal{F}_\zeta\big) \to (\mathcal{X}, \mathcal{B}(\mathcal{X}))$$

is a measurable mapping. It is straightforward to see that, for any $T > 0$, the restriction of $Z(\zeta)$ to $[0, T]$ depends only on $\zeta_{[0,T]}$. Finally, if

$$\mathfrak{B}_1 = \sum_{j=1}^{\infty} \alpha_j b_j^2 < \infty,$$

where $\{\alpha_j\}$ are the eigenvalues of the Stokes operator, then $\zeta \in C(\mathbb{R}_+; V)$ almost surely. The restriction of Z to $C(\mathbb{R}_+; V)$ defines a continuous linear mapping from $C(\mathbb{R}_+; V)$ to $\mathcal{X}$.

Exercise 2.4.4 Without assuming that $\{e_j\}$ is an eigenbasis for L, prove that the stochastic convolution (2.101) defines a random process that has continuous trajectories and satisfies relation (2.103).

We now show how to reduce the stochastic Navier–Stokes system (2.98) to an equation with random coefficients. As before, we assume that the deterministic and random forces h and η satisfy the hypotheses formulated in the beginning of this subsection. From now on, we change slightly the definition of the space $\mathcal{H}$ and denote

$$\mathcal{H} = \{u \in L^2(0, T; V) : \dot{u} \in L^2(0, T; V^*)\};$$

cf. Section 2.1.3. We seek a solution in the form $u = z + v$, where $z(t)$ is the process constructed in Proposition 2.4.2. The function v must satisfy the equations

$$\dot{v} + \nu L v + B(v + z) = h, \tag{2.111}$$

$$v(0) = u_0. \tag{2.112}$$

This is a Navier–Stokes-type system with random coefficients entering through the stochastic process z.

Proposition 2.4.5 *Let ν and T be some positive constants and let $h \in H$ be a given function. Then for any $u_0 \in H$ and $z \in \mathcal{X}_T$ problem* (2.111), (2.112) *has a unique solution $v \in \mathcal{H}$. Moreover, the operator $\mathcal{R} : H \times \mathcal{X}_T \to \mathcal{H}$ taking a pair (u_0, z) to the solution $v \in \mathcal{H}$ is locally Lipschitz continuous. That is, for any $R > 0$ there is a constant $C_R > 0$ such that*

$$\|\mathcal{R}(u_{01}, z_1) - \mathcal{R}(u_{02}, z_2)\|_{\mathcal{H}} \le C_R\big(|u_{01} - u_{02}|_2 + \|z_1 - z_2\|_{\mathcal{X}_T}\big), \tag{2.113}$$

where $u_{0i} \in H$ and $z_i \in \mathcal{X}_T$ are arbitrary functions whose norms do not exceed R.

In the case $z \equiv 0$, this result is a straightforward consequence of Theorem 2.1.13 and Proposition 2.1.25. The case of a non-zero function z can be handled with similar arguments, and therefore we omit the proof; cf. Exercise 2.1.27. We are now ready to establish the main theorem on the existence and uniqueness of a solution for the stochastic Navier–Stokes equation (2.98).

Theorem 2.4.6 *For any $\nu > 0$ and any $\mathcal{G}_0$-measurable random variable $u_0(x)$ the Navier–Stokes system* (2.98) *has a solution $u(t)$, $t \ge 0$, satisfying the initial* *condition*

$$u(0) = u_0 \quad \textit{almost surely.} \tag{2.114}$$

Moreover, if $\tilde{u}(t)$ is another solution of (2.98), (2.114)*, then $u(t) \equiv \tilde{u}(t)$ almost surely. Furthermore, the solution $u(t)$ possesses the following properties.*

(i) *Almost all trajectories of $u(t)$ are continuous with range in H and locally square integrable with range in V.*
(ii) *The process $u(t)$ can be written in the form*

$$u(t) = u_0 + \int_0^t f(s)\,ds + \zeta(t), \quad t \geq 0, \tag{2.115}$$

where $f(t)$ is a V^*-valued $\mathcal{G}_t$-progressively measurable process such that

$$\mathbb{P}\left\{\int_0^T \|f(t)\|_{V^*}^2 dt < \infty \textit{ for any } T > 0\right\} = 1. \tag{2.116}$$

Let $\mathcal{C}_T = C(0,T;H)$ and $m_{\zeta,T} = \mathcal{D}\big(\zeta|_{[0,T]}\big)$. Then $\operatorname{supp} m_{\zeta,T} = C(0,T;H_\zeta)$; see (2.110). Let $\mathcal{F}_{\zeta,T}$ be the $m_{\zeta,T}$-completion of $\mathcal{B}(\mathcal{C}_T)$. Remark 2.4.3 and Proposition 2.4.5 imply the following result.

Proposition 2.4.7 *There exists a measurable mapping*

$$U : \big(H \times \mathcal{C}_T, \mathcal{B}(H) \otimes \mathcal{F}_{\zeta,T}\big) \to \big(\mathcal{X}_T, \mathcal{B}(\mathcal{X}_T)\big),$$

which is locally Lipschitz continuous in $u \in H$ for $m_{\zeta,T}$-almost every $\omega \in \mathcal{C}_T$, such that

$$u^\omega = U(u_0, \zeta^\omega) \quad \textit{for } m_{\zeta,T}\textit{-almost every } \omega \textit{ and all } u_0 \in H. \tag{2.117}$$

Moreover, the restriction of U to $H \times C(0,T;V)$ is locally Lipschitz continuous in both variables.

In particular, if we denote by U_t the restriction of U at time $t \in [0,T]$, then for any (random) initial function u_0^ω which is independent of ζ^ω the law of $u^\omega(t)$ can be written as the image of the product measure $\mathcal{D}(u_0) \otimes m_{\zeta,t}$ under the mapping U_t:

$$\mathcal{D}(u^\omega(t)) = U_{t*}\big(\lambda \otimes m_{\zeta,t}\big), \quad \lambda = \mathcal{D}(u_0). \tag{2.118}$$

Proof of Theorem 2.4.6 We first prove uniqueness. If u and $\tilde{u}$ are two solutions, then for almost every $\omega \in \Omega$ the difference $w = u - \tilde{u}$ belongs to $\mathcal{X}$, vanishes at $t = 0$, and satisfies the equation

$$\dot{w} + \nu L w + B(w, u) + B(\tilde{u}, w) = 0.$$

This equation implies that $\dot{w} \in L^2(0,T;V^*)$ for any $T > 0$, and as in the case of the deterministic equation (see the proof of Theorem 2.1.13), it follows that $w \equiv 0$ almost surely.

To prove existence, we denote by $z(t)$ the solution of Eq. (2.100) vanishing at zero; see Proposition 2.4.2. Let $\Omega_0 \subset \Omega$ be a set of full measure that consists of those $\omega \in \Omega$ for which $z \in \mathcal{X}$. We define $v \in \mathcal{X}$ as the solution of problem (2.111), (2.112) for $\omega \in \Omega_0$ and set $v = 0$ on the complement of Ω_0. Then the random process $u = v + z$ is a solution of problem (2.98), (2.114).

We now prove assertions (i) and (ii). The continuity of trajectories of u in H follows from a similar property for z and v. Relation (2.99) implies that $u(t)$ can be written in the form (2.115) with $f(t) = h - \nu L u(t) - B(u(t))$. Finally, it follows from Propositions 2.4.2 and 2.4.5 that (2.116) holds. □

Theorem 2.4.6 established above not only gives the existence and uniqueness of a solution for the stochastic Navier–Stokes equation (2.98), but also ensures that the solution meets Condition A.7.4 from Section A.7. Therefore the infinite-dimensional Itô formula proved in the appendix applies to the constructed solution. This provides us with a convenient tool for deriving a priori estimates without resorting to finite-dimensional approximations of the Navier–Stokes system.

In conclusion, let us note that the family of solutions corresponding to all possible initial data form a Markov family; see Section 1.3.1. This fact is true for "any" PDE perturbed by the time derivative of a stochastic process with independent increments, provided that the corresponding Cauchy problem is well posed. We refer the reader to chapter 7 of [Øks03] for the proof of the Markov property in the case of SDEs (see also section V.5 of [Kry02]) and to section 9.2 of [DZ92] where this property is proved for various classes of SPDEs. In the context of the Navier–Stokes equations perturbed by spatially regular white noise, the Markov property can be established with the help of a similar argument, using relation (2.118) for the law of a solution. We also note that, as will be shown in Section 2.4.4, the family of solutions for (2.98) forms a Markov RDS, and as was explained in Section 1.3.3, to each Markov RDS one can associate, in a canonical way, a Markov process. This provides a rigorous proof of the above-mentioned claim.

For the reader's convenience, we now briefly describe the construction of a Markov process associated with the stochastic Navier–Stokes system. Let $\widetilde{\Omega} = H \times \Omega$, where $(\Omega, \mathcal{F}, \mathcal{G}_t, \mathbb{P})$ is the filtered probability space defined in the beginning of this section (see page 69), and let

$$\widetilde{\mathcal{F}}_t = \mathcal{B}(H) \otimes \mathcal{G}_t, \quad \mathbb{P}_v = \delta_v \otimes \mathbb{P},$$

where $v \in H$. Writing $\widetilde{\omega} = (v, \omega)$ for points of $\widetilde{\Omega}$, we now define the process $t \mapsto u_t(\widetilde{\omega}) \equiv u^\omega(t; v)$, where $u^\omega(t; v)$ denotes the solution of (2.98) constructed in Theorem 2.4.6 with a deterministic initial function $u_0 = v$. We thus obtain a family of Markov processes $(u_t, \mathbb{P}_v)$, $v \in H$, associated with (2.98). All concepts and properties that are valid in the general setting of Markov processes apply in the case under study. In particular, a measure $\mu \in \mathcal{P}(H)$ is said to be *stationary* for Eq. (2.98) if $\mathfrak{P}_t^* \mu = \mu$ for all $t \geq 0$, where $\mathfrak{P}_t^* : \mathcal{P}(H) \to \mathcal{P}(H)$ denotes the Markov semigroup for the family $(u_t, \mathbb{P}_v)$, and a solution $u(t)$ of Eq. (2.98) is called *stationary* if its law $\mathcal{D}(u(t))$ coincides with a stationary measure for all $t \geq 0$. Let us also note that $\mathfrak{P}_t^* \lambda$ coincides with the right-hand side of (2.118).

2.4.2 Additional results: energy balance, higher Sobolev norms, and time averages

Our first result concerns the *energy balance* for solutions of the Navier–Stokes system (2.98). It is an analogue of relation (2.39) in the stochastic case.

Proposition 2.4.8 *Under the hypotheses of Theorem 2.4.6, assume that the initial function u_0 has a finite second moment, $\mathbb{E}\,|u_0|_2^2 < \infty$. Then for any $\nu > 0$ the following relation holds for a solution $u(t)$ of problem* (2.98), (2.114):

$$\mathbb{E}\,|u(t)|_2^2 + 2\nu \mathbb{E} \int_0^t |\nabla u(s)|_2^2 ds = \mathbb{E}\,|u_0|_2^2 + \mathfrak{B} t + 2\mathbb{E} \int_0^t \langle u, h \rangle ds, \quad t \geq 0. \tag{2.119}$$

Moreover,

$$\mathbb{E}\,|u(t)|_2^2 \leq e^{-\nu\alpha_1 t} \mathbb{E}\,|u_0|_2^2 + (\nu\alpha_1)^{-1} \mathfrak{B} + (\nu\alpha_1)^{-2} |h|_2^2, \quad t \geq 0, \tag{2.120}$$

where $\alpha_1 > 0$ stands for the first eigenvalue[8] *of the Stokes operator L.*

Proof We wish to apply Theorem A.7.5 to the functional $F(u) = |u|_2^2$. Since

$$\partial_u F(u; v) = 2\langle u, v \rangle, \quad \partial_u^2 F(u; v) = 2|v|_2^2,$$

relation (A.24) takes the form

$$\begin{aligned} |u(t \wedge \tau_n)|_2^2 = |u_0|_2^2 &+ 2 \int_0^{t\wedge\tau_n} \big(\langle u, h - \nu L u \rangle + \mathfrak{B} \big)\, ds \\ &+ 2 \sum_{j=1}^{\infty} b_j \int_0^{t\wedge\tau_n} \langle u, e_j \rangle d\beta_j(s), \end{aligned} \tag{2.121}$$

where we set

$$\tau_n = \inf\{t \geq 0 : |u(t)|_2 > n\}.$$

[8] Recall that, in the case of periodic boundary conditions, we consider the problem in the spaces of functions with zero mean value, and therefore the first eigenvalue of L is positive.

Taking the mean value in (2.121) and using Doob's optional sampling theorem, we obtain

$$\mathbb{E}\,|u(t\wedge\tau_n)|_2^2 + 2\nu\mathbb{E}\int_0^{t\wedge\tau_n}|\nabla u(s)|_2^2 ds = \mathbb{E}\,|u_0|_2^2 + \mathfrak{B}\mathbb{E}\,(t\wedge\tau_n) + 2\mathbb{E}\int_0^{t\wedge\tau_n}\langle u,h\rangle ds. \quad (2.122)$$

Since the trajectories of $u(t)$ are continuous H-valued functions of time, we have $\tau_n \to \infty$ as $n \to \infty$. Passing to the limit in (2.122) as $n \to \infty$ and using the monotone convergence theorem, we arrive at (2.119).

We now note that

$$|\nabla u|_2^2 \geq \alpha_1 |u|_2^2, \quad 2|\langle u,h\rangle| \leq \nu\alpha_1|u|_2^2 + (\nu\alpha_1)^{-1}|h|_2^2.$$

Substituting these estimates into (2.119) and using Gronwall's inequality, we arrive at (2.120). □

The technique used in the previous proposition applies to higher moments of the L^2-norm of solutions and gives some a priori estimates for $\mathbb{E}\,|u(t)|_2^{2m}$ in terms of $E\,|u_0|_2^{2m}$, where $m \geq 2$ is an arbitrary integer. We shall not give either a proof or a formulation of the corresponding results, referring the reader to the book [VF88]. Instead we now establish an estimate for the second exponential moment that will imply the above-mentioned results under a stronger assumption on the initial function.

Proposition 2.4.9 *Under the hypotheses of Theorem 2.4.6, there is a constant $c > 0$ not depending on ν, h, and $\{b_j\}$ such that if $\varkappa > 0$ satisfies the condition*

$$\varkappa \sup_{j\geq 1} b_j^2 \leq c, \quad (2.123)$$

then the following assertion holds: for any $\mathcal{G}_0$-measurable H-valued random variable $u_0(x)$ such that

$$\mathbb{E}\exp\big(\varkappa\nu|u_0|_2^2\big) < \infty \quad (2.124)$$

the corresponding solution of (2.98), (2.114) *satisfies the inequality*

$$\mathbb{E}\exp\big(\varkappa\nu|u(t)|_2^2\big) \leq e^{-\varkappa\nu^2 t}\mathbb{E}\exp\big(\varkappa\nu|u_0|_2^2\big) + K(\nu,\varkappa,\mathfrak{B},h), \quad (2.125)$$

where we set

$$K(\nu,\varkappa,\mathfrak{B},h) = \nu^{-1}R\exp(\varkappa R/\alpha_1), \quad R = C\nu^{-1}|h|_2^2 + \mathfrak{B} + \nu, \quad (2.126)$$

and $C > 0$ is an absolute constant.

Proof We wish to apply Theorem A.7.5 to a functional $F : H \to \mathbb{R}$ defined as $F(u) = \exp(\varkappa\nu|u|_2^2)$. It is straightforward to verify that

$$\partial_u F(u; v) = 2\varkappa\nu F(u)\langle u, v\rangle,$$
$$\partial_u^2 F(u; v) = 2\varkappa\nu F(u)\big(|v|_2^2 + 2\varkappa\nu\langle u, v\rangle^2\big).$$

It follows that relation (A.24) can be rewritten as

$$F(u(t\wedge\tau_n)) = F(u_0) + \int_0^{t\wedge\tau_n} A(s)\,ds + M(t\wedge\tau_n), \tag{2.127}$$

where $M(t)$ is a stochastic integral (cf. (2.121)), and

$$A(t) = 2\varkappa\nu F(u(t))\Big(-\nu\langle u(t), Lu(t)\rangle + \langle u(t), h\rangle + \frac{1}{2}\mathfrak{B} + \varkappa\nu\sum_{j=1}^{\infty} b_j^2\langle u(t), e_j\rangle^2\Big).$$

Taking the mean value in (2.127) and using Doob's optional sampling theorem, we derive

$$\mathbb{E}\,F(u(t\wedge\tau_n)) = \mathbb{E}\,F(u_0) + \mathbb{E}\int_0^{t\wedge\tau_n} A(s)\,ds. \tag{2.128}$$

Now note that if (2.123) holds with $c \le (8\alpha_1)^{-1}$, then

$$A(t) \le \varkappa\nu F(u(t))\big(-\nu\|u(t)\|_V^2 + C_1\nu^{-1}|h|_2^2 + \mathfrak{B}\big)$$
$$\le -\varkappa\nu^2 F(u(t)) + \varkappa\nu^2 K(\nu, \varkappa, \mathfrak{B}, h),$$

where $\|u\|_V^2 = \langle u, Lu\rangle$. Substituting this inequality into (2.128) and using Fatou's lemma to pass to the limit as $n \to \infty$, we obtain

$$\mathbb{E}\exp(\varkappa\nu|u(t)|_2^2) + \varkappa\nu^2\int_0^t \mathbb{E}\exp(\varkappa\nu|u(s)|_2^2)\,ds \le \varkappa\nu^2 K(\nu, \varkappa, \mathfrak{B}, h).$$

Application of Gronwall's inequality completes the proof of the proposition. □

The next result gives an estimate for the rate of growth of solutions as $t \to \infty$. Let us set

$$\mathcal{E}_u(t) = |u(t)|_2^2 + \nu\int_0^t |\nabla u(s)|_2^2 ds. \tag{2.129}$$

Proposition 2.4.10 *Assume that the hypotheses of Theorem 2.4.6 are fulfilled, and u_0 is an H-valued random variable with a finite second moment. Then for any $\rho > 0$ the solution $u(t)$ of problem* (2.98), (2.114) *satisfies the inequality*

$$\mathbb{P}\Big\{\sup_{t\ge 0}\big(\mathcal{E}_u(t) - (\mathfrak{B} + 2(\alpha_1\nu)^{-1}|h|_2^2)\,t\,\big) \ge |u_0|_2^2 + \rho\Big\} \le e^{-\gamma\nu\rho}, \tag{2.130}$$

where we set $\gamma = \frac{1}{4}\alpha_1\big(\sup_j b_j^2\big)^{-1}$.

Proof Let us recall that relation (2.121) was established in the proof of Proposition 2.4.8. In view of (2.119), we have

$$\mathbb{E}\sum_{j=1}^{\infty} b_j^2 \int_0^t \langle u(s), e_j\rangle^2 ds \le \Big(\sup_{j\ge 1} b_j^2\Big)\int_0^t |u(s)|_2^2 ds < \infty \quad \text{for any } t \ge 0.$$

So the stochastic integral

$$M_t = 2\sum_{j=1}^{\infty} b_j \int_0^t \langle u(s), e_j\rangle d\beta_j(s)$$

defines a martingale (see Section A.11), and relation (2.121) implies that, with probability 1, we have

$$\mathcal{E}_u(t) = |u_0|_2^2 + \mathfrak{B}t + \nu\int_0^t |\nabla u(s)|_2^2 ds + 2\int_0^t \langle u(s), h\rangle ds + M_t, \qquad t \ge 0. \tag{2.131}$$

To estimate the right-hand side of this relation, we apply the classical method of exponential supermartingales; e.g., see problem 5 in section 2.9 of [McK69] for the one-dimensional case. Namely, noting that the quadratic variation of M_t is equal to

$$\langle M\rangle_t = 4\sum_{j=1}^{\infty} b_j^2 \int_0^t \langle u(s), e_j\rangle^2 ds$$

(see Section A.11 and [Kry02]), we rewrite (2.131) in the form

$$\mathcal{E}_u(t) - \mathfrak{B}t = |u_0|_2^2 + \Big(M_t - \frac{1}{2}\gamma\nu\langle M\rangle_t\Big) + K_t, \tag{2.132}$$

where we set

$$K_t = 2\int_0^t \langle u(s), h\rangle ds + \frac{1}{2}\gamma\nu\langle M\rangle_t - \nu\int_0^t |\nabla u(s)|_2^2 ds.$$

Since

$$\left|\int_0^t \langle u(s), h\rangle ds\right| \le \frac{\nu}{4}\int_0^t |\nabla u(s)|_2^2 ds + (\alpha_1\nu)^{-1}|h|_2^2 t,$$

$$\langle M\rangle_t \le 4\alpha_1^{-1}\sup_{j\ge 1} b_j^2 \int_0^t |\nabla u(s)|_2^2 ds = \frac{1}{\gamma}\int_0^t |\nabla u(s)|_2^2 ds,$$

we see that

$$K_t \le 2(\alpha_1\nu)^{-1}|h|_2^2 t.$$

Substituting this inequality into (2.132), we obtain

$$\mathcal{E}_u(t) - \big(\mathfrak{B} + 2(\alpha_1\nu)^{-1}|h|_2^2\big)\, t \le |u_0|_2^2 + \Big(M_t - \frac{1}{2}\gamma\nu\langle M\rangle_t\Big). \tag{2.133}$$

Consider the exponential function $\exp(\gamma\nu M_t - \frac{1}{2}(\gamma\nu)^2\langle M\rangle_t)$. Applying the exponential supermartingale result given in Section A.11, we see that

$$\mathbb{P}\Big\{\sup_{t\ge 0}\big(M_t - \tfrac{1}{2}\gamma\nu\langle M\rangle_t\big) \ge \rho\Big\} = \mathbb{P}\Big\{\sup_{t\ge 0}\exp\big(\gamma\nu M_t - \tfrac{1}{2}(\gamma\nu)^2\langle M\rangle_t\big) \ge e^{\gamma\nu\rho}\Big\}$$
$$\le e^{-\gamma\nu\rho}.$$

Combining this with (2.133), we arrive at the required inequality (2.130). □

Corollary 2.4.11 *Under the hypotheses of Proposition 2.4.10, assume that the random variable u_0 satisfies the condition*

$$\mathbb{E}\,|u_0|_2^{2m} < \infty, \tag{2.134}$$

where $m \ge 1$ is an integer. Then there is a constant $C_m > 0$ depending only on m such that for any $T \ge 1$ we have

$$\mathbb{E}\sup_{0\le t\le T}\mathcal{E}_u(t)^m \le C_m\big(\mathbb{E}\,|u_0|_2^{2m} + T^m\mathfrak{B}^m + \nu^{-m}(\gamma^{-m} + \alpha_1^{-m}T^m|h|_2^{2m})\big), \tag{2.135}$$

where $\gamma > 0$ is the constant defined in Proposition 2.4.10.

Proof Inequality (2.130) implies that

$$\mathbb{P}\Big\{\sup_{0\le t\le T}\mathcal{E}_u(t) \ge |u_0|_2^2 + C\,T + \rho\Big\} \le e^{-\gamma\nu\rho}, \tag{2.136}$$

where we set $C = \mathfrak{B} + 2(\alpha_1\nu)^{-1}|h|_2^2$. Now note that if ξ and η are non-negative random variables, then

$$\mathbb{E}\,\xi^m \le 2^{m-1}\big(\mathbb{E}(\xi-\eta)^m\mathbb{I}_{\{\xi>\eta\}} + \mathbb{E}\,\eta^m\big)$$
$$= 2^{m-1}\int_0^\infty \mathbb{P}\{\xi - \eta > \lambda^{1/m}\}\,d\lambda + 2^{m-1}\mathbb{E}\,\eta^m.$$

Applying this inequality to $\xi = \sup_t \mathcal{E}_u(t)$ and $\eta = |u_0|_2^2 + C\,T$ and using (2.136), we derive

$$\mathbb{E}\sup_{0\le t\le T}\mathcal{E}_u(t)^m \le 2^{m-1}\int_0^\infty \exp\big(-\gamma\nu\lambda^{1/m}\big)\,d\lambda + 2^{m-1}\mathbb{E}\,(|u_0|_2^2 + C\,T)^m.$$

It remains to note that the right-hand side can be estimated by that of (2.135) with a constant C_m depending only on m. □

Let us emphasise that Theorem 2.4.6, Propositions 2.4.8–2.4.10, and Corollary 2.4.11 are valid both for periodic and Dirichlet boundary conditions, and the same proof works in these two cases. In the next result concerning higher Sobolev norms, it is essential that the problem is considered under periodic boundary conditions. To simplify the presentation, we shall assume that the sequence $\{e_j\}$ entering the right-hand side of (2.98) coincides with the family of normalised eigenfunctions of the Stokes operator L and denote by $\alpha_1 \le \alpha_2 \le \cdots$ the corresponding eigenvalues. For any integer $k \ge 0$, we set

$$\mathcal{E}_u(k,t) = t^k \|u(t)\|_k^2 + \nu \int_0^t s^k \|u(s)\|_{k+1}^2 ds,$$

so that $\mathcal{E}_u(0,t)$ coincides with the function $\mathcal{E}_u(t)$ defined by (2.129).

Proposition 2.4.12 *Let $k \ge 1$ be an integer, let $h \in V^k$, and let the sequence $\{b_j\}$ satisfy the condition*

$$\mathfrak{B}_k := \sum_{j=1}^{\infty} \alpha_j^k b_j^2 < \infty. \tag{2.137}$$

Let u_0 be a $\mathcal{G}_0$-measurable H-valued random variable satisfying (2.134) *for any integer $m \ge 1$. Then for any $m \ge 1$ and $T \ge 1$ there is a $C(k,m,T) > 0$ such that*

$$\mathbb{E} \sup_{0 \le t \le T} \mathcal{E}_u(k,t)^m \le C(k,m,T)\big(1 + \nu^{-m(7k+2)}\big(\mathbb{E}\,|u_0|_2^{4m(k+1)} + 1\big)\big). \tag{2.138}$$

In what follows, this result is only used in Exercise 2.5.8 (for proving similar bounds for stationary solutions) and in Section 3.5.3 (devoted to convergence of mean values of functionals on higher Sobolev space). Its proof is based on the standard idea of differentiating higher Sobolev norms along solutions. Accurate calculations, which involve Itô's formula for infinite-dimensional diffusions and some interpolation inequalities, are rather technical, and we present them in Section 2.6. The reader not interested in these details may safely skip them, together with the rest of this subsection, and jump directly to the final part of this section, which is devoted to the construction of a Markov RDS corresponding to Navier–Stokes equations with white noise; see page 85.

Corollary 2.4.13 *Under the conditions of Proposition 2.4.12, assume that inequality* (2.124) *holds for the initial function u_0 with a constant $\varkappa > 0$ satisfying* (2.123). *Then for any integer $m \ge 1$ and any constants $T \ge 1$ and $t_0 \ge 1$ we have*

$$\begin{aligned} &\mathbb{E} \sup_{t_0 \le t \le t_0+T} \|u(t)\|_k^{2m} + \nu^m \mathbb{E}\left(\int_{t_0}^{t_0+T} \|u(s)\|_{k+1}^2 ds\right)^m \\ &\quad \le C\left(1 + \nu^{-m(9k+4)}\left(e^{-\varkappa\nu t_0}\mathbb{E}\exp(\varkappa\nu|u_0|_2^2) + K\right) + \nu^{-m(7k+2)}\right), \end{aligned} \tag{2.139}$$

where $K = K(\nu, \varkappa, \mathfrak{B}, h)$ is defined by (2.126), *and $C > 0$ is a constant depending on k, m, T, and $\varkappa$.*

Proof Let us denote by $P_{km}(t_0, T)$ the left-hand side of (2.139). Applying Proposition 2.4.12 in which the interval $[0, T]$ is replaced by $[t_0 - 1, t_0 + T]$, we conclude that

$$P_{km}(t_0,T)\le C_1\bigl(1+\nu^{-m(7k+2)}\bigl(\mathbb{E}\,|u(t_0-1)|_2^{4m(k+1)}+1\bigr)\bigr). \tag{2.140}$$

Now note that

$$\nu^{-m(7k+2)}\mathbb{E}\,|u(t_0-1)|_2^{4m(k+1)}\le C_2\nu^{-m(9k+4)}\mathbb{E}\exp\bigl(\varkappa\nu|u(t_0-1)|_0^2\bigr),$$

where C_2 depends only on k, m, and $\varkappa$. Using Proposition 2.4.9 to bound the quantity $\mathbb{E}\exp\bigl(\varkappa\nu|u(t_0-1)|_2^2\bigr)$ and substituting the resulting estimate into (2.140), we arrive at (2.139). □

We now state an exercise summarising some further properties of solutions for the stochastic Navier–Stokes system on the torus. They can be established with the help of the methods developed above.

Exercise 2.4.14 Let us consider the stochastic Navier–Stokes system (2.11) in which $h\in H$ is a deterministic function and $\{e_j\}$ is the family of normalised eigenfunctions of the Stokes operator L.

(i) Assume that $\mathfrak{B}_1<\infty$ (see (2.137)) and the initial function u_0 is such that

$$\mathbb{E}\exp\bigl(\varkappa\nu\|u_0\|_1^2\bigr)<\infty,$$

where $\varkappa>0$ satisfies inequality (2.123) with a sufficiently small $c>0$. Show that

$$\mathbb{E}\exp\bigl(\varkappa\nu\|u(t)\|_1^2\bigr)\le e^{-\varkappa\nu^2t}\mathbb{E}\exp\bigl(\varkappa\nu\|u_0\|_1^2\bigr)+\nu^{-1}R_1e^{\varkappa R_1/\alpha_1}, \tag{2.141}$$

where $R_1=C\nu^{-1}|h|_2^2+\mathfrak{B}_1+\nu$, and $C>0$ is a constant not depending on other parameters. *Hint:* Repeat the argument used in the proof of Proposition 2.4.9.

(ii) In addition to the above conditions, assume that

$$\sum_{j=1}^{\infty}e^{\rho\sqrt{\alpha_j}}\bigl(\langle h,e_j\rangle^2+b_j^2\bigr)<\infty, \tag{2.142}$$

where $\rho>0$ is a constant. Prove that there are positive constants p and q such that, for any any integer $m\ge 1$ and any $t_0\ge 1$ and $T\ge 1$, we have

$$\mathbb{E}\sup_{t_0\le t\le t_0+T}\biggl(\sum_{j=1}^{\infty}\exp\bigl(\sqrt{\alpha_j}r_\nu\nu^p\bigr)\langle u(t),e_j\rangle^2\biggr)^m\le C_m\nu^{-q},\quad \mathbb{E}\,r_\nu^{-m}\le C_m,$$

where $C_m>0$ is a constant not depending on ν, and r_ν is a random variable not depending on p, q, and m. *Hint:* A proof can be found in [Shi02].

2.4.3 Universality of white-noise forces

Theorem 2.2.2 and Exercise 2.2.4 show that a spatially regular white-noise force is universal in the sense that an appropriately normalised high-frequency random kick force converges to it in distribution. In this section, we shall prove that this property is inherited by solutions of the corresponding equations. Namely, together with Eq. (2.98), let us consider the Navier–Stokes system perturbed by the high-frequency kick force (2.69):

$$\dot{u} + \nu L u + B(u) = h + \eta_\varepsilon, \quad \eta_\varepsilon(t,x) = \sqrt{\varepsilon}\sum_{k=1}^{\infty}\eta_k(x)\delta(t-k\varepsilon), \tag{2.143}$$

where $\{\eta_k\}$ is a sequence of i.i.d. random variables of the form (2.70) and $0 < \varepsilon \le 1$. As was explained in Section 2.3, Eq. (2.143) is equivalent to the discrete-time system

$$u_k = S_\varepsilon(u_{k-1}) + \sqrt{\varepsilon}\,\eta_k, \quad k \ge 1, \tag{2.144}$$

where S_t denotes the time shift along trajectories of Eq. (2.143) with $\eta_\varepsilon \equiv 0$. With a slight abuse[9] of notation, we denote by $P_t^\varepsilon(u,\cdot)$, $t \in \varepsilon\mathbb{Z}_+$, the transition function associated with (2.144) and by $\mathfrak{P}_t(\varepsilon)$ and $\mathfrak{P}_t^*(\varepsilon)$ the corresponding Markov operators. We extend the transition function $P_t^\varepsilon(u,\cdot)$ (as well as the corresponding Markov semigroups) to the half-line $\mathbb{R}_+$ by defining it as the law of the solution for (2.143) issued from $u \in H$.

Theorem 2.4.15 *In addition to the above hypotheses, assume that $\mathfrak{B}_1 < \infty$. Then for any $T > 0$ and $R > 0$ we have*

$$\sup_{0\le t\le T}\sup_{|v|_2\le R}\left\|P_t^\varepsilon(v,\cdot) - P_t(v,\cdot)\right\|_L^* \to 0 \quad as \quad \varepsilon \to 0^+. \tag{2.145}$$

Proof We need to prove that if $u(t;v)$ and $u_\varepsilon(t;v)$ are solutions issued from $v \in H$ for the Navier–Stokes system perturbed by a high-frequency kick force or spatially regular white noise, then for any sequence $\{\varepsilon_n > 0\}$ going to zero we have

$$\sup_{t,v,f}\left|\mathbb{E}\big(f(u_{\varepsilon_n}(t;v)) - f(u(t;v))\big)\right| \to 0 \quad \text{as } n \to \infty, \tag{2.146}$$

where the supremum is taken over $t \in [0,T]$, $v \in B_H(R)$, and $f \in L_b(H)$ with $\|f\|_L \le 1$. This is done in two steps.

Step 1. We first note that it suffices to prove convergence (2.146) in which u_ε is replaced by the solution $\tilde{u}_\varepsilon$ of Eq. (4.110) with $\eta = \partial_t\tilde{\zeta}_\varepsilon$, where $\tilde{\zeta}_\varepsilon$

[9] In Section 2.3, the discrete time varies in $\mathbb{Z}_+$ independently of the length T of the interval between two consecutive kicks. Since we wish to compare solutions for systems perturbed by white noise and random kicks, we have to take here the same scale of time.

is the continuous process defined in Section 2.2. Indeed, it follows from the condition $\mathfrak{B}_1 < \infty$ that, with probability 1, the random variables η_k belong to a bounded ball in V. Therefore the definition of $\tilde{\zeta}_\varepsilon$ implies that (cf. (2.74))

$$\|\tilde{\zeta}_\varepsilon - \zeta_\varepsilon\|_{L^\infty(0,T;V)} \le C_1\sqrt{\varepsilon}.$$

Recalling Exercise 2.1.27, we see that, on the set $\Gamma_r = \{\|\zeta_\varepsilon\|_{L^\infty(0,T;V)} \le r\}$, we have the estimate

$$\sup_{0\le t\le T} \|u_\varepsilon(t;v) - u(t;v)\|_1 \le C_2(r)\sqrt{\varepsilon}.$$

On the other hand, applying the Doob–Kolmogorov inequality (A.55) to ζ_ε, we derive

$$\mathbb{P}(\Gamma_r^c) = \mathbb{P}\Big\{\sup_{0\le t\le T} \|\zeta_\varepsilon(t)\|_1 > r\Big\} \le r^{-2}\mathbb{E}\,\|\zeta_\varepsilon(T)\|_1^2 \le C_3(T)r^{-2}.$$

Combining these estimates, for any $t \in [0,T]$, $v \in B_H(R)$, and $f \in L_b(H)$ with $\|f\|_L \le 1$, we obtain

$$\begin{aligned}\big|\mathbb{E}\big(f(u_{\varepsilon_n}(t,v)) - f(u(t,v))\big)\big| &\le \mathbb{E}\left(\mathbb{I}_{\Gamma_r}\|u_\varepsilon(t;v) - u(t;v)\|_1\right) + 2\,\mathbb{P}(\Gamma_r^c)\\ &\le C_2(r)\sqrt{\varepsilon} + 2\,C_3(T)r^{-2}.\end{aligned}$$

The right-hand side of this inequality can be made arbitrarily small by choosing $R \gg 1$ and $\varepsilon \ll 1$.

Step 2. We now prove (2.146) with u_{ε_n} replaced by $\tilde{u}_{\varepsilon_n}$. To this end, note that the expression in (2.146) depends only on the laws of u_{ε_n} and u. In other words, if we replace $\tilde{\zeta}_{\varepsilon_n}$ and ζ by some other random variables ξ_n and ξ valued in $C(0,T;V)$ such that

$$\mathcal{D}(\xi_n) = \mathcal{D}(\tilde{\zeta}_{\varepsilon_n}), \quad \mathcal{D}(\xi) = \mathcal{D}(\zeta), \tag{2.147}$$

then the corresponding solutions $w_n(t;v)$ and $w(t;v)$ of Eq. (4.110) will satisfy the relation

$$\mathbb{E}\big(f(\tilde{u}_{\varepsilon_n}(t;v)) - f(u(t;v))\big) = \mathbb{E}\big(f(w_n(t;v)) - f(w(t;v))\big), \quad t \in [0,T].$$

In view of Exercise 2.2.4, we have $\mathcal{D}(\tilde{\zeta}_{\varepsilon_n}) \to \mathcal{D}(\zeta)$ in the weak topology of the space $\mathcal{P}(C(0,T;V))$. Therefore, by Skorohod's embedding theorem (see theorem 11.7.2 in [Dud02]), there are random variables ξ_n and ξ valued in $C(0,T;V)$ such that (2.147) holds and, for any $\delta > 0$, we have

$$\mathbb{P}\big\{\|\xi_n - \xi\|_{C(0,T;V)} < \delta\big\} \to 0 \quad \text{as } n \to \infty.$$

On the other hand, the Doob–Kolmogorov inequality (A.55) implies that

$$\mathbb{P}\{\|\xi\|_{C(0,T;V)} > r\} \to 0 \quad \text{as } r \to \infty.$$

Combining these two relations with the argument used in Step 1, we easily prove the required convergence. The proof of the theorem is complete. □

The following exercise shows that convergence of solutions remains true in the case when the initial function is random, and moreover, it is uniform with respect to measures almost entirely concentrated on bounded parts of the phase space.

Exercise 2.4.16 Under the hypotheses of Theorem 2.4.15, prove that, for any $T > 0$ and any subset $\Lambda \subset \mathcal{P}(H)$ satisfying the condition

$$\sup_{\lambda\in\Lambda} \int_H |u|_2^2 \lambda(du) < \infty,$$

we have

$$\sup_{0\le t\le T} \sup_{\lambda\in\Lambda} \left\| \mathfrak{P}_t^*(\varepsilon)\lambda - \mathfrak{P}_t^*\lambda \right\|_L^* \to 0 \quad \text{as} \quad \varepsilon \to 0^+. \tag{2.148}$$

Hint: Use the idea applied in Step 3 of the proof of Theorem 4.3.1.

2.4.4 RDS associated with Navier–Stokes equations

We conclude this section with a construction of a Markov RDS associated with the Navier–Stokes system perturbed by spatially regular white noise. For any $b > 0$, denote by $(\Omega_0, \mathcal{A}_0, \mathbb{P}_b)$ the probability space associated with the two-sided centred Brownian motion with variance b^2. Namely, Ω_0 is the space of continuous functions $\omega_t : \mathbb{R} \to \mathbb{R}$ vanishing at $t = 0$ with the metric of uniform convergence on bounded intervals (see (1.9)), $\mathcal{A}_0$ is the Borel σ-algebra, and $\mathbb{P}_b$ is a Gaussian measure on $(\Omega, \mathcal{A}_0)$ such that, for any $t_0 < t_1 < \cdots < t_n$ and any Borel subsets $\Gamma_1, \ldots, \Gamma_n \subset \mathbb{R}$, we have

$$\mathbb{P}_b\big(\{\omega \in \Omega_0 : \omega_{t_1} - \omega_{t_0} \in \Gamma_1, \ldots, \omega_{t_n} - \omega_{t_{n-1}} \in \Gamma_n\}\big) = \prod_{k=1}^n \mathcal{N}_{b(t_k - t_{k-1})}(\Gamma_k),$$

where $\mathcal{N}_\sigma$ stands for the normal distribution with zero mean value and variance σ^2. Given a random process ζ of the form (2.98), we denote

$$\Lambda = \{j \ge 1 : b_j \ne 0\}.$$

Define $(\Omega, \mathcal{F}, \mathbb{P})$ as the completion of the direct product of the probability spaces $(\Omega_0, \mathcal{A}_0, \mathbb{P}_{b_j})$ with $j \in \Lambda$. Elements of Ω are continuous functions $\omega : \mathbb{R} \to \mathbb{R}^\Lambda$, where $\mathbb{R}^\Lambda$ is endowed with the Tikhonov topology; see formula (1.9) and the discussion after it.

Exercise 2.4.17 Let $\omega^{(j)}$ be the jth component of $\omega \in \Omega$. Show that the series

$$\tilde{\zeta}(t, x) = \sum_{j\in\Lambda} \omega_t^{(j)} e_j(x), \quad t \in \mathbb{R}, \tag{2.149}$$

defines an H-valued random process with almost surely continuous trajectories. Show also that the law of the restriction of $\tilde{\zeta}$ to the positive half-line coincides with that of ζ.

Let us define a family of shifts on the probability space $(\Omega, \mathcal{F}, \mathbb{P})$ by the relation

$$(\theta_t\omega)_s = \omega_{t+s} - \omega_t, \quad t, s \in \mathbb{R}.$$

Exercise 2.4.18 Show that $\boldsymbol{\theta} = \{\theta_t : \Omega \to \Omega, t \in \mathbb{R}\}$ is a group of measure-preserving transformations on $(\Omega, \mathcal{F}, \mathbb{P})$.

Everything is now ready for the construction of an RDS associated with the stochastic Navier–Stokes system (2.98). We first explain the construction for the simpler case of a more regular force. Namely, assume that $\mathfrak{B}_1 < \infty$. In this case, the trajectories of $\tilde{\zeta}$ belong to $C(\mathbb{R}; V)$ on a set $\Omega_* \in \mathcal{F}$ of full measure, and solutions of (2.98) with $\zeta = \tilde{\zeta}$ can be written as $u = \tilde{\zeta} + v$, where v is a solution of the problem

$$\dot{v} + \nu L(v + \tilde{\zeta}) + B(v + \tilde{\zeta}) = h, \quad v(0) = u_0.$$

It is clear that Ω_* can be chosen to be invariant with respect to the group $\{\theta_t\}$. Let us define $\varphi_t^\omega u_0 = \tilde{\zeta}^\omega(t) + v^\omega(t)$ for $\omega \in \Omega_*$ and $\varphi_t^\omega u_0 = u_0$ for $\omega \notin \Omega_*$. The co-cycle property is an obvious consequence of the uniqueness of solutions for the above problem, while the Markov property follows from the fact that $\varphi_t^{\theta_s\omega} u_0$ depends only on the increments $\{\omega_{s+r} - \omega_s, r \in [0, t]\}$.

In the general case, the construction of an RDS associated with (2.98) is similar. However, we need to use the Ornstein–Uhlenbeck process because the regularity of $\tilde{\zeta}$ is not sufficient. Let us consider the stochastic Stokes equation (2.100) in which $\eta = \partial_t\tilde{z}$. Its solution issued from zero can be written as

$$\tilde{z}(t) = \int_0^t e^{-\nu(t-s)L} d\tilde{\zeta}(s), \quad t \geq 0. \tag{2.150}$$

Let $\widetilde{\Omega}$ be the set of those $\omega \in \Omega$ for which the corresponding trajectory $\tilde{z}^\omega$ belongs to the space $\mathcal{X} = C(\mathbb{R}_+; H) \cap L^2_{\text{loc}}(\mathbb{R}_+; V)$. Then $\widetilde{\Omega} \in \mathcal{F}$, and by Proposition 2.4.2, we have $\mathbb{P}(\widetilde{\Omega}) = 1$. It follows that $\Omega_* = \bigcap_{n\in\mathbb{Z}} \theta_n(\widetilde{\Omega})$ is a set of full measure. We now denote

$$\varphi_t^\omega u_0 = \begin{cases} \tilde{z}^\omega(t) + \tilde{v}^\omega(t), & \text{for } \omega \in \Omega_*, \\ u_0, & \text{for } \omega \notin \Omega_*, \end{cases} \tag{2.151}$$

where $\tilde{v}^\omega$ stands for the solution of (2.111), (2.112) with $z = \tilde{z}^\omega$. The proof of the following result on the existence of an RDS associated with the Navier–Stokes system is given in Section 2.6.

Theorem 2.4.19 *The family* $\boldsymbol{\Phi} = \{\varphi_t^\omega, t \ge 0, \omega \in \Omega\}$ *is a continuous Markov RDS in* H *over* $\boldsymbol{\theta}$*. Moreover, for any* $u_0 \in H$ *the law of the random variable* $\{\varphi_t u_0, t \ge 0\}$ *in the space* $C(\mathbb{R}_+; H)$ *coincides with that of the solution for* (2.98), (2.114).

2.5 Existence of a stationary distribution

There is a general and simple method for constructing stationary distributions for Markov processes that satisfy a compactness condition. It was introduced by Bogoliouboff and Kryloff in [KB37] and enables one to prove the existence of a stationary measure for a large class of nonlinear stochastic PDEs. In this section, we first use the Bogoliouboff–Kryloff method to establish an abstract result and then apply it to the Markov process associated with the 2D Navier–Stokes system with random perturbation.

2.5.1 The Bogolyubov–Krylov argument

Let $(\Omega, \mathcal{F})$ be a measurable space with a filtration $\{\mathcal{F}_t\}$, let X be a Polish space, and let $(u_t, \mathbb{P}_v)$, $v \in X$, be a family of X-valued Markov processes on $(\Omega, \mathcal{F})$ adapted to $\mathcal{F}_t$. We assume that the corresponding transition function $P_t(u, \cdot)$ associated with $(u_t, \mathbb{P}_v)$ possesses the Feller property and that the trajectories $u_t(\omega)$ are continuous in time for any $\omega \in \Omega$. Denote by $\mathfrak{P}_t$ and $\mathfrak{P}_t^*$ the Markov semigroups associated with $P_t(u, \cdot)$. Let us fix $\lambda \in \mathcal{P}(X)$ and define a family of measures $\bar{\lambda}_t$, $t \in \mathcal{T}_+$, by the relation

$$\bar{\lambda}_t(\Gamma) = \frac{1}{t}\int_0^t (\mathfrak{P}_s^*\lambda)(\Gamma)\, ds = \frac{1}{t}\int_0^t \int_X P_s(u, \Gamma)\, \lambda(du)\, ds, \quad \Gamma \in \mathcal{B}(X), \tag{2.152}$$

where the integral in $s \in [0, t]$ should be replaced by the sum over $s = 0, \dots, t-1$ if the time parameter is discrete. The Feller property implies that $P_s(u, \Gamma)$ is a measurable function of $(s, u) \in \mathcal{T}_+ \times X$, and therefore the integrals in (2.152) are well defined. Recall that a family of measures $\{\mu_\alpha, \alpha \in \mathcal{A}\} \subset \mathcal{P}(X)$ is said to be *tight* if for any $\varepsilon > 0$ there is a compact subset $K \subset X$ such that $\mu_\alpha(K) \ge 1 - \varepsilon$ for any $\alpha \in \mathcal{A}$.

Theorem 2.5.1 *Under the above condition, assume that* $\lambda \in \mathcal{P}(X)$ *is a measure for which the family* $\{\bar{\lambda}_t, t \ge 0\}$ *is tight. Then the Markov family* $(u_t, \mathbb{P}_v)$ *has at least one stationary measure.*

Proof In view of the Prokhorov compactness criterion (see Theorem 1.2.14), there is a subsequence $\{t_n\} \subset \mathcal{T}_+$ going to $+\infty$ such that $\{\bar{\lambda}_{t_n}\}$ converges weakly to a measure $\mu \in \mathcal{P}(X)$. That is,

$$(f, \bar{\lambda}_{t_n}) \to (f, \mu) \quad \text{as } n \to \infty \text{ for any } f \in C_b(X).$$

We claim that μ is a stationary measure for $(u_t, \mathbb{P}_v)$. Indeed, in view of this convergence, Exercise 1.3.23, and the definition of $\bar{\lambda}_t$, for any $f \in C_b(X)$ and $r \in \mathcal{T}_+$ we have

$$
\begin{aligned}
(f, \mathfrak{P}_r^* \mu) &= (\mathfrak{P}_r f, \mu) = \lim_{n\to\infty} (\mathfrak{P}_r f, \bar{\lambda}_{t_n}) \\
&= \lim_{n\to\infty} \frac{1}{t_n} \int_0^{t_n} (\mathfrak{P}_r f, \mathfrak{P}_s^* \lambda)\, ds = \lim_{n\to\infty} \frac{1}{t_n} \int_0^{t_n} (f, \mathfrak{P}_{s+r}^* \lambda)\, ds \\
&= \lim_{n\to\infty} \frac{1}{t_n} \left(\int_0^{t_n} (f, \mathfrak{P}_s^* \lambda)\, ds + \int_{t_n}^{t_n + r} (f, \mathfrak{P}_s^* \lambda)\, ds - \int_0^r (f, \mathfrak{P}_s^* \lambda)\, ds \right) \\
&= (f, \mu).
\end{aligned}
$$

Since this is true for any $f \in C_b(X)$, we see that $\mathfrak{P}_r^* \mu = \mu$ for each $r \in \mathcal{T}_+$. So, μ is a stationary measure. □

Exercise 2.5.2 Let $(u_t, \mathbb{P}_v)$, $v \in H$, be a family of Markov processes in X satisfying the above conditions. Suppose that there is an initial point $v \in X$, an increasing sequence of compact subsets $K_m \subset X$, and finite times $t_m \in \mathcal{T}_+$ such that

$$\sup_{t \geq t_m} \mathbb{P}_v\{u_t \notin K_m\} \to 0 \quad \text{as } m \to \infty.$$

Show that the hypotheses of Theorem 2.5.1 are satisfied.

2.5.2 Application to Navier–Stokes equations

Let us consider the problem of the existence of a stationary measure for the kick-forced Navier–Stokes system (2.76).

Theorem 2.5.3 *Let $h \in H$ be a deterministic function and let $\{\eta_k\}$ be a sequence of i.i.d. random variables in H such that $\mathbb{E}\,|\eta_1|_2 < \infty$. Then Eq. (2.76) possesses at least one stationary measure. Moreover, every stationary measure $\mu \in \mathcal{P}(H)$ satisfies*

$$\int_H |u|_2 \mu(du) < \infty. \tag{2.153}$$

Proof

Step 1. To prove the existence of a stationary measure, we shall show that the hypotheses of Exercise 2.5.2 hold for the initial point $v = 0$. In view of relation (2.79), it suffices to construct, for any given $\varepsilon > 0$, two compact subsets $K^{(1)}, K^{(2)} \subset H$ such that

$$\mathbb{P}\{S_T(u_{k-1}) \notin K^{(1)}\} \leq \varepsilon, \quad \mathbb{P}\{\eta_k \notin K^{(2)}\} \leq \varepsilon,$$

where $\{u_k\}$ is the trajectory of (2.79) with $u_0 = 0$. Taking then $K = K^{(1)} + K^{(2)}$, we see that $\mathbb{P}\{u_k \notin K\} \le 2\varepsilon$ for any $k \ge 1$.

The existence of a compact set $K^{(2)}$ with the required property follows immediately from Ulam's theorem on regularity of probability measures on a Polish space; see Section 1.2.1. To construct $K^{(1)}$, we note that (see (2.81))

$$\mathbb{E}\,|u_k|_2 \le C \quad \text{for all } k \ge 0.$$

Hence, by Chebyshev's inequality, for any $\varepsilon > 0$ there is $R_\varepsilon > 0$ such that

$$\mathbb{P}\{|u_{k-1}| > R_\varepsilon\} \le R_\varepsilon^{-1}\mathbb{E}\,|u_{k-1}|_2 \le \varepsilon \quad \text{for } k \ge 1. \tag{2.154}$$

By part (ii) of Proposition 2.1.25, the operator S_T is continuous from H to V. Denoting by $B_H(r)$ the closed ball in H of radius r centred at zero, we conclude that the set $K^{(1)} := S_T(B_H(R_\varepsilon))$ is compact in H. Inequality (2.154) implies that the probability of the event $\{S_T(u_{k-1}) \notin K^{(1)}\}$ does not exceed ε.

Step 2. We now prove (2.153) for any stationary measure $\mu \in \mathcal{P}(H)$. Let us fix a constant $R > 0$ and consider a function $f_R : H \to \mathbb{R}$ defined by

$$f_R(u) = \begin{cases} |u|_2, & |u|_2 \le R, \\ R, & |u|_2 > R. \end{cases}$$

Then, for any $k \ge 0$, we have

$$\int_H f_R(u)\,\mu(du) = \int_H \int_H P_k(u, dv) f_R(v)\,\mu(du). \tag{2.155}$$

Let us estimate the right-hand side of (2.155). If $|u|_2 \le \rho$, then by inequality (2.81) with $m = 1$ we have

$$\int_H P_k(u, dv) f_R(v) \le \mathbb{E}\,|u_k|_2 \le q^k \rho + C, \tag{2.156}$$

where $\{u_k\}$ is the trajectory of (2.79) with $u_0 = u$, and $C > 0$ and $q < 1$ are constants not depending on u, k, and R. Substituting (2.156) into (2.155) and using the inequality $f_R \le R$, for any $R > 0$ and $\rho > 0$, we derive

$$\int_H f_R(u)\,\mu(du) \le R\,\mu\big(H \setminus B_H(\rho)\big) + q^k \rho + C.$$

Passing to the limit as $k \to \infty$ and $\rho \to \infty$, we see that

$$\int_H f_R(u)\,\mu(du) \le C.$$

Fatou's lemma now implies the required inequality (2.153). □

Exercise 2.5.4 Show that if the law of the random variables η_k satisfies the hypotheses of part (i) of Proposition 2.3.5 with some integer $s \geq 0$, then for any stationary measure $\mu \in \mathcal{P}(H)$ of (2.76) we have

$$\int_H \|u\|_s^m \,\mu(du) < \infty \quad \text{for any } m \geq 1.$$

In particular, if (2.88) holds for any $s > 0$, then any stationary measure is concentrated on the space $C^\infty(\mathbb{T}^2;\mathbb{R}^2)$. Similarly, show that, under the hypotheses of part (ii) of Proposition 2.3.5, any stationary measure $\mu \in \mathcal{P}(H)$ satisfies the inequality

$$\int_H \exp\bigl(\varkappa_s \|u\|_s^{p_s}\bigr)\,\mu(du) < \infty,$$

where $\varkappa_s$ and p_s are some positive constants, and $p_0 = p_1 = 2$. *Hint:* Use the argument of the proof of inequality (2.153).

Theorem 2.5.3 is valid for the Navier–Stokes equations in a bounded domain, and its proof remains literally the same. Moreover, the results announced in Exercise 2.5.4 are also true in this case, except that $p_1 < 2$.

We now turn to the existence of a stationary measure and some estimates for its moments in the case of the Navier–Stokes system with spatially regular white noise. Since the proofs are based on similar ideas, we shall mostly confine ourselves to formulating corresponding results and outlining their proofs.

Theorem 2.5.5 *Under the hypotheses of Theorem 2.4.6, the stochastic Navier–Stokes system* (2.98) *with an arbitrary* $\nu > 0$ *has a stationary measure. Moreover, any stationary measure* $\mu_\nu \in \mathcal{P}(H)$ *satisfies the relations*

$$\int_H \Bigl(\nu\,\|u\|_1^2 + \exp\bigl(\varkappa\nu|u|_2^2\bigr)\Bigr)\,\mu_\nu(du) \leq C\bigl(\mathfrak{B} + \nu^{-1}|h|_2^2\bigr), \tag{2.157}$$

$$\nu\int_H \|u\|_1^2\mu_\nu(du) = \frac{\mathfrak{B}}{2} + \int_H \langle u,h\rangle\,\mu_\nu(du), \tag{2.158}$$

where the positive constants $\varkappa$ *and* C *do not depend on* ν.

The left-hand side of (2.158) is called the *rate of dissipation of energy*; cf. (2.119).

Proof We first prove the existence of a stationary measure. Let us denote by $u(t,x)$ the solution of (2.98) issued from $u_0 = 0$ and by λ_t the law of $u(t)$ regarded as a random variable in H. Thus, we have $\lambda_t = \mathfrak{P}_t^*\delta_0$ for $t \geq 0$, where $\mathfrak{P}_t^* : \mathcal{P}(H) \to \mathcal{P}(H)$ stands for the Markov semigroup associated with the equation and δ_0 denotes the Dirac measure concentrated at zero. By Theorem 2.5.1, the existence of a stationary measure will be established if we show that the family $\{\bar\lambda_t, t \geq 0\}$ is tight, where $\bar\lambda_t$ is defined by relation (2.152) with $X = H$. Since the embedding $V \subset H$ is compact (see Property 1.1.2), it suffices to prove that

$$\sup_{t\ge 0} \bar\lambda_t\big(H \setminus B_V(R)\big) \to 0 \quad \text{as } R \to \infty, \tag{2.159}$$

where $B_V(R)$ stands for the ball in V of radius R centred at zero. To this end, let us note that relation (2.119) with $u_0 = 0$ implies that

$$\mathbb{E}\int_0^t |\nabla u(s)|_2^2 ds \le t\,(\mathfrak{B} + \nu^{-1}|h|_2^2), \quad t \ge 0.$$

Combining this with Chebyshev's inequality, we derive

$$\begin{aligned}\bar\lambda_t\big(H \setminus B_V(R)\big) &= \frac{1}{t}\int_0^t \lambda_s\big(H \setminus B_V(R)\big)\,ds = \frac{1}{t}\int_0^t \mathbb{P}\big\{\|u(s)\|_1 > R\big\}\,ds\\ &\le \frac{1}{t}R^{-2}\,\mathbb{E}\int_0^t |\nabla u(s)|_2^2\,ds \le R^{-2}(\mathfrak{B} + \nu^{-1}|h|_2^2),\end{aligned}$$

whence follows convergence (2.159).

To prove that any stationary measure satisfies (2.157) and (2.158), we first note that inequality (2.125) and the argument used in the derivation of (2.153) enable one to prove that

$$\int_H \exp\big(\varkappa\nu|u|_2^2\big)\,\mu_\nu(du) \le C_1\big(\mathfrak{B} + \nu^{-1}|h|_2^2\big).$$

Let u_0 be an H-valued random variable independent of ζ whose law coincides with μ and let $u(t)$ be a solution of (2.98) issued from u_0. Then $u(t)$ is a stationary process, and it follows from (2.119) that

$$2\nu t\,\mathbb{E}\|u\|_1^2 = \mathfrak{B}t + 2t\,\mathbb{E}\langle u, h\rangle.$$

Dividing the above relation by $2t$, we obtain (2.158). Combining (2.158) with Schwarz's and Friedrichs' inequalities, we derive the required upper bound for the first term on the left-hand side of (2.157). We leave the details to the reader as an exercise. □

Exercise 2.5.6 Under the hypotheses of Theorem 2.5.5, show that any stationary solution $u(t, x)$ of (2.98) satisfies the inequality

$$\begin{aligned}\mathbb{E}\sup_{t_0\le t\le t_0+T}&\left(|u(t)|_2^2 + \nu\int_{t_0}^t \|u(s)\|_1^2 ds\right)^m\\ &\le C_m\big(T^m\mathfrak{B}^m + \nu^{-m}(\gamma^{-m} + \alpha_1^{-m}T^m|h|_2^{2m})\big),\end{aligned}$$

for any constants $t_0 \ge 0$, $T \ge 1$, $\nu > 0$ and any integer $m \ge 1$. Here C_m is a constant depending only on m, and $\gamma > 0$ is defined in Proposition 2.4.10. *Hint:* Use inequalities (2.125), (2.135), and (2.157).

The above theorem and exercise are true both for periodic and Dirichlet boundary conditions. The following exercise gives some extra a priori estimates for stationary measures for the Navier–Stokes system on the torus.

Exercise 2.5.7 In addition to the hypotheses of Theorem 2.5.5, assume that $\mathfrak{B}_1 < \infty$ and $h \in V$. Show that there are positive constants $\varkappa$ and C not depending on ν such that any stationary measure $\mu_\nu \in \mathcal{P}(H)$ for (2.98) satisfies the relations

$$\int_H \exp\big(\varkappa\nu\|u\|_1^2\big)\,\mu_\nu(du) \le C\,\big(\mathfrak{B}_1 + \nu^{-1}\|h\|_1^2\big), \tag{2.160}$$

$$\nu\int_H |Lu|_2^2\mu_\nu(du) = \frac{\mathfrak{B}_1}{2} + \int_H \langle\nabla u, \nabla h\rangle\mu_\nu(du). \tag{2.161}$$

Combine (2.160) with an analogue of (2.125) and (2.135) for $u_0 \in V$ to obtain some bounds for $\mathbb{E}\sup_t \mathcal{E}_u(1,t)^m$, where $u(t,x)$ is a stationary solution for (2.98). *Hint:* To prove (2.160), combine (2.141) with the idea of the proof of (2.153).

The left-hand side of (2.161) is called the *rate of dissipation of enstrophy*; cf. (2.158).

In the case of higher regularity of the deterministic and random forces, some further bounds for various moments of stationary solutions in terms of the viscosity can be found in [BKL00; Shi02; KS03]. In Chapter 5, we shall need the result announced in the following exercise.

Exercise 2.5.8 Let the hypotheses of Proposition 2.4.12 be fulfilled for some integer $k \ge 1$. Show that, for any $m \ge 1$ and $T > 0$, there is a constant $C(k,m,T) > 0$ such that any stationary solution of (2.98) satisfies the inequality

$$\mathbb{E}\sup_{0\le t\le T} \mathcal{E}_u(k,t)^m \le C(k,m,T)\big(1+\nu^{-m(7k+2)}\big). \tag{2.162}$$

In particular, if $h \in C^\infty(\mathbb{T}^2;\mathbb{R}^2)$ and (2.137) holds for all $k \ge 1$, then any stationary measure is concentrated on the space $C^\infty(\mathbb{T}^2;\mathbb{R}^2)$. *Hint:* Use inequality (2.138) and the scheme of the proof of Theorem 2.5.3.

We conclude this chapter by the following result on the uniqueness of a stationary measure for the Navier–Stokes system with small noise.

Exercise 2.5.9 Show that for any $\nu > 0$ there is an $\varepsilon > 0$ such that if $|h|_2 + \mathfrak{B} \le \varepsilon$, then the RDS Φ associated with the stochastic Navier–Stokes system (2.98) satisfies inequality (1.75). In particular, Φ has a unique stationary measure μ_ν. Show also that the transition function converges to μ_ν exponentially fast in the dual-Lipschitz distance. *Hint:* Use Propositions 1.3.31 and 2.4.10, as well as the proof of (2.54).

2.6 Appendix: some technical proofs

Proof of Proposition 2.1.25 We shall confine ourselves to the formal derivation of (2.54) and (2.55). The calculations can be justified with the help of Galerkin approximations.

(i) Let us set $u_i(t) = S_t(u_{0i}, f_i)$, $i = 1, 2$. Then the difference $u = u_1 - u_2$ satisfies the equation

$$\dot{u} + Lu + B(u, u_1) + B(u_2, u) = f_1 - f_2. \tag{2.163}$$

Let us take the scalar product of this equation with $2u(t)$ in H. Using (2.11) and the inequalities

$$\big|(B(u, u_1), u)\big| \le c_1 \|u\|_{1/2}^2 \|u_1\|_1 \le c_2 |u|_2 \|u\|_1 \|u_1\|_1 \le \frac{1}{4}\|u\|_1^2 + c_2^2 \|u_1\|_1^2 \, |u|_2^2,$$

$$\big|(f_1 - f_2, u)\big| \le \|f_1 - f_2\|_{-1} \|u\|_1 \le \frac{1}{4}\|u\|_1^2 + \|f_1 - f_2\|_{-1}^2,$$

we derive from (2.163) the differential inequality

$$\partial_t\left(|u|_2^2 + \int_0^t \|u\|_1^2 ds\right) \le 2c_2^2 \|u_1\|_1^2 \left(|u|_2^2 + \int_0^t \|u\|_1^2 ds\right) + 2\|f_1 - f_2\|_{-1}^2. \tag{2.164}$$

Applying the Gronwall inequality, we arrive at

$$\begin{aligned} |u(t)|_2^2 + \int_0^t \|u\|_1^2 ds \le \exp\left(2c_2^2 \int_0^t \|u_1\|_1^2 dr\right) |u_{01} - u_{02}|_2^2 \\ + 2\int_0^t \exp\left(2c_2^2 \int_s^t \|u_1\|_1^2 dr\right) \|f_1(s) - f_2(s)\|_{-1}^2 ds. \end{aligned} \tag{2.165}$$

This implies, in particular, inequality (2.54).

(ii) We now take the scalar product of (2.163) with $2tLu$ in H:

$$\begin{aligned} \partial_t\big(t\|u\|_1^2\big) + 2t|Lu|_2^2 = \|u\|_1^2 - 2t\big(B(u, u_1), Lu\big) - 2t\big(B(u_2, u), Lu\big) \\ + 2t(f_1 - f_2, Lu). \end{aligned} \tag{2.166}$$

Let us use the inequalities

$$\|v\|_\infty^2 \le c_3 |v|_2 \, |Lv|_2, \quad \|v\|_1^2 \le |v|_2 \, |Lv|_2$$

to estimate the second and third terms on the right-hand side of (2.166):

$$\begin{aligned}\big|(B(u,u_1),Lu)\big| &\le c_4\|u\|_\infty\|u_1\|_1\,|Lu|_2 \le c_5|u|_2^{1/2}|Lu|_2^{3/2}|u_1|_2^{1/2}|Lu_1|_2^{1/2}\\ &\le \frac{1}{8}|Lu|_2^2 + c_6|u|_2^2|u_1|_2^2|Lu_1|_2^2, \qquad (2.167)\\ \big|(B(u_2,u),Lu)\big| &\le c_4\|u_2\|_\infty\|u\|_1\,|Lu|_2 \le c_5|u_2|_2^{1/2}|Lu_2|_2^{1/2}|u|_2^{1/2}|Lu|_2^{3/2}\\ &\le \frac{1}{8}|Lu|_2^2 + c_6|u|_2^2|u_2|_2^2|Lu_2|_2^2. \qquad (2.168)\end{aligned}$$

Furthermore, by the Cauchy inequality,

$$|(f_1 - f_2, Lu)| \le \frac{1}{4}|Lu|_2^2 + |f_1 - f_2|_2^2. \qquad (2.169)$$

Substituting (2.167)–(2.169) into (2.166) and integrating in time, we derive

$$\begin{aligned} t\|u\|_1^2 + \int_0^t s|Lu|_2^2 ds \le \int_0^t \|u\|_1^2 ds + c_7\int_0^t s|u|_2^2\big(|u_1|_2^2\,|Lu_1|_2^2\\ + |u_2|_2^2\,|Lu_2|_2^2\big)\,ds + 2\int_0^t s\,|f_1 - f_2|_2^2 ds.\end{aligned}$$

Combining this with (2.165), we see that the required inequality will be established if we show that

$$\int_0^t s|u_i|_2^2|Lu_i|_2^2\,ds \le c_8\,t^{-2}\exp\Big(c_8\int_0^t \big(\|u_i\|_1^2 + |f_i|_2^2\big)\,ds\Big), \qquad (2.170)$$

where $i = 1, 2$ and $0 < t \le 1$.

To prove (2.170), we first note that, by (2.43),

$$\int_0^t s|Lu_i|_2^2 ds \le |u_{0i}|_2^2 + 2\int_0^t |f|_2^2 ds \quad \text{for } 0 \le t \le 1. \qquad (2.171)$$

Furthermore, taking the scalar product of (2.19) with $2u$ and integrating in time, we derive

$$|u_i(t)|_2^2 + 2\int_0^t \|u_i\|_1^2 ds = |u_{0i}|_2^2 + 2\int_0^t \langle u_i, f_i\rangle\,ds.$$

Combining this with (2.51), we conclude that

$$\big(1 - e^{-\alpha_1 t}\big)|u_{0i}|^2 \le c_9\int_0^t \big(\|u_i\|_1^2 + \|f_i\|_{-1}^2\big)\,ds. \qquad (2.172)$$

It follows from (2.51), (2.171), and (2.172) that

$$\int_0^t s|u_i|_2^2|Lu_i|_2^2\,ds \le c_{10}\left(|u_{0i}|_2^2 + \int_0^t |f_i|_2^2 ds\right)^2$$
$$\le c_{11}\left((1-e^{-\alpha_1 t})^{-1}\int_0^t \left(\|u_i\|_1^2 + \|f_i\|_{-1}^2\right) ds + \int_0^t |f_i|_2^2 ds\right)^2.$$

This implies the required inequality (2.170), and the proof of the proposition is complete. □

Proof of Proposition 2.4.12 The proof is by induction in k. For $k = 0$, a stronger result is established in Corollary 2.4.11. We now assume that $k = n \ge 1$ and that for $k \le n-1$ inequality (2.138) is already proved. We wish to apply Itô's formula to the functional

$$F_n(t,u) = t^n \|u\|_n^2 = \langle L^n u, u\rangle.$$

To this end, we use Theorem A.7.5 with $H = V^n$, $V = V^{n+1}$, and $V^* = V^{n-1}$. The fact that the hypotheses of Theorem A.7.5 are fulfilled can be checked with the help of an argument similar to that used in the proof of Propositions 2.4.8 and 2.4.10 (which correspond to the case $n = 0$). Therefore we shall confine ourselves to a formal derivation of estimates.

Let us set $f_n(t) = F_n(t, u(t))$. By Itô's formula (A.37), we have (cf. (2.131))

$$f_n(t) = \int_0^t \left(ns^{n-1}\|u\|_n^2 + 2s^n\langle L^n u, h - \nu Lu - B(u)\rangle + \mathfrak{B}_n s^n\right) ds + \sum_{j=1}^{\infty} 2b_j \int_0^t s^n \langle L^n u, e_j\rangle\, d\beta_j(s). \quad (2.173)$$

We now distinguish between two cases: $n = 1$ and $n \ge 2$. In the first case, by Lemma 2.1.16, we have $\langle Lu, B(u)\rangle = 0$. Combining this with (2.173) and the relations

$$2s\,|\langle Lu, h\rangle| \le \|u\|_1^2 + s^2\|h\|_1^2, \quad \langle Lu, e_j\rangle = \alpha_j\langle u, e_j\rangle,$$

we derive

$$\mathcal{E}_u(1,t) \le 2\int_0^t \|u\|_1^2\,ds + \frac{1}{2}\mathfrak{B}_1 t^2 + \frac{1}{3}\|h\|_1^2 t^3 + K_1(t),$$

where we set

$$K_1(t) = \sum_{j=1}^{\infty} 2b_j\alpha_j \int_0^t s\langle u, e_j\rangle\, d\beta_j(s) - \nu\int_0^t s\|u\|_2^2\,ds. \quad (2.174)$$

It follows that

$$\sup_{0\le t\le T} \mathcal{E}_u(1,t) \le 2\int_0^T \|u\|_1^2\,ds + \mathfrak{B}_1 T^2 + \|h\|_1^2 T^3 + \sup_{0\le t\le T} K_1(t). \quad (2.175)$$

The first term on the right-hand side of (2.175) can be estimated with the help of (2.135). To bound the last term, we repeat the argument used in the proof of (2.130). Namely, we denote by $M(t)$ the martingale defined by the sum in (2.174) and remark that its quadratic variation is equal to

$$\langle M\rangle_t = \sum_{j=1}^{\infty} 4b_j^2\alpha_j^2 \int_0^t s^2\langle u, e_j\rangle^2 ds \le 4T\Bigl(\sup_{j\ge 1} b_j^2\Bigr)\int_0^t s\,\|u(s)\|_2^2\,ds.$$

Therefore, setting $\gamma = (2T\sup_j b_j^2)^{-1}$, we obtain

$$K_1(t) \le M(t) - \frac{\gamma\nu}{2}\langle M\rangle_t.$$

By the supermartingale inequality (A.57), we have

$$\mathbb{P}\Bigl\{\sup_{0\le t\le T} K_1(t)\ge\rho\Bigr\} = \mathbb{P}\Bigl\{\sup_{0\le t\le T}\exp\bigl(\gamma\nu M(t) - \tfrac{(\gamma\nu)^2}{2}\langle M\rangle_t\bigr)\ge e^{\gamma\nu\rho}\Bigr\} \le e^{-\gamma\nu\rho},$$

whence it follows that (cf. proof of Corollary 2.4.11)

$$\begin{aligned}\mathbb{E}\sup_{0\le t\le T}\mathcal{E}_u(1,t)^m &\le C_1\bigl(\mathfrak{B}_1^m T^{2m} + \|h\|_1^{2m}T^{3m}\bigr)\\ &\quad + C_1\Bigl(\nu^{-m}\mathbb{E}\,\mathcal{E}_u(T)^m + \int_0^\infty e^{-\nu\gamma\lambda^{1/m}}\,d\lambda\Bigr). \quad (2.176)\end{aligned}$$

Combining this with (2.135), we obtain the inequality

$$\begin{aligned}\mathbb{E}\sup_{0\le t\le T}\mathcal{E}_u(1,t)^m &\le C_2\bigl(\mathfrak{B}_1^m T^{2m} + \|h\|_1^{2m}T^{3m}\bigr)\\ &\quad + C_2\nu^{-m}\bigl(1 + \mathbb{E}\,|u_0|_2^{2m} + T^m\mathfrak{B}^m\bigr)\\ &\quad + C_2\nu^{-2m}\bigl(1 + T^m\alpha_1^{-m}|h|_2^{2m}\bigr), \quad (2.177)\end{aligned}$$

which gives the required estimate (2.138) with $k=1$.

We now assume that $n\ge 2$. In this case, the scalar product $\langle L^n u, B(u)\rangle$ does not vanish, and we need to estimate it. In view of Lemma 2.1.20, we have

$$\begin{aligned}|\langle L^n u, B(u)\rangle| &\le C_2\|u\|_{n+1}^{(4n-1)/2n}\|u\|_1^{(n+1)/2n}|u|_2^{1/2}\\ &\le \frac{\nu}{4}\|u\|_{n+1}^2 + C_2^2\nu^{-4n}\|u\|_1^{2(n+1)}|u|_2^{2n}.\end{aligned}$$

Combining this with (2.173) and the relations

$$2s\,|\langle L^n u, h\rangle| \le s^{n-1}\|u\|_n^2 + s^{n+1}\|h\|_n^2, \quad \langle L^n u, e_j\rangle = \alpha_j^n\langle u, e_j\rangle,$$

we derive

$$\mathcal{E}_u(n,t) \le (n+1)\int_0^t s^{n-1}\|u\|_n^2\,ds + \frac{1}{n+1}\mathfrak{B}_n t^{n+1} + \frac{1}{n+2}\|h\|_n^2\,t^{n+2}$$
$$+ C_3\nu^{-4n}\int_0^t s^n\|u\|_1^{2(n+1)}|u|_2^{2n}ds + K_n(t),$$

where we set

$$K_n(t) = \sum_{j=1}^{\infty} 2b_j\alpha_j^n\int_0^t s^n\langle u, e_j\rangle\,d\beta_j(s) - \frac{\nu}{2}\int_0^t s^n\|u\|_{n+1}^2\,ds.$$

Since

$$\sup_{0\le t\le T}\int_0^t s^n\|u\|_1^{2(n+1)}|u|_2^{2n}ds \le \nu^{-1}\sup_{0\le t\le T}\mathcal{E}_u(t)^{n+1}\mathcal{E}_u(1,t)^n,$$

we conclude that

$$\sup_{0\le t\le T}\mathcal{E}_u(n,t) \le C_4\big(\nu^{-1}\mathcal{E}_u(n-1,T) + \mathfrak{B}_n T^{n+1} + \|h\|_n^2 T^{n+2}\big)$$
$$+ C_4\nu^{-(4n+1)}\sup_{0\le t\le T}\big(\mathcal{E}_u(t)^{n+1}\,\mathcal{E}_u(1,t)^n\big) + \sup_{0\le t\le T}K_n(t).$$

Setting $P(m,n) = \mathbb{E}\sup\mathcal{E}_u(n,t)^m$ and repeating the argument used in the derivation of (2.176), we prove that

$$P(m,n) \le C_5\nu^{-m}\big(P(m,n-1) + \nu^{-4mn}\big(P(2m(n+1),0)\,P(2mn,1)\big)^{1/2}\big)$$
$$+ C_5(1+\nu^{-m}). \tag{2.178}$$

Recalling that (see (2.135) and (2.177))

$$P(2m(n+1),0) \le C_6\big(\mathbb{E}\,|u_0|_2^{4m(n+1)} + \nu^{-2m(n+1)} + 1\big),$$
$$P(2mn,1) \le C_7\big(1 + \nu^{-2mn}\mathbb{E}\,|u_0|_2^{4mn} + \nu^{-4mn}\big),$$

and using the induction hypothesis, we see that (2.178) implies the required estimate (2.138) with $k = n$. This completes the proof of the proposition. □

Proof of Theorem 2.4.19

Step 1. Suppose we have shown that

$$\theta_s(\Omega_*) = \Omega_* \quad \text{for any } s \in \mathbb{R}. \tag{2.179}$$

To prove that $\boldsymbol{\Phi}$ is an RDS, we need to check the continuity and cocylcle properties of Definition 1.3.14. The continuity of $\varphi_t^\omega(\cdot)$ is trivial for $\omega \notin \Omega_*$. To prove it for $\omega \in \Omega_*$, let us note that, in view of (2.151), we have

$$\varphi_t^\omega u_0 = \tilde{z}^\omega(t) + \mathcal{R}_t(u_0, \tilde{z}^\omega(\cdot)) \quad \text{for } t \ge 0,\ \omega \in \Omega_*, \tag{2.180}$$

where $\mathcal{R}_t : H \times \mathcal{X} \to H$ stands for the operator that takes (u_0, z) to the solution at time t of problem (2.111), (2.112). Proposition 2.4.5 now implies the required continuity $\omega \in \Omega_*$.

Let us prove the cocycle property (1.62). The first relation is obvious, as is the second one with $\omega \notin \Omega_*$ (in view of (2.179)). We need to show that

$$\varphi_{t+s}^{\omega} u_0 = \varphi_t^{\theta_s \omega}(\varphi_s^{\omega} u_0) \quad \text{for } t, s \geq 0, \omega \in \Omega_*, u_0 \in H. \tag{2.181}$$

To this end, note that, in view of (2.150) and (2.102), we have

$$\tilde{z}^{\theta_s \omega}(t) = \tilde{z}_s^{\omega}(s+t) \quad \text{for } t \geq 0, s \in \mathbb{R}, \omega \in \Omega_*, \tag{2.182}$$

where $\tilde{z}_s(t)$ is the solution of the problem

$$\dot{z} + \nu L z = \partial_t \tilde{\zeta}, \quad z(s) = 0.$$

Combining (2.180) and (2.182), we write

$$\varphi_t^{\theta_s \omega}(\varphi_s^{\omega} u_0) = \tilde{z}_s^{\omega}(t+s) + \mathcal{R}_t(\varphi_s^{\omega} u_0, \tilde{z}_s^{\omega}(s, \cdot)). \tag{2.183}$$

For any $s \in \mathbb{R}$ and $\omega \in \Omega_*$, the right-hand side of (2.183) regarded as a function of t is a solution of the problem

$$\dot{u} + \nu L u + B(u) = h + \partial_t \tilde{\zeta}^{\omega}, \quad u(s) = \varphi_s^{\omega} u_0. \tag{2.184}$$

On the other hand, the function $\varphi_{\cdot+s}^{\omega} u_0$ also satisfies (2.184). By uniqueness, the two functions must coincide, and we obtain (2.181).

Step 2. Let us prove (2.179). To this end, it suffices to show that $\theta_s \omega \in \widetilde{\Omega}$ for any $s \in \mathbb{R}$. Let $\omega \in \Omega_*$ and let $n \in \mathbb{Z}$ be such that $n + s \geq 0$. In view of (2.182), the difference

$$w(t) = \tilde{z}^{\theta_{-n}\omega}(s+n+t) - \tilde{z}^{\theta_s \omega}(t) = \tilde{z}^{\theta_{-n}\omega}(s+n+t) - \tilde{z}_s^{\omega}(s+t)$$

regarded as a function of t satisfies the equations

$$\dot{w} + \nu L w = 0, \quad w(0) = \tilde{z}^{\omega}(s+n) \in H.$$

It follows that $w \in \mathcal{X}$. Since $\tilde{z}^{\theta_{-n}\omega}(s+n+\cdot) \in \mathcal{X}$, we conclude that $\tilde{z}^{\theta_s \omega} = \tilde{z}^{\omega}(s+n+\cdot) - w \in \mathcal{X}$ and, hence, $\theta_s \omega \in \widetilde{\Omega}$ for any $s \in \mathbb{R}$.

Step 3. We now prove that $\boldsymbol{\Phi}$ is Markov. In view of Definition 1.3.18, we need to show for any finite sets $u_i^{\pm} \in H$, $t_i^{\pm}, s_i^{\pm} \in \mathbb{R}$, $i = 1, \ldots, N^{\pm}$, such that

$$s_i^- < 0, \quad 0 < t_i^- \leq -s_i^-, \quad s_i^+ \geq 0, \quad t_i^+ > 0,$$

and any Borel subsets $\Gamma_i^{\pm} \subset H$, the events[10]

$$A^{\pm} = \{\omega \in \Omega : \varphi(t_i^{\pm}, \theta_{s_i^{\pm}}\omega)u_i^{\pm} \in \Gamma_i^{\pm}, i = 1, \dots, N^{\pm}\}$$

are independent. To this end, note that (cf. (2.180) and (2.102))

$$\varphi_t^{\theta_s\omega} u_0 = z_s^{\omega}(s+t) + \mathcal{R}_t(u_0, z_s^{\omega}(s+\cdot)),$$

$$z_s^{\omega}(s+t) = \zeta^{\omega}(s+t) - e^{-\nu t L}\zeta^{\omega}(s) - \nu \int_s^{s+t} Le^{-\nu(t-s)L}\zeta^{\omega}(s)\,ds,$$

where $s \in \mathbb{R}$, $t \geq 0$, and $\omega \in \Omega_*$. It follows that, up to an event of zero measure, A^- depends only on the path $\{\omega_s, s \leq 0\}$, whereas A^+ depends only on $\{\omega_s, s \geq 0\}$. Since the process ω_t has independent increments and vanishes at $t = 0$, we conclude that A^- and A^+ are independent.

Finally, the fact that the law of $\{\varphi_t u_0, t \geq 0\}$ is the same as that of the solution $u(t)$ for (2.98), (2.114) follows immediately from (2.151) and the construction of u (see Theorem 2.4.6). The proof of the theorem is now complete. □

Notes and comments

The results presented in Section 2.1 are very well known. The well-posedness of the 2D Navier–Stokes system in a bounded domain and the regularity of solutions was proved due to contributions of many researchers, starting from the pioneering paper of Leray [Ler34], followed by Ladyzhenskaya [Lad59; Lad63], Lions and Prodi [LP59; Lio69], Temam [Tem68], and others. We refer the reader to the books [Tem79; CF88; Soh01] for a comprehensive study of the initial-boundary value problem for Navier–Stokes equations.

The Foiaş–Prodi estimates presented in Section 2.1.8 are of fundamental importance for this book. They show that the dynamics of the 2D Navier–Stokes system in a bounded domain is determined by finitely many modes. This type of result was first established by Foiaş and Prodi [FP67]. It is used in various problems related to the large-time behaviour of solutions, including the theory of attractors and inertial manifolds; see the books [Tem88; CFNT89; BV92] and references therein.

The theory of well-posedness for Navier–Stokes equations with various types of additive noise is a rather straightforward consequence of deterministic results, due to the simple reductions described in Sections 2.3 and 2.4. The study of the Navier–Stokes system perturbed by spatially regular white noise was initiated by Bensoussan and Temam [BT73], Viot [Vio75; Vio76], and Vishik, Komech, and Fursikov [VKF79; VF88]. They proved that the 2D

10 We write here $\varphi(t, \theta_s\omega)$ instead of $\varphi_t^{\theta_s\omega}$ to avoid triple subscripts.

Navier–Stokes system perturbed by a white-noise force defines a Markov process in a suitable function space and that Itô's formula applies to its solutions and results in the a priori estimates (2.119) and (2.135) with $m = 1$. These inequalities with $m \geq 2$ and other estimates for more regular random forces were derived by E, Mattingly, and Sinai [EMS01; Mat02b] and Kuksin and Shirikyan [KS02a; KS03]. The case of an analytic random force was investigated by Bricmont, Kupiainen, and Lefevere [BKL00], Mattingly [Mat02a], and Shirikyan [Shi02]. A more delicate task is to prove the existence and uniqueness of a solution for white forces of low spatial smoothness. This situation was studied, for instance, by Flandoli [Fla94], Ferrario [Fer03], Da Prato and Debussche [DD02], and Brzeźniak and Ferrario [BF09]. Navier–Stokes equations perturbed by random kicks are less traditional, but their investigation is technically simpler. The estimates described in Section 2.3 are taken from [KS01b; Shi04]. Theorem 2.4.15 and relation (2.148) belong to a large group of results in stochastic PDEs motivated by numerical methods, since a popular way to calculate solutions of a PDE perturbed by a white force is to replace it by high-frequency normalised random kicks. This approach is called the *splitting up method* in numerical analysis. We refer the reader to the paper of Gyöngy and Krylov [GK03] for some results in this direction that concern, in particular, the rate of convergence.

Finally, the idea of studying limit points of the averaged laws of solutions to construct a stationary distribution of Markov processes goes back to Kryloff and Bogoliouboff [KB37]. It is a simple, but effective tool for producing stationary distributions in various problems; see the books [VF88; KH95] and references therein. The question of uniqueness of a stationary distribution is, in general, much more complicated. However, in the laminar case, the RDS generated by the 2D Navier–Stokes system is globally asymptotically stable, which implies the uniqueness and exponential mixing of a stationary measure (see Exercise 2.5.9). This fact was proved by Mattingly [Mat99].

3

Uniqueness of stationary measure and mixing

This chapter contains some results on uniqueness of a stationary distribution and the property of exponential mixing. Their proofs are based on a development of the classical coupling argument introduced by Doeblin in the late 1930s; see [Doe38; Doe40]. Without going into details, let us describe two essentially equivalent versions of Doeblin's approach to study ergodic properties of Markov chains.

Let X be a complete metric space and let $(u_k, \mathbb{P}_u)$, $k \in \mathbb{Z}_+$, be a Feller family of Markov chains in X parametrised by the initial point $u \in X$. We shall denote by $P_k(u, \Gamma)$, $u \in X$, $\Gamma \in \mathcal{B}(X)$, the transition function associated with the Markov family and by $\mathfrak{P}_k : C_b(X) \to C_b(X)$ and $\mathfrak{P}_k^* : \mathcal{P}(X) \to \mathcal{P}(X)$ the corresponding Markov operators (see Section 1.3.3). Let us assume that

$$\|P_1(u, \cdot) - P_1(u', \cdot)\|_{\text{var}} \le \gamma \tag{3.1}$$

for any $u, u' \in X$, where $\|\cdot\|_{\text{var}}$ denotes the total variation distance, and $\gamma < 1$ is a constant not depending on u and u'. In this case, the uniqueness of a stationary distribution and exponential mixing for the family $(u_k, \mathbb{P}_u)$ can be proved using the coupling lemma in one of the two equivalent forms given in Section 1.2.4.

Contraction of the space of measures

Let us endow $\mathcal{P}(X)$ with the total variation distance and consider the operator $\mathfrak{P}_1^* : \mathcal{P}(X) \to \mathcal{P}(X)$. We claim that this is a contraction:

$$\|\mathfrak{P}_1^*\lambda - \mathfrak{P}_1^*\lambda'\|_{\text{var}} \le \gamma\, \|\lambda - \lambda'\|_{\text{var}} \quad \text{for any } \lambda, \lambda' \in \mathcal{P}(X). \tag{3.2}$$

Indeed, let us use Corollary 1.2.25 to write

$$\lambda = (1-\delta)\nu + \delta\tilde{\lambda}, \quad \lambda' = (1-\delta)\nu + \delta\tilde{\lambda}',$$

where $\delta = \|\lambda - \lambda'\|_{\mathrm{var}}$, $\nu \in \mathcal{P}(X)$, and $\tilde{\lambda}$ and $\tilde{\lambda}'$ are mutually singular measures. Using (3.1) and (1.22), for any $\Gamma \in \mathcal{B}(X)$ we obtain

$$\begin{aligned}\mathfrak{P}_1^*\lambda(\Gamma) - \mathfrak{P}_1^*\lambda'(\Gamma) &= \delta\left(\mathfrak{P}_1^*\tilde{\lambda}(\Gamma) - \mathfrak{P}_1^*\tilde{\lambda}'(\Gamma)\right)\\ &= \delta \iint_{X\times X} \big(P_1(u,\Gamma) - P_1(u',\Gamma)\big)\tilde{\lambda}(du)\tilde{\lambda}'(du')\\ &\le \delta \iint_{X\times X} \gamma\, \tilde{\lambda}(du)\tilde{\lambda}'(du')\\ &= \|\lambda - \lambda'\|_{\mathrm{var}}\,\gamma.\end{aligned}$$

Using symmetry and taking the supremum over $\Gamma \in \mathcal{B}(X)$, we arrive at (3.2).

Since $\mathfrak{P}_1^*$ is a contraction of the complete metric space $(\mathcal{P}(X), \|\cdot\|_{\mathrm{var}})$, there is a unique measure $\mu \in \mathcal{P}(X)$ such that $\mathfrak{P}_1^*\mu = \mu$, and for any $\lambda \in \mathcal{P}(X)$ we have

$$\|\mathfrak{P}_k^*\lambda - \mu\|_{\mathrm{var}} \le \gamma^k, \quad k \ge 0. \tag{3.3}$$

When (3.3) holds, we say that the Markov process (which defines the semi-group) is *exponentially mixing* in the total variation norm. Similarly, if inequality (3.3) holds with the total variation distance replaced by the dual-Lipschitz distance, then we say that the Markov process is *exponentially mixing* in the dual-Lipschitz norm.

Coupling argument

Let $(\mathcal{R}(u,u',\cdot), \mathcal{R}'(u,u',\cdot))$ be a pair of random variables in X that depend on $u, u' \in X$ and form a maximal coupling for the measures $P_1(u,\cdot)$ and $P_1(u',\cdot)$. That is, $\mathcal{D}(\mathcal{R}) = P_1(u,\cdot)$, $\mathcal{D}(\mathcal{R}') = P_1(u',\cdot)$, and

$$\mathbb{P}\big\{\mathcal{R}(u,u') \ne \mathcal{R}'(u,u')\big\} = \|P_1(u,\cdot) - P_1(u',\cdot)\|_{\mathrm{var}} \quad \text{for all } u, u' \in X; \tag{3.4}$$

cf. Section 1.2.4. In particular, $\mathcal{R}(u,u) = \mathcal{R}'(u,u)$ for any $u \in X$. Such random variables exist in view of Theorem 1.2.28. Let us denote by Ω the direct product of countably many copies of the probability space on which $\mathcal{R}$ and $\mathcal{R}'$ are defined and consider a family of Markov chains $\{\boldsymbol{u}_k\}$ in $\boldsymbol{X} = X \times X$ given by the rule

$$\boldsymbol{u}_0(\omega) = \boldsymbol{u}, \qquad \boldsymbol{u}_k(\omega) = (\mathcal{R}(\boldsymbol{u}_{k-1},\omega_k), \mathcal{R}'(\boldsymbol{u}_{k-1},\omega_k)) \quad \text{for } k \ge 1. \tag{3.5}$$

Here $\omega = (\omega_j, j \ge 1) \in \Omega$ denotes the random parameter and $\boldsymbol{u} \in \boldsymbol{X}$ is an initial point. Writing $\boldsymbol{u} = (u, u')$ and $\boldsymbol{u}_k = (u_k, u_k')$, we derive from inequalities (3.1), (3.4) and the Markov property that

$$\mathbb{P}_{\boldsymbol{u}}\{u_{k+1} \ne u'_{k+1} \mid \mathcal{F}_k\} \le \gamma \quad \text{for any } \boldsymbol{u} \in \boldsymbol{X}, k \ge 0, \tag{3.6}$$

where $\mathcal{F}_k$ denotes the σ-algebra generated by $\boldsymbol{u}_1, \dots, \boldsymbol{u}_k$. Iterating inequality (3.6) and using the fact that $u_{k-1} = u'_{k-1}$ implies $u_k = u'_k$, we obtain by induction

$$\mathbb{P}_{\boldsymbol{u}}\{u_k \neq u'_k\} = \mathbb{E}_{\boldsymbol{u}}\big(\mathbb{I}_{\{u_{k-1}\neq u'_{k-1}\}}\mathbb{P}_{\boldsymbol{u}}\{u_k \neq u'_k \mid \mathcal{F}_{k-1}\}\big)$$
$$\leq \gamma\, \mathbb{P}_{\boldsymbol{u}}\{u_{k-1} \neq u'_{k-1}\} \leq \gamma^k \tag{3.7}$$

for any $\boldsymbol{u} \in \boldsymbol{X}$, $k \geq 0$. So,

$$\|P_k(u, \cdot) - P_k(u', \cdot)\|_{\text{var}} \leq \gamma^k. \tag{3.8}$$

Combining this with the Kolmogorov–Chapman relation, we see that for any two measures λ and λ' on X the total variation distance between $\mathfrak{P}_k^*\lambda$ and $\mathfrak{P}_k^*\lambda'$ goes to zero exponentially fast. In particular, there is at most one stationary distribution. Moreover, the sequence $\{P_k(u, \cdot)\}$ converges to a limiting measure μ, which is stationary for $(u_k, \mathbb{P}_u)$. Finally, inequality (3.3) also follows from (3.8).

Inequality (3.1) is satisfied for a number of problems with compact phase space, e.g., for stochastic differential equations (SDEs) with non-degenerate diffusion on a compact manifold. On the other hand, condition (3.1) is rather restrictive if the phase space is not compact. For instance, in the case of SDEs in $\mathbb{R}^n$, it is fulfilled only if there is a strong nonlinear drift towards a bounded ball. However, sometimes one can overcome this difficulty with the help of the following modification of the coupling argument.

Let X be a Polish space and let $(u_k, \mathbb{P}_u)$ be a Feller family of Markov chains in X. Retaining the notation used above, suppose we can find a closed subset $B \subset X$ for which the two properties below are satisfied:

Recurrence. The first hitting time τ_B of the set B is almost surely finite for any initial point $u \in X$, and there is a $\delta > 0$ such that

$$\mathbb{E}_u \exp\big(\delta\tau_B\big) < \infty \quad \text{for all } u \in X. \tag{3.9}$$

Squeezing. Inequality (3.1) with a constant $\gamma < 1$ holds for any $u, u' \in B$.

Let $(\mathcal{R}, \mathcal{R}')$ be the family of random variables in $\boldsymbol{X}$ defined above and let $\{\boldsymbol{u}_k\}$ be the family of Markov chains given by (3.5). Denote by ρ_n the nth instant when the trajectory $\boldsymbol{u}_k$ enters the set $\boldsymbol{B} := B \times B$. Then, using (3.1), (3.4), and the strong Markov property, we get (cf. (3.6))

$$\mathbb{P}\{u_{\rho_n+1} \neq u'_{\rho_n+1} \mid \mathcal{F}_{\rho_n}\} \leq \gamma \quad \text{for any } \boldsymbol{u} \in \boldsymbol{X}, n \geq 1, \tag{3.10}$$

where $\mathcal{F}_{\rho_n}$ denotes the σ-algebra associated with the Markov time ρ_n. Iteration of (3.10) results in (cf. (3.7))

$$\mathbb{P}_{\boldsymbol{u}}\{u_{\rho_n+1} \neq u'_{\rho_n+1}\} \leq \gamma^n \quad \text{for any } \boldsymbol{u} \in \boldsymbol{X}, n \geq 1.$$

Combining this with (3.9), one can prove inequality (3.8) with a larger constant $\gamma < 1$, and this implies all the properties established for the case in which (3.1) holds uniformly. Thus, Doeblin's method applies also in the case of an unbounded phase space, provided that inequality (3.1) is satisfied on a subset that can be reached from any initial point at a random time with finite exponential moment.

Application of the above technique to stochastic PDEs encounters one essential difficulty: inequality (3.1) cannot be true (even for close u and u'), unless some very restrictive conditions are imposed on the diffusion; see Section 3.5.1 below for the case of Navier–Stokes equations. In this chapter, we show how to develop Doeblin's approach to be able to treat the 2D stochastic Navier–Stokes equations and other dissipative SPDEs with degenerate diffusion. In Section 3.1, some general criteria are established for the uniqueness of a stationary distribution and mixing in the dual-Lipshcitz norm. In Sections 3.2–3.4 we apply these criteria to the Navier–Stokes system with various random perturbations. Section 3.5 is devoted to some further results on uniqueness and mixing, a discussion of the Navier–Stokes system perturbed by a compound Poisson process, and a description of an alternative proof for mixing of the model with random kicks. Finally, in Section 3.7, we clarify the importance of the results of this chapter for physics.

3.1 Three results on uniqueness and mixing

In this section, we establish some sufficient conditions for uniqueness of a stationary measure for a Markov process. We begin with the case in which there is a Kantorovich functional (see Section 1.2.5) decaying along any pair of trajectories. This property immediately implies that the Markov process in question defines a contraction in the space of measures, and therefore has a unique stationary measure, which is exponentially mixing. We next discuss a simple criterion for uniqueness and mixing. Roughly speaking, we prove that if a Markov process is recurrent, and the dual-Lipschitz distance between the laws of trajectories issued from close points remains small (without any contraction!), then there is at most one stationary measure, and if it exists, then the solution converges to it in distribution. Our third result gives a sufficient condition for uniqueness and exponential mixing. Its proof is based on a stopping-time technique well known in the theory of renewal processes; see chapters XI in [Fel71] or chapter 5 in [PSS89].

3.1.1 Decay of a Kantorovich functional

Let X be a Polish space, let $\mathcal{T} = \mathbb{R}$ or $\mathbb{Z}$, let $\mathcal{T}_+ = \{t \in \mathcal{T} : t \geq 0\}$, and let $(u_t, \mathbb{P}_u)$ be a Feller family of Markov processes in X. Recall that, given a measurable symmetric function $F : X \times X \to \mathbb{R}_+$ satisfying (1.40), we define the Kantorovich functional $\mathcal{K}_F$ with the density F by relation (1.41). In what follows, we suppose that there is a constant $C > 0$ and a point $u_0 \in X$ such that

$$F(u_1, u_2) \le C\left(1 + \mathrm{dist}_X(u_1, u_0) + \mathrm{dist}_X(u_2, u_0)\right) \quad \text{for any } u_1, u_2 \in X. \tag{3.11}$$

Recall that $\mathcal{P}_1(X)$ is the class of measures $\mu \in \mathcal{P}(X)$ such that

$$\mathfrak{m}_1(\mu) = \int_X \mathrm{dist}_X(u, u_0)\mu(du) < \infty.$$

Inequality (3.11) implies that if ξ_1 and ξ_2 are two random variables whose laws belong to $\mathcal{P}_1(X)$, then $\mathbb{E}\, F(\xi_1, \xi_2) < \infty$.

Theorem 3.1.1 *Let $(u_t, \mathbb{P}_u)$ be a Feller family of Markov processes in X satisfying the following two conditions.*

A priori estimate: *There is a constant $C > 0$ and a point $u_0 \in X$ such that*

$$\int_X \mathrm{dist}_X(z, u_0) P_t(u, dz) \le C\left(1 + \mathrm{dist}_X(u, u_0)\right) \quad \textit{for any } u \in X,\ t \in \mathcal{T}_+. \tag{3.12}$$

Contraction: *There is a time $s \in \mathcal{T}_+$, a constant $\gamma < 1$, and a Kantorovich density $F : X \times X \to \mathbb{R}_+$ satisfying (3.11) such that*

$$\mathcal{K}_F(\mathfrak{P}_s^*\lambda, \mathfrak{P}_s^*\lambda') \le \gamma\, \mathcal{K}_F(\lambda, \lambda') \quad \textit{for any } \lambda, \lambda' \in \mathcal{P}_1(X). \tag{3.13}$$

Then the Markov family $(u_t, \mathbb{P}_u)$ has a unique stationary distribution $\mu \in \mathcal{P}_1(X)$. Moreover, there are positive constants C and α such that

$$\|\mathfrak{P}_t^*\lambda - \mu\|_L^* \le C\, e^{-\alpha t}\left(1 + \mathfrak{m}_1(\lambda)\right) \quad \textit{for } t \in \mathcal{T}_+, \tag{3.14}$$

where $\lambda \in \mathcal{P}_1(X)$ is an arbitrary measure.

In what follows, we shall say that a Markov process is *exponentially mixing* in the dual-Lipschitz norm if it satisfies (3.14).

Proof

Step 1. Let us endow $\mathcal{P}(X)$ with the dual-Lipschitz distance. By Theorem 1.2.15, this is a complete metric space. Integrating (3.12) with respect to $\lambda(du)$, we see that

$$\mathfrak{m}_1(\mathfrak{P}_t^*\lambda) \le C\,(1 + \mathfrak{m}_1(\lambda)) \quad \text{for } t \in \mathcal{T}_+. \tag{3.15}$$

Inequalities (1.42) and (3.13) imply that

$$\|\mathfrak{P}_{ks}^*\lambda - \mathfrak{P}_{ks}^*\lambda'\|_L^* \le \mathcal{K}_F(\mathfrak{P}_{ks}^*\lambda, \mathfrak{P}_{ks}^*\lambda') \le \gamma^k \mathcal{K}_F(\lambda, \lambda') \quad \text{for } k \ge 0. \tag{3.16}$$

Let us show that $\{P_{ks}(u, \cdot), k \ge 1\}$ is a Cauchy sequence in $\mathcal{P}(X)$ for any $u \in X$. Indeed, applying the Kolmogorov–Chapman relation, for $u, u' \in X$, $l \ge k$, and for an arbitrary function $f \in L_b(X)$ with $\|f\|_L \le 1$, we obtain

$$\begin{aligned}
&(f, P_{ls}(u', \cdot)) - (f, P_{ks}(u, \cdot)) \\
&\quad = \int_X P_{(l-k)s}(u', dz) \int_X \big(P_{ks}(z, dw) - P_{ks}(u, dw)\big) f(w) \\
&\quad \le \gamma^k \int_X P_{(l-k)s}(u', dz) \mathcal{K}_F(\delta_z, \delta_u).
\end{aligned}$$

Using the fact that $\mathcal{K}_F(\delta_{u_1}, \delta_{u_2}) = F(u_1, u_2)$ and recalling inequality (3.11), we see that

$$\begin{aligned}
&(f, P_{ls}(u', \cdot)) - (f, P_{ks}(u, \cdot)) \\
&\quad \le C_1 \gamma^k \int_X P_{(l-k)s}(u', dz)\big(1 + \mathrm{dist}_X(z, u_0) + \mathrm{dist}_X(u, u_0)\big) \\
&\quad \le C_2 \gamma^k \big(1 + \mathrm{dist}_X(u', u_0) + \mathrm{dist}_X(u, u_0)\big).
\end{aligned}$$

Integrating this inequality with respect to $\lambda(du)\,\lambda'(du')$, where $\lambda, \lambda' \in \mathcal{P}_1(X)$ are arbitrary measures, using the symmetry with respect to λ and λ', and taking the supremum over f, for $l \ge k$ we derive

$$\begin{aligned}
\|\mathfrak{P}_{ls}^* \lambda' - \mathfrak{P}_{ks}^* \lambda\|_L^* &\le C_2 \gamma^k \int_X \big((1 + \mathrm{dist}_X(u', u_0) + \mathrm{dist}_X(u, u_0)\big)\lambda(du)\lambda(du') \\
&\le C_2 \gamma^k \big(1 + \mathfrak{m}_1(\lambda) + \mathfrak{m}_1(\lambda')\big). \qquad (3.17)
\end{aligned}$$

In particular, taking $\lambda = \lambda' = \delta_u$, we see that $\{P_{ks}(u, \cdot)\}$ is a Cauchy sequence in $\mathcal{P}(X)$, and therefore it converges to a limit $\mu \in \mathcal{P}(X)$. It follows from (3.17) that μ does not depend on $u \in X$. Since $\mu = \lim_{k\to\infty} (\mathfrak{P}_s^*)^k \delta_u$, applying $\mathfrak{P}_s^*$ to this relation, we see that $\mathfrak{P}_s^* \mu = \mu$.

Step 2. Let us show that $\mu \in \mathcal{P}_1(X)$. We choose any increasing sequence $\{f_n\} \subset C_b(X)$ converging pointwise to the function $\mathrm{dist}_X(\cdot, u_0)$. Then, using (3.12), we can write

$$(f_n, \mu) = \lim_{k\to\infty} (f_n, P_{ks}(u_0, \cdot)) \le \limsup_{k\to\infty} \int_X \mathrm{dist}_X(z, u_0) P_{ks}(u_0, dz) \le C.$$

Applying the monotone convergence theorem and passing to the limit as $n \to \infty$, we see that $\mathfrak{m}_1(\mu) \le C$.

Step 3. Applying inequality (3.17) with $l = k$, $\lambda' = \mathfrak{P}_t^* \mu$, $\lambda = \mu$, and using (3.15), we obtain

$$\|\mathfrak{P}_t^* \mu - \mu\|_L^* = \|\mathfrak{P}_{ks}^* \mathfrak{P}_t^* \mu - \mathfrak{P}_{ks}^* \mu\|_L^* \le C_3 \gamma^k \big(1 + \mathfrak{m}_1(\mu)\big).$$

Letting $k \to \infty$, we obtain $\mathfrak{P}_t^* \mu = \mu$ for each $t \in \mathcal{T}_+$, and we see that μ is a stationary distribution. If $\lambda \in \mathcal{P}_1(X)$ is another stationary measure, then by (3.17) we have $\lambda = \mu$. So μ is a unique stationary distribution for the Markov family $(u_t, \mathbb{P}_u)$ in the space $\mathcal{P}_1(X)$.

Step 4. To complete the proof of the theorem, it remains to establish (3.14). To this end, take any $t \in \mathcal{T}_+$ and write $t = ks + r$, where $k \geq 0$ is an integer and $0 \leq r < s$. Using (3.15) with $t = r$ and inequality (3.17) in which $\lambda' = \mu$ and λ is replaced by $\mathfrak{P}_r^*\lambda$, we derive

$$\|\mathfrak{P}_t^*\lambda - \mu\|_L^* = \|\mathfrak{P}_{ks}^*(\mathfrak{P}_r^*\lambda) - \mu\|_L^* \leq C_3\,\gamma^k\big(1 + \mathfrak{m}_1(\mathfrak{P}_r^*\lambda)\big) \leq C_4\,\gamma^k\big(1 + \mathfrak{m}_1(\lambda)\big).$$

This inequality readily implies (3.14). □

3.1.2 Coupling method: uniqueness and mixing

As before, we denote by $(u_t, \mathbb{P}_u)$, $t \in \mathcal{T}_+$, a Feller family of Markov processes in X. Let $\boldsymbol{X} = X \times X$ be the direct product of two copies of X. We write $\boldsymbol{u} = (u, u') \in \boldsymbol{X}$ and denote by $\Pi : \boldsymbol{u} \mapsto u$ and $\Pi' : \boldsymbol{u} \mapsto u'$ the natural projections. Let us consider a family of strong Markov processes $(\boldsymbol{u}_t, \mathbb{P}_{\boldsymbol{u}})$, $t \in \mathcal{T}_+$, in $\boldsymbol{X}$. We denote by $\boldsymbol{P}_t(\boldsymbol{u}, \boldsymbol{\Gamma})$ its transition function and by $\boldsymbol{P}_t$ and $\boldsymbol{P}_t^*$ the corresponding Markov operators. We shall write $\boldsymbol{u}_t = (u_t, u_t')$.

Definition 3.1.2 The Markov family $(\boldsymbol{u}_t, \mathbb{P}_{\boldsymbol{u}})$ is called a *coupling of two copies of* $(u_t, \mathbb{P}_u)$ if for any $\boldsymbol{u} = (u, u') \in \boldsymbol{X}$ the laws under $\mathbb{P}_{\boldsymbol{u}}$ of the processes $\Pi\boldsymbol{u}_t$ and $\Pi'\boldsymbol{u}_t$ (regarded as measures on the space of functions from $\mathcal{T}_+$ to X) coincide with those of u_t under $\mathbb{P}_u$ and $\mathbb{P}_{u'}$, respectively. In this case, we shall also say that $(\boldsymbol{u}_t, \mathbb{P}_{\boldsymbol{u}})$ is an *extension* of $(u_t, \mathbb{P}_u)$.

That is, if $(\boldsymbol{u}_t, \mathbb{P}_{\boldsymbol{u}})$ is an extension of $(u_t, \mathbb{P}_u)$, then

$$\Pi_*\boldsymbol{P}_t(\boldsymbol{u}, \cdot) = P_t(u, \cdot), \qquad \Pi'_*\boldsymbol{P}_t(\boldsymbol{u}, \cdot) = P_t(u', \cdot). \tag{3.18}$$

For a closed subset $B \subset X$, we denote by $\boldsymbol{B}$ the direct product $B \times B$ and by $\boldsymbol{\tau}(B)$ the first hitting time of $\boldsymbol{B}$ for $\boldsymbol{u}_t$:

$$\boldsymbol{\tau}(B) = \min\{t \geq 0 : u_t \in B, u_t' \in B\} = \min\{t \geq 0 : \boldsymbol{u}_t \in \boldsymbol{B}\}.$$

The following result gives a sufficient condition for the uniqueness of a stationary distribution and mixing in the dual-Lipschitz distance.

Theorem 3.1.3 *Let $(u_t, \mathbb{P}_u)$ be a Markov process and let $(\boldsymbol{u}_t, \mathbb{P}_{\boldsymbol{u}})$ be its extension. Let us assume that for any integer $m \geq 1$ there is a closed subset $B_m \subset X$ and a constant $\delta_m > 0$ such that $\delta_m \to 0$ as $m \to \infty$, and the following two properties hold.*

Recurrence: *For any $\boldsymbol{u} = (u, u') \in \boldsymbol{X}$ and $m \geq 1$, we have*

$$\mathbb{P}_{\boldsymbol{u}}\{\boldsymbol{\tau}(B_m) < \infty\} = 1. \tag{3.19}$$

Stability: *There is a constant $T_m \in \mathcal{T}_+$ such that*

$$\sup_{t \geq T_m} \left\| P_t(u, \cdot) - P_t(u', \cdot) \right\|_L^* \leq \delta_m \quad \textit{for any } \boldsymbol{u} \in \boldsymbol{B}_m = B_m \times B_m. \tag{3.20}$$

Then, for any $u, u' \in X$, we have

$$\left\| P_t(u, \cdot) - P_t(u', \cdot) \right\|_L^* \to 0 \quad \textit{as } t \to \infty. \tag{3.21}$$

Moreover, if μ is a stationary distribution for the family $(u_t, \mathbb{P}_u)$, then it is unique, and for any $\lambda \in \mathcal{P}(X)$ we have

$$\left\| \mathfrak{P}_t^* \lambda - \mu \right\|_L^* \to 0 \quad \textit{as } t \to \infty. \tag{3.22}$$

That is, the Markov process is *mixing* in the dual-Lipschitz norm.

Proof

Step 1. We first assume that (3.21) is proved and establish the claims concerning the stationary distribution. The uniqueness is a straightforward consequence of (3.22). To prove (3.22), we take any function $f \in L_b(X)$ and write

$$\begin{aligned} |(f, \mathfrak{P}_t^* \lambda - \mu)| &= |(f, \mathfrak{P}_t^* \lambda - \mathfrak{P}_t^* \mu)| \\ &= \left| \iint_{X \times X} \big(\mathfrak{P}_t f(u) - \mathfrak{P}_t f(u') \big) \, \lambda(du) \, \mu(du') \right| \\ &\leq \iint_{X \times X} \left| \mathfrak{P}_t f(u) - \mathfrak{P}_t f(u') \right| \lambda(du) \, \mu(du'). \end{aligned} \tag{3.23}$$

Taking into account (3.21) and using the Lebesgue theorem on dominated convergence, we see that the right-hand side converges to zero as $t \to +\infty$ for any $f \in L_b(X)$. In view of assertion (ii) of Theorem 1.2.15, this is equivalent to (3.22).

Step 2. We now prove (3.21). For any $m \geq 1$ and $t \geq T_m$, write

$$\boldsymbol{\tau}(m, t) = \boldsymbol{\tau}(B_m) \wedge t, \quad p(\boldsymbol{u}, m, t) = \mathbb{P}_{\boldsymbol{u}}\{\boldsymbol{\tau}(B_m) + T_m > t\}.$$

Applying the strong Markov property (1.57) to the family $(\boldsymbol{u}_t, \mathbb{P}_{\boldsymbol{u}})$ and taking into account the first relation in (3.18), for any integer $m \geq 1$ we obtain

$$\begin{aligned} P_t(u, \Gamma) = \boldsymbol{P}_t(\boldsymbol{u}, \Gamma \times X) &= \mathbb{E}_{\boldsymbol{u}} \boldsymbol{P}_{t - \boldsymbol{\tau}(m,t)}(\boldsymbol{u}_{\boldsymbol{\tau}(m,t)}, \Gamma \times X) \\ &= \mathbb{E}_{\boldsymbol{u}} P_{t - \boldsymbol{\tau}(m,t)}(u_{\boldsymbol{\tau}(m,t)}, \Gamma). \end{aligned}$$

A similar relation holds for $P_t(u', \Gamma)$. It follows that

$$\left\| P_t(u, \cdot) - P_t(u', \cdot) \right\|_L^* \leq \mathbb{E}_{\boldsymbol{u}} g(t - \boldsymbol{\tau}(m, t), \boldsymbol{u}_{\boldsymbol{\tau}(m,t)}), \tag{3.24}$$

where for $s \ge 0$ and $\boldsymbol{z} = (z, z') \in \boldsymbol{X}$ we set

$$g(s, \boldsymbol{z}) = \| P_s(z, \cdot) - P_s(z', \cdot) \|_L^*.$$

Now note that, by the stability condition (3.20), we have

$$\sup_{s \ge T_m} g(s, \boldsymbol{z}) \le \delta_m \quad \text{for } \boldsymbol{z} \in \boldsymbol{B}_m. \tag{3.25}$$

Combining inequalities (3.24) and (3.25) and taking into account that the relation $\boldsymbol{\tau}(B_m) + T_m \le t$ implies $g(t - \boldsymbol{\tau}(m, t), \boldsymbol{u}_{\tau(m,t)}) \le \delta_m$, we get

$$\| P_t(u, \cdot) - P_t(u', \cdot) \|_L^* \le \delta_m + p(\boldsymbol{u}, m, t) \quad \text{for } t \ge T_m.$$

In view of (3.19), the right-hand side of this inequality can be made arbitrarily small by choosing m first and then taking t to be sufficiently large. This completes the proof of Theorem 3.1.3. □

Analysing the above proof, it is easy to see that if $\boldsymbol{Y} \subset \boldsymbol{X}$ is a Borel subset such that

$$\sup_{\boldsymbol{u} \in \boldsymbol{Y}} \mathbb{P}_{\boldsymbol{u}}\{\boldsymbol{\tau}(B_m) > t\} \le p(m, t), \tag{3.26}$$

where $p(m, t) \to 0$ as $t \to \infty$ for any integer $m \ge 1$, then

$$\sup_{\boldsymbol{u} \in \boldsymbol{Y}} \| P_t(u, \cdot) - P_t(u', \cdot) \|_L^* \to 0 \quad \text{as } t \to \infty. \tag{3.27}$$

This simple observation enables one to prove the following result on uniform convergence to the stationary distribution.

Theorem 3.1.4 *Under the hypotheses of Theorem 3.1.3, assume that* (3.26) *holds for any compact set $\boldsymbol{Y} \subset \boldsymbol{X}$ and any integer $m \ge 1$, where $p(m, t) \to 0$ as $t \to \infty$. Then for any compact subset Λ of the space $\mathcal{P}(H)$ endowed with the dual-Lipschitz metric convergence* (3.22) *holds uniformly with respect to $\lambda \in \Lambda$.*

Proof Let us fix a constant $\varepsilon > 0$ and use Prokhorov's theorem to find a compact subset $Y \subset X$ such that $\lambda(Y) \ge 1 - \varepsilon$ for any $\lambda \in \Lambda \cup \{\mu\}$. Let $f \in L_b(X)$ be an arbitrary function such that $\| f \|_L \le 1$. Then it follows from (3.23) that

$$|(f, \mathfrak{P}_t^* \lambda - \mu)| \le \sup_{u, u' \in Y} \left| \mathfrak{P}_t f(u) - \mathfrak{P}_t f(u') \right| + 4\varepsilon.$$

Using inequality (3.27) with $\boldsymbol{Y} = Y \times Y$ and recalling that ε and f are arbitrary, we arrive at the required uniform convergence. □

Let us also mention that the rate of convergence in (3.22) can be estimated in terms of the mean value of the hitting time $\boldsymbol{\tau}(B)$ and the sequence $\{\delta_m\}$ (see the recurrence and stability conditions of Theorem 3.1.3). In particular, if the Markov process in question depends on a parameter θ in such a way that the above-mentioned quantities are estimated uniformly in θ, then the rate of convergence to the stationary measure is also uniform in θ.

An application of Theorem 3.1.3 to the Navier–Stokes system is presented in Section 3.3. Here we discuss the case of the viscous *Burgers equation*, for which the stability holds uniformly with respect to the viscosity.

Example 3.1.5 Let us consider the Burgers equation with periodic boundary conditions:

$$\dot{u} - \nu\partial_x^2 u + \partial_x u^2 = \frac{\partial}{\partial t}\zeta^\omega(t,x), \quad x \in \mathbb{T}, \quad \int_{\mathbb{T}} u\,dx \equiv \int_{\mathbb{T}} \zeta\,dx \equiv 0, \tag{3.28}$$

where $\mathbb{T} = \mathbb{R}/2\pi\mathbb{Z}$. We abbreviate $\dot{H}^m(\mathbb{T};\mathbb{R}) = \dot{H}^m$ for $m \in \mathbb{R}$ (see Section 1.1.1) and denote by $\{e_j, j \in \mathbb{Z}_0\}$ the usual L^2-normalised trigonometric basis of these spaces. As for the white-forced Navier–Stokes system, we choose $\zeta^\omega(t,x) = \sum_j b_j\beta_j^\omega(t)e_j(x)$. For simplicity, assume that all quantities $\mathfrak{B}_m = \sum_j j^{2m}b_j^2$ are finite. Then Eq. (3.28) is well posed in each space $\dot{H}^m$, $m \geq 1$, and defines there a Markov process; see [DZ96]. Consider the RDS $\{\varphi_t^\omega, t \geq 0\}$ defined by (3.28). It is known that for a.e. ω the maps φ_t^ω extend by continuity to non-expanding transformations $\dot{L}^1 \to \dot{L}^1$, where $\dot{L}^1$ denotes the space of integrable functions on $\mathbb{T}$ with zero mean value; e.g., see lemma 3.2.2 in [Hör97]. So (3.28) defines a Markov process in $\dot{L}^1$. Theorem 3.1.3 applies to it. Indeed, choose $B_m = B_{\dot{L}^1}(1/m)$. Then the recurrence follows from the same simple argument as for the Navier–Stokes system (see below Section 3.3.2 and the discussion after Theorem 3.3.1). The stability with $T_m = 0$ and $\delta_m = 1/m$ immediately follows from the fact that the maps $\varphi_t^\omega : \dot{L}^1 \to \dot{L}^1$ are non-expanding.

If all numbers b_j are non-zero, then the rate of recurrence may be chosen to be independent of ν (cf. Section 3.3.2 below). Since the characteristics of stability are also independent of ν, in this case the rate of mixing for (3.28) does not depend on the viscosity. See [Bor12] for further properties of the Burgers equation (3.28) and [BK07] for its relevance as a physical model.

3.1.3 Coupling method: exponential mixing

This subsection can be omitted at first reading, since its results are used only in the proof of exponential mixing for equations with unbounded noise. To simplify the presentation, we assume here that X is a separable Banach space with a norm $\|\cdot\|$. The main result of this section – Theorem 3.1.7 – gives a sufficient condition for exponential mixing for Markov processes in X. In addition to the coupling, the proof uses some techniques of the theory of renewal processes; cf. [Fel71, chapter XI] or [PSS89, chapter 5].

Let $(u_t, \mathbb{P}_u)$ be a Feller family of Markov processes in X, let $(\boldsymbol{u}_t, \mathbb{P}_{\boldsymbol{u}})$ be its extension, and let B be a closed subset in X. Recall that $\boldsymbol{\tau}(B)$ stands for the first hitting time of the set $\boldsymbol{B} = B \times B$ for $\boldsymbol{u}_t$. We also introduce the stopping time

$$\sigma = \inf\{t \in \mathcal{T}_+ : \|u_t - u'_t\| \geq C e^{-\beta t}\}, \tag{3.29}$$

where C and β are some fixed positive constants. In other words, σ is the first instant when the curves u_t and u'_t "stop converging" to each other exponentially fast. In particular, if $\sigma = \infty$, then

$$\|u_t - u'_t\| \leq C e^{-\beta t} \quad \text{for } t \geq 0. \tag{3.30}$$

Definition 3.1.6 We shall say that the family $(u_t, \mathbb{P}_u)$ satisfies the *coupling hypothesis* if there is an extension $(\boldsymbol{u}_t, \mathbb{P}_{\boldsymbol{u}})$, a closed set $B \subset X$, and an increasing function $g(r) \geq 1$ of the variable $r \geq 0$ such that the following two properties hold.

Recurrence: There is $\delta > 0$ such that

$$\mathbb{E}_{\boldsymbol{u}} \exp(\delta \boldsymbol{\tau}(B)) \leq G(\boldsymbol{u}) \quad \text{for all } \boldsymbol{u} = (u, u') \in \boldsymbol{X}, \tag{3.31}$$

where we set $G(\boldsymbol{u}) = g(\|u\|) + g(\|u'\|)$.

Exponential squeezing: There are positive constants δ_1, δ_2, c, K, and $q > 1$ such that, for any $\boldsymbol{u} \in \boldsymbol{B}$, we have

$$\mathbb{P}_{\boldsymbol{u}}\{\sigma = \infty\} \geq \delta_1, \tag{3.32}$$

$$\mathbb{E}_{\boldsymbol{u}}\{\mathbb{I}_{\{\sigma<\infty\}} \exp(\delta_2 \sigma)\} \leq c, \tag{3.33}$$

$$\mathbb{E}_{\boldsymbol{u}}\{\mathbb{I}_{\{\sigma<\infty\}} G(\boldsymbol{u}_\sigma)^q\} \leq K. \tag{3.34}$$

Any extension of $(u_t, \mathbb{P}_u)$ satisfying the above properties will be called a *mixing extension*.

Note that $\sigma \equiv 0$ for any $\boldsymbol{u} \in \boldsymbol{X}$ with $\|u - u'\| > C$. It follows from (3.32) that the set $\boldsymbol{B}$ must belong to the C-neighbourhood of the diagonal, that is, $\boldsymbol{B} \subset \{\boldsymbol{u} : \|u - u'\| \leq C\}$.

Before formulating the main result of this subsection, we make some comments on the above definition. Let us take an arbitrary initial point $\boldsymbol{u} \in \boldsymbol{B}$. Then, in view of (3.32), with $\mathbb{P}_u$-probability $\geq \delta_1$, we have $\sigma = \infty$, and therefore, with the same probability, the trajectories u_t and u'_t converge exponentially fast (see (3.30)). On the other hand, if they do not, inequality (3.33) says that the first instant σ when the trajectories "stop converging" is not very large. Moreover, by (3.34), we have some control over $\boldsymbol{u}_t$ at the instant $t = \sigma$. If the initial point $\boldsymbol{u} \in \boldsymbol{X}$ does not belong to $\boldsymbol{B}$, we cannot claim that the above properties hold. However, we know that, with probability 1, any trajectory hits the set $\boldsymbol{B}$, and by (3.31), the first hitting time $\boldsymbol{\tau}(B)$ has a finite exponential moment.

These observations make it plausible that, for any initial point $\boldsymbol{u} \in \boldsymbol{X}$, the trajectories u_t and u'_t converge exponentially fast. In fact, we have the following result.

Theorem 3.1.7 *Let $(u_t, \mathbb{P}_u)$ be a Feller family of Markov processes that possesses a mixing extension $(\boldsymbol{u}_t, \mathbb{P}_{\boldsymbol{u}})$. Then there is a random time $\ell \in \mathcal{T}_+$ such that, for any $\boldsymbol{u} \in \boldsymbol{X}$, with $\mathbb{P}_{\boldsymbol{u}}$-probability 1, we have*

$$\|u_t - u'_t\| \le C_1 e^{-\beta(t-\ell)} \quad \text{for } t \ge \ell, \tag{3.35}$$

$$\mathbb{E}_{\boldsymbol{u}} e^{\alpha\ell} \le C_1 G(\boldsymbol{u}), \tag{3.36}$$

where $\boldsymbol{u} \in \boldsymbol{X}$ is an arbitrary initial point, $g(r)$ is the function in Definition 3.1.6, and C_1, α, and β are positive constants not depending on $\boldsymbol{u}$ and t. If, in addition, there is an increasing function $\tilde{g}(r) \ge 1$ such that

$$\mathbb{E}_u \, g\big(\|u_t\|\big) \le \tilde{g}(\|u\|) \quad \text{for } u \in X,\ t \ge 0, \tag{3.37}$$

then the family $(u_t, \mathbb{P}_u)$ has a unique stationary measure $\mu \in \mathcal{P}(X)$, and there is a constant $\gamma > 0$ such that

$$\| P_t(u, \cdot) - \mu\|_L^* \le V(\|u\|) e^{-\gamma t} \quad \text{for } t \ge 0,\ u \in X, \tag{3.38}$$

where V is given by the relation

$$V(r) = 3C_1\big(g(r) + \tilde{g}(0)\big). \tag{3.39}$$

Proof We first show that inequalities (3.35)–(3.37) imply the existence and uniqueness of a stationary measure and the mixing property (3.38). Namely, we shall derive from these relations that, for any $u, u' \in X$,

$$\big\| P_t(u, \cdot) - P_t(u', \cdot)\big\|_L^* \le 3C_1 G(\boldsymbol{u})\, e^{-\gamma t}, \quad t \ge 0. \tag{3.40}$$

To this end, we fix an arbitrary functional $f \in L_b(X)$ with $\|f\|_L \le 1$ and note that

$$\begin{aligned} \big|\big(f, P_t(u, \cdot) - P_t(u', \cdot)\big)\big| &\le \mathbb{E}_{\boldsymbol{u}}\big|f(u_t) - f(u'_t)\big| \\ &\le 2\mathbb{P}_{\boldsymbol{u}}\big\{\ell > \tfrac{t}{2}\big\} + \mathbb{E}_{\boldsymbol{u}}\big\{\mathbb{I}_{\{\ell \le \frac{t}{2}\}}\big|f(u_t) - f(u'_t)\big|\big\}. \end{aligned} \tag{3.41}$$

In view of (3.36) and Chebyshev's inequality, we have

$$\mathbb{P}_{\boldsymbol{u}}\big\{\ell > \tfrac{t}{2}\big\} \le C_1 G(\boldsymbol{u})\, e^{-\frac{\alpha t}{2}}.$$

Furthermore, it follows from the assumption $\|f\|_L \le 1$ and inequality (3.35) that the second term on the right-hand side of (3.41) does not exceed

$$\mathbb{E}_{\boldsymbol{u}}\big\{\mathbb{I}_{\{\ell \le \frac{t}{2}\}}\|u_t - u'_t\|\big\} \le C_1 e^{-\frac{\beta t}{2}}.$$

Substituting the last two estimates into (3.41) and using that $G \geq 2$, we obtain the required inequality (3.40) with $\gamma = \frac{1}{2}(\alpha \wedge \beta)$.

We now use (3.40) to prove the existence and uniqueness of a stationary measure and inequality (3.38). Let us fix arbitrary points $u, u' \in X$ and a functional $f \in L_b(X)$ such that $\|f\|_L \leq 1$. By the Kolmogorov–Chapman relation and inequality (3.40), for $t \leq s$ we have

$$\begin{aligned}
\big|\big(f, P_t(u,\cdot) - P_s(u',\cdot)\big)\big| &= \left|\int_X P_{s-t}(u', dz)\int_X \big(P_t(u,dv) - P_t(z,dv)\big) f(v)\right| \\
&\leq 3C_1 e^{-\gamma t}\int_X P_{s-t}(u', dz)\big[g(\|u\|) + g(\|z\|)\big] \\
&= 3C_1 e^{-\gamma t}\big[g(\|u\|) + \mathbb{E}_{u'}\, g(\|u_{s-t}\|)\big].
\end{aligned}$$

Taking into account (3.37), we conclude that

$$\big\|P_t(u,\cdot) - P_s(u',\cdot)\big\|_L^* \leq 3C_1 e^{-\gamma t}\big(g(\|u\|) + \tilde{g}(\|u'\|)\big). \tag{3.42}$$

Since $\mathcal{P}(X)$ is a complete metric space with respect to the dual-Lipschitz distance, we conclude that $P_t(u,\cdot)$ converges, as $t \to +\infty$, to a measure $\mu \in \mathcal{P}(X)$, which does not depend on u and is stationary. Setting $u' = 0$ in (3.42) and passing to the limit as $s \to +\infty$, we obtain inequality (3.38) with V given by (3.39).

Thus, we need to establish inequalities (3.35) and (3.36). Their proof is divided into four steps.

Step 1. We introduce the stopping time

$$\rho = \inf\{t \in \mathcal{T}_+ : t \geq \sigma, \boldsymbol{u}_t \in \boldsymbol{B}\}. \tag{3.43}$$

In other words, we wait until the first instant σ when the trajectories u_t and u_t' "stop converging" and denote by ρ the first hitting time of $\boldsymbol{B}$ after σ. Let δ, δ_1, and δ_2 be the constants in (3.31), (3.32), and (3.33). We claim that, for any $\boldsymbol{u} \in \boldsymbol{B}$,

$$\mathbb{P}_{\boldsymbol{u}}\{\rho = \infty\} \geq \delta_1, \tag{3.44}$$

$$\mathbb{E}_{\boldsymbol{u}}\big\{\mathbb{I}_{\{\rho<\infty\}} e^{\alpha\rho}\big\} \leq f, \tag{3.45}$$

where $\alpha \leq \delta_2 \wedge \delta$ and $f < 1$ are positive constants not depending on $\boldsymbol{u}$. Indeed, the definition of ρ and (3.31) imply that $\{\rho = \infty\} = \{\sigma = \infty\}$, so (3.44) is an immediate consequence of (3.32).

To prove (3.45), we first show that

$$\mathbb{E}_{\boldsymbol{u}}\big\{\mathbb{I}_{\{\rho<\infty\}} e^{\delta_3\rho}\big\} \leq M \quad \text{for any } \boldsymbol{u} \in \boldsymbol{B}, \tag{3.46}$$

where $\delta_3 = \frac{(q-1)(\delta_2\wedge\delta)}{q}$ and $M > 0$ is a constant not depending on $\boldsymbol{u}$. Indeed, using relation (3.43), the strong Markov property, and inequality (3.31), we derive

$$\mathbb{E}_{\boldsymbol{u}}\big\{\mathbb{I}_{\{\rho<\infty\}}e^{\delta_3\rho}\big\} = \mathbb{E}_{\boldsymbol{u}}\big\{\mathbb{I}_{\{\sigma<\infty\}}e^{\delta_3\sigma}\big(\mathbb{E}_{\boldsymbol{u}_\sigma}\, e^{\delta_3\tau_{\boldsymbol{B}}}\big)\big\} \le \mathbb{E}_{\boldsymbol{u}}\big\{\mathbb{I}_{\{\sigma<\infty\}}e^{\delta_3\sigma}\,G(\boldsymbol{u}_\sigma)\big\}.$$

Combining this with (3.33) and (3.34), we conclude that

$$\begin{aligned}\mathbb{E}_{\boldsymbol{u}}\big\{\mathbb{I}_{\{\rho<\infty\}}e^{\delta_3\rho}\big\} &\le \big(\mathbb{E}_{\boldsymbol{u}}\big\{\mathbb{I}_{\{\sigma<\infty\}}e^{\delta_2\sigma}\big\}\big)^{\frac{q-1}{q}}\big(\mathbb{E}_{\boldsymbol{u}}\,\big\{\mathbb{I}_{\{\sigma<\infty\}}G(\boldsymbol{u}_\sigma)^q\big\}\big)^{\frac{1}{q}}\\ &\le (c^{q-1}K)^{\frac{1}{q}} =: M.\end{aligned}$$

To derive (3.45), let us set $\alpha = \varepsilon\delta_3$ and note that, in view of (3.44) and (3.46), we have

$$\mathbb{E}_{\boldsymbol{u}}\big\{\mathbb{I}_{\{\rho<\infty\}}e^{\alpha\rho}\big\} \le \big(\mathbb{P}_{\boldsymbol{u}}\{\rho<\infty\}\big)^{1-\varepsilon}\big(\mathbb{E}_{\boldsymbol{u}}\big\{\mathbb{I}_{\{\rho<\infty\}}e^{\delta_3\rho}\big\}\big)^{\varepsilon} \le (1-\delta_1)^{1-\varepsilon}M^{\varepsilon}.$$

The right-hand side of this inequality is less than 1 if $\varepsilon > 0$ is sufficiently small.

Step 2. We now consider the iterations of ρ. Namely, we define two sequences of stopping times ρ_k and ρ_k' by the formulas

$$\begin{aligned}\rho_0 &= \tau_{\boldsymbol{B}},\\ \rho_k' &= \inf\{t\in\mathcal{T}_+ : t\ge\rho_{k-1},\ \|u_t-u_t'\|\ge C\,e^{-\beta(t-\rho_{k-1})}\}, \quad k\ge 1,\\ \rho_k &= \inf\{t\in\mathcal{T}_+ : t\ge\rho_k',\ \boldsymbol{u}_t\in\boldsymbol{B}\}, \quad k\ge 1.\end{aligned}$$

That is, ρ_k is the first occurrence of ρ after ρ_{k-1}. We claim that

$$\mathbb{E}_{\boldsymbol{u}}\big\{\mathbb{I}_{\{\rho_k<\infty\}}e^{\alpha\rho_k}\big\} \le f^k G(\boldsymbol{u}) \quad \text{for any } \boldsymbol{u}\in\boldsymbol{X}. \tag{3.47}$$

Indeed, inequality (3.47) is true for $k = 0$ in view of (3.31). Let us assume that (3.47) holds for $\rho_0, \ldots, \rho_{k-1}$, where $k \ge 1$. Since $\boldsymbol{u}_{\rho_k} \in \boldsymbol{B}$, inequality (3.45) and the strong Markov property imply that

$$\begin{aligned}\mathbb{E}_{\boldsymbol{u}}\big\{\mathbb{I}_{\{\rho_k<\infty\}}e^{\alpha\rho_k}\big\} &\le \mathbb{E}_{\boldsymbol{u}}\Big\{\mathbb{I}_{\{\rho_{k-1}<\infty\}}e^{\alpha\rho_{k-1}}\sup_{\boldsymbol{v}\in\boldsymbol{B}}\mathbb{E}_{\boldsymbol{v}}\big(\mathbb{I}_{\{\rho<\infty\}}e^{\alpha\rho}\big)\Big\}\\ &\le f\,\mathbb{E}_{\boldsymbol{u}}\big\{\mathbb{I}_{\{\rho_{k-1}<\infty\}}e^{\alpha\rho_{k-1}}\big\}.\end{aligned}$$

Combining this with the induction hypothesis, we arrive at (3.47).

Step 3. We now note that, if $\rho_k < \infty$ and $\rho_{k+1} = \infty$ for an integer $k \ge 0$, then

$$\|u_t-u_t'\| \le C\,e^{-\beta(t-\rho_k)} \quad \text{for } t\ge\rho_k. \tag{3.48}$$

For any $\boldsymbol{u}\in\boldsymbol{X}$, let us set $\bar{k} = \sup\{k\ge 0 : \rho_k<\infty\}$. We wish to show that, for any $\boldsymbol{u}\in\boldsymbol{X}$,

$$\bar{k}<\infty \quad \mathbb{P}_{\boldsymbol{u}}\text{-almost surely.} \tag{3.49}$$

To this end, note that, in view of (3.44) and the strong Markov property,

$$\mathbb{P}_{\boldsymbol{u}}\{\rho_k<\infty\} \le (1-\delta_1)\mathbb{P}_{\boldsymbol{u}}\{\rho_{k-1}<\infty\} \le (1-\delta_1)^k\mathbb{P}_{\boldsymbol{u}}\{\rho_0<\infty\} \le (1-\delta_1)^k.$$

Hence, the Borel–Cantelli lemma implies (3.49).

Step 4. Let us set

$$\ell = \begin{cases} \rho_{\bar{k}} & \text{if } \bar{k} < \infty, \\ +\infty & \text{if } \bar{k} = \infty. \end{cases}$$

Inequality (3.35) follows immediately from (3.48), the definition of ρ_k, and the fact that $\rho_{\bar{k}+1} = \infty$. To prove (3.36), we write

$$\mathbb{E}_{\boldsymbol{u}} e^{\alpha\ell} = \sum_{k=0}^{\infty} \mathbb{E}_{\boldsymbol{u}} \{\mathbb{I}_{\{\bar{k}=k\}} e^{\alpha\rho_k}\} \le \sum_{k=0}^{\infty} \mathbb{E}_{\boldsymbol{u}} \{\mathbb{I}_{\{\rho_k<\infty\}} e^{\alpha\rho_k}\} \le (1-f)^{-1} G(\boldsymbol{u}),$$

where we used inequality (3.47) and the fact that $\ell < \infty$ with $\mathbb{P}_{\boldsymbol{u}}$-probability 1 for any $\boldsymbol{u} \in \boldsymbol{X}$. This completes the proof of Theorem 3.1.7. □

Remark 3.1.8 Analysing the proof given above, it is not difficult to see that Theorem 3.1.7 remains valid if σ is replaced by any other stopping time $\tilde{\sigma}$ such that

$$\mathbb{P}_{\boldsymbol{u}}\{\tilde{\sigma} \le \sigma\} = 1 \quad \text{for any } \boldsymbol{u} \in \boldsymbol{B}.$$

In other words, if inequalities (3.32)–(3.35) hold with σ replaced by $\tilde{\sigma}$, then the conclusion of Theorem 3.1.7 is true. To see this, it suffices to repeat the arguments above, replacing σ by $\tilde{\sigma}$ everywhere.

3.2 Dissipative RDS with bounded kicks

3.2.1 Main result

Let H be a separable Banach space with norm $\|\cdot\|$ and let $S : H \to H$ be a continuous (nonlinear) operator. We consider an RDS defined by the relation

$$u_k = S(u_{k-1}) + \eta_k, \quad k \ge 1, \tag{3.50}$$

where $\{\eta_k\}$ is a sequence of independent identically distributed (i.i.d.) random variables in H whose law $\mathcal{D}(\eta_1)$ has bounded support. As explained in Example 1.3.15, one can regard (3.50) as a Markov RDS in H. Our goal in this section is to study the Markov chain defined by (3.50) in H (see Example 1.3.6). In what follows, we assume that the following four conditions are satisfied.

Condition 3.2.1 For any $R > r > 0$ there are positive constants $C = C(R)$, $a = a(R, r) < 1$ and an integer $n_0 = n_0(R, r) \ge 1$ such that

$$\|S(u_1) - S(u_2)\| \le C(R)\|u_1 - u_2\| \quad \text{for all} \quad u_1, u_2 \in B_H(R), \tag{3.51}$$

$$\|S^n(u)\| \le \max\{a\|u\|, r\} \quad \text{for} \quad u \in B_H(R), \quad n \ge n_0, \tag{3.52}$$

where S^n stands for the nth iteration of S.

Let us denote by $\mathcal{K}$ the support of the measure $\mathcal{D}(\eta_1)$ and define the sets of attainability from $B \subset H$ by the relations

$$\mathcal{A}_0(B) = B, \quad \mathcal{A}_k(B) = S\big(\mathcal{A}_{k-1}(B)\big) + \mathcal{K} \quad \text{for} \quad k \geq 1.$$

Condition 3.2.2 There is a constant $\rho > 0$ and a non-decreasing continuous function $k_0 = k_0(R) > 0$ such that

$$\mathcal{A}_k(B_H(R)) \subset B_H(\rho) \quad \text{for any } R > 0 \text{ and } k \geq k_0(R), \tag{3.53}$$

where $B_H(r)$ stands for the closed ball of radius r centred at the origin.

Let us introduce the set

$$\mathcal{A} = \overline{\bigcup_{k \geq 0} \mathcal{A}_k\big(B_H(\rho)\big)}, \tag{3.54}$$

where ρ is the constant defined in Condition 3.2.2. A straightforward consequence of (3.51) and (3.53) is that $\mathcal{A}$ is a bounded invariant absorbing set for the RDS (3.50); see Section 1.3.4.

Condition 3.2.3 There is a finite-dimensional subspace $E \subset H$, a continuous linear projection operator $\mathsf{P} : H \to H$ onto E, and a constant $\gamma < 1$ such that

$$\big\|\mathsf{Q}\big(S(u_1) - S(u_2)\big)\big\| \leq \gamma \, \|u_1 - u_2\| \quad \text{for all} \quad u_1, u_2 \in \mathcal{A}, \tag{3.55}$$

where $\mathsf{Q} = I - \mathsf{P}$.

Note that, in view of the Hahn–Banach theorem, a continuous projection onto a finite-dimensional subspace of a Banach space always exists. The last condition concerns the law of random variables η_k.

Condition 3.2.4 The support of $\mathcal{D}(\eta_1)$ contains the origin, and the random variables $\mathsf{P}_*\eta_1$ and $\mathsf{Q}_*\eta_1$ are independent. Furthermore, the measure $\mathsf{P}_*\mathcal{D}(\eta_1)$ has a density $p(x)$ with respect to the Lebesgue measure on E, and there is a constant C such that

$$\int_E |p(x+y) - p(x)| \, dx \leq C \, \|y\| \quad \text{for } y \in E. \tag{3.56}$$

Before formulating the main result of this section, we make some comments on the above conditions. Inequality (3.51) is nothing else but the Lipschitz continuity of S on bounded subsets of H, while (3.52) expresses the property of dissipativity for S. In particular, it follows from (3.52) that $u = 0$ is a stable fixed point for the deterministic dynamical system generated by S. It is easy to see that inequality (3.52) of Condition 3.2.1 is fulfilled if S satisfies the inequality

$$\|S(u)\| \leq q \, \|u\| \quad \text{for all } u \in H,$$

where $q < 1$ and C are positive constants not depending on u. Condition 3.2.2 means that the RDS (3.50) has a bounded absorbing set, and the function $k_0(R)$ provides an upper bound for the time of absorption. In applications to evolutionary PDEs, Condition 3.2.3 (sometimes referred to as the *squeezing property*) manifests itself as a smoothing property of flow maps. It is usually satisfied for quasilinear parabolic PDEs, such as the Ginzburg–Landau and reaction–diffusion equations. It also holds for the 2D Navier–Stokes system, which can be written as a nonlocal parabolic PDE; see (2.19). Finally, Condition 3.2.4 expresses the non-degeneracy of the random perturbation and the continuous dependence on the shifts of the law of its projection to E. The latter property is certainly satisfied if the density p belongs to the Sobolev class $W^{1,1}(E)$.[1]

We now formulate the main result of this section. Its proof is based on Theorem 3.1.1 and is given in the next two subsections. An application to the randomly forced Navier–Stokes system is discussed in Section 3.2.4.

Theorem 3.2.5 *Suppose that Conditions 3.2.1–3.2.4 are satisfied. Then the RDS* (3.50) *has a unique stationary distribution* $\mu \in \mathcal{P}(H)$. *Moreover, there are positive constants* C *and* α_0 *such that, for any* $\alpha \in (0, \alpha_0]$, *we have*

$$\|\mathfrak{P}_k^*\lambda - \mu\|_L^* \le Ce^{-\alpha k} \int_H \exp\bigl(\alpha k_0(\|u\|)\bigr)\,\lambda(du), \quad k \ge 0, \tag{3.57}$$

where $\lambda \in \mathcal{P}(H)$ *is an arbitrary measure for which the right-hand side of* (3.57) *is finite, and* k_0 *is defined in Condition 3.2.2.*

3.2.2 Coupling

Let us denote by $\chi(du)$ the law of the random variable η_1, by $\chi_E(dx)$ its image under the projection P, and by $\chi_E(u, dx)$, $u \in H$, the law of $\mathsf{P}(S(u) + \eta_1)$. We shall need the following auxiliary result.

Lemma 3.2.6 *Let* $S : H \to H$ *be a continuous operator, and suppose that Condition 3.2.4 is satisfied. Then there is a probability space* $(\Omega, \mathcal{F}, \mathbb{P})$ *such that for any pair* $v, v' \in H$ *there are two* H*-valued random variables* $\zeta = \zeta(v, v', \omega)$ *and* $\zeta' = \zeta'(v, v', \omega)$ *possessing the following properties.*

(i) *The laws of* ζ *and* ζ' *coincide with* χ.
(ii) *The random variables* $(\mathsf{P}\zeta, \mathsf{P}\zeta')$ *and* $(\mathsf{Q}\zeta, \mathsf{Q}\zeta')$ *are independent. Furthermore, the projections* $\mathsf{Q}\zeta$ *and* $\mathsf{Q}\zeta'$ *coincide for all* $\omega \in \Omega$ *and do not depend on* (v, v').
(iii) *The pair*

$$V = \mathsf{P}\bigl(S(v) + \zeta\bigr), \quad V' = \mathsf{P}\bigl(S(v') + \zeta'\bigr)$$

[1] Inequality (3.56) means precisely that the density p belongs to the Nikolski–Besov space $\Lambda_1^{1,\infty}(E)$; see chapter V in [Ste70].

is a maximal coupling for $(\chi_E(v,\cdot),\chi_E(v',\cdot))$. *Moreover,*

$$\mathbb{P}\{V\neq V'\}\le C_1\|S(v)-S(v')\| \quad \text{for any } v,v'\in H, \tag{3.58}$$

where $C_1>0$ *is a constant not depending on* v *and* v'.

(iv) *The functions* ζ *and* ζ' *are measurable with respect to* (v,v',ω).

Proof For any $v,v'\in H$, let (V,V') be a maximal coupling for the pair of measures $(\chi_E(v,\cdot),\chi_E(v',\cdot))$. By Theorem 1.2.28, we can assume that V and V' are defined on the same probability space $(\Omega_1,\mathcal{F}_1,\mathbb{P}_1)$ for all $v,v'\in H$ and are measurable functions of (v,v',ω_1). Let $(\Omega_2,\mathcal{F}_2,\mathbb{P}_2)$ be the probability space on which the random variables η_k are defined. We denote by $(\Omega,\mathcal{F},\mathbb{P})$ the direct product of these two spaces and, for any $\omega=(\omega_1,\omega_2)\in\Omega$, set

$$\begin{aligned}\zeta(v,v',\omega)&=V(v,v',\omega_1)-\mathsf{P}S(v)+\mathsf{Q}\eta_1(\omega_2),\\ \zeta'(v,v',\omega)&=V'(v,v',\omega_1)-\mathsf{P}S(v')+\mathsf{Q}\eta_1(\omega_2).\end{aligned} \tag{3.59}$$

Assertions (i), (ii), (iv), and the first part of (iii) are straightforward consequences of the construction. To prove inequality (3.58), we note that $\chi_E(v,dy)$ is absolutely continuous with respect to the Lebesgue measure on E, and the corresponding density is given by $p(x-\mathsf{P}S(v))$. Since (V,V') is a maximal coupling, inequality (3.56) and relation (1.23) imply that

$$\begin{aligned}\mathbb{P}\{V\neq V'\}&=\big\|\chi_E(v,\cdot)-\chi_E(v',\cdot)\big\|_{\mathrm{var}}\\ &=\frac12\int_E\big|p(x-\mathsf{P}S(v))-p(x-\mathsf{P}S(v'))\big|\,dx\\ &\le\frac12 C\,\|\mathsf{P}(S(v)-S(v'))\|\le C_1\|S(v)-S(v')\|,\end{aligned}$$

where we used the continuity of the projection P. This completes the proof of Lemma 3.2.6. □

We now fix a constant $d>0$ and use the following rules to define *coupling operators* $(\mathcal{R},\mathcal{R}')$ on the probability space $(\Omega,\mathcal{F},\mathbb{P})$ (constructed in the proof of Lemma 3.2.6): for $v,v'\in\mathcal{A}$ with $\|v-v'\|\le d$, we set

$$\mathcal{R}(v,v',\omega)=S(v)+\zeta(v,v',\omega),\quad \mathcal{R}'(v,v',\omega)=S(v')+\zeta'(v,v',\omega); \tag{3.60}$$

for $v,v'\in\mathcal{A}$ with $\|v-v'\|>d$, we set

$$\mathcal{R}(v,v',\omega)=S(v)+\eta_1(\omega_2),\quad \mathcal{R}'(v,v',\omega)=S(v')+\eta_1(\omega_2). \tag{3.61}$$

Let us recall that the concept of extension for a family of Markov chains was introduced in Section 3.1.2. We now define an extension of the Markov chain associated with the RDS (3.50). Denote by $(\widehat{\Omega},\widehat{\mathcal{F}},\widehat{\mathbb{P}})$ the direct product of countably many copies of the probability space $(\Omega,\mathcal{F},\mathbb{P})$ in Lemma 3.2.6. We shall write $\widehat{\omega}=(\omega^1,\omega^2,\dots)$ for the points of $\widehat{\Omega}$.

Let us consider the following RDS $\boldsymbol{\Phi}_k = (\Phi_k, \Phi'_k)$ on the product space $H \times H$:

$$\Phi_0(\widehat{\omega})(v, v') = v, \quad \Phi'_0(\widehat{\omega})(v, v') = v',$$
$$\Phi_k(\widehat{\omega})(v, v') = \mathcal{R}(\Phi_{k-1}(v, v'), \Phi'_{k-1}(v, v'), \omega^k),$$
$$\Phi'_k(\widehat{\omega})(v, v') = \mathcal{R}'(\Phi_{k-1}(v, v'), \Phi'_{k-1}(v, v'), \omega^k).$$

Note that $\boldsymbol{\Phi}_k$ depends only on $(\omega^1, \dots, \omega^k)$. Arguing by induction and using the definition of the coupling operators $\mathcal{R}$ and $\mathcal{R}'$, we see that the Markov chain $(\boldsymbol{u}_k, \mathbb{P}_{\boldsymbol{u}})$ associated with $\boldsymbol{\Phi}_k$ is an extension[2] for the restriction to $\mathcal{A}$ of the Markov chain $(u_k, \mathbb{P}_u)$ corresponding to (3.50). In what follows, we shall drop the hat from the notation and write $\mathbb{P}$, ω, etc. The following two propositions are crucial points of the proof of Theorem 3.2.5.

Proposition 3.2.7 *Suppose that Conditions 3.2.1–3.2.4 are satisfied. Then, for sufficiently small $d > 0$ there is a constant $K > 0$ such that, for any pair of vectors $v, v' \in \mathcal{A}$ with $\|v - v'\| \le d$, we have*

$$\mathbb{P}\big\{\|\mathcal{R}(v, v') - \mathcal{R}'(v, v')\| \le \gamma\, \|v - v'\|\big\} \ge 1 - K\, \|v - v'\|, \tag{3.62}$$

$$\mathbb{P}\big\{\|\Phi_k(v, v') - \Phi'_k(v, v')\| \le \gamma^k \|v - v'\| \text{ for } k \ge 0\big\} \ge 1 - K_1 \|v - v'\|, \tag{3.63}$$

where $K_1 = \frac{K}{1-\gamma}$.

Proof We first prove (3.62). The definition of the operators $\mathcal{R}$ and $\mathcal{R}'$ implies that if $v, v' \in \mathcal{A}$ and $\|v - v'\| \le d$, then

$$\|\mathcal{R}(v, v') - \mathcal{R}'(v, v')\| \le \|V(v, v') - V'(v, v')\| + \|\mathsf{Q}(S(v) - S(v'))\|. \tag{3.64}$$

In view of (3.51) and (3.58), the first term on the right-hand side of (3.64) vanishes with probability no less than $1 - K\, \|v - v'\|$. Furthermore, it follows from (3.55) that

$$\|\mathsf{Q}(S(v) - S(v'))\| \le \gamma\, \|v - v'\|.$$

Combining this with (3.64), we arrive at (3.62).

We now iterate (3.62) to prove (3.63). Let us assume that $d > 0$ is so small that $K_1 d \le 1/2$. For any integer $k \ge 1$, we set

$$G_k = \big\{\|\Phi_k(v, v') - \Phi'_k(v, v')\| \le \gamma\, \|\Phi_{k-1}(v, v') - \Phi'_{k-1}(v, v')\|\big\},$$

$$\overline{G}_k = \bigcap_{l=1}^{k} G_l, \quad k \le +\infty.$$

[2] In other words, the laws of the components of the random variables $\boldsymbol{u}_k = \boldsymbol{\Phi}_k(\omega)(v, v')$ coincide with $P_k(v, \cdot)$ and $P_k(v', \cdot)$ for any $v, v' \in \mathcal{A}$.

It is clear that the event on the left-hand side of (3.63) coincides with $\overline{G}_\infty$. Since $\overline{G}_1 \supset \overline{G}_2 \supset \cdots$, inequality (3.63) will be established once we show that

$$\mathbb{P}(\overline{G}_k) \geq 1 - K\,\|v - v'\| \sum_{l=0}^{k-1} \gamma^l =: \varkappa_k \quad \text{for each } k \geq 1. \tag{3.65}$$

The proof of (3.65) is by induction in k. For $k = 1$, inequality (3.65) coincides with (3.62). Assume now that (3.65) is established for some k. Let us write $\omega = (\omega_k, \omega_k')$, where $\omega_k := (\omega^1, \dots, \omega^k)$ and $\omega_k' := (\omega^j, j \geq k+1)$, and denote by $\mathbb{P}_k$ the projections of $\mathbb{P}$ to the first k components of ω. Since Φ_k depends only on ω_k, the event $\overline{G}_k$ can be written as

$$\overline{G}_k = \{\omega = (\omega_k, \omega_k') \in \Omega : \omega_k \in g_k\},$$

where g_k is a measurable subset of the direct product of k copies of the probability space constructed in Lemma 3.2.6. By the induction hypothesis, we have $\mathbb{P}^k(g_k) \geq \varkappa_k$, and by (3.62), for each $\omega_k \in g_k$ the probability of the event $\{\omega^{k+1} : (\omega_k, \omega^{k+1}) \in g_{k+1}\}$ is minorised by $1 - K\gamma^k\|v - v'\|$. Therefore,

$$\mathbb{P}_{k+1}(g_{k+1}) \geq \varkappa_k(1 - K\gamma^k\|v - v'\|) > \varkappa_{k+1},$$

whence we conclude that (3.65) holds with k replaced by $k+1$. This completes the proof of the proposition. □

Proposition 3.2.8 *Suppose that Conditions 3.2.1–3.2.4 are satisfied and that* $2K_1 d \leq 1$, *where* K_1 *is the constant constructed in Proposition 3.2.7. Then for any* $\delta > 0$ *there is an integer* $\ell = \ell_\delta \geq 1$ *and a constant* $\theta_\delta > 0$ *such that*

$$\mathbb{P}\big\{\|\Phi_\ell(v, v') - \Phi_\ell'(v, v')\| \leq \delta\big\} \geq \theta_\delta \quad \textit{for any } v, v' \in \mathcal{A}. \tag{3.66}$$

Proof We fix $\delta > 0$ and assume, without loss of generality, that $\delta \leq d \wedge (2K_1)^{-1}$. Let us introduce the stopping time

$$\tau = \tau(v, v') = \min\{k \geq 0 : \|\Phi_k(v, v') - \Phi_k'(v, v')\| \leq d\}.$$

Choose $m \geq 0$ so large that $d\gamma^m \leq \delta$. In view of the strong Markov property and inequality (3.63), for any integer $\ell \geq 1$ we have

$$\begin{aligned}
\mathbb{P}\{\|\Phi_{\ell+m} - \Phi_{\ell+m}'\| \leq \delta\} &\geq \mathbb{P}\big(\{\tau \leq \ell\} \cap \{\|\Phi_{\ell+m} - \Phi_{\ell+m}'\| \leq \delta\}\big) \\
&= \mathbb{E}\big(\mathbb{I}_{\{\tau \leq \ell\}} \mathbb{P}\big(\{\|\Phi_{\ell+m} - \Phi_{\ell+m}'\| \leq \delta\} \,|\, \mathcal{F}_\tau\big)\big) \\
&\geq (1 - K_1 d)\,\mathbb{P}\{\tau \leq \ell\} \\
&\geq \tfrac{1}{2}\,\mathbb{P}\{\tau \leq \ell\}.
\end{aligned}$$

Thus, inequality (3.66) will be established if we show that

$$\mathbb{P}\{\tau(v, v') > \ell\} \leq p \quad \text{for all } v, v' \in \mathcal{A}, \tag{3.67}$$

where $\ell \geq 1$ is a suitable integer and $p < 1$.

To prove (3.67), we use inequality (3.52) with $r = d/4$ to find an integer $\ell \geq 1$ such that

$$\|S^\ell(u)\| \leq \tfrac{d}{4} \quad \text{for all } u \in \mathcal{A}.$$

The Lipschitz continuity of S on bounded subsets implies that if $\varepsilon > 0$ is sufficiently small and $\|\eta_k\| \leq \varepsilon$ for $k = 1, \dots, \ell$, then the trajectory $\{u_k\}$ of (3.50) issued from any point $u \in \mathcal{A}$ satisfies the inequality $\|u_\ell\| \leq d/2$. By Condition 3.2.4, we have $\nu = \mathbb{P}\{\|\eta_1\| \leq \varepsilon\} > 0$. Let us define random variables $\zeta_k(v, v')$ and $\zeta_k'(v, v')$ by the relations

$$\zeta_k = \Phi_k - S(\Phi_{k-1}), \quad \zeta_k' = \Phi_k' - S(\Phi_{k-1}'), \quad k \geq 1.$$

The construction implies that each of the sequences $\{\zeta_k\}$ and $\{\zeta_k'\}$ consists of i.i.d. random variables whose distributions coincide with that of η_1. Let us consider the events

$$G = \{\|\zeta_k\| \leq \varepsilon, 1 \leq k \leq \ell\}, \quad G' = \{\|\zeta_k'\| \leq \varepsilon, 1 \leq k \leq \ell\}.$$

What has been said above implies that $\mathbb{P}(G) = \mathbb{P}(G') = \nu^\ell$ and

$$\|\Phi_\ell(v, v')\| \vee \|\Phi_\ell'(v, v')\| \leq d/2 \quad \text{for } \omega \in G \cap G',$$

whence it follows that

$$\|\Phi_\ell(v, v') - \Phi_\ell'(v, v')\| \leq d \quad \text{for } \omega \in G \cap G'.$$

Thus, we see that

$$\tau(v, v') \leq \ell \quad \text{for } \omega \in G \cap G' \text{ and any } v, v' \in \mathcal{A}. \tag{3.68}$$

On the other hand, the definition of Φ_k and Φ_k' implies that if $\omega \in \{\tau > \ell\}$, then $\zeta_k = \zeta_k'$ for $1 \leq k \leq \ell$. It follows that

$$G \cap G' \cap \{\tau > \ell\} = G \cap \{\tau > \ell\} = G' \cap \{\tau > \ell\}. \tag{3.69}$$

Assume that $\mathbb{P}\{\tau > \ell\} > 1 - \nu^\ell$. Then the event in (3.69) is not empty, which contradicts (3.68). So (3.67) is proved with $p = 1 - \nu^\ell$. □

3.2.3 Proof of Theorem 3.2.5

Step 1. Let us recall that the set $\mathcal{A}$ defined by (3.54) is invariant and absorbing for the RDS (3.50). Therefore, by Lemma 1.3.30, the support of any stationary measure is contained in $\mathcal{A}$. Let us fix any stationary measure μ and suppose we have proved that

$$\|P_k(u, \cdot) - \mu\|_L^* \leq C\, e^{-\alpha_0 k} \quad \text{for } k \geq 0, u \in \mathcal{A}, \tag{3.70}$$

where C and α_0 are some positive constants. In this case, if μ' is another stationary measure, then by the Lebesgue theorem on dominated convergence, for any $f \in L_b(H)$ we have

$$(f,\mu') = (f,\mathfrak{P}_k^*\mu') = (\mathfrak{P}_k f,\mu') = \int_H (f,P_k(u,\cdot))\,\mu'(du) \to (f,\mu)$$
$$\text{as } k\to\infty.$$

Thus, $(f,\mu) = (f,\mu')$ for any $f \in L_b(H)$, and therefore $\mu' = \mu$.

We now prove that (3.70) implies (3.57). By Condition 3.2.2, we have

$$u_k \in \mathcal{A} \quad \text{for any } u\in H \text{ and } k \ge k_0(\|u\|).$$

Therefore, the support of the measure $P_k(u,\cdot)$ is contained in $\mathcal{A}$ for $k \ge k_0(\|u\|)$. Combining this with inequality (3.70) and the Kolmogorov–Chapman relation, we obtain

$$\begin{aligned} |(f,P_k(u,\cdot)) - (f,\mu)| &= \left|\int_H P_{k_0}(u,dz)\big((f,P_{k-k_0}(z,\cdot)) - (f,\mu)\big)\right| \\ &\le \int_H P_{k_0}(u,dz)\sup_{z\in\mathcal{A}}\big|(f,P_{k-k_0}(z,\cdot)) - (f,\mu)\big| \\ &\le C\,e^{-\alpha_0(k-k_0)}, \end{aligned} \tag{3.71}$$

where $f \in L_b(H)$ is any function with $\|f\|_L \le 1$ and $k \ge k_0 = k_0(\|u\|)$. Furthermore, inequality (3.71) remains valid for $0 \le k < k_0$, provided that $C \ge 2$. It follows that, for any $u \in H$, $k \ge 0$, and $\alpha \in (0,\alpha_0]$, we have

$$|(f,P_k(u,\cdot)) - (f,\mu)| \le C\,e^{-\alpha(k-k_0)}.$$

Integrating this inequality with respect to $\lambda(du)$ and taking the supremum over $f \in L_b(H)$ with $\|f\|_L \le 1$, we arrive at (3.57).

Step 2. We now prove (3.70). To this end, we apply Theorem 3.1.1, in which $X = \mathcal{A}$, and the Markov family in question is the one generated by the restriction of the RDS (3.50) to $\mathcal{A}$. Since the metric space $\mathcal{A}$ is bounded, inequality (3.12) is trivial, and we only need to check (3.13).

Let us choose a small constant $d > 0$ and set $d_{-1} = +\infty$ and $d_m = \gamma^m d$ for $m \ge 0$, where $\gamma > 0$ is defined in Proposition 3.2.7. We now define a Kantorovich density F by the relation

$$F(u_1,u_2) = \begin{cases} R & \text{for } \|u_1-u_2\| > d_0, \\ R^2 d_m & \text{for } d_{m+1} < \|u_1-u_2\| \le d_m, \end{cases}$$

where $m \ge 0$, and $R \ge 1$ stands for a constant that will be chosen later in such a way that $Rd \le 1/2$, so that $F(u_1,u_2) \le R$. We claim that if $d > 0$ is sufficiently small, then

$$\mathcal{K}_F(\mathfrak{P}_\ell^*\lambda,\mathfrak{P}_\ell^*\lambda') \le \gamma'\mathcal{K}_F(\lambda,\lambda') \quad \text{for any } \lambda,\lambda' \in \mathcal{P}(\mathcal{A}), \tag{3.72}$$

where $\ell \geq 1$ is the integer defined in Proposition 3.2.8 for $\delta = d$, and $\gamma' < 1$ is a constant not depending on the measures λ and λ'. Once this inequality is established, the required result will follow from Theorem 3.1.1.

For any pair of vectors $(v, v') \in H \times H$, we set

$$f(v, v') = \mathbb{E}\, F\big(\Phi_l(v, v'), \Phi'_l(v, v')\big),$$

where the RDS $\boldsymbol{\Phi}_k = (\Phi_k, \Phi'_k)$ is defined in Section 3.2.2. Define the subsets

$$\mathcal{A}_m = \{(v, v') \in \mathcal{A} \times \mathcal{A} : d_{m+1} < \|v - v'\| \leq d_m\}, \quad m \geq -1.$$

Since $F \leq R$, for $(v, v') \in \mathcal{A}_m$ we have

$$\begin{aligned} f(v, v') &\leq R^2 d_{m+1} \mathbb{P}\big(Q_m(v, v')\big) + R\big(1 - \mathbb{P}\big(Q_m(v, v')\big)\big) \\ &= R^2 d_{m+1} P_m(v, v') + R\big(1 - P_m(v, v')\big), \end{aligned}$$

where $m \geq -1$, $Q_m(v, v') = \{\omega \in \Omega : \|\Phi_\ell(v, v') - \Phi'_\ell(v, v')\| \leq d_{m+1}\}$, and $P_m(v, v') = \mathbb{P}\big(Q_m(v, v')\big)$. It follows that if (u, u') is a coupling for the pair of measures $\lambda, \lambda' \in \mathcal{P}(\mathcal{A})$ which is independent of $\{\boldsymbol{\Phi}_k, k \geq 1\}$, then

$$\begin{aligned} &\mathbb{E}\, F(\Phi_\ell(u, u'), \Phi'_\ell(u, u')) \\ &\quad = \sum_{m=-1}^{\infty} \mathbb{E}\left\{\mathbb{I}_{\{(u,u') \in \mathcal{A}_m\}} f(u, u')\right\} \\ &\quad \leq \sum_{m=-1}^{\infty} \mathbb{E}\left\{\mathbb{I}_{\{(u,u') \in \mathcal{A}_m\}} \Big(R^2 d_{m+1} P_m(u, u') + R\big(1 - P_m(u, u')\big)\Big)\right\} \\ &\quad = R\, \mathbb{E}\left\{\mathbb{I}_{\{(u,u') \in \mathcal{A}_{-1}\}} \big(1 - P_{-1}(u, u')(1 - Rd)\big)\right\} \\ &\qquad + \sum_{m=0}^{\infty} R^2 d_m \mathbb{E}\left\{\mathbb{I}_{\{(u,u') \in \mathcal{A}_m\}} \big(\gamma P_m(u, u') + (R d_m)^{-1}(1 - P_m(u, u'))\big)\right\}. \end{aligned}$$

By (3.66) and (3.63) for $(u, u') \in \mathcal{A}_{-1}$ and $(u, u') \in \mathcal{A}_m, m \geq 0$, we have $P_{-1} \geq \theta_d$ and $P_m \geq 1 - K_1 d$. So choosing R and d so that $\gamma + R^{-1} K_1 < 1$ and $2Rd \leq 1$, we derive

$$\begin{aligned} \mathbb{E}\, F(\Phi_\ell(u, u'), \Phi'_\ell(u, u')) &\leq R\big(1 - \theta_d(1 - Rd)\big) \mathbb{P}\{(u, u') \in \mathcal{A}_{-1}\} \\ &\quad + \sum_{m=0}^{\infty} R^2 d_m (\gamma + R^{-1} K_1)\, \mathbb{P}\big\{(u, u') \in \mathcal{A}_m\big\} \\ &\leq \gamma' \mathbb{E}\, F(u, u'), \end{aligned} \tag{3.73}$$

where θ_d is constructed in Proposition 3.2.8, and $\gamma' < 1$ is a constant not depending on u and u'. Since inequality (3.73) is true for some coupling (u, u') of the pair of measures (λ, λ'), taking the infimum over all couplings, we arrive at (3.72). The proof of the theorem is complete.

3.2.4 Application to Navier–Stokes equations

Let us consider the homogeneous Navier–Stokes system perturbed by a random kick force (cf. (2.76)):

$$\dot{u} + \nu L u + B(u) = \sum_{k=1}^{\infty} \eta_k \delta(t - kT). \tag{3.74}$$

Here $\{\eta_k\}$ is a sequence of i.i.d. random variables in H of the form

$$\eta_k(x) = \sum_{j=1}^{\infty} b_j \xi_{jk} e_j(x), \tag{3.75}$$

where $\{e_j\}$ stands for an orthonormal basis in H consisting of the eigenfunctions of the Stokes operator L, ξ_{jk} are independent scalar random variables whose ranges are included in $[-1, 1]$ and laws are independent of k, and $b_j \geq 0$ are some constants such that

$$\mathfrak{B} := \sum_{j=1}^{\infty} b_j^2 < \infty. \tag{3.76}$$

As was explained in Section 2.3, the restriction of trajectories for (3.74) to $T\mathbb{Z}_+$ satisfies relation (3.50), where $u_k = u(kT)$, and $S : H \to H$ stands for the operator that takes u_0 to the solution at time T of the homogeneous Navier–Stokes system. We denote by $\boldsymbol{\Phi} = \{\varphi_k, k \geq 0\}$ the Markov RDS associated with (3.50) and by $\mathfrak{P}_k$ and $\mathfrak{P}_k^*$ the corresponding Markov semigroups; see Section 1.3.3. Recall that a function $p : [-1, 1] \to \mathbb{R}$ is said to have *bounded variation* if[3]

$$\mathrm{Var}(p) = \sup_{\{r_j\}} \sum_{j=1}^{m} |p(r_j) - p(r_{j-1})|, \tag{3.77}$$

where the supremum is taken over all partitions $r_0 = -1 < r_1 < \cdots < r_m = 1$ of the interval $[-1, 1]$.

Theorem 3.2.9 *Under the above conditions, assume that the laws of ξ_{jk} are absolutely continuous with respect to the Lebesgue measure, and the corresponding densities p_j are functions of bounded variation such that*

$$\int_{-\varepsilon}^{\varepsilon} p_j(r)\,dr > 0 \quad \textit{for any } \varepsilon > 0. \tag{3.78}$$

In this case, there is an integer $N = N(\mathfrak{B}, \nu) \geq 1$ such that if

$$b_j \neq 0 \quad \textit{for } j = 1, \ldots, N, \tag{3.79}$$

[3] Equivalently, the derivative p' of the function $p(x)$ in the sense of distributions is a signed measure, and $\mathrm{Var}(p) = 2\|p'\|_{\mathrm{var}}$ (see (1.13)).

then the RDS $\boldsymbol{\Phi}$ *has a unique stationary measure* $\mu \in \mathcal{P}(H)$. *Moreover, there are positive constants* C *and* α *such that*

$$\|\mathfrak{P}_k^*\lambda - \mu\|_L^* \le C\, e^{-\alpha k}\Big(1 + \int_H |u|_2 \lambda(du)\Big), \quad k \ge 0, \tag{3.80}$$

where $\lambda \in \mathcal{P}(H)$ *is an arbitrary measure with finite first moment.*

For example, the assumptions of Theorem 3.2.9 are met for any $\nu > 0$ if the random variables ξ_{jk} are independent, each ξ_{jk} is uniformly distributed on a non-degenerate segment $[r_j', r_j''] \subset [-1, 1]$ containing the point $r = 0$, all numbers b_j are non-zero, and $\mathfrak{B} < \infty$.

Proof We shall show that the Markov family generated by $\boldsymbol{\Phi}$ satisfies the hypotheses of Theorem 3.2.5. It follows from assertion (i) of Proposition 2.1.25 and inequality (2.24) that the operator $S : H \to H$ is Lipschitz continuous on bounded subsets. Moreover, inequality (2.51) implies that

$$|S(u)|_2 \le q|u|_2 \quad \text{for any } u \in H, \tag{3.81}$$

where $q = \exp(-\alpha_1 T/2)$. Thus, S satisfies Condition 3.2.1.

We now check Condition 3.2.2. It follows from (3.76) that $|\eta_k|_2 \le \sqrt{\mathfrak{B}}$ almost surely, and therefore the support of the law for η_k is contained in a ball $B_H(r)$. Combining this with inequality (3.81), we see that

$$|\varphi_1 u|_2 \le q\,|u|_2 + r \quad \text{for any } u \in H \text{ and almost every } \omega \in \Omega.$$

It follows that

$$\sup_{u \in \mathcal{A}_k(B_H(R))} |u|_2 \le q^k R + (1-q)^{-1} r.$$

We conclude that Condition 3.2.2 holds with

$$\rho = 2r(1-q)^{-1}, \quad k_0(R) = \big[\big(\ln q^{-1}\big)^{-1} \ln\big(\tfrac{2R}{r} + 1\big)\big] + 1, \tag{3.82}$$

where $[a]$ stands for the integer part of a.

Let us check Condition 3.2.3. To this end, we use the regularising property of S. Inequality (2.55) implies that

$$\|S(u_1) - S(u_2)\|_1 \le C_\nu |u_1 - u_2|_2 \quad \text{for any } u_1, u_2 \in \mathcal{A},$$

where $\mathcal{A}$ is defined by (3.54). Combining this with Poincaré's inequality, we see that

$$\begin{aligned} |(I - \mathsf{P}_N)(S(u_1) - S(u_2))|_2 &\le \frac{1}{\sqrt{\alpha_{N+1}}} \|S(u_1) - S(u_2)\|_1 \\ &\le \frac{C_\nu}{\sqrt{\alpha_{N+1}}}\, |u_1 - u_2|_2, \end{aligned}$$

where α_j denotes the eigenvalue of L corresponding to an eigenvector e_j, and P_N stands for the orthogonal projection to the space $E :=$

span$\{e_1, \dots, e_N\}$. Since $\alpha_j \to \infty$ as $j \to \infty$, we conclude that Condition 3.2.3 is satisfied for $\mathsf{P} = \mathsf{P}_N$ with a sufficiently large $N \geq 1$.

It remains to verify Condition 3.2.4. We first show that the support of $\mathcal{D}(\eta_1)$ contains the origin. To this end, it suffices to prove that

$$\mathbb{P}\{|\eta_1|_2 < \delta\} > 0 \quad \text{for any } \delta > 0.$$

But this relation readily follows from (3.76) and (3.78). Indeed, if m is sufficiently large, then (3.76) implies that $|(I - \mathsf{P}_m)\eta_1|_2 \leq \delta/2$ for almost all ω, and it follows from (3.78) that $\mathbb{P}\{|\mathsf{P}_m \eta_1|_2 < \delta/2\} > 0$ for any $m \geq 1$.

To check the second part of Condition 3.2.4, we note that the law of $\mathsf{P}_N \eta_1$ possesses a density (with respect to the Lebesgue measure on E) given by the relation

$$p(x) = \prod_{j=1}^{N} q_j(x_j), \quad q_j(x_j) = b_j^{-1} p_j(x_j/b_j), \quad x = (x_1, \dots, x_N) \in E. \tag{3.83}$$

To prove (3.56), we first assume that p_j are C^1-smooth functions. In this case, we have

$$\begin{aligned}\int_E |p(x+y) - p(x)|\, dx &\leq |y| \int_E \int_0^1 \big|(\nabla p)(x + \theta y)\big|\, d\theta dx \\ &= |y| \int_E \big|(\nabla p)(x)\big|\, dx \leq |y| \sum_{j=1}^{N} \int_{\mathbb{R}} \big|\partial_{x_j} q_j(x_j)\big|\, dx_j \\ &= |y| \sum_{j=1}^{N} \mathrm{Var}(q_j) = |y| \sum_{j=1}^{N} b_j^{-1}\, \mathrm{Var}(p_j).\end{aligned}$$

In the general case, inequality (3.56) can be established by a standard approximation argument.

We have thus shown that the conclusion of Theorem 3.2.5 holds for the RDS (3.50) corresponding to the kicked Navier–Stokes system (3.74). This proves the uniqueness of a stationary measure and inequality (3.57). To establish (3.80), we choose $\alpha \in (0, \alpha_0]$ such that $\alpha \leq \ln q^{-1}$. In this case, taking into account the second relation in (3.82), we see that

$$\exp\big(\alpha k_0(R)\big) \leq C_1(R+1).$$

Substituting this inequality into (3.57), we arrive at (3.80). The proof of Theorem 3.2.9 is complete. □

3.3 Navier–Stokes system perturbed by white noise

In this section, we consider the stochastic Navier–Stokes system in a bounded domain $Q \subset \mathbb{R}^2$ (or on a torus $\mathbb{R}^2/(a\mathbb{Z} \oplus b\mathbb{Z})$), assuming that the space-mean values of the force and solutions vanish). Namely, we study the problem

$$\dot{u} + \nu L u + B(u) = h + \eta(t), \tag{3.84}$$

$$u(0) = u_0, \tag{3.85}$$

where $\nu > 0$, $h \in H$ is a deterministic function, and η is a random process of the form (2.66), i.e.,

$$\eta(t, x) = \frac{\partial}{\partial t} \sum_{j=1}^{\infty} b_j \beta_j(t) e_j(x),$$

where $\mathfrak{B} = \sum_j b_j^2 < \infty$. We shall denote by $u(t; u_0)$ the unique solution of (3.84), (3.85) constructed in Theorem 2.4.6. The main result of this section claims that, under some non-degeneracy assumptions, the Markov process associated with (3.84) has a unique stationary measure μ, and all the trajectories of (3.84) converge to μ in law. Among a number of theorems of this kind, we have chosen the one with the simplest proof. Stronger results and those with different assumptions on the random force η are discussed in Section 3.5.

3.3.1 Main result and scheme of its proof

Let us denote by $(u_t, \mathbb{P}_u)$ the Markov family in H associated with the Navier–Stokes system (3.84) and denote by $\mathfrak{P}_t$ and $\mathfrak{P}_t^*$ the corresponding Markov semigroups. Recall that if u_0 is an H-valued random variable independent of $\{\beta_j\}$, then the law of the solution for (3.84), (3.85) coincides with $\mathfrak{P}_t^* \lambda$, where λ is the distribution of u_0.

Theorem 3.3.1 *Let us assume that the basis $\{e_j\}$ in the definition of the process η consists of the eigenfunctions for the Stokes operator L and the coefficients b_j satisfy* (2.67). *Then the Markov family $(u_t, \mathbb{P}_u)$ has a unique stationary measure $\mu \in \mathcal{P}(H)$, provided that $b_j \neq 0$ for all $j \geq 1$. Moreover, for any compact subset Λ of the space $\mathcal{P}(H)$ endowed with the dual-Lipschitz metric we have*

$$\sup_{\lambda \in \Lambda} \|\mathfrak{P}_t^* \lambda - \mu\|_L^* \to 0 \quad \textit{as } t \to \infty. \tag{3.86}$$

This theorem remains true under the weaker hypothesis that $b_j \neq 0$ for $j = 1, \dots, N$, with a sufficiently large N. Indeed, the only point where we use that all the coefficients b_j (rather than finitely many of them) are non-zero is Lemma 3.3.11. However, we need the conclusion of that lemma to be true only for the family of balls B centred at a given point $\hat{u} \in H$. In this situation, finitely many non-zero components of the noise η are sufficient to be able to prove that, for an appropriately chosen function $\hat{u} \in H$, the probability of transition from a given point to any neighbourhood of $\hat{u}$ is positive. For instance, if $h = 0$, then the function $\hat{u} = 0$ is a globally exponentially stable fixed point for the unperturbed dynamics (see Corollary 2.1.24), and it is not difficult to see that the probability of transition from any point to an arbitrary small neighbourhood of the origin is positive. We refer the reader to Exercise 3.3.12 for some more hints in the case of a non-zero function h and finitely many non-zero coefficients.

Convergence (3.86) and the a priori estimates for solutions of the Navier–Stokes system imply that for a large class of functionals $f : H \to \mathbb{R}$ the following assertion holds: the mean value of f calculated at a solution at time t converges, as $t \to \infty$, to the mean value of f with respect to the stationary distribution. Namely, denote by $\mathcal{W}$ the space of continuous functions $w(r) > 0$ that are defined and non-decreasing on the positive half-line. In what follows, the elements of $\mathcal{W}$ will be called *weight functions*. For any $w \in \mathcal{W}$, let $C(H, w)$ be the space of continuous functionals $f : H \to \mathbb{R}$ such that

$$|f|_w := \sup_{u \in H} \frac{|f(u)|}{w(|u|_2)} < \infty. \tag{3.87}$$

Corollary 3.3.2 *Under the hypotheses of Theorem 3.3.1, there is a* $\varkappa > 0$ *such that if* $f \in C(H, w)$ *with* $w(r) = \exp(\varkappa \nu r^2)$*, then*

$$\mathbb{E}\, f\big(u(t; v)\big) \to \int_H f(z)\mu(dz) \quad \textit{as } t \to +\infty \textit{ for any } v \in H. \tag{3.88}$$

In particular, this convergence holds for the energy functional $E(u) = \frac{1}{2}|u|_2^2$.

Proof Let us denote by $\varkappa_1$ and $\varkappa_2$ the constants $\varkappa > 0$ defined in Proposition 2.4.9 and Theorem 2.5.5, respectively, and let $\varkappa = (\varkappa_1/2) \wedge \varkappa_2$. Then for $f \in C(H, w)$ the quantities $\mathbb{E} f(u(t; v))$ and (f, μ) are well defined in view of (2.125) and (2.157).

Convergence (3.86) with $\lambda = \delta_v$ and assertion (ii) of Theorem 1.2.15 imply that

$$\mathfrak{P}_t g(v) = \mathbb{E}\, g\big(u(t; v)\big) \to (g, \mu) \quad \text{as } t \to +\infty, \tag{3.89}$$

where $g \in C_b(H)$ is an arbitrary function. For any $R > 0$, let $\chi_R \in C_b(H)$ be a function such that $0 \le \chi_R \le 1$, $\chi_R(u) = 1$ for $|u|_2 \le R$, and $\chi_R(u) = 0$ for $|u|_2 \ge R + 1$. Then for any $f \in C(H, w)$ we can write

$$\mathbb{E}\, f\big(u(t; v)\big) = (\mathfrak{P}_t f_{\le R})(u) + \mathbb{E}\, f_{\ge R}\big(u(t; v)\big), \tag{3.90}$$

where we set $f_{\le R} = \chi_R f$ and $f_{\ge R} = (1 - \chi_R) f$. Note that $f_{\le R} \in C_b(H)$. By (3.89), the first term in the right-hand side of (3.90) satisfies the inequality

$$|(\mathfrak{P}_t f_{\le R})(u) - (f_{\le R}, \mu)| \le \varepsilon_1(t, R, u),$$

where $\varepsilon_1(t, R, u) \to 0$ as $t \to \infty$ for any $R > 0$ and $u \in H$. Furthermore, it follows from Proposition 2.4.9 and Chebyshev's inequality that

$$\begin{aligned}|\mathbb{E}\, f_{\geq R}\big(u(t;v)\big)| &\leq \big|\mathbb{E}\,\big\{\big(1 - \chi_R(u(t;v))\big)\exp\big(\varkappa\nu|u(t;v)|_2^2\big)\big\}\big| \\ &\leq \big(\mathbb{P}\{|u(t;v)|_2 > R\}\big)^{1/2}\big(\mathbb{E}\,\exp(2\varkappa\nu|u(t;v)|_2^2)\big)^{1/2} \\ &\leq C(|u|_2)\varepsilon_2(R),\end{aligned}$$

where $t \geq 0$ is arbitrary, and $\varepsilon_2(R) \to 0$ as $R \to \infty$. We have thus shown that

$$\left|\mathbb{E}\, f(u(t;v)) - \int_H f(z)\mu(dz)\right| \leq |(f_{\geq R}, \mu)| + \varepsilon_1(t, R, u) + C(|u|_2)\varepsilon_2(R).$$

This inequality implies the required convergence, since $(f_{\geq R}, \mu) \to 0$ as $R \to \infty$ in view of Theorem 2.5.5 and Lebesgue's theorem on dominated convergence. □

Theorem 3.3.1 applies to the Navier–Stokes system (3.84) in a bounded domain or on the torus. In the latter case, we can consider the unique stationary distribution μ as a measure on the space of locally square-integrable functions on $\mathbb{R}^2$ that are 2π-periodic in both variables. Hence, one may ask a question about the invariance of μ with respect to translations in space. It turns out that this is indeed true, provided that the random force is invariant. Namely, let us consider the white-forced Navier–Stokes system (3.84), in which[4] $h \equiv 0$ and the coefficients of the random process

$$\eta(t, x) = \frac{\partial}{\partial t}\zeta(t, x), \quad \zeta(t, x) = \sum_{s \in \mathbb{Z}_0^2} b_s \beta_s(t) e_s(x),$$

satisfy the relations

$$b_s = b_{-s} \quad \text{for all } s \in \mathbb{Z}_0^2. \tag{3.91}$$

We claim that in this case the law of the process $\zeta(t, x)$ is invariant under translations with respect to x. Indeed, if $a \in \mathbb{R}^2$ and $T_a : \mathbb{T}^2 \to \mathbb{T}^2$ denotes the translation operator taking x to $x + a \pmod{2\pi}$, then

$$\begin{aligned}\zeta(t, T_a x) &= \sum_{s \in \mathbb{Z}_+^2} c_s b_s s^{\perp}\big(\beta_s(t)\cos\langle s, T_a x\rangle + \beta_{-s}(t)\sin\langle s, T_a x\rangle\big) \\ &= \sum_{s \in \mathbb{Z}_+^2} c_s b_s s^{\perp}\big(\tilde\beta_s(t)\cos\langle s, x\rangle + \tilde\beta_{-s}(t)\sin\langle s, x\rangle\big),\end{aligned}$$

[4] Note that the Dirac measure concentrated at a function $h(x)$ is invariant with respect to translations if and only if h is constant. Since the mean value of h over the torus must be zero, the assumption that $h \equiv 0$ is necessary for the law of the right-hand side of (3.84) to be invariant.

where c_s and $\mathbb{Z}^2_+$ are defined in Section 2.1.5,

$$\tilde\beta_s(t) = \beta_s(t)\cos\langle s, a\rangle + \beta_{-s}(t)\sin\langle s, a\rangle,$$
$$\tilde\beta_{-s}(t) = -\beta_s(t)\sin\langle s, a\rangle + \beta_{-s}(t)\cos\langle s, a\rangle.$$

The following property of the multidimensional Brownian motion is a simple consequence of the definition.

Exercise 3.3.3 Let $B(t)$ be a d-dimensional Brownian motion, that is, an $\mathbb{R}^d$-valued stochastic process whose components are independent standard Brownian motions. Then for any orthogonal $d \times d$ matrix U the process $UB(t)$ is also a Brownian motion. *Hint:* Show that $UB(t)$ is a Gaussian process with the same correlations as $B(t)$.

Applying this property with $d = 2$, we see that $\tilde\beta_s$ and $\tilde\beta_{-s}$ are independent Brownian motions. Thus, the processes $\zeta(t, T_a x)$ and $\zeta(t, x)$ have exactly the same form and, hence, the same laws. A similar argument proves that the laws of $\zeta(t, -x)$ and $\zeta(t, x)$ also coincide. The following assertion shows that the same properties hold for the stationary measure.

Corollary 3.3.4 *Let us assume that $h \equiv 0$ and the coefficients $b_s \neq 0$ are such that $\sum_s b_s^2 < \infty$ and* (3.91) *holds. Then for any $\nu > 0$ the unique stationary measure μ_ν of* (3.84) *is invariant under translations of $\mathbb{R}^2$ and the reflection $x \mapsto -x$.*

Proof We first prove the invariance of μ_ν under translations. Let us consider Eq. (3.84) in which η is replaced by $\eta_a(t, x) = \eta(t, T_a x)$. Then, by Theorem 3.3.1 applied to the torus, the new equation has a unique stationary distribution, which must coincide with μ_ν, because the laws of η and η_a are the same. On the other hand, the process $u^a_\nu(t, x) = u_\nu(t, T_a x)$ is a stationary solution of Eq. (3.84) with η replaced by η_a. By the uniqueness of a stationary measure, the law of $u^a_\nu(t)$ coincides with μ_ν. A similar argument shows that μ_ν is invariant under the reflection $x \mapsto -x$. □

Example 3.3.5 Let us assume that the hypotheses of Corollary 3.3.4 are satisfied and the unique stationary measure $\mu \in \mathcal{P}(H)$ for (3.84) is concentrated on V^{k+1} with some integer $k \geq 1$. (This is indeed the case if $h \in V^k$ and (2.137) holds; see Exercise 2.5.8.) Consider a functional of the form $f_x(u) = g(u(x))$, where $g : \mathbb{R}^2 \to \mathbb{R}$ is a continuous function and $x \in \mathbb{T}^2$ is a fixed point. The functional f_x is well defined on V^2, and therefore its mean value (f_x, μ) has a sense. Moreover, the invariance of μ with respect to space translations implies that the function $x \mapsto (f_x, \mu)$ is constant. On the other hand, by Fubini's theorem, we have

$$\langle\!\langle (f_x, \mu)\rangle\!\rangle = (2\pi)^{-2}\int_{\mathbb{T}^2}\int_{V^{k+1}} g(u(x))\,\mu(du)\,dx = \int_{V^{k+1}} \langle g(u(\cdot))\rangle\,\mu(du).$$

In particular, taking $g(v^1, v^2) = v^j$ with $j = 1$ or 2, we see that $\langle u^j(\cdot)\rangle = 0$ for any $u \in H$ and, hence,

$$\int_{V^{k+1}} u^j(x)\mu(du) = 0 \quad \text{for } x \in \mathbb{T}^2,\ j = 1, 2.$$

Thus, the mean velocity in the unique stationary regime is equal to zero. Let us also note that we did not use the uniqueness, and the same conclusion is true for any space-invariant stationary distribution concentrated[5] on V^{k+1} with $k \geq 1$.

Let us outline now the proof of Theorem 3.3.1. We wish to apply Theorem 3.1.4. To this end, we shall construct an extension $(\boldsymbol{u}_t, \mathbb{P}_{\boldsymbol{u}})$ for the family $(u_t, \mathbb{P}_u)$ and a decreasing sequence of closed subsets $B_m \subset H$ such that the recurrence and stability properties of Theorem 3.1.3 hold, together with inequality (3.26) in which $\boldsymbol{Y} \subset \boldsymbol{H}$ is an arbitrary compact subset.

Let us first describe a trivial extension for $(u_t, \mathbb{P}_u)$ that will be called a *pair of independent copies*. We denote by $(\Omega, \mathcal{F})$ the measurable space on which the family $(u_t, \mathbb{P}_u)$ is defined and by $(\boldsymbol{\Omega}, \boldsymbol{\mathcal{F}})$ the direct product of two copies $(\Omega, \mathcal{F})$:

$$\boldsymbol{\Omega} = \Omega \times \Omega, \quad \boldsymbol{\mathcal{F}} = \mathcal{F} \otimes \mathcal{F}.$$

The points of $\boldsymbol{\Omega}$ will be denoted by $\boldsymbol{\omega} = (\omega, \omega')$. We now define a process $\boldsymbol{u}_t(\boldsymbol{\omega})$ in the space $\boldsymbol{H} = H \times H$ by the formula

$$\boldsymbol{u}_t(\boldsymbol{\omega}) = (u_t(\omega), u_t(\omega'))$$

and, for any $\boldsymbol{u} = (u, u') \in \boldsymbol{H}$, we denote $\mathbb{P}_{\boldsymbol{u}} = \mathbb{P}_u \times \mathbb{P}_{u'}$. Clearly, the family $(\boldsymbol{u}_t, \mathbb{P}_{\boldsymbol{u}})$ is an extension for $(u_t, \mathbb{P}_u)$, and its transition function $\boldsymbol{P}_t$ satisfies the relation

$$\boldsymbol{P}_t(\boldsymbol{u}, \Gamma \times \Gamma') = P_t(u, \Gamma)\, P_t(u', \Gamma') \quad \text{for } \boldsymbol{u} = (u, u') \in \boldsymbol{H},\ \Gamma, \Gamma' \in \mathcal{B}(H), \tag{3.92}$$

where $P_t(u, \Gamma)$ stands for the transition function of $(u_t, \mathbb{P}_u)$.

Let us denote by $B_m \subset H$ the closed ball of radius $1/m$ centred at zero and define the stopping time

$$\boldsymbol{\tau} = \boldsymbol{\tau}(m) = \min\{t \geq 0 : \boldsymbol{u}_t \in B_m \times B_m\}.$$

We shall show in the next two subsections that the following propositions are true.

[5] The fact that μ is concentrated on V^{k+1} is not really important: one can give a meaning to the mean value (f_x, μ) for any space-invariant measure $\mu \in \mathcal{P}(H)$, and the same conclusion will be valid; see [VF88].

Proposition 3.3.6 *Under the hypotheses of Theorem 3.3.1, for any integer $m \geq 1$ there are two positive constants C and α such that*

$$\mathbb{E}_u \exp(\alpha\boldsymbol{\tau}) \leq C\big(1 + |u|_2^2 + |u'|_2^2\big) \quad \textit{for all } u, u' \in H. \tag{3.93}$$

In particular, the stopping time $\boldsymbol{\tau}$ is almost surely finite for any initial point $\boldsymbol{u} \in \boldsymbol{H}$.

Proposition 3.3.7 *Under the hypotheses of Theorem 3.3.1, there is a sequence $\delta_m > 0$ going to zero such that*

$$\sup_{t\geq 0} \| P_t(u, \cdot) - P_t(u', \cdot)\|_L^* \leq \delta_m \quad \textit{for any } u, u' \in B_m. \tag{3.94}$$

Once these two propositions are established, the required results will follow from Theorem 3.1.3.

Analysing the proof of the theorem, it is not difficult to see that convergence is uniform with respect to the random perturbations η that are bounded and "uniformly non-degenerate". Namely, we have the following result that can be established by repeating step by step the proof of Theorem 3.3.1.

Exercise 3.3.8 Let $\nu > 0$ be a constant, let $h \in H$ be a deterministic function, and let $\{\hat{b}_j\}$ be a sequence of positive numbers such that $\sum_j \hat{b}_j^2 < \infty$. Prove that for any $\mathfrak{B} > 0$ there is $\delta > 0$ such that if

$$\sum_{j=1}^{\infty} b_j^2 \leq \mathfrak{B}, \quad \sup_{j\geq 1} |b_j - \hat{b}_j| \leq \delta,$$

then the Markov family $(u_t, \mathbb{P}_u)$ associated with Eq. (3.84) has a unique stationary measure $\mu \in \mathcal{P}(H)$, and for any compact subset $\Lambda \subset \mathcal{P}(H)$ we have

$$\sup_{\lambda\in\Lambda} \|\mathfrak{P}_t^*\lambda - \mu\|_L^* \leq \alpha_\Lambda(t) \to 0 \quad \text{as } t \to \infty,$$

where α_Λ is a function depending only on $\mathfrak{B}$ and $\{\hat{b}_j\}$.

Finally, it is sometimes useful to know that laws of solutions in the space of trajectories also converge to a limiting measure. Namely, let $\tilde{u}(t)$, $t \geq 0$, be a stationary solution of (3.84) and let $\tilde{\mu} \in \mathcal{P}\big(C(\mathbb{R}_+; H)\big)$ be its law.

Exercise 3.3.9 Under the hypotheses of Theorem 3.3.1, show that for any solution $u(t)$ of (3.84) we have

$$\mathcal{D}\big(u(t + \cdot)\big) \to \tilde{\mu} \quad \text{as } t \to \infty,$$

where the convergence holds in the weak topology of the space $\mathcal{P}\big(C(\mathbb{R}_+; H)\big)$.

3.3.2 Recurrence: proof of Proposition 3.3.6

For any $R > 0$, define the stopping time

$$T_R = \min\{t \geq 0 : \|\boldsymbol{u}_t\| \leq R\},$$

where we set

$$\|\boldsymbol{u}\| = |\nabla \boldsymbol{u}|, \quad |\boldsymbol{u}|^2 = |u|_2^2 + |u'|_2^2.$$

The proof of Proposition 3.3.6 is based on the following two lemmas.

Lemma 3.3.10 *For any $\nu > 0$, $h \in H$, and $\mathfrak{B} \geq 0$ there is an $R > 0$ such that if the coefficients $b_j \geq 0$ satisfy the inequality $\sum_j b_j^2 \leq \mathfrak{B}$, then*

$$\mathbb{E}_{\boldsymbol{u}} \exp(\gamma T_R) \leq 1 + K\, |\boldsymbol{u}|^2 \quad \textit{for any } \boldsymbol{u} \in \boldsymbol{H}, \tag{3.95}$$

where K and γ are positive constants. In particular, $\mathbb{P}_{\boldsymbol{u}}\{T_R < \infty\} = 1$ for any $\boldsymbol{u} \in \boldsymbol{H}$.

Lemma 3.3.11 *Under the hypotheses of Theorem 3.3.1, for any $R > 0$ and any non-degenerate ball $B \subset H$ there is a $p > 0$ such that*

$$\boldsymbol{P}_1(\boldsymbol{u}, B \times B) \geq p \quad \textit{for any } u, u' \in B_V(R). \tag{3.96}$$

Let us emphasise that, in the first lemma, some of the constants b_j may be zero, whereas in the second lemma, the condition that all of them are positive is crucial for the result to be true with an *arbitrary* non-degenerate ball. Note, however, that if we wish (3.96) to be true for any ball *centred at a fixed point*, then it suffices to assume that a large, but finite number of coefficients b_j are non-zero; see below Exercise 3.3.12. Taking the above lemmas for granted, let us complete the proof of the proposition. It is divided into three steps.

Step 1. Let us introduce an increasing sequence of stopping times σ_n' by the rule

$$\sigma_0' = T_R, \quad \sigma_n' = \min\{t \geq \sigma_{n-1}' + 1 : \|\boldsymbol{u}_t\| \leq R\}, \quad n \geq 1.$$

We also set $\sigma_n = \sigma_n' + 1$. In order to estimate the stopping time $\boldsymbol{\tau} = \boldsymbol{\tau}(m)$ (which is the first instant when $|u_t| \vee |u_t'| \leq 1/m$), we start with estimating the first integer $n = n(m)$ for which $|\boldsymbol{u}_{\sigma_n}| \leq 1/m$. The probability of the event $\{n(m) > k\}$ is equal to

$$P_{\boldsymbol{u}}(k) = \mathbb{P}_{\boldsymbol{u}}\biggl(\bigcap_{n=1}^{k}\{|\boldsymbol{u}_{\sigma_n}| > 1/m\}\biggr).$$

By the construction of σ_n' and the strong Markov property, we have

$$P_{\boldsymbol{u}}(k) \le (1-p)^k \qquad \text{for any } \boldsymbol{u} \in \boldsymbol{H},\ k \ge 1, \tag{3.97}$$

where $p = p(m) > 0$ is the constant constructed in Lemma 3.3.11 for the ball $B = B_m$.

Step 2. We now show that

$$\mathbb{E}_{\boldsymbol{u}} \exp(\gamma\sigma_k) \le C_1^k\big(1+|\boldsymbol{u}|^2\big), \quad k \ge 1, \tag{3.98}$$

where $C_1 > 0$ depends only on R. Indeed, define $T_R' = \min\{t \ge 1 : \|\boldsymbol{u}_t\| \le R\}$. Then, by the Markov property and inequalities (3.95) and (2.120), for any $\boldsymbol{u}$ such that $|\boldsymbol{u}| \le R$ we have

$$\mathbb{E}_{\boldsymbol{u}} e^{\gamma T_R'} = \mathbb{E}_{\boldsymbol{u}}\mathbb{E}_{\boldsymbol{u}}\big(e^{\gamma T_R'} \mid \mathcal{F}_1\big) = e^{\gamma}\mathbb{E}_{\boldsymbol{u}}\left(\mathbb{E}_{\boldsymbol{u}_1} e^{\gamma T_R}\right) \le e^{\gamma}\mathbb{E}_{\boldsymbol{u}}\big(1 + K\,|\boldsymbol{u}_1|^2\big) \le C_2,$$

where $\mathcal{F}_t$ is the filtration corresponding to the process $\boldsymbol{u}_t$ and $C_2 > 0$ is a constant depending only on R. Applying now the strong Markov property, for $k \ge 1$ we obtain[6]

$$\mathbb{E}_{\boldsymbol{u}} e^{\gamma\sigma_k'} = \mathbb{E}_{\boldsymbol{u}}\mathbb{E}_{\boldsymbol{u}}\big(e^{\gamma\sigma_k'} \mid \mathcal{F}_1\big) = \mathbb{E}_{\boldsymbol{u}}\big(e^{\gamma\sigma_{k-1}'}\mathbb{E}_{\boldsymbol{u}(\sigma_{k-1}')} e^{\gamma T_R'}\big) \le C_2\,\mathbb{E}_{\boldsymbol{u}} e^{\gamma\sigma_{k-1}'},$$

where we used again the fact that $\|\boldsymbol{u}_{\sigma_n'}\| \le R$ for any $n \ge 0$. Iteration of this inequality results in

$$\mathbb{E}_{\boldsymbol{u}} e^{\gamma\sigma_k'} \le C_2^k \mathbb{E}_{\boldsymbol{u}} e^{\gamma\sigma_0'}.$$

Combining this with (3.95), we arrive at (3.98).

Step 3. We can now prove (3.93). For any initial point $\boldsymbol{u} \in \boldsymbol{H}$, any constant $M > 0$, and any integer $k \ge 1$, we have

$$\begin{aligned} \mathbb{P}_{\boldsymbol{u}}\{\boldsymbol{\tau} \ge M\} &= \mathbb{P}_{\boldsymbol{u}}\{\boldsymbol{\tau} \ge M, \sigma_k < M\} + \mathbb{P}_{\boldsymbol{u}}\{\boldsymbol{\tau} \ge M, \sigma_k \ge M\} \\ &\le \mathbb{P}_{\boldsymbol{u}}\{\boldsymbol{\tau} > \sigma_k\} + \mathbb{P}_{\boldsymbol{u}}\{\sigma_k \ge M\}. \end{aligned} \tag{3.99}$$

It follows from (3.97), (3.98), and Chebyshev's inequality that

$$\begin{aligned} &\mathbb{P}_{\boldsymbol{u}}\{\boldsymbol{\tau} > \sigma_k\} \le P_{\boldsymbol{u}}(k) \le (1-p)^k, \\ &\mathbb{P}_{\boldsymbol{u}}\{\sigma_k \ge M\} \le e^{-\gamma M}\mathbb{E}_{\boldsymbol{u}}\exp(\gamma\sigma_k) \le C_1^k e^{-\gamma M}\big(1+|\boldsymbol{u}|^2\big). \end{aligned}$$

Substitution of these estimates into (3.99) results in

$$\mathbb{P}_{\boldsymbol{u}}\{\boldsymbol{\tau} \ge M\} \le (1-p)^k + C_1^k e^{-\gamma M}\big(1+|\boldsymbol{u}|^2\big).$$

Choosing $k \sim \varepsilon M$, where $\varepsilon > 0$ is sufficiently small, we obtain

$$\mathbb{P}_{\boldsymbol{u}}\{\boldsymbol{\tau} \ge M\} \le C_3 e^{-\gamma' M}\big(1+|\boldsymbol{u}|^2\big),$$

[6] We shall sometimes write $u(\sigma_n)$ instead of u_{σ_n} to avoid double subscripts.

where C_3 and $\gamma' < \gamma$ are some positive constants. This inequality immediately implies (3.93), where $\alpha < \gamma'$. To complete the proof of Proposition 3.3.6, it remains to establish the two lemmas above.

Proof of Lemma 3.3.10 If all the coefficients b_j are zero, then we get a deterministic Navier–Stokes system, and the lemma is a straightforward consequence of (2.44) (with $t = 1$ and $m = 1$) and (2.52). Therefore, we shall assume that there are non-zero coefficients.

Recall that relation (2.131) was established in the proof of Proposition 2.4.10 as a result of application of Itô's formula to the functional $F(u) = |u|_2^2$. A similar argument applied to $F(t, \boldsymbol{u}) = e^{\alpha_1 \nu t}|\boldsymbol{u}|^2$ implies that

$$e^{\alpha_1 \nu t}|\boldsymbol{u}_t|^2 + 2\nu \int_0^t e^{\alpha_1 \nu s}|\nabla \boldsymbol{u}_s|^2 ds$$

$$= |\boldsymbol{u}_0|^2 + \int_0^t e^{\alpha_1 \nu s}\big(\alpha_1 \nu |\boldsymbol{u}_s|^2 + 2\mathfrak{B} + 2\langle u_s + u_s', h\rangle\big)\, ds + M_t,$$

where M_t is the corresponding stochastic integral. Let us fix a parameter $N > 0$, take $t = T_R \wedge N =: T$, and apply $\mathbb{E}_{\boldsymbol{u}}$. Using Doob's optional sampling theorem and the inequalities

$$|\boldsymbol{u}_s|^2 \le \alpha_1^{-1}|\nabla \boldsymbol{u}_s|^2, \quad |\langle u_s + u_s', h\rangle| \le \frac{\alpha_1 \nu}{4}|\boldsymbol{u}_t|^2 + \frac{2}{\alpha_1 \nu}|h|_2^2,$$

we derive

$$\mathbb{E}_{\boldsymbol{u}}\big(e^{\alpha_1 \nu T}|\boldsymbol{u}_T|^2\big) + \mathbb{E}_{\boldsymbol{u}} \int_0^T e^{\alpha_1 \nu s}\Big(\frac{\nu}{2}|\nabla \boldsymbol{u}_s|^2 - \big(2\mathfrak{B} + \tfrac{4}{\alpha_1 \nu}|h|_2^2\big)\Big) ds \le |\boldsymbol{u}|^2. \tag{3.100}$$

Now note that $\|\boldsymbol{u}_s\|^2 \ge R^2$ for $0 \le t \le T_R$. Hence, if

$$R^2 \ge 8\nu^{-1}\big(\mathfrak{B} + \tfrac{2}{\alpha_1 \nu}|h|_2^2\big),$$

then (3.100) implies that

$$\mathbb{E}_{\boldsymbol{u}}\left(\frac{\nu R^2}{4}\int_0^T e^{\alpha_1 \nu s} ds\right) \le |\boldsymbol{u}|^2.$$

Therefore,

$$\mathbb{E}_{\boldsymbol{u}}\big(\exp\{\alpha_1 \nu (T_R \wedge N)\} - 1\big) \le \frac{4\alpha_1}{R^2}|\boldsymbol{u}|^2.$$

Passing to the limit as $N \to \infty$ and using Fatou's lemma, we obtain the required inequality (3.95) with $\gamma = \alpha_1 \nu$ and $K = 4\alpha_1 R^{-2}$. □

Proof of Lemma 3.3.11 In view of (3.92), it suffices to show that

$$\inf_{v \in B_V(R)} P_1(v, B) > 0.$$

In other words, we must prove that

$$\inf_{v\in B_V(R)} \mathbb{P}\{|u(1;v)-\hat{u}|_2<\varepsilon\}>0 \quad \text{for any } \hat{u}\in H, \varepsilon>0. \tag{3.101}$$

Before giving a rigorous proof of (3.101), we describe the main idea. We wish to show that the probability of transition from any state $v\in B_V(R)$ to the ε-neighbourhood of a given point $\hat{u}\in H$ is separated from zero. Since our noise is non-degenerate in all Fourier modes, it is straightforward to find a path $\hat{\zeta}$ that belongs to the support of the noise and stirs the solution from v to an arbitrarily small neighbourhood of $\hat{u}$. By continuity of the resolving operator, we can find $\delta>0$ such that for any path ζ from the δ-tube $\mathcal{O}_\delta$ around $\hat{\zeta}$ the corresponding trajectory ends up in the ε-neighbourhood of $\hat{u}$. Since $\hat{\zeta}$ is in the support of the noise, the probability of the event $\{\zeta\in\mathcal{O}_\delta\}$ is positive, and therefore so is the probability of $\{|u(1;v)-\hat{u}|_2<\varepsilon\}$. The fact that these probabilities are separated from zero uniformly in $v\in B_V(R)$ follows from some compactness and continuity arguments.

The accurate realisation of this scheme is rather simple when the external force is H^1-regular. To simplify the presentation, we give here the proof in this particular case; the general situation is considered in Section 3.6.1.

Let us assume that $h\in V$ and $\mathfrak{B}_1<\infty$. In view of Proposition 2.4.7, solutions of the stochastic Navier–Stokes systems are locally Lipschitz continuous with respect to the trajectories of the noise. In particular, a solution at time 1 can be written as $u(1)=U_1(v,\zeta)$, where $U_1: H\times C(0,T;V)\to H$ is a locally Lipschitz continuous mapping. We claim that, for any $\varepsilon>0$ and $v\in B_V(R)$, we can find $\hat{\zeta}_v\in C(0,T;V)$ such that

$$\sup_{v\in B_V(R)} |U_1(v,\hat{\zeta}_v)-\hat{u}|_2\le\varepsilon/2, \tag{3.102}$$

and the mapping $v\mapsto\hat{\zeta}_v$ is continuous from H to $C(0,T;V)$. Indeed, let $\chi(t)$ be a smooth function equal to 1 for $t\le 0$ and to 0 for $t\ge 1$. For $\delta>0$ and $0\le t\le 1$, we set

$$u_\delta(t)=\chi(t)e^{(\nu t+\delta)L}v+(1-\chi(t))e^{\delta L}\hat{u},$$

$$\hat{\zeta}_v(t)=u_\delta(t)-u_\delta(0)+\int_0^t\big(\nu Lu_\delta+B(u_\delta)\big)ds-th.$$

Then $u_\delta(0)=e^{\delta L}v$ and $U_1(e^{\delta L}v,\zeta_\delta)=e^{\delta L}\hat{u}$. Choosing $\delta>0$ sufficiently small and using the uniform continuity of U_1, as well as the compactness of $B_V(R)$ in H, we conclude that (3.102) holds. The continuous dependence of $\hat{\zeta}_v$ on v is a straightforward consequence of the above relations.

We now choose $\delta>0$ so small that for any $\xi\in C(0,T;V)$ satisfying the inequality $\|\xi-\hat{\zeta}_v\|_{C(0,T;V)}<\delta$ we have (cf. (3.102))

$$\sup_{v\in B_V(R)} |U_1(v,\xi)-\hat{u}|_2<\varepsilon.$$

It follows that

$$\{|u(1;v)-\hat u|_2<\varepsilon\}\supset\{\|\zeta-\hat\zeta_v\|_{C(0,T;V)}<\delta\}=:\Gamma(v)\quad\text{for } v\in B_V(R),$$

whence we conclude that

$$\inf_{v\in B_V(R)}\mathbb{P}\{|u(1;v)-\hat u|_2<\varepsilon\}\ge\inf_{v\in B_V(R)}\mathbb{P}\big(\Gamma(v)\big). \tag{3.103}$$

Since the support of the law for the restriction of ζ to $[0,T]$ coincides with $C(0,T;V)$, we have $\mathbb{P}\big(\Gamma(v)\big)>0$ for any $v\in B_V(R)$. Furthermore, in view of the portmanteau theorem, the function $v\mapsto\mathbb{P}\big(\Gamma(v)\big)$ is lower semicontinuous. It remains to note that the image of $B_V(R)$ by the mapping $v\mapsto\hat\zeta_v$ is a compact subset of $C(0,T;V)$, and therefore the right-hand side of (3.103) must be positive. This completes the proof of the lemma in the case of an H^1-regular external force. □

Exercise 3.3.12 Prove that, for any $R,\nu>0$ and $h\in H$, there is a function $\hat u\in H$ and an integer $N\ge 1$ such that, if $b_j\ne 0$ for $j=1,\dots,N$, then

$$\boldsymbol{P}_T(\boldsymbol{u},B\times B)\ge p\quad\text{for any } u,u'\in B_V(R), \tag{3.104}$$

where $B\subset H$ is any ball centred at $\hat u$, and $T>0$ and $p\in(0,1]$ are some constants depending on B. *Hint:* If $h=0$, then the claim is true with $\hat u=0$ for any coefficients $b_j\ge 0$. Similarly, if $|h|_2\ll 1$, then the unperturbed problem has a unique stationary point $\hat u\in V$, which is globally exponentially stable as $t\to\infty$, and the claim is valid again for any $b_j\ge 0$. In the case of an arbitrary $h\in H$, one can use finitely many non-zero components of the noise to annihilate the low modes of h, while the contribution of high modes will be small.

3.3.3 Stability: proof of Proposition 3.3.7

We first outline the scheme of the proof, which is based on the Foiaş–Prodi estimate and Girsanov's theorem. Without loss of generality, we can assume that $u=0$, because the general case can be easily derived from this one with the help of the triangle inequality. Thus, we need to show that

$$\|\mathcal{D}(u_t)-\mathcal{D}(u_t')\|_L^*\le\delta_m\quad\text{for all } t\ge 0, \tag{3.105}$$

where u_t and u_t' stand for trajectories of the Navier–Stokes system (3.84) that are issued from $u=0$ and $u'\in B_m=\{u\in H:|u|_2\le m^{-1}\}$, respectively, and $\delta_m\to 0$ as $m\to\infty$. Let us define an auxiliary process v_t as a solution of the problem

$$\dot v+\nu Lv+B(v)+\lambda\mathsf{P}_N(v-u_t')=h+\eta(t), \tag{3.106}$$

$$v_0=0, \tag{3.107}$$

where $\mathsf{P}_N:H\to H$ denotes the orthogonal projection on the subspace $H_{(N)}$ spanned by e_j, $j=1,\dots N$, and $\lambda>0$ and $N\ge 1$ are large parameters that will be chosen later. Using the Foiaş–Prodi estimate (see Theorem 2.1.28), we shall show that, with high probability,

$$|v_t - u'_t|_2 \le Ce^{-t}|u'|_2 \quad \text{for } t \ge 0, \tag{3.108}$$

where $C > 0$ is a deterministic constant. This inequality will imply that

$$\|\mathcal{D}(v_t) - \mathcal{D}(u'_t)\|_L^* \le \delta_m^{(1)} \quad \text{for all } t \ge 0, \tag{3.109}$$

where $\delta_m^{(1)} \to 0$ as $m \to \infty$. Thus, it suffices to compare the laws of u_t and v_t.

We shall prove that if $|u'|_2 \le \frac{1}{m}$, then

$$\|\mathcal{D}(u_t) - \mathcal{D}(v_t)\|_L^* \le \delta_m^{(2)} \quad \text{for all } t \ge 0, \tag{3.110}$$

where $\delta_m^{(2)} \to 0$ as $m \to \infty$. Note that since inequality (3.110) concerns the law of solutions (and not the solutions themselves), we can choose the underlying probability space $(\Omega, \mathcal{F}, \mathbb{P})$ at our convenience. We assume that it coincides with the canonical space of the Wiener process $\{\zeta(t), t \ge 0\}$; see (2.71). Namely, Ω is the space of continuous functions $\omega : \mathbb{R}_+ \to H$ endowed with the metric of uniform convergence on bounded intervals, $\mathbb{P}$ is the law of ζ, and $\mathcal{F}$ is the completion of the Borel σ-algebra with respect to $\mathbb{P}$ (cf. Example 1.3.13 and section 3.4 in [Str93]). In this case, ζ is the *canonical process* given by $\zeta(t) = \omega_t$ for $t \ge 0$. We now define a transformation $\Phi : \Omega \to \Omega$ by the relation

$$\Phi(\omega)_t = \omega_t - \lambda \int_0^t \mathsf{P}_N(v_s - u'_s)\,ds, \tag{3.111}$$

where v_t and u'_t are solutions of the corresponding equations with right-hand side $\eta(t) = \eta^\omega(t) = \partial_t \omega_t$. We shall show that

$$\mathbb{P}\big\{u_t(\Phi(\omega)) = v_t(\omega) \text{ for all } t \ge 0\big\} = 1. \tag{3.112}$$

In view of Exercise 1.2.13 (ii) and inequality (1.16), to prove (3.110) it suffices to estimate the total variation distance between $\mathbb{P}$ and $\Phi_*(\mathbb{P})$. This will be done with the help of Girsanov's theorem. We now turn to the accurate proof, which is divided into several steps.

Step 1. Let us prove (3.109). To this end, we fix any integer $\varepsilon > 0$ and use Proposition 2.4.10 to find positive constants K and ρ_ε such that

$$\mathbb{P}\left\{\int_0^t \|u'_s\|_1^2 ds \le \rho_\varepsilon + Kt \text{ for all } t \ge 0\right\} \ge 1 - \varepsilon \quad \text{for any } u' \in B_1.$$

Let $\Omega_{K,\rho_\varepsilon}$ be the event in the left-hand side of this inequality. By Theorem 2.1.28 with $M = 1$ and Remark 2.1.29, there is a constant $\lambda > 0$ and an integer $N \ge 1$ such that if v is a solution of (3.106), (3.107), then inequality (3.108) with

$C = C_\varepsilon = e^{c\rho_\varepsilon}$ holds for $\omega \in \Omega_{K,\rho_\varepsilon}$. In this case, for any $t \geq 0$, $u' \in B_1$, and $f \in L_b(H)$ with $\|f\|_L \leq 1$ we have

$$\left|\mathbb{E}\big(f(v_t) - f(u_t')\big)\right| \leq 2\mathbb{P}(\Omega^c_{K,\rho_\varepsilon}) + \left|\mathbb{E}\,\mathbb{I}_{\Omega_{K,\rho_\varepsilon}}\big(f(v_t) - f(u_t')\big)\right| \leq 2\varepsilon + C_\varepsilon|u'|_2.$$

Since f was arbitrary, we conclude that (3.109) holds.

Step 2. We now prove (3.112). Let us set $\tilde{v}_t(\omega) = u_t(\Phi(\omega))$. Since

$$\frac{\partial}{\partial t}\Phi(\omega) = \frac{\partial \zeta}{\partial t} - \lambda \mathsf{P}_N(v_t - u_t'),$$

the process $\tilde{v}$ is a solution of problem (3.106), (3.107). By the uniqueness of the solution (see Theorem 2.4.6), we conclude that

$$\mathbb{P}\big\{\tilde{v}_t(\omega) = v_t(\omega) \text{ for all } t \geq 0\big\} = 1.$$

This relation coincides with (3.112).

Step 3. We have thus shown that, for any $t \geq 0$, the random variables u_t and v_t satisfy the hypotheses of Exercise 1.2.13(ii). Therefore, to establish (3.110), it suffices to prove that the transformation $\Phi : \Omega \to \Omega$ defined by (3.111) satisfies the inequality

$$\|\mathbb{P} - \Phi_*(\mathbb{P})\|_{\mathrm{var}} \leq \delta_m^{(2)} \to 0 \quad \text{as } m \to \infty. \tag{3.113}$$

The proof of this fact is based on Girsanov's theorem. The space $\Omega = C(\mathbb{R}_+, H)$ can be written as the direct sum of the closed subspaces $\Omega_N = C(\mathbb{R}_+, H_{(N)})$ and $\Omega_N^\perp = C(\mathbb{R}_+, H_{(N)}^\perp)$. We shall accordingly write $\Omega \ni \omega = (\omega^{(1)}, \omega^{(2)})$. In this notation, the transformation Φ takes the form

$$\Phi(\omega^{(1)}, \omega^{(2)}) = \big(\Psi(\omega^{(1)}, \omega^{(2)}), \omega^{(2)}\big),$$

where $\Psi : \Omega \to \Omega_N$ is defined by the relation

$$\Psi(\omega^{(1)}, \omega^{(2)})_t = \omega_t^{(1)} + \int_0^t a\big(s; \omega^{(1)}, \omega^{(2)}\big)\, ds, \quad a(t) = -\lambda \mathsf{P}_N\big(v_t - u_t'\big). \tag{3.114}$$

We now need the following lemma established at the end of this subsection.

Lemma 3.3.13 *Let $\mathbb{P}_N$ and $\mathbb{P}_N^\perp$ be the images of $\mathbb{P}$ under the natural projections*

$$\mathsf{P}_N : \Omega \to \Omega_N, \quad \mathsf{Q}_N : \Omega \to \Omega_N^\perp.$$

Then

$$\|\Phi_*(\mathbb{P}) - \mathbb{P}\|_{\mathrm{var}} \leq \int_{\Omega_N^\perp} \big\|\Psi_*\big(\mathbb{P}_N, \omega^{(2)}\big) - \mathbb{P}_N\big\|_{\mathrm{var}}\, \mathbb{P}_N^\perp(d\omega^{(2)}), \tag{3.115}$$

where $\Psi_*\big(\mathbb{P}_N, \omega^{(2)}\big)$ stands for the image of $\mathbb{P}_N$ under the mapping $\Psi(\cdot, \omega^{(2)})$.

Thus, to prove the required result, it suffices to derive an appropriate estimate for the right-hand side of (3.115). To do this, we may try to apply Theorem A.10.1 to the processes $y = \omega^{(1)}$ and $\tilde{y} = \Psi(\omega^{(1)}, \omega^{(2)})$. However, we cannot do it directly because the process $a(t)$ does not necessarily satisfy Novikov's condition (A.51). To overcome this difficulty, we fix any $\varepsilon > 0$ and find a constant $C = C_\varepsilon > 0$ such that (3.108) holds with probability no less than $1 - \varepsilon$ (cf. Step 1). We now introduce an auxiliary process defined as

$$\tilde{a}(t) = \chi\Bigl(\bigl(C|u'|_2\bigr)^{-1} \sup_{0\le s\le t} \bigl(e^s|v_s - u'_s|_2\bigr)\Bigr)\, a(t),$$

where χ is the indicator function of the interval $[0, 1]$. In other words, the process $\tilde{a}$ coincides with a if $v - u'$ satisfies (3.108) and is zero starting from the first instant $t \ge 0$ when (3.108) fails. This construction implies that

$$\mathbb{P}\{\tilde{a}(t) = a(t) \text{ for all } t \ge 0\} \ge 1 - \varepsilon. \tag{3.116}$$

We now define $\widetilde{\Phi}(\omega^{(1)}, \omega^{(2)}) = \bigl(\widetilde{\Psi}(\omega^{(1)}, \omega^{(2)}), \omega^{(2)}\bigr)$, where $\widetilde{\Psi} : \Omega \to \Omega_N$ is given by relation (3.114) with a replaced by $\tilde{a}$. Inequality (3.116) implies that Φ and $\widetilde{\Phi}$ coincide with probability $\ge 1 - \varepsilon$, whence it follows that

$$\|\Phi_*(\mathbb{P}) - \widetilde{\Phi}_*(\mathbb{P})\|_{\text{var}} \le \varepsilon. \tag{3.117}$$

We now estimate the total variation distance between $\widetilde{\Phi}_*(\mathbb{P})$ and $\mathbb{P}$. In view of (3.115), it suffices to compare $\widetilde{\Psi}_*(\mathbb{P}, \omega^{(2)})$ and $\mathbb{P}_N$. To this end, we apply Theorem A.10.1. Define the processes

$$y(t) = \omega_t^{(1)}, \quad \tilde{y}(t) = \omega_t^{(1)} + \int_0^t \tilde{a}\bigl(s; \omega^{(1)}, \omega^{(2)}\bigr)\, ds$$

and note that $\mathbb{P}_N = \mathcal{D}(y)$ and $\widetilde{\Psi}_*(\mathbb{P}, \omega^{(2)}) = \mathcal{D}(\tilde{y})$. Furthermore, the definition of $\tilde{a}$ implies that

$$|\tilde{a}(t)| \le C\, e^{-t} |u'|_2 \quad \text{for all } t \ge 0 \text{ and a.e. } \omega \in \Omega, \tag{3.118}$$

and therefore Novikov's condition (A.51) holds for $\tilde{a}$. Hence, recalling that $b_j > 0$ for all $j \ge 1$ and using (A.52), we derive

$$\|\widetilde{\Psi}_*(\mathbb{P}, \omega^{(2)}) - \mathbb{P}_N\|_{\text{var}} \le \frac{1}{2}\bigl(\sqrt{Q_N} - 1\bigr)^{1/2}, \tag{3.119}$$

where we set

$$Q_N = \mathbb{E} \exp\Bigl(6\, \bar{b}_N^{-2} \int_0^\infty |\tilde{a}(t)|^2 dt\Bigr), \quad \bar{b}_N = \min_{1\le j\le N} b_j.$$

It follows from (3.118) that $Q_N \le \exp(C_{N,\varepsilon}|u'|_2^2)$, where $C_{N,\varepsilon} > 0$ does not depend on u'. Combining this with (3.117) and (3.119), we conclude that

$$\|\Phi_*(\mathbb{P}) - \mathbb{P}\|_{\text{var}} \le \varepsilon + C'_{N,\varepsilon}|u'|_2 \quad \text{for any } u' \in B_1.$$

Since $\varepsilon > 0$ was arbitrary, we arrive at (3.113). To complete the proof of Proposition 3.3.7, it remains to establish Lemma 3.3.13.

Step 4: Proof of Lemma 3.3.13. Let $f : \Omega \to \mathbb{R}$ be a bounded continuous function with $\|f\|_\infty \le 1$. Then

$$
\begin{aligned}
\left|\mathbb{E}\left(f(\Phi(\omega)) - f(\omega)\right)\right| &= \left|\int_\Omega \left(f(\Psi(\omega), \omega^{(2)}) - f(\omega^{(1)}, \omega^{(2)})\right) \mathbb{P}(d\omega)\right| \\
&\le \int_{\Omega_N^\perp} \left|\int_{\Omega_N} \left(f(\Psi(\omega^{(1)}, \omega^{(2)}), \omega^{(2)}) - f(\omega^{(1)}, \omega^{(2)})\right) \mathbb{P}_N(d\omega^{(1)})\right| \mathbb{P}_N^\perp(d\omega^{(2)}) \\
&\le \int_{\Omega_N^\perp} \|\Psi_*(\mathbb{P}_N, \omega^{(2)}) - \mathbb{P}_N\|_{\text{var}}\, \mathbb{P}_N^\perp(d\omega^{(2)}).
\end{aligned}
$$

Since f was arbitrary, we arrive at (3.115). The proof of Proposition 3.3.7 is complete.

3.4 Navier–Stokes system with unbounded kicks

In Section 3.2.4, we proved that the homogeneous Navier–Stokes system perturbed by a sufficiently non-degenerate bounded kick force has a unique stationary distribution, and the laws of all other solutions converge to it exponentially fast. The proof of this result was based on the exponential decay of a Kantorovich functional (see Theorem 3.1.1). In this section we establish these results for the case of unbounded kicks. To do this, we use Theorem 3.1.7, whose proof combines the coupling with a stopping-time technique.

3.4.1 Formulation of the result

Let us consider the Navier–Stokes system (3.84) in a bounded domain $Q \subset \mathbb{R}^2$. We assume that $h \in H$ is a deterministic function and η is the random process given by (2.65), (3.75):

$$
\eta(t, x) = \sum_{k=1}^{\infty} \eta_k(x)\delta(t - kT), \quad \eta_k(x) = \sum_{j=1}^{\infty} b_j \xi_{jk} e_j(x),
$$

where the kicks $\{\eta_k\}$ form a sequence of i.i.d. random variables in H. As was explained in Section 2.3, the problem in question is equivalent to the discrete-time RDS (3.50), in which $S = S_T : H \to H$ stands for the operator that takes a function $u_0 \in H$ to the value at time T of the solution of the deterministic Navier–Stokes system (2.77) supplemented with the initial condition $u(0) = u_0$.

We now formulate the main conditions imposed on the kicks. We assume that $\{e_j\}$ is an orthonormal basis in H formed of the eigenfunctions of L, $b_j \ge 0$ are some constants satisfying (3.76), and ξ_{jk} are independent scalar random variables satisfying the inequality

$$\mathbb{E}\exp\bigl(\varkappa\,|\xi_{jk}|^2\bigr) \le C \tag{3.120}$$

with some positive constants C and $\varkappa$ not depending on j and k. It is straightforward to see that if this inequality holds, then $\mathbb{E}\,\exp(\varkappa_0|\eta_1|_2^2) < \infty$ for some $\varkappa_0 \in (0, \varkappa]$. In particular, the conclusions of Propositions 2.3.4 and 2.3.8 hold.

Recall that we denote by $\{\varphi_k, k \ge 0\}$ the Markov RDS associated with (3.50); that is, $\varphi_k u_0 = u_k$ for $k \ge 0$, where $u_0, u_1, u_2, \dots$ are defined by (3.50). Let $\mathfrak{P}_k$ and $\mathfrak{P}_k^*$ be the corresponding Markov operators. The following theorem is the main result of this section (cf. Theorem 3.2.9).

Theorem 3.4.1 *Under the above conditions, assume that the laws of the random variables ξ_{jk} are absolutely continuous with respect to the Lebesgue measure, and the corresponding densities p_j are functions of bounded total variation that are positive almost everywhere. In this case, there is an integer $N = N(\mathfrak{B}, \nu) \ge 1$ such that if*

$$b_j \ne 0 \quad \textit{for } j = 1, \dots, N, \tag{3.121}$$

then the RDS $\{\varphi_k, k \ge 0\}$ has a unique stationary measure $\mu \in \mathcal{P}(H)$. Moreover, there are positive constants C and α such that

$$\|\mathfrak{P}_k^*\lambda - \mu\|_L^* \le C\,e^{-\alpha k}\Bigl(1 + \int_H |u|_2\lambda(du)\Bigr), \quad k \ge 0, \tag{3.122}$$

where $\lambda \in \mathcal{P}(H)$ is an arbitrary measure with finite first moment.

The scheme of the proof of this result is described in the next subsection, and the details are given in Sections 3.6.2 and 3.6.3. Here we make a comment on the conditions imposed on the external force of (3.84) and establish a corollary of Theorem 3.4.1.

In the case of bounded kicks, we assumed that the function h is zero, and the support of the law for the kicks contains zero. This hypothesis ensured that, with positive probability, any two solutions approach each other arbitrarily closely. In the present section, the function $h \in H$ is arbitrary, however, we do assume that the noise is very efficient in the sense that the support of its law contains a *subspace* of high dimension (rather than a ball in a subspace of high dimension, as was the case in Theorem 3.2.9).

As in the case of white noise, the mixing and a priori estimates imply the convergence of the mean value of a functional $f : H \to \mathbb{R}$, calculated on solutions, to the mean value of f with respect to the stationary distribution. Namely, for any weight function $w \in \mathcal{W}$ and a constant $\gamma \in (0, 1]$, we denote by $C^\gamma(H, w)$ the space of Hölder-continuous functions $f : H \to \mathbb{R}$

with finite norm

$$|f|_{w,\gamma} = \sup_{u\in H} \frac{|f(u)|}{w(|u|_2)} + \sup_{0<|u-v|_2\le 1} \frac{|f(u)-f(v)|}{|u-v|_2^\gamma\big(w(|u|_2)+w(|v|_2)\big)}.$$

Note that in the case $\gamma = 1$ and $w \equiv 1$ we obtain the space $L_b(H)$.

Corollary 3.4.2 *Under the conditions of Theorem 3.4.1, for any $\gamma \in (0, 1]$, $\nu > 0$, and $\mathfrak{B} > 0$ there are positive constants C, β, and $\varkappa$ such that, for any function $f \in C^\gamma(H, w)$ with $w(r) = \exp(\varkappa r^2)$, we have*

$$\left|\mathbb{E}\, f(\varphi_k u) - \int_H f(v)\mu(dv)\right| \le C|f|_{w,\gamma}\, w(|u|_2)e^{-\beta k} \quad \textit{for } k \ge 0,\ u \in H. \tag{3.123}$$

Proof We shall apply the scheme used in the proof of Corollary 3.3.2. The only difference is that now the estimates are more quantitative.

We first note that both terms on the left-hand side of (3.123) are well defined. Indeed, the fact that $\mathbb{E}\, f(\varphi_k u)$ is finite was proved in Proposition 2.3.4. Furthermore, repeating the argument of the proof of Theorem 2.5.3 and using inequality (2.87) instead of (2.81), for $\varkappa \ll 1$ one can show that

$$\int_H \exp(2\varkappa|v|_2^2)\mu(dv) < \infty. \tag{3.124}$$

This implies, in particular, that (f, μ) is finite for any $f \in C^\gamma(H, w)$.

To prove (3.123), we fix any function $f \in C^\gamma(H, w)$ and assume, without loss of generality, that $|f|_{w,\gamma} = 1$. Using representation (3.90) with $t = k$, we estimate the left-hand side Δ_k of (3.123) as follows:

$$\Delta_k \le |\mathfrak{P}_k f_{\le R}(u) - (f_{\le R}, \mu)| + |\mathbb{E}\, f_{\ge R}(\varphi_k u)| + |(f_{\ge R}, \mu)|. \tag{3.125}$$

Let us estimate each term of the right-hand side. Inequality (3.122) and Lemma 1.2.6 imply that

$$|\mathfrak{P}_k g(u) - (g, \mu)| \le C_1|g|_\gamma e^{-\gamma\alpha k}\big(1 + |u|_2\big), \quad k \ge 0, \tag{3.126}$$

where $g \in C_b^\gamma(H)$ is an arbitrary function. Taking $g = f_{\le R}$, we get

$$|\mathfrak{P}_k f_{\le R}(u) - (f_{\le R}, \mu)| \le C_2 e^{-\gamma\alpha k} w(R+1)\big(1 + |u|_2\big).$$

Furthermore, it follows from inequality (2.87) with $\varkappa$ replaced by $2\varkappa$ that

$$\begin{aligned}
|\mathbb{E}\, f_{\ge R}(\varphi_k u)| &= \big|\mathbb{E}\big(\mathbb{I}_{\{|\varphi_k u|_2 \ge R\}} f_{\ge R}(\varphi_k u)\big)\big| \\
&\le \big(\mathbb{P}\{|\varphi_k u|_2 \ge R\}\big)^{1/2}\big(\mathbb{E}\, f_{\ge R}^2(\varphi_k u)\big)^{1/2} \\
&\le \big(\mathbb{P}\{w(\varphi_k u) \ge w(R)\}\big)^{1/2}\big(\mathbb{E}\exp(2\varkappa|\varphi_k u|_2^2)\big)^{1/2} \\
&\le \frac{C_3 w(|u|_2)}{\sqrt{w(R)}}.
\end{aligned}$$

Finally, inequality (3.124) implies that

$$|(f_{\geq R},\mu)| \leq \frac{C_4}{w(R)}.$$

Substituting the above estimates into (3.125), we obtain

$$\Delta_k \leq C_5\Big(e^{-\gamma\alpha k}w(R+1)\big(1+|u|_2\big)+\frac{1+w(|u|_2)}{\sqrt{w(R)}}\Big).$$

Choosing R such that $R^2=\frac{\gamma\alpha k}{2\varkappa}$, we arrive at the required inequality (3.123) with $\beta=\frac{\gamma\alpha}{4}$. □

3.4.2 Proof of Theorem 3.4.1

Let $(u_k,\mathbb{P}_u)$ be the Markov family that is associated with the RDS $\{\varphi_k, k\geq 0\}$ corresponding to (3.50). It follows from Proposition 2.3.3 that

$$\mathbb{E}\,|u_k|_2 \leq C_1\big(1+|u|_2\big) \quad \text{for all } u\in H, k\geq 0$$

where $C_1>0$ does not depend on u and k. By Theorem 3.1.7, the required result will be established if we construct a mixing extension $(\boldsymbol{u}_k,\mathbb{P}_{\boldsymbol{u}})$ for the family $(u_k,\mathbb{P}_u)$ and show that relations (3.31)–(3.34) hold with a function of the form

$$g(r)=C(1+r). \tag{3.127}$$

It turns out that one can construct a suitable extension in the same way as for the case of bounded kicks (see Section 3.2.2). Namely, we fix an integer $N\geq 1$ and denote by $\mathsf{P}_N: H\to H$ and $\mathsf{Q}_N: H\to H$ the orthogonal projections onto the subspaces $H_{(N)}$ and $H_{(N)}^{\perp}$, respectively. The proof of Lemma 3.2.6 does not use the boundedness of the kick η_k and its assertions remain valid in the context of this section with $E=H_{(N)}$, $\mathsf{P}=\mathsf{P}_N$, and $\mathsf{Q}=\mathsf{Q}_N$. Let us denote by $\zeta=\zeta(v,v',\omega)$ and $\zeta'=\zeta'(v,v',\omega)$ the random variables constructed in Lemma 3.2.6 and define *coupling operators* $(\mathcal{R},\mathcal{R}')$ by relations (3.60). Repeating the construction of Section 3.2.2, we obtain an RDS $\boldsymbol{\Phi}_k=(\Phi_k,\Phi_k')$ in the product space $H\times H$ such that the corresponding Markov family $(\boldsymbol{u}_k,\mathbb{P}_{\boldsymbol{u}})$ is an extension for $(u_k,\mathbb{P}_u)$. Let us set

$$\sigma_0=\min\{k\geq 0: |u_k-u_k'|_2\geq c\,e^{-k}\}, \tag{3.128}$$

where $c>0$ is a constant that will be chosen later. In Section 3.6, we shall show that the following two propositions are true.

Proposition 3.4.3 *Suppose that condition* (3.121) *is satisfied with a sufficiently large $N\geq 1$. Then for any $d>0$ the recurrence property of Definition 3.1.6 holds for the extension $(\boldsymbol{u}_k,\mathbb{P}_{\boldsymbol{u}})$ with $\boldsymbol{B}=\boldsymbol{B}_d=B_H(d)\times B_H(d)$ and a function g of the form* (3.127).

Proposition 3.4.4 *For sufficiently large integers $N \geq 1$ there is a stopping time σ and a constant $d > 0$ such that if condition* (3.121) *is satisfied, then*

$$\mathbb{P}_{\boldsymbol{u}}\{\sigma \leq \sigma_0\} = 1 \quad \text{for any } \boldsymbol{u} \in \boldsymbol{B}_d, \tag{3.129}$$

and relations (3.32)–(3.34) *hold for* $(\boldsymbol{u}_k, \mathbb{P}_{\boldsymbol{u}})$ *with* $\boldsymbol{B} = \boldsymbol{B}_d$.

Combining these results with Theorem 3.1.7 and Remark 3.1.8, we easily derive the uniqueness of the stationary distribution and inequality (3.122). Indeed, in view of the propositions, we can choose the parameters $N \geq 1$ and $d > 0$ so that, if condition (3.121) is fulfilled, then the Markov family $(\boldsymbol{u}_k, \mathbb{P}_{\boldsymbol{u}})$ possesses the recurrence and exponential squeezing properties of Definition 3.1.6 with $\boldsymbol{B} = \boldsymbol{B}_d$, a stopping time σ satisfying (3.129), and a function g of the form (3.127). Therefore, by Remark 3.1.8, the conclusions of Theorem 3.1.7 are true for the Markov family $(u_k, \mathbb{P}_u)$. This completes the proof of the theorem.

3.5 Further results and generalisations

3.5.1 The Flandoli–Maslowski theorem

The very first result on the uniqueness of a stationary distribution for the 2D Navier–Stokes system perturbed by a random force was obtained by Flandoli and Maslowski [FM95]. In this subsection we formulate their result and give some references.

Let us consider again the Navier–Stokes system (3.84), in which $h \in H$ is a deterministic function and η is a random process of the form (2.66). In contrast to the situation studied in Section 3.3, we now assume that *all* coefficients b_j are non-zero and satisfy the two-sided inequality

$$c\alpha_j^{-\frac{1}{2}} \leq b_j \leq C\alpha_j^{-\frac{3}{8}-\varepsilon} \quad \text{for any } j \geq 1, \tag{3.130}$$

where C, c, and ε are positive constants and $\{\alpha_j\}$ is the non-decreasing sequence of eigenvalues for the Stokes operator L. Note that, by the well-known spectral asymptotics for L (e.g., see chapter 4 in [CF88]), we have $\alpha_j \sim j$ as $j \to \infty$, whence it follows that condition (2.67) is not satisfied, so that the random force $\eta(t, x)$ is very rough as a function of the space variable x. Nevertheless, one can show that the initial-boundary value problem for (3.84) is well posed. We consider the corresponding Markov process in H. The main result of [FM95] is the following theorem.

Theorem 3.5.1 *Assume that the coefficients b_j satisfy condition* (3.130). *Then the following assertions hold.*

Uniqueness: *The Markov process possesses a unique stationary distribution $\mu \in \mathcal{P}(H)$.*

Mixing: *For any $\lambda \in \mathcal{P}(H)$, we have*

$$\|\mathfrak{P}_t^*\lambda - \mu\|_{\mathrm{var}} \to 0 \quad \text{as } t \to \infty.$$

Ergodicity: *Let $f : H \to \mathbb{R}$ be a Borel-measurable function integrable with respect to μ and let $u(t, x)$ be any solution with a deterministic initial datum $u(0) \in H$. Then*

$$\lim_{T\to\infty} \frac{1}{T} \int_0^T f(u(t))\,dt = \int_H f\,d\mu \quad \text{almost surely.}$$

Without going into details, let us outline the main ideas of the proof.[7] It is based on Doob's theorem which provides sufficient conditions for uniqueness and mixing; see [Doo48; DZ96; Ste94; Sei97]. According to that result, if the transition function of the Markov RDS $\boldsymbol{\Phi}$ is *regular* in the sense that the probability measures $P_t(u, \cdot)$, $u \in H$, are absolutely continuous with respect to one another for some $t > 0$, then the conclusions of Theorem 3.5.1 are valid. The regularity of the transition function is, in turn, a consequence of the following two properties.

Irreducibility: There is a $t_1 > 0$ such that

$$P_{t_1}\big(u, B_H(v, r)\big) > 0 \quad \text{for all } u, v \in H, r > 0.$$

Strong Feller property: There is a $t_2 > 0$ such that the function $u \mapsto P_{t_2}(u, \Gamma)$ is continuous on H for any $\Gamma \in \mathcal{B}(H)$.

It is not difficult to show that if these properties hold, then the transition probabilities are mutually absolutely continuous for $t = t_1 + t_2$. Thus, it suffices to establish the irreducibility and strong Feller property.

The proof of irreducibility can be carried out by repeating the argument used to establish Lemma 3.3.11. The hard point of the Flandoli–Maslowski analysis is the verification of the strong Feller property. The latter is based on an application of the Bismut–Elworthy formula [Bis81; Elw92] to Galerkin's approximations of the Navier–Stokes system and derivation of a uniform estimate for the derivative of the Markov semigroup.

The Flandoli–Maslowski result was developed and extended by many authors. In particular, Ferrario [Fer99] proved that the non-degeneracy condition (3.130) can be relaxed to allow random forces that are regular with respect to the space variables, Bricmont, Kupiainen, and Lefevere [BKL01] established the exponential convergence to the unique stationary distribution in the total variation norm for the Navier–Stokes system perturbed by a kick force, and Goldys and Maslowski [GM05] obtained a similar result in the case

[7] The scheme presented below is not entirely accurate. We refer the reader to the original work [FM95] for the precise argument, which is more complicated.

of white-noise perturbations. We refer the reader to those papers for further references on this subject.

3.5.2 Exponential mixing for the Navier–Stokes system with white noise

Let us recall that Theorem 3.3.1 establishes the uniqueness of a stationary measure and the mixing property for the Navier–Stokes system (3.84) with a spatially regular white noise η. On the other hand, Theorems 3.2.9 and 3.4.1 show that, in the case of a non-degenerate kick force, the corresponding RDS is exponentially mixing. It turns out that a similar result is true for white noise perturbations. Namely, the following theorem was proved in [BKL02; Mat02b; KS02a; Oda08] at various levels of generality; see *Notes and comments* at the end of this chapter.

Theorem 3.5.2 *In the setting of Section 3.3, assume that* $\mathfrak{B} = \sum_j b_j^2 < \infty$. *Then there is an integer* $N \geq 1$ *depending on* ν, $\mathfrak{B}$, *and* $|h|_2$ *such that the following properties hold, provided that*

$$b_j \neq 0 \quad \textit{for } j = 1, \ldots, N. \tag{3.131}$$

Uniqueness: *Equation* (2.98) *has a unique stationary measure* $\mu \in \mathcal{P}(H)$.

Exponential mixing: *There are positive constants* C *and* α *such that*

$$\|\mathfrak{P}_t^*\lambda - \mu\|_L^* \leq C\, e^{-\alpha t}\Big(1 + \int_H |u|_2^2 \lambda(du)\Big), \quad t \geq 0, \tag{3.132}$$

where $\lambda \in \mathcal{P}(H)$ *is an arbitrary measure for which the right-hand side is finite.*

Let us also mention that this theorem remains true for some classes of multiplicative noise (see [Oda08]). The proof is based on the ideas used to establish the exponential mixing in the case of a kick force, and therefore we confine ourselves to a brief description of the coupling construction. The interested reader will be able to fill in the details repeating the scheme applied for the kick force model (cf. the paper [Shi08] in which the details are carried out for the complex Ginzburg–Landau equation (4)).

Outline of the proof of Theorem 3.5.2 Let $(u_t, \mathbb{P}_v)$ be the Markov process associated with (2.98) and let $\{u(t; v), t \geq 0\}$ be a solution of (2.98) issued from $v \in H$ at $t = 0$. We wish to construct a mixing extension for $(u_t, \mathbb{P}_v)$; cf. Definition 3.1.6. Let us fix a time $T > 0$, an integer $N \geq 1$, and an arbitrary function $\chi \in C^\infty(\mathbb{R})$ such that $\chi(t) = 0$ for $t \leq 0$ and $\chi(t) = 1$ for $t \geq T$. Recall that P_N and Q_N stand for the orthogonal projections in H onto the subspaces H_N and $H_N^\perp$, respectively, where H_N is the vector span of the first N eigenfunctions of the Stokes operator. Given two points $v, v' \in H$, we denote by $\lambda(v), \lambda'(v,v') \in \mathcal{P}(C(0,T;H\times H))$ the laws of the random variables

$$\big((u(t;v),\zeta(t)), t\in[0,T]\big), \quad \big((u(t;v')-\chi(t)\mathsf{P}_N(v'-v),\zeta(t)), t\in[0,T]\big).$$

Let us define a mapping $f: C(0,T;H\times H)\to C(0,T;H_N\times H_N^\perp)$ by the formula

$$f\big((u(t),\xi(t), t\in[0,T])\big) = \big((\mathsf{P}_N u(t), \mathsf{Q}_N\xi(t), t\in[0,T]\big)).$$

In view of Exercise 1.2.30, there is a coupling $(\boldsymbol{\Upsilon}(v,v'), \boldsymbol{\Upsilon}'(v,v'))$ for the pair of measures $(\lambda(v), \lambda'(v,v'))$ such that $\big(f(\boldsymbol{\Upsilon}(v,v')), f(\boldsymbol{\Upsilon}'(v,v'))\big)$ is a maximal coupling for $\big(f_*(\lambda(v)), f_*(\lambda'(v,v'))\big)$. Thus, $\boldsymbol{\Upsilon}$ and $\boldsymbol{\Upsilon}'$ are $H\times H$-valued stochastic processes on the interval $[0,T]$, and we denote by $\Upsilon_t(v,v')$ and $\Upsilon_t'(v,v')$ the restriction of their first components to time t. We now define coupling operators on the interval $[0,T]$ by the relations (cf. (3.60))

$$\mathcal{R}_t(v,v') = \Upsilon_t(v,v'), \quad \mathcal{R}_t'(v,v') = \Upsilon_t'(v,v') + \chi(t)\mathsf{P}_N(v'-v), \quad 0\le t\le T.$$

Iteration of $\mathcal{R}$ and $\mathcal{R}'$ enables one to construct a Markov RDS $\boldsymbol{\Phi}_t = (\Phi_t, \Phi_t')$ in the product space $H\times H$; cf. Section 3.2.2. Namely, we define $(\Omega,\mathcal{F},\mathbb{P})$ as the direct product of countably many copies of the probability space for the coupling operators and denote by $\omega = (\omega^1,\omega^2,\dots)$ points of Ω. Given $v, v'\in H$, we now set

$$\left.\begin{aligned}\Phi_t(\omega)(v,v') &= \mathcal{R}_t(v,v',\omega^1)\\ \Phi_t'(\omega)(v,v') &= \mathcal{R}_t'(v,v',\omega^1)\end{aligned}\right\} \quad \text{for } 0\le t\le T,$$

$$\left.\begin{aligned}\Phi_t(\omega)(v,v') &= \mathcal{R}_t(\boldsymbol{\Phi}_{(k-1)T}(\omega)(v,v'),\omega^k)\\ \Phi_t'(\omega)(v,v') &= \mathcal{R}_t'(\boldsymbol{\Phi}_{(k-1)T}(\omega)(v,v'),\omega^k)\end{aligned}\right\} \quad \text{for } (k-1)T\le t\le kT,$$

where $k\ge 2$. It is straightforward to see that the Markov process $(\boldsymbol{u}_t, \mathbb{P}_{\boldsymbol{v}})$ associated with the RDS $\boldsymbol{\Phi}_t$ is an extension for $(u_t,\mathbb{P}_v)$. It turns out that the recurrence and exponential squeezing properties of Definition 3.1.6 hold for it, so that we have a mixing extension for $(u_t,\mathbb{P}_v)$. Application of Theorem 3.1.7 enables one to conclude the proof. □

As in the case of the Navier–Stokes system perturbed by random kicks, Theorem 3.5.2 implies the exponential convergence of the probabilistic average for an observable to its mean value with respect to the stationary distribution. Namely, we have the following result that can be established by repeating the arguments of the proof of Corollary 3.4.2.

Corollary 3.5.3 *Under the hypotheses of Theorem 3.5.2, for any* $\gamma\in(0,1]$ *and sufficiently small* $\varkappa>0$ *there are positive constants* C *and* β *such that, for any* $f\in C^\gamma(H,w)$ *with* $w(r)=e^{\varkappa r^2}$, *we have*

$$\big|\mathbb{E}_u f(u_t) - (f,\mu)\big| \le Ce^{-\beta t}|f|_{\gamma,w}w(|u|_2), \quad t\ge 0, \quad u\in H. \tag{3.133}$$

All the results of this chapter that concern the Navier–Stokes system were established for a general bounded domain with smooth boundary. In the case of periodic boundary conditions and white-noise perturbations, one can go further and prove the uniqueness and exponential mixing for *any* $\nu > 0$ *with a fixed finite-dimensional external force*. More precisely, let us consider the Navier–Stokes equation (2.98) on the standard 2D torus. The following result was established by Hairer and Mattingly [HM06; HM08; HM11]; see also the papers [AS05; AS06] by Agrachev and Sarychev dealing with the closely related question of controllability of the Navier–Stokes system by a finite-dimensional external force and the review [Kup10] by Kupiainen.

Theorem 3.5.4 *In the setting of Section 3.3, assume that* $x \in \mathbb{T}^2$ *and* $h = 0$. *Then there is an integer* $N \geq 1$ *not depending on* $\nu > 0$ *such that the conclusions of Theorem 3.5.2 are true, provided that condition* (3.131) *holds.*

The proof of this theorem is based on the Kantorovich functional technique (see Sections 3.1 and 3.2) and some gradient estimates for the Markov semigroup $\mathfrak{P}_t$. The latter, in turn, uses a version of Malliavin calculus developed by Mattingly and Pardoux [MP06]. We refer the reader to the above-cited original works for more details of the proof. Let us also mention the paper [AKSS07], which proves that, for a large class of stochastic nonlinear systems with finite-dimensional noises, the corresponding solutions are such that projections of their distributions to any finite-dimensional subspace are absolutely continuous with respect to Lebesgue measure.

3.5.3 Convergence for functionals on higher Sobolev spaces

Theorems 3.2.9, 3.3.1, 3.4.1, and 3.5.2 show that if a random perturbation acting on the Navier–Stokes system is sufficiently non-degenerate, then for a large class of continuous functionals on H, the ensemble average converges, as $t \to \infty$, to the space average with respect to the stationary distribution. At the same time, many important characteristics of fluid flows, such as the enstrophy or correlation tensors, are continuous functions on higher Sobolev spaces, but not on H. It is therefore desirable to have similar convergence results for that type of functional. In this subsection, we prove that convergence of functionals defined on higher Sobolev spaces can easily be obtained from what we have already established, provided that the external force is regular in the space variables. To simplify the presentation, we confine ourselves to the case of a torus with a spatially regular white force.

We thus consider Eq. (2.98) on $\mathbb{T}^2$ and assume that, for some integer $k \geq 1$, we have

$$h \in V^k, \quad \mathfrak{B}_k = \sum_{j=1}^{\infty} \alpha_j^k b_j^2 < \infty. \tag{3.134}$$

For any integer $m \geq 0$ and $\varkappa > 0$, we define weight functions $w^{(m)}(r) = (1 + r)^m$ and $w_\rho(r) = e^{\rho r^2}$.

Theorem 3.5.5 *In addition to the hypotheses of Theorem 3.5.2, let us assume that* (3.134) *holds for some integer $k \geq 1$. Then for any $\gamma \in (0, 1]$ and $\rho > 0$, any integer $m \geq 0$, and any functional $f \in C^\gamma(V^k, w^{(m)})$ we have*

$$\left|\mathbb{E}_u f(u_t) - (f, \mu)\right| \leq C e^{-\beta t} |f|_{\gamma, w^{(m)}} w_\rho(|u|_2), \quad t \geq 0, \quad u \in H, \tag{3.135}$$

where C and β are positive constants not depending on f.

For instance, if (3.134) holds with $k = 1$, then the result applies to the enstrophy functional $\Omega(u) = \frac{1}{2}|\operatorname{curl} u|_2^2$. Furthermore, if (3.134) holds with $k = 2$, then convergence (3.135) is valid for the *velocity correlation tensors*

$$f^n(u) = u^{i_1}(x_1) \cdots u^{i_n}(x_n), \tag{3.136}$$

where $i_l \in \{1, 2\}$ and $x_1, \ldots, x_n \in \mathbb{T}^2$. Or one can take for f a structure function $f(u) = |u(x + \ell) - u(x)|^p$ with $x \in \mathbb{T}^2$, $\ell \in \mathbb{R}^2$, and $p \in \mathbb{N}$.

To prove Theorem 3.5.5, we shall need the following lemma, which is of independent interest and will also be used in Chapter 4.

Lemma 3.5.6 *Suppose that* (3.134) *holds for some integer $k \geq 0$. Then the following assertions hold.*

(i) *For any $\gamma \in (0, 1]$ and sufficiently small $\varkappa > 0$ there exist positive constants $\alpha \leq \gamma$ and ρ such that $\mathfrak{P}_t f \in C^\alpha(H, w_\rho)$ for any $f \in C^\gamma(H, w_\varkappa)$, and the norm of the linear operator*

$$\mathfrak{P}_t : C^\gamma(H, w_\varkappa) \to C^\alpha(H, w_\rho) \tag{3.137}$$

is bounded uniformly on any compact interval $[0, T]$. *Moreover, the constants α and ρ can be chosen in such a way that $\rho \to 0$ as $\varkappa \to 0$.*

(ii) *For any integer $m \geq 0$ and any $\gamma \in (0, 1]$ and $\rho > 0$, there is an $\alpha \in (0, \gamma]$ such that $\mathfrak{P}_t f \in C^\alpha(H, w_\rho)$ for any $f \in C^\gamma(V^k, w^{(m)})$ and $t > 0$. Moreover, the norm of the linear operator*

$$\mathfrak{P}_t : C^\gamma(V^k, w^{(m)}) \to C^\alpha(H, w_\rho) \tag{3.138}$$

is bounded uniformly on any compact interval $[\tau, T]$, where $\tau = 0$ for $k = 0$ and $\tau > 0$ for $k \geq 1$.

Proof of Theorem 3.5.5 Let us note that, by the semigroup property, for $t \geq 1$ we have

$$\mathbb{E}_u f(u_t) = \mathfrak{P}_t f(u) = \big(\mathfrak{P}_{t-1}(\mathfrak{P}_1 f)\big)(u).$$

In view of Lemma 3.5.6(ii), the function $\mathfrak{P}_1 f$ belongs to $C^\alpha(H, w_\rho)$, where $\rho > 0$ can be chosen arbitrarily small. Applying Corollary 3.5.3 to $\mathfrak{P}_1 f$, we see that

$$\begin{aligned}\left|\mathbb{E}_u f(u_t) - (f, \mu)\right| &= \left|\left(\mathfrak{P}_{t-1}(\mathfrak{P}_1 f)\right)(u) - (\mathfrak{P}_1 f, \mu)\right| \\ &\le C e^{-\beta(t-1)} |\mathfrak{P}_1 f|_{\alpha, w_\rho} w_\rho(|u|_2).\end{aligned}$$

Since the operator (3.138) is continuous, the right-hand side of this inequality can be estimated by that of (3.135). □

Proof of Lemma 3.5.6 (i) Let us fix arbitrary constants $\gamma \in (0, 1]$ and $\varkappa > 0$ and a functional $f \in C^\gamma(H, w_\varkappa)$ with norm $|f|_{w_\varkappa, \gamma} \le 1$. The continuity of operator (3.137) and the uniform boundedness of its norm will be established if we show that

$$\left|\mathfrak{P}_t f(u)\right| \le C\, e^{\rho |u|_2^2}, \tag{3.139}$$

$$\left|\mathfrak{P}_t f(u) - \mathfrak{P}_t f(v)\right| \le C\, |u - v|_2^\alpha\, e^{bt + \rho(|u|_2^2 + |v|_2^2)}, \tag{3.140}$$

where $u, v \in H$, $0 \le t \le T$, and b, C are some positive constants depending only on γ and $\varkappa$. Inequality (3.139) with $\rho = \varkappa$ follows immediately from (2.125). To prove (3.140), let us denote by u_t and v_t the trajectories of (2.98) issued from the initial points u and v, respectively. Then, by (2.54), the difference $u_t - v_t$ satisfies the inequality

$$|u_t - v_t|_2 \le |u - v|_2 \exp\Big(C_1 \int_0^t \|u_s\|_1^2 ds\Big), \quad t \ge 0. \tag{3.141}$$

Now note that, for any $\alpha \in (0, \gamma]$, we have

$$\begin{aligned}\left|f(u_t) - f(v_t)\right| &\le |u_t - v_t|_2^\gamma \left(w_\varkappa(|u_t|_2) + w_\varkappa(|v_t|_2)\right) \\ &\le C_2\, |u_t - v_t|_2^\alpha \big(w_{2\varkappa}(|u_t|_2) + w_{2\varkappa}(|v_t|_2)\big).\end{aligned} \tag{3.142}$$

Combining inequalities (3.141) and (3.142) and choosing $\alpha \le \gamma$ so small that $\alpha C_1 \le 2\varkappa\nu$, we derive

$$\left|\mathfrak{P}_t f(u) - \mathfrak{P}_t f(v)\right| \le \mathbb{E}\left|f(u_t) - f(v_t)\right| \le C_3\, |u - v|_2^\alpha\, \mathbb{E} e^{2\varkappa \xi_t}, \tag{3.143}$$

where we set

$$\xi_t = |u_t|_2^2 + |v_t|_2^2 + \nu \int_0^t \|u_s\|_1^2 ds.$$

Inequality (2.130) implies that

$$\mathbb{P}_u\{\xi_t \ge |u|_2^2 + |v|_2^2 + 2\mathfrak{B}_0 t + z\} \le e^{-cz}, \quad t \ge 0, \quad z \in \mathbb{R},$$

where $c > 0$ does not depend on t, z and $u, v \in H$. Therefore, if $0 < \varepsilon < c$, then

$$\mathbb{E}\, e^{\varepsilon \xi_t} \le C_4 \exp\bigl(2c\mathfrak{B}_0 t + c(|u|_2^2 + |v|_2^2)\bigr).$$

Hence, if $2\varkappa \le \varepsilon$, then

$$\mathbb{E}\, e^{2\varkappa \xi_t} \le \bigl(\mathbb{E}\, e^{\varepsilon \xi_t}\bigr)^{2\varkappa/\varepsilon} \le C_4 \exp\bigl(bt + \rho(|u|_2^2 + |v|_2^2)\bigr),$$

where $b = 4c\varkappa\mathfrak{B}_0/\varepsilon$ and $\rho = 2\varkappa c/\varepsilon$. Comparing this with (3.143), we arrive at (3.140). The explicit form of the constant ρ implies that it goes to zero with $\varkappa$.

(ii) We shall confine ourselves to proving the required properties under a slightly more restrictive condition. Namely, we shall assume that (3.134) holds with k replaced by $k+1$. The proof in the general case is based on an estimate for the V^k-norm of the difference between two solutions; cf. (2.55) for the case $k = 1$. That estimate can be obtained by taking the scalar product in H of (2.163) with the function $2tL^k u$ and carrying out some arguments similar to those in Section 2.6. Since the corresponding estimates are technically rather complicated, we do not present them here, leaving the proof of assertion (ii) in full generality to the reader as an exercise.

We need to show that if $f \in C^\gamma(V^k, w^{(m)})$ is a functional whose norm does not exceed 1, then inequalities (3.139) and (3.140) hold for $t \in [\tau, T]$ with some positive constants C and b. The first of them follows immediately from (2.138). To prove the second, we use the interpolation inequality (1.7) and the Hölder continuity of f to write

$$\begin{aligned}\bigl|f(u_t) - f(v_t)\bigr| &\le \|u_t - v_t\|_k^\gamma \bigl(w^{(m)}(\|u_t\|_k) + w^{(m)}(\|v_t\|_k)\bigr)\\ &\le C_5\, |u_t - v_t|_2^\alpha \bigl(1 + \|u_t\|_{k+1} + \|v_t\|_{k+1}\bigr)^{m+1},\end{aligned}$$

where $\alpha \in (0, \frac{\gamma}{k+1}]$ is arbitrary, and we keep the notation used in the proof of (i). Combining this with (3.141), we derive

$$\begin{aligned}\bigl|\mathfrak{P}_t f(u) - \mathfrak{P}_t f(v)\bigr| &\le \mathbb{E}\,\bigl|f(u_t) - f(v_t)\bigr|\\ &\le C_5\, |u - v|_2^\alpha\, \mathbb{E}\Bigl\{\bigl(1 + \|u_t\|_{k+1} + \|v_t\|_{k+1}\bigr)^{m+1}\\ &\quad \times \exp\Bigl(\alpha C_1 \int_0^t \|u_s\|_1^2 ds\Bigr)\Bigr\}.\end{aligned}$$

Applying Schwarz's inequality to the expectation, using (2.138) with k and m replaced by $k+1$ and $m+1$, and repeating the argument used in the proof of (i), we see that (3.140) holds for sufficiently small $\alpha > 0$. This completes the proof of the lemma. □

3.5.4 Mixing for Navier–Stokes equations perturbed by a compound Poisson process

In this section, we consider the Navier–Stokes system (3.84) in which η is the time derivative of a compound Poisson process (2.68). It is straightforward to check that a solution of (3.84), (2.68) satisfies a deterministic Navier–Stokes system between two consecutive random times t_{k-1} and t_k, while at the point $t = t_k$ it has a jump of size η_k. Let us set $\mathcal{T}_\omega = \{t_k(\omega), k \geq 1\}$. The following result is a consequence of the existence and uniqueness of a solution for the Navier–Stokes system and the well-known property of independence of increments for a compound Poisson process.

Exercise 3.5.7 Show that, for any $u_0 \in H$, problem (3.84), (2.68) has a unique solution $u(t, x)$ issued from u_0 whose almost every trajectory belongs to the space $C(\mathbb{R}_+ \setminus \mathcal{T}_\omega; H) \cap L^2_{\text{loc}}(\mathbb{R}_+; V)$ and can be chosen to be right-continuous with left limits existing at the points t_k. Prove also that the corresponding solutions form a Markov process in H.

Let us denote by $\mathfrak{P}_t$ and $\mathfrak{P}_t^*$ the Markov semigroups associated with (3.84), (2.68). The following result on existence, uniqueness, and polynomial mixing of a stationary measure was established by Nersesyan [Ner08] in the case of the complex Ginzburg–Landau equation, which is technically more complicated than the 2D Navier–Stokes system.

Theorem 3.5.8 *Assume that the random kicks η_k satisfy the hypotheses imposed in Theorem 3.2.9. Then for any $\nu > 0$ there is an integer $N \geq 1$ such that if $b_j \neq 0$ for $j = 1, \dots, N$, then the following assertions hold.*

Existence and uniqueness: *The semigroup $\mathfrak{P}_t^*$ has a unique stationary distribution μ.*

Polynomial mixing: *For any $p \geq 1$ there is a constant $C_p > 0$ such that for any function $f \in L_b(H)$ we have*

$$|\mathfrak{P}_t f(u) - (f, \mu)| \leq C_p t^{-p}(1 + |u|_2^2)\|f\|_L, \quad t \geq 0.$$

We confine ourselves to outlining the main idea of the proof, referring the reader to the original work [Ner08] for the details. Let us set $\tilde{u}_k = u(t_k)$ and note that

$$\tilde{u}_k = S_{\tau_k}(\tilde{u}_{k-1}) + \eta_k, \quad k \geq 1. \tag{3.144}$$

It turns out that the trajectories $\{\tilde{u}_k\}$ form a Markov chain in H, whose properties are rather similar to those of the RDS (3.50) with unbounded[8] kicks. Using the

[8] Note that even though the kicks η_k are bounded, in the case when they are non-zero almost all trajectories of (3.144) are unbounded in H. This is due to the fact that on any finite interval the system can receive an arbitrarily large number of kicks.

ideas of Section 3.4, it is possible to prove that (3.144) has a unique stationary measure $\tilde{\mu}$, which is mixing faster than any negative degree of k. Once this property is established, one can go back to the original problem and prove the uniqueness and polynomial mixing of a stationary measure μ for it. Moreover, the measures μ and $\tilde{\mu}$ are linked by the following *Khas'minskii relation*:

$$(f, \mu) = (\mathbb{E}\,\tau_1)^{-1}\,\mathbb{E}_{\tilde{\mu}} \int_0^{\tau_1} f(u_t)\,dt, \qquad f \in C_b(H). \tag{3.145}$$

Exercise 3.5.9 Prove that if $\tilde{\mu}$ is a stationary measure for (3.144), then the measure μ defined by (3.145) is stationary for the Markov process associated with (3.84), (2.68).

3.5.5 Description of some results on uniqueness and mixing for other PDEs

The methods developed in this chapter for proving the uniqueness of a stationary measure and exponential mixing apply, roughly speaking, to random perturbations of any dissipative PDE with a finite-dimensional attractor. Let us describe briefly some results concerning the complex Ginzburg–Landau equation:

$$\dot{u} - (\nu + i)\Delta u + (\alpha + i\lambda)|u|^{2p}u = h(x) + \eta(t, x), \quad u\big|_{\partial Q} = 0. \tag{3.146}$$

Here $\nu > 0$, $\alpha \geq 0$, $\lambda > 0$, and $p > 0$ are some parameters, $Q \subset \mathbb{R}^d$ is a bounded domain with smooth boundary, h is a deterministic function, and η is a random force from one of the classes described in Section 2.2. We thus consider only the case when the conservative term $i|u|^{2p}u$ is defocusing. In this case, the Cauchy problem for (3.146) is well posed in $H_0^1 = H_0^1(Q; \mathbb{C})$ under the following assumptions on the parameter p:

$$p < \infty \quad \text{for } d = 1, 2, \qquad p \leq \frac{2}{d-2} \quad \text{for } d \geq 3. \tag{3.147}$$

The behaviour of solutions for (3.146) heavily depends on whether the parameter α is positive or not. Roughly speaking, when $\alpha > 0$, the dissipative term $\alpha|u|^{2p}u$ compensates for the strong nonlinear effects due to the conservative term, and the uniqueness of a stationary measure and exponential mixing can be established under essentially the same hypotheses. On the other hand, without strong nonlinear dissipation (i.e., for $\alpha = 0$), these properties are not known to hold, unless either the random perturbation is bounded or the parameter p satisfies much more restrictive conditions. Let us discuss the latter situation in more detail, referring the reader to the papers [Hai02b; Oda08] for the easier case of equations with strong nonlinear dissipation. Note that Eq. (3.146) with $\alpha = 0$ is a stochastic PDE with conservative nonlinearity and linear damping. So this is an analogue of the main topic of this book – the stochastic Navier–Stokes system.

We assume that p satisfies the following inequalities (cf. (3.147)):

$$p \le 1 \quad \text{for } d = 1, \qquad p < 1 \quad \text{for } d = 2, \qquad p \le \frac{2}{d} \quad \text{for } d \ge 3. \tag{3.148}$$

Concerning the external force, we suppose that $h \in H^1(Q;\mathbb{C})$ and $\eta(t,x)$ is a random process of the form (2.66), where $\{b_j\} \subset \mathbb{R}$ is a sequence decaying faster than every negative degree of j, $\{e_j\}$ is an eigenbasis for the Dirichlet Laplacian in Q, and $\{\beta_j = \beta_j^1 + i\beta_j^2\}$ is a sequence of independent complex Brownian motions. A proof of the following result can be found in [Shi08].

Theorem 3.5.10 *Under the above hypotheses, assume that $\alpha = 0$ and $b_j \ne 0$ for all $j \ge 1$. Then for any $\nu > 0$ and $\lambda > 0$ the Markov process associated with* (3.146) *has a unique stationary measure $\mu \in \mathcal{P}(H_0^1)$. Moreover, for any function $f \in L_b(L^2(Q;\mathbb{R}))$ and any solution $u(t)$ for* (3.146), *we have*

$$|\mathbb{E} f(u(t)) - (f,\mu)| \le C e^{-\gamma t} \|f\|_L \big(1 + \mathbb{E}|u(0)|_2^2\big), \quad t \ge 0,$$

where C and $\gamma > 0$ are universal constants.

A brief discussion on other results concerning the uniqueness and mixing for various classes of randomly forced PDEs can be found in the section *Notes and comments* at the end of this chapter.

3.5.6 An alternative proof of mixing for kick force models

In Section 3.2, we established the uniqueness of a stationary distribution and the exponential mixing for the Markov chain defined by (3.50). The proof presented there is based on a coupling argument which enabled one to establish a contraction in a space of measures. In this section, we outline an alternative proof of the uniqueness and mixing, which is based on a Lyapunov–Schmidt reduction and a version of the Ruelle–Perron–Frobenius theorem. To simplify the presentation, we impose more restrictive hypotheses on the system than in Section 3.2; however, essentially the same proof works in the general case.

We consider the discrete-time RDS (3.50), in which $S : H \to H$ is a locally Lipschitz compact mapping in a Hilbert space H with norm $\|\cdot\|$ and $\{\eta_k\}$ is a sequence of i.i.d. random variables in H supported by a compact subset. We assume that

$$\|S(u)\| \le q\,\|u\| \quad \text{for all } u \in H, \tag{3.149}$$

where $q < 1$ does not depend on u. Then, for sufficiently large $R > 0$, the ball $B_H(R)$ is an invariant absorbing set for (3.50). It follows that the RDS (3.50) has a compact absorbing set $X \subset B_H(R)$, and we can confine our consideration to X. We assume, in addition, that there is a constant $\gamma < 1$ and a projection $\mathsf{P} : H \to H$ with a finite-dimensional range $E \subset H$ such that inequality (3.55) holds with $\mathcal{A} = B_H(r)$ and $r = R/(1-q)$, the random variables $(I - \mathsf{P})\eta_k$ are almost surely zero, and the law of $\mathsf{P}\eta_k$ has a C^1-smooth density ρ with respect to the Lebesgue measure on E such that $\rho(0) > 0$. Let us denote by $P_k(u, \Gamma)$ the transition function of the Markov chain (3.50) restricted to X and by $\mathfrak{P}_k$ and $\mathfrak{P}_k^*$ the corresponding Markov semigroups. Our aim is to establish the following result.

Theorem 3.5.11 *Under the above hypotheses, there is a unique stationary distribution $\mu \in \mathcal{P}(X)$ for $\mathfrak{P}_k^*$, and for any $\lambda \in \mathcal{P}(X)$ we have*

$$\|\mathfrak{P}_k^*\lambda - \mu\|_L^* \to 0 \quad \text{as } k \to \infty. \tag{3.150}$$

The existence of a stationary measure is a consequence of the Bogolyubov-Krylov argument (see Section 2.5.2). We now outline the proof of the uniqueness under the additional hypothesis that $(I - \mathsf{P})\eta_k = 0$ almost surely. It is divided into three steps.

Step 1: Lyapunov–Schmidt reduction. Let us consider the projection of Eq. (3.50) to the orthogonal complement of E:

$$w_k = \mathsf{Q}S(v_{k-1} + w_{k-1}). \tag{3.151}$$

Here $\mathsf{Q} = I - \mathsf{P}$, $v_k = \mathsf{P}u_k$, and $w_k = \mathsf{Q}u_k$. We denote by $\boldsymbol{X}$ and $\boldsymbol{Y}$ the spaces of sequences $\boldsymbol{v} = (v_k, k \in \mathbb{Z}_-)$ and $\boldsymbol{w} = (w_k, k \in \mathbb{Z}_-)$ with $v_k \in B_E(R)$ and $w_k \in B_{E^\perp}(qr)$, endowed with the metric

$$d_{\boldsymbol{X}}(\boldsymbol{u}^1, \boldsymbol{u}^2) = \sup_{k \le 0}\bigl(e^{\alpha k}\|u_k^1 - u_k^2\|\bigr), \quad \boldsymbol{u}^i = (u_k^i, k \in \mathbb{Z}_-),$$

where $\alpha > 0$ is such that $\gamma e^{\alpha} < 1$. Using the contraction mapping principle, it is straightforward to prove that for any $\boldsymbol{v} \in \boldsymbol{X}$ there is a unique $\boldsymbol{w} := \boldsymbol{\mathcal{W}}(\boldsymbol{v}) \in \boldsymbol{Y}$ satisfying (3.151) for each $k \le 0$. Moreover, the mapping $\boldsymbol{\mathcal{W}} : \boldsymbol{X} \to \boldsymbol{Y}$ is globally Lipschitz continuous. (These properties are essentially a consequence of the Foiaş–Prodi estimates discussed in Section 2.1.8.)

Let us define a Markov RDS in $\boldsymbol{X}$ by the formula

$$\boldsymbol{v}^k = \bigl(\boldsymbol{v}^{k-1}, \mathsf{P}S(v_0^{k-1} + \mathcal{W}(\boldsymbol{v}^{k-1})) + \eta_k\bigr), \tag{3.152}$$

where $\mathcal{W} = \mathcal{W}(\boldsymbol{v})$ stands for the zeroth component of $\boldsymbol{\mathcal{W}}(\boldsymbol{v})$, and given two elements $\boldsymbol{v} = (v_k, k \in \mathbb{Z}_-) \in \boldsymbol{X}$ and $v \in B_E(R)$, we write $(\boldsymbol{v}, v)$ for the sequence $(\dots, v_{-1}, v_0, v) \in \boldsymbol{X}$. Given a stationary measure μ for $\mathfrak{P}_k^*$, we denote by $\boldsymbol{\mu} \in \mathcal{P}(\boldsymbol{X})$ the unique law satisfying the relation

$$\begin{aligned}&\boldsymbol{\mu}\bigl(\{(v_k, k \in \mathbb{Z}_-) \in \boldsymbol{X} : (v_{-l}, \dots, v_0) \in \Gamma\}\bigr)\\ &\quad = \int_{\widetilde{\Gamma}} \mu(du_{-l})P_1(u_{-l}, du_{1-l}) \dots P_1(u_1, du_0) \quad \text{for any } \Gamma \in \mathcal{B}(H^{l+1}).\end{aligned}$$

Here $P_k(u, \cdot)$ is the transition function of the family of Markov chains corresponding to (3.50), and

$$\widetilde{\Gamma} = \{(u_{-l}, \dots, u_0) \in H^{l+1} : (\mathsf{P}u_{-l}, \dots, \mathsf{P}u_0) \in \Gamma\}.$$

In other words, $\boldsymbol{\mu}$ is the restriction to $\mathbb{Z}_-$ of the law for the projection to E of a stationary trajectory of (3.50) with distribution μ. The following proposition establishes a relationship between (3.50) and (3.152). Its proof can be found in [KS00].

Proposition 3.5.12 *Let $\mu \in \mathcal{P}(H)$ be a stationary measure for $\mathfrak{P}_k^*$. Then $\boldsymbol{\mu}$ is a stationary measure for* (3.152).

Thus, the uniqueness of a stationary measure for $\mathfrak{P}_k^*$ will be established if we establish this property for the Markov chain (3.152). In the next step, we formulate an abstract result that gives a sufficient condition for the uniqueness and mixing of a stationary measure for a Markov semigroup on a compact space.

Step 2: Ruelle–Perron–Frobenius type theorem. Let $\boldsymbol{X}$ be a compact metric space and let $\boldsymbol{P}_k(\boldsymbol{v}, \Gamma)$ be a Markov transition function on $\boldsymbol{X}$. We denote by $\boldsymbol{P}_k : C(\boldsymbol{X}) \to C(\boldsymbol{X})$ and $\boldsymbol{P}_k^* : \mathcal{P}(\boldsymbol{X}) \to \mathcal{P}(\boldsymbol{X})$ the corresponding Markov semigroups. Recall that a family $\mathcal{C} \subset C(\boldsymbol{X})$ is said to be *determining* if any two measures $\boldsymbol{\mu}, \boldsymbol{\nu} \in \mathcal{P}(\boldsymbol{X})$ satisfying the relation $(f, \boldsymbol{\mu}) = (f, \boldsymbol{\nu})$ for all $f \in \mathcal{C}$ coincide. The following result gives a sufficient condition for the uniqueness and mixing for a stationary measure.

Theorem 3.5.13 *Suppose that the transition function satisfies the following two hypotheses.*

Uniform Feller property: *There is a determining family $\mathcal{C}$ such that for any $f \in \mathcal{C}$ the sequence $\{\boldsymbol{P}_k f, k \geq 0\}$ is uniformly equicontinuous.*

Uniform recurrence: *There is a $\hat{\boldsymbol{v}} \in \boldsymbol{X}$ such that for any $r > 0$ one can find an integer $m \geq 1$ and a constant $p > 0$ for which*

$$\boldsymbol{P}_k(\boldsymbol{u}, B_{\boldsymbol{X}}(\hat{\boldsymbol{v}}, r)) \geq p \quad \text{for } k \geq m \text{ and all } \boldsymbol{u} \in \boldsymbol{X}. \tag{3.153}$$

Then $\boldsymbol{P}_k^*$ has a unique stationary measure $\boldsymbol{\mu} \in \mathcal{P}(\boldsymbol{X})$, and for any $f \in C(\boldsymbol{X})$, we have

$$\boldsymbol{P}_k f \to (f, \boldsymbol{\mu}) \quad \text{as } k \to \infty \text{ uniformly on } \boldsymbol{X}. \tag{3.154}$$

This theorem has independent interest, and its proof is given at the end of this subsection. We now show how to apply it in our situation.

Step 3: Verification of the hypotheses. The uniform recurrence condition is a straightforward consequence of the fact that $u = 0$ is a globally stable fixed point for the deterministic (nonlinear) semigroup obtained by iterations of S and the fact that the random variables η_k are small with a positive probability. The latter follows from the inequality $\rho(0) > 0$. To prove the uniform Feller property, we note that the transition function $\boldsymbol{P}_k(\boldsymbol{v}, \cdot)$ for the Markov chain associated with (3.152) is such that

$$\boldsymbol{P}_k(\boldsymbol{v}, \boldsymbol{\Gamma}) = \int_\Gamma \prod_{l=1-k}^{0} \rho(z_l - F(\boldsymbol{v}, z_{1-k}, \dots, z_{-l-1})\, dz_{1-k} \dots dz_0, \qquad (3.155)$$

where $\Gamma \subset E^{k+1}$ is an arbitrary Borel subset, $F(\boldsymbol{v}) = \mathsf{P}S(v_0 + \mathcal{W}(\boldsymbol{v}))$ for $\boldsymbol{v} = (v_m, m \in \mathbb{Z}_-)$, and

$$\boldsymbol{\Gamma} = \{\boldsymbol{v} = (v_m, m \in \mathbb{Z}_-) : (v_{-k}, \dots, v_0) \in \Gamma\}.$$

Let us denote by $\mathcal{C}$ the space of functions $f \in C(\boldsymbol{X})$ that depend on finitely many components. It follows from (3.155) that if a function $f \in \mathcal{C}$ depends only on $v_{-N}, \dots, v_0$, then

$$\boldsymbol{P}_k f(\boldsymbol{v}) = \int_{E^k} \prod_{l=1-k}^{0} \rho(z_l - F(\boldsymbol{v}, z_{1-k}, \dots, z_{l-1})) f(z_{-N}, \dots, z_0)\, dz_{1-k} \dots dz_0.$$

Since the mapping $\mathcal{W}$ is Lipschitz continuous, so is F, and the above formula readily gives uniform equicontinuity of the family $\{\boldsymbol{P}_k f, k \geq 0\}$ for any $f \in \mathcal{C}$.

Further analysis shows that the original and the reduced systems are equivalent, and therefore the convergence to the stationary measure for (3.152) implies (3.150). We refer the reader to the original work [KS00] for more details on this subject.

Proof of Theorem 3.5.13 The existence of a stationary measure follows immediately from the compactness of $\boldsymbol{X}$ and the Bogolyubov–Krylov argument. To prove uniqueness, it suffices to show that if $\boldsymbol{\mu} \in \mathcal{P}(\boldsymbol{X})$ is a stationary distribution, then (3.154) holds for any $f \in \mathcal{C}$. Indeed, if (3.154) is proved and $\boldsymbol{\mu}, \boldsymbol{\nu} \in \mathcal{P}(\boldsymbol{X})$ are two stationary measures, then $(f, \boldsymbol{\mu}) = (f, \boldsymbol{\nu})$ for any $f \in \mathcal{C}$. Since $\mathcal{C}$ is a determining family, we conclude that $\boldsymbol{\mu} = \boldsymbol{\nu}$.

Suppose now that $\boldsymbol{\mu} \in \mathcal{P}(\boldsymbol{X})$ is a stationary measure. Let $f \in \mathcal{C}$ be an arbitrary function. We wish to prove (3.154). There is no loss of generality in assuming that $0 \leq f \leq 1$. It follows from the uniform Feller property that $\boldsymbol{P}_{k_n} f$ converges uniformly to a limit $g \in C(\boldsymbol{X})$ for some sequence $k_n \to \infty$. We claim that g must be constant. If this assertion is proved, then the obvious relation $(\boldsymbol{P}_1 h, \boldsymbol{\mu}) = (h, \boldsymbol{\mu})$, which holds for any $h \in C(\boldsymbol{X})$, implies that the constant must be equal to $(f, \boldsymbol{\mu})$, whence we conclude that the whole sequence $\{\boldsymbol{P}_k f\}$ converges to $(f, \boldsymbol{\mu})$.

Suppose that $g \not\equiv C$. We can assume that $m_n = k_{n+1} - k_n \to \infty$ as $n \to \infty$. It is easy to see that $\boldsymbol{P}_{m_n} g \to g$ uniformly on $\boldsymbol{X}$. Since g is not constant, then $g(\hat{\boldsymbol{v}})$ (where $\hat{\boldsymbol{v}}$ is defined in the uniform recurrence condition) is either smaller than the maximum of g or greater than the minimum of g. Assume that the first case is true. Denoting by $\boldsymbol{u} \in \boldsymbol{X}$ a point where g attains its maximum, we can find positive constants δ and r such that

$$g(\boldsymbol{u}) \geq g(z) + \delta \quad \text{for} \quad z \in B_X(\hat{\boldsymbol{v}}, r). \tag{3.156}$$

We now write

$$\begin{aligned} \boldsymbol{P}_{m_n} g(\boldsymbol{u}) &= \int_X P_{m_n}(\boldsymbol{u}, dz) g(z) \\ &= \int\limits_{B_X(\hat{\boldsymbol{v}},r)} P_{m_n}(\boldsymbol{u}, dz) g(z) + \int\limits_{B_X^c(\hat{\boldsymbol{v}},r)} P_{m_n}(\boldsymbol{u}, dz) g(z). \end{aligned}$$

Using (3.156), we derive

$$\begin{aligned} \boldsymbol{P}_{m_n} g(\boldsymbol{u}) &\leq P_{m_n}(\boldsymbol{u}, B_X(\hat{\boldsymbol{v}}, r))(g(\boldsymbol{u}) - \delta) + P_{m_n}(\boldsymbol{u}, B_X^c(\hat{\boldsymbol{v}}, r)) g(\boldsymbol{u}) \\ &\leq g(\boldsymbol{u}) - \delta P_{m_n}(\boldsymbol{u}, B_X(\hat{\boldsymbol{v}}, r)). \end{aligned} \tag{3.157}$$

Since $m_n \to \infty$, inequality (3.153) implies that

$$\boldsymbol{P}_{m_n} g(\boldsymbol{u}) \leq g(\boldsymbol{u}^+) - \delta p \quad \text{for } n \gg 1. \tag{3.158}$$

Passing to the limit as $n \to \infty$, we arrive at a contradiction. This completes the proof of the theorem. □

3.6 Appendix: some technical proofs

In this section, we have compiled some relatively difficult proofs that are not of primary importance. The reader not interested in the technical details may safely skip this section.

3.6.1 Proof of Lemma 3.3.11

Compared to the case of an H^1-regular external force, the additional difficulty is that the resolving operator is not continuous as a mapping defined on the space of H-valued paths. To overcome this problem, we first rewrite the Navier–Stokes system as an equation with random coefficients and then repeat essentially the same scheme, using the continuous dependence of solutions on coefficients. The proof is divided into three steps.

Step 1. Let us recall the decomposition of $u(t; v)$ described in Section 2.4. We write $u(t; v) = z(t) + \tilde{v}(t)$, where $z(t)$ is the solution of Stokes' equation (2.100) with zero initial condition and

$$\tilde{v}(t) = \mathcal{R}_t(v, z)$$

is the solution of problem (2.111), (2.112); see Proposition 2.4.5. Recall the space $\mathcal{X}_T = C(0, T; H) \cap L^2(0, T; V)$ and denote by $\dot{\mathcal{X}}_T$ the space of functions in $\mathcal{X}_T$ vanishing at $t = 0$. We claim that the following two properties hold.

(a) For any $\hat{u} \in V$ there is a continuous operator $Z_{\hat{u}} : H \to \dot{\mathcal{X}}_1$, $v \mapsto Z_{\hat{u}}(v; t)$, that is uniformly Lipschitz on bounded subsets and is such that the solution of (2.100) with $\eta = \partial_t Z_{\hat{u}} + \nu L Z_{\hat{u}}$ equals $\hat{u}$ at $t = 1$:

$$\mathcal{R}_1(v, Z_{\hat{u}}(v)) + Z_{\hat{u}}(v; 1) = \hat{u}. \tag{3.159}$$

(b) Let $\mathcal{O} \subset \dot{\mathcal{X}}_1$ be any neighbourhood of the origin and let $\mathcal{K} \subset \dot{\mathcal{X}}_1$ be a compact set. Then

$$\inf_{\hat{z} \in \mathcal{K}} \mathbb{P}\{z - \hat{z} \in \mathcal{O}\} > 0. \tag{3.160}$$

Taking these properties for granted, let us complete the proof of (3.101). Without loss of generality, we can assume that $\hat{u} \in V$. Since the operator Z is continuous, there is a constant $\rho > 0$ such that the image of the compact set $B_V(R) \subset H$ under Z is contained in the ball $B_{\dot{\mathcal{X}}_1}(\rho)$. The Lipschitz property on bounded subsets implies that

$$|\mathcal{R}_1(v, z_1) - \mathcal{R}_1(v, z_2)|_2 \le C \, \|z_1 - z_2\|_{\mathcal{X}_1}$$

for any $v \in B_V(R)$ and $z_1, z_2 \in B_{\dot{\mathcal{X}}_1}(\rho)$, where $C \ge 1$ is a constant depending only on R. In particular, taking $z_2 = Z_{\hat{u}}(v)$, we see that

$$|\mathcal{R}_1(v, z_1) - \mathcal{R}_1(v, Z_{\hat{u}}(v))|_2 \le \frac{\varepsilon}{2} \quad \text{for } \|z_1 - Z_{\hat{u}}(v)\|_{\mathcal{X}_1} \le \frac{\varepsilon}{2C}.$$

It follows from property (a) that

$$|u(1; v) - \hat{u}|_2 \le |z(1) - Z_{\hat{u}}(v; 1)|_2 + |\mathcal{R}_1(v, z_1) - \mathcal{R}_1(v, Z_{\hat{u}}(v))|_2.$$

Combining the last two inequalities, we conclude that

$$\mathbb{P}\{|u(1; v) - \hat{u}|_2 < \varepsilon\} \ge \mathbb{P}\Big\{\|z - Z_{\hat{u}}(v)\|_{\mathcal{X}_1} < \frac{\varepsilon}{2C}\Big\} \quad \text{for any } v \in B_V(R). \tag{3.161}$$

Applying property (b) in which $\mathcal{K} = Z_{\hat{u}}(B_V(R))$ and $\mathcal{O} \subset \dot{\mathcal{X}}_1$ is the open ball of radius $\frac{\varepsilon}{2C}$ centred at zero, we see that the right-hand side of (3.161) is separated from zero uniformly in $v \in B_V(R)$. Thus, to complete the proof of (3.101), it suffices to establish properties (a) and (b).

Step 2: Proof of (a). Let $\widetilde{w} \in \mathcal{H}$ be the solution of the problem

$$\partial_t \widetilde{w} + \nu L \widetilde{w} = 0, \quad \widetilde{w}(0) = v.$$

For any $v \in H$, define a function $w \in \mathcal{H}$ by the relation $w(t) = \chi(t)\widetilde{w}(t) + t\hat{u}$, where $\chi \in C^\infty(\mathbb{R})$ is an arbitrary function such that $\chi(t) = 1$ for $t \le 0$ and $\chi(t) = 0$ for $t \ge 1$. Then $w(0) = v$, $w(1) = \hat{u}$, and

$$\eta := \partial_t w + \nu L w + B(w) - h \in L^2(0, 1; V^*).$$

Moreover, η is uniformly Lipschitz continuous in $v \in H$ on bounded subsets as a function with range in $L^2(0, 1; V^*)$. Let us define $Z_{\hat{u}}(v) \in \dot{\mathcal{X}}_1$ as the unique solution of the problem

$$\dot{Z} + \nu L Z = \eta(t), \quad Z(0) = 0.$$

Then

$$w(t) = Z_{\hat{u}}(v; t) + \mathcal{R}_t(v, Z_{\hat{u}}(v)),$$

whence we conclude that (3.159) holds. The Lipschitz continuity of $Z_{\hat{u}}(v)$ follows from a similar property for the mapping $v \mapsto \eta$.

Step 3: Proof of (b). Let us note that the process $z(t)$ can be written in the form (cf. (2.101))

$$z(t) = \sum_{j=1}^{\infty} b_j \int_0^t e^{-\alpha_j \nu (t-s)} d\beta_j(s)\, e_j(x), \quad 0 \le t \le 1,$$

where $\{e_j\}$ stands for the family of normalised eigenfunctions of the Stokes operator, α_j denotes the eigenvalue corresponding to e_j, and the series above converges in $L^2(\Omega; \dot{\mathcal{X}}_1)$. We claim that the support of the law for the process $\{z(t), 0 \le t \le 1\}$ regarded as a random variable in $\dot{\mathcal{X}}_1$coincides with the entire space. To prove this, it suffices to show that, for any $\hat{z} \in C(0, 1; V)$ vanishing at zero and any $\varepsilon > 0$, we have

$$\mathbb{P}\big\{\|z - \hat{z}\|_{\mathcal{X}_1} < \varepsilon\big\} > 0. \tag{3.162}$$

Let us fix a function $\hat{z}$ and write it in the form

$$\hat{z}(t) = \sum_{j=1}^{\infty} \hat{z}_j(t) e_j(x),$$

where $\hat{z}_j(t) = \langle z(t), e_j \rangle$, and the series converges in V uniformly in $t \in [0, 1]$. For any integer $N \ge 1$, we have

$$z(t) - \hat{z}(t) = D_N(t) + r_N(t) + \hat{r}_N(t), \tag{3.163}$$

where we set

$$D_N(t) = \sum_{j=1}^{N} \Big(b_j \int_0^t e^{-\alpha_j \nu(t-s)} d\beta_j(s) - \hat{z}_j(t) \Big) e_j(x),$$

$$r_N(t) = \sum_{j=N+1}^{\infty} b_j \int_0^t e^{-\alpha_j \nu(t-s)} d\beta_j(s)\, e_j(x),$$

$$\hat{r}_N(t) = \sum_{j=N+1}^{\infty} \hat{z}_j(t) e_j(x).$$

Let us estimate each term on the right-hand side of (3.163). Since $\{e_j\}$ is an orthogonal system with respect to the scalar product $\langle Lu, u\rangle^{1/2}$ and $\|e_j\|_V^2 = \alpha_j$ for any $j \geq 1$, we have

$$\|D_N(t)\|_V \leq \sqrt{\alpha_N} \max_{1\leq j\leq N} \Big(\sup_{0\leq t\leq 1} b_j \Big| \int_0^t e^{-\alpha_j \nu(t-s)} d\beta_j(s) - \hat{z}_j(t) \Big| \Big).$$

Since $\hat{z}_j$ is a continuous function on [0, 1] vanishing at zero, combining this inequality with Proposition A.4.2, we see that

$$\mathbb{P}\Big\{ \sup_{0\leq t\leq 1} \|D_N(t)\|_V < \frac{\varepsilon}{3} \Big\} > 0 \quad \text{for any } N \geq 1. \tag{3.164}$$

Now let $\mathsf{P}_N : H \to H$ be the orthogonal projection onto the vector span of $e_1, \dots, e_N$. Then one can find an integer $N_1(\varepsilon) \geq 1$ such that

$$\sup_{0\leq t\leq 1} \|\hat{r}_N(t)\|_V = \sup_{0\leq t\leq 1} \|(I - \mathsf{P}_N)\hat{z}(t)\|_V \leq \frac{\varepsilon}{3} \quad \text{for } N \geq N_1(\varepsilon).$$

Furthermore, the technique developed in the proof of Proposition 2.4.2 (see the derivation of (2.109)) enables one to show that

$$\mathbb{E}\, \|r_N\|_{\mathcal{X}_1}^2 = \mathbb{E}\, \|(I - \mathsf{P}_N)z\|_{\mathcal{X}_1}^2 \to 0 \quad \text{as } N \to \infty.$$

Therefore there is an integer $N_2(\varepsilon) \geq 1$ such that

$$\mathbb{P}\Big\{ \|r_N\|_{\mathcal{X}_1} \leq \frac{\varepsilon}{3} \Big\} > 0 \quad \text{for } N \geq N_2(\varepsilon).$$

Combining the above estimates and noting that the random variables D_N and r_N are independent, we arrive at the required inequality (3.162).

We are now ready to prove (3.160). Since $\operatorname{supp} \mathcal{D}(z) = \mathcal{X}_1$, for any neighbourhood of zero $\mathcal{O} \subset \dot{\mathcal{X}}_1$ the function $p(\hat{z}) = \mathbb{P}\{z - \hat{z} \in \mathcal{O}\}$ is positive at any point of $\dot{\mathcal{X}}_1$. Furthermore, in view of a well-known property of the weak convergence of measures (e.g., see theorem 11.1.1 in [Dud02]), p is lower semicontinuous. It follow that p is separated from zero on any compact subset of $\dot{\mathcal{X}}_1$. This completes the proof of Lemma 3.3.11.

3.6.2 Recurrence for Navier–Stokes equations with unbounded kicks

This section is devoted to the proof of Proposition 3.4.3. For any $d > 0$, we introduce the stopping time

$$\boldsymbol{\tau}_d = \inf\{k \geq 0 : \boldsymbol{u}_k \in \boldsymbol{B}_d\}.$$

We wish to show that, for any $\boldsymbol{u} \in \boldsymbol{H}$ and $d > 0$, the random variable $\boldsymbol{\tau}_d$ is $\mathbb{P}_{\boldsymbol{u}}$-almost surely finite, and there are positive constants δ and $C > 0$ such that

$$\mathbb{E}_{\boldsymbol{u}} \exp(\delta \boldsymbol{\tau}_d) \leq C(1 + |\boldsymbol{u}|), \tag{3.165}$$

where $|\boldsymbol{u}| = |u|_2 + |u'|_2$.

As in the case of the Navier–Stokes system with spatially regular white noise, inequality (3.165) is a consequence of the following two lemmas (cf. Section 3.3.2).

Lemma 3.6.1 *There is an* $R_* > 0$ *such that*

$$\mathbb{P}_{\boldsymbol{u}}\{\boldsymbol{\tau}_R < \infty\} = 1 \quad \textit{for any } \boldsymbol{u} \in \boldsymbol{H},\ R \geq R_*. \tag{3.166}$$

Moreover, there are positive constants K *and* γ *such that*

$$\mathbb{E}_{\boldsymbol{u}} \exp(\gamma \boldsymbol{\tau}_R) \leq 1 + K R^{-1} |\boldsymbol{u}| \quad \textit{for any } \boldsymbol{u} \in \boldsymbol{H},\ R \geq R_*. \tag{3.167}$$

Lemma 3.6.2 *For any positive constants* $\mathfrak{B}$, R, d, *and* ν *there is a* $p > 0$ *and integers* $N_0 \geq 1$ *and* $m \geq 1$ *such that if* (3.121) *holds, then*

$$\mathbb{P}_{\boldsymbol{u}}\{\boldsymbol{u}_m \in \boldsymbol{B}_d\} \geq p \quad \textit{for any } \boldsymbol{u} \in \boldsymbol{B}_R. \tag{3.168}$$

Once these lemmas are established, inequality (3.165) will follow by exactly the same argument as in Section 3.3.2. Thus, we shall confine ourselves to the proof of the above lemmas.

Proof of Lemma 3.6.1 We use a well-known argument based on the existence of a Lyapunov function. It follows from Proposition 2.3.3 that

$$\mathbb{E}_{\boldsymbol{u}} |\boldsymbol{u}_k| \leq q\big(|\boldsymbol{u}| \vee R_*\big) \quad \text{for any } \boldsymbol{u} \in \boldsymbol{H}, \tag{3.169}$$

where $q < 1$ and $R_* > 0$ are some constants not depending on $\boldsymbol{u}$. Let us fix any $R \geq R_*$ and set

$$p_k(\boldsymbol{u}) = \mathbb{E}_{\boldsymbol{u}}\big(\mathbb{I}_{\{\boldsymbol{\tau}_R > k\}} |\boldsymbol{u}_k|\big).$$

The Markov property and inequality (3.169) imply that

$$\begin{aligned} p_{k+1}(\boldsymbol{u}) &\leq \mathbb{E}_{\boldsymbol{u}}\big(\mathbb{I}_{\{\boldsymbol{\tau}_R > k\}} (\mathbb{E}_{\boldsymbol{u}} |\boldsymbol{u}_{k+1}| \mid \mathcal{F}_k)\big) = \mathbb{E}_{\boldsymbol{u}}\big(\mathbb{I}_{\{\boldsymbol{\tau}_R > k\}} (\mathbb{E}_{\boldsymbol{v}} |\boldsymbol{u}_1|)\big|_{\boldsymbol{v}=\boldsymbol{u}_k}\big) \\ &\leq \mathbb{E}_{\boldsymbol{u}}\big(\mathbb{I}_{\{\boldsymbol{\tau}_R > k\}}\, q(|\boldsymbol{u}_k| \vee R_*)\big) = q\, p_k(\boldsymbol{u}), \end{aligned}$$

where we used the fact that $|\boldsymbol{u}_k| > R \geq R_*$ on the set $\{\boldsymbol{\tau}_R > k\}$. Iterating the above inequality and noting that $p_0(\boldsymbol{u}) \leq |\boldsymbol{u}|$, we obtain

$$p_k(\boldsymbol{u}) \leq q^k |\boldsymbol{u}| \quad \text{for all } k \geq 0, \boldsymbol{u} \in \boldsymbol{H}.$$

It follows that

$$\mathbb{P}_{\boldsymbol{u}}\{\boldsymbol{\tau}_R > k\} \leq R^{-1}\mathbb{E}_{\boldsymbol{u}}\big(\mathbb{I}_{\{\boldsymbol{\tau}_R > k\}}|\boldsymbol{u}_k|\big) \leq R^{-1}q^k|\boldsymbol{u}|, \tag{3.170}$$

whence we conclude that (3.166) holds.

We now prove (3.167). Let $\gamma > 0$ be so small that $e^\gamma q < 1$. Then, using (3.166) and (3.170), we derive

$$\begin{aligned}\mathbb{E}_{\boldsymbol{u}} \exp(\gamma \boldsymbol{\tau}_R) &= \sum_{k=0}^{\infty} \mathbb{E}_{\boldsymbol{u}}\big(\mathbb{I}_{\{\boldsymbol{\tau}_R = k\}} \exp(\gamma \boldsymbol{\tau}_R)\big) \\ &\leq 1 + \sum_{k=1}^{\infty} e^{\gamma k}\mathbb{P}_{\boldsymbol{u}}\big\{\boldsymbol{\tau}_R > k-1\big\} \\ &\leq 1 + R^{-1}|\boldsymbol{u}| \sum_{k=1}^{\infty} e^{\gamma k} q^{k-1}.\end{aligned}$$

The series on the right-hand side of this inequality converges in view of the choice of γ, and we obtain (3.167). □

Proof of Lemma 3.6.2

Step 1. Inequality (3.168) will be established if for any constants $R > d > 0$ we find $\varepsilon > 0$ and $q < 1$ such that

$$\mathbb{P}\big\{|\mathcal{R}(v, v')|_2 \vee |\mathcal{R}'(v, v')|_2 \leq \big(q(|v| \vee |v'|)\big) \vee d\big\} \geq \varepsilon \quad \text{for } (v, v') \in \boldsymbol{B}_R. \tag{3.171}$$

Indeed, let us denote by $\boldsymbol{G}(v, v')$ the event on the left-hand side of this inequality and choose an integer $m \geq 1$ so that $qd < q^m R \leq d$. Denoting $\boldsymbol{B}^{(k)} = \boldsymbol{B}_{q^k R}$, $k \geq 0$, we have $\boldsymbol{B}^{(m)} \subset \boldsymbol{B}_d$. Combining (3.171) with the Markov property, for $1 \leq k \leq m-1$ we derive

$$\begin{aligned}\mathbb{P}_{\boldsymbol{u}}\{\boldsymbol{u}_k \in \boldsymbol{B}^{(k)}\} &\geq \mathbb{P}_{\boldsymbol{u}}\big(\boldsymbol{G}(\boldsymbol{u}_{k-1}) \cap \{\boldsymbol{u}_{k-1} \in \boldsymbol{B}^{(k-1)}\}\big) \\ &= \mathbb{E}_{\boldsymbol{u}}\big(\mathbb{I}_{\boldsymbol{B}^{(k-1)}}(\boldsymbol{u}_{k-1})\,\mathbb{P}_{\boldsymbol{u}}\{\boldsymbol{G}(\boldsymbol{u}_{k-1}) \mid \mathcal{F}_{k-1}\}\big) \\ &= \mathbb{E}_{\boldsymbol{u}}\Big(\mathbb{I}_{\boldsymbol{B}^{(k-1)}}(\boldsymbol{u}_{k-1})\,\mathbb{P}\big(\boldsymbol{G}(v, v')\big)\big|_{(v,v')=\boldsymbol{u}_{k-1}}\Big) \\ &\geq \varepsilon\,\mathbb{P}_{\boldsymbol{u}}\{\boldsymbol{u}_{k-1} \in \boldsymbol{B}^{(k-1)}\}.\end{aligned}$$

A similar argument shows that

$$\mathbb{P}_{\boldsymbol{u}}\{\boldsymbol{u}_m \in \boldsymbol{B}_d\} \geq \varepsilon\,\mathbb{P}_{\boldsymbol{u}}\{\boldsymbol{u}_{m-1} \in \boldsymbol{B}^{(m-1)}\}.$$

Combining the above inequalities, we see that

$$\mathbb{P}_{\boldsymbol{u}}\{\boldsymbol{u}_m \in \boldsymbol{B}_d\} \geq \varepsilon^m \mathbb{P}_{\boldsymbol{u}}\{\boldsymbol{u}_0 \in \boldsymbol{B}^{(0)}\} = \varepsilon^m.$$

This implies inequality (3.168) with $p = \varepsilon^m$.

Step 2. We now prove (3.171). For $\delta > 0$, we set

$$A_\delta = \big\{\mathsf{P}_N\mathcal{R}(v, v') = \mathsf{P}_N\mathcal{R}'(v, v'), |\mathsf{P}_N\mathcal{R}(v, v')|_2 \leq \delta, |\mathsf{Q}_N\zeta(v, v')|_2 \leq \delta\big\},$$

where $\zeta(v, v') = \mathcal{R}(v, v') - S(v)$. Suppose that for any $\delta > 0$ and a constant $\varepsilon' = \varepsilon'(\delta) > 0$ we have

$$\mathbb{P}(A_\delta) \geq \varepsilon' \quad \text{for } (v, v') \in \boldsymbol{B}_R. \tag{3.172}$$

In this case, using the regularising property of S and the Foiaş–Prodi estimate (see Theorems 2.1.18 and 2.1.30), for any $\omega \in A_\delta$ we derive

$$\begin{aligned}
|\mathcal{R}(v, v')|_2 &\leq |\mathsf{P}_N\mathcal{R}(v, v')|_2 + |\mathsf{Q}_N\mathcal{R}(v, v')|_2 \\
&\leq \delta + C_1\alpha_N^{-1/2}\|S(v)\|_1 + |\mathsf{Q}_N\zeta(v, v')|_2 \\
&\leq 2\delta + C_2\alpha_N^{-1/2}|v|_2, \\
|\mathcal{R}'(v, v')|_2 &\leq |\mathcal{R}(v, v')|_2 + |\mathcal{R}'(v, v') - \mathcal{R}(v, v')|_2 \\
&\leq 2\delta + C_2\alpha_N^{-1/2}|v|_2 + C_3(R)\alpha_N^{-1/2}|v - v'|_2.
\end{aligned}$$

Combining these two inequalities, we obtain

$$|\mathcal{R}(v, v')|_2 \vee |\mathcal{R}'(v, v')|_2 \leq 2\delta + C_4(R)\alpha_N^{-1/2}\big(|v|_2 \vee |v'|_2\big).$$

Choosing $N \geq 1$ and $\delta > 0$ so that $C_4(R)\alpha_N^{-1/2} \leq 1/2$ and $\delta = d/6$, we see that the right-hand side of this inequality does not exceed $(\frac{3}{4}(|v|_2 \vee |v'|_2)) \vee d$. We thus arrive at (3.171) with $q = \frac{3}{4}$.

Step 3. It remains to establish (3.172). The construction of $(\mathcal{R}, \mathcal{R}')$ implies that $\mathsf{Q}_N\zeta$ is independent of $(\mathsf{P}_N\mathcal{R}, \mathsf{P}_N\mathcal{R}')$. Hence,

$$\mathbb{P}(A_\delta) \geq \mathbb{P}(A_\delta^{(1)})\,\mathbb{P}(A_\delta^{(2)}), \tag{3.173}$$

where we set

$$\begin{aligned}
A_\delta^{(1)} &= \big\{|\mathsf{Q}_N\zeta(v, v')|_2 \leq \delta\big\}, \\
A_\delta^{(2)} &= \big\{|\mathsf{P}_N\mathcal{R}(v, v')|_2 \leq \delta, \mathsf{P}_N\mathcal{R}(v, v') = \mathsf{P}_N\mathcal{R}'(v, v')\big\}.
\end{aligned}$$

Since $\mathcal{D}(\zeta(v, v')) = \mathcal{D}(\eta_1)$, the conditions imposed on b_j and $\mathcal{D}(\xi_{jk})$ imply that

$$\mathbb{P}(A_\delta^{(1)}) \geq \varepsilon_1(\delta) > 0 \quad \text{for any } \delta > 0.$$

Furthermore, recalling that the pair $\big(\mathsf{P}_N\mathcal{R}(v,v'),\mathsf{P}_N\mathcal{R}'(v,v')\big)$ is a maximal coupling for $(\chi_N^v,\chi_N^{v'})$, where $\chi_N^u=(\mathsf{P}_N)_*\mathcal{D}(S(u)+\eta_1)$, and using Lemma 1.2.26, we derive

$$\mathbb{P}(A_\delta^{(2)})=\big(\chi_N^v\wedge\chi_N^{v'}\big)\big(B_{H_{(N)}}(\delta)\big).$$

Now note that $\chi_N^u(dx)=p(x-\mathsf{P}_NS(u))\,dx$, where p stands for the density of the law for $\mathsf{P}_N\eta_1$ with respect to the Lebesgue measure on $H_{(N)}$ (see (3.83)). Since $p(x)>0$ almost everywhere, we conclude that

$$\inf_{(v,v')\in \boldsymbol{B}_R}\big(\chi_N^v\wedge\chi_N^{v'}\big)\big(B_{H_{(N)}}(\delta)\big)\ge\varepsilon_2(\delta)>0\quad\text{for any }\delta>0.$$

The required inequality (3.172) follows now from (3.173). □

3.6.3 Exponential squeezing for Navier–Stokes equations with unbounded kicks

In this section, we establish Proposition 3.4.4. Its proof is divided into three steps.

Step 1. For any $M>0$, we define the stopping times

$$\boldsymbol{\tau}(M)=\min\Big\{k\ge1:\sum_{j=0}^k\int_0^T\big(\|S_t(u_j)\|_1^2+\|S_t(u_j')\|_1^2\big)\,dt>M(k+1)\Big\},$$
$$\sigma(M)=\boldsymbol{\tau}(M)\wedge\min\{k\ge1:\mathsf{P}_Nu_k\ne\mathsf{P}_Nu_k'\}.$$

We claim that for any $M>0$ there is an integer $N\ge1$ such that, for any $d>0$ and an appropriate choice of the parameter $c>0$ entering the definition of σ_0, relation (3.129) holds, that is, $\sigma\le\sigma_0$ almost surely. Indeed, by the Foiaş–Prodi estimate (2.62), on the set $\{\sigma(M)>k\}$ we have

$$|u_k-u_k'|_2\le\big(C\alpha_N^{-1/2}\big)^k\exp\big(C(M+1)k\big)|u_0-u_0'|_2\le2d\big(Ce^{C(M+1)}\alpha_N^{-1/2}\big)^k.$$

Choosing $c=3d$ and $N\ge1$ so large that $Ce^{C(M+1)}\alpha_N^{-1/2}\le e^{-1}$, we see that (3.129) is true.

Step 2. We now prove that, for sufficiently large $M>0$ and $N\ge1$, the stopping time $\sigma(M)$ satisfies (3.32)–(3.34). To this end, it suffices to show that

$$\mathbb{P}_{\boldsymbol{u}}\{\sigma(M)=k\}\le2e^{-2k}\quad\text{for }k\ge1,\ \boldsymbol{u}\in\boldsymbol{B}_d. \tag{3.174}$$

Indeed, if these relations are established, then

$$\mathbb{P}_{\boldsymbol{u}}\{\sigma(M)=\infty\}=1-\sum_{k=1}^{\infty}\mathbb{P}_{\boldsymbol{u}}\{\sigma(M)=k\}\geq 1-2\sum_{k=1}^{\infty}e^{-2k}>0,$$

$$\mathbb{E}_{\boldsymbol{u}}\big(\mathbb{I}_{\{\sigma(M)<\infty\}}e^{\sigma(M)}\big)=\sum_{k=1}e^{k}\mathbb{P}_{\boldsymbol{u}}\{\sigma(M)=k\}=2\sum_{k=1}^{\infty}e^{-k}<\infty.$$

Furthermore, by inequality (2.82), we have

$$\mathbb{E}_{\boldsymbol{u}}\big(\mathbb{I}_{\{\sigma(M)<\infty\}}|\boldsymbol{u}_{\sigma(M)}|^2\big)\leq \mathbb{E}_{\boldsymbol{u}}|\boldsymbol{u}|^2+C_1\leq 2d^2+C_1,\quad \boldsymbol{u}\in\boldsymbol{B}_d,$$

where $C_1>0$ is a constant depending on $\mathbb{E}\,|\eta_1|_2^4$ and $\sum_k(\mathbb{P}_{\boldsymbol{u}}\{\sigma(M)=k\})^{1/2}$. Thus, it remains to prove (3.174).

Step 3. Let us introduce the events

$$A(k)=\{\mathsf{P}_N u_k=\mathsf{P}_N u_k'\},\quad \bar{A}(k)=\bigcap_{l=1}^{k}A(l),$$

$$D(k)=\Big\{\sum_{j=0}^{k}\int_0^T\big(\|S_t(u_j)\|_1^2+\|S_t(u_j')\|_1^2\big)\,dt\leq M(k+1)\Big\}.$$

We argue by induction on $k\geq 1$. For $k=1$, we have

$$\{\sigma(M)=1\}\subset\{\boldsymbol{\tau}(M)=1\}\cup A(1)^c. \tag{3.175}$$

Proposition 2.3.8 implies that if $M>0$ is sufficiently large, then

$$\mathbb{P}_{\boldsymbol{u}}\big\{\boldsymbol{\tau}(M)=1\big\}\leq e^{-2}. \tag{3.176}$$

Furthermore, since $S:H\to H$ is Lipschitz continuous on bounded subsets (see Proposition 2.1.25), we have

$$|S(v)-S(v')|_2\leq C_2|v-v'|\leq 2C_2d\quad\text{for }(v,v')\in\boldsymbol{B}_d.$$

Assertion (iii) of Lemma 3.2.6 now implies that

$$\mathbb{P}_{\boldsymbol{u}}\{A(1)^c\}=\mathbb{P}_{\boldsymbol{u}}\{\mathsf{P}_N u_k\neq\mathsf{P}_N u_k'\}\leq C_3(N)d\leq e^{-2}\quad\text{for }d\ll 1.$$

Combining this with (3.175) and (3.176), we obtain (3.174) with $k=1$.

We now assume that $k=m\geq 2$ and for $1\leq k\leq m-1$ inequality (3.174) is already established. It follows from the definition of $\sigma(M)$ that (cf. (3.175))

$$\{\sigma(M)=m\}\subset\{\boldsymbol{\tau}(M)=m\}\cup B(m), \tag{3.177}$$

where $B(m)=\bar{A}(m-1)\cap A(m)^c\cap\{\boldsymbol{\tau}(M)>m\}$. Let us estimate the probability of the two events on the right-hand side of (3.177). Proposition 2.3.8

implies that, for $M \gg 1$, we have

$$\mathbb{P}_{\boldsymbol{u}}\{\boldsymbol{\tau}(M) = m\} \le C_4 Q^m e^{-\delta M(m+1)} \le e^{-2m}. \tag{3.178}$$

Furthermore, using (3.174) with $1 \le k \le m-1$, for any $\boldsymbol{u} \in \boldsymbol{B}_d$ we derive

$$\mathbb{P}_{\boldsymbol{u}}\{\bar{A}(m-1) \cap D(m-1)\} \ge \mathbb{P}_{\boldsymbol{u}}\{\sigma(M) \ge m\} \ge 1 - 2\sum_{k=1}^{m-1} e^{-2k} \ge \frac{1}{2}. \tag{3.179}$$

Therefore, since $\{\boldsymbol{\tau}(M) > m\} \subset D(m-1)$, we can write

$$\begin{aligned}\mathbb{P}_{\boldsymbol{u}}(B(m)) &\le \mathbb{P}_{\boldsymbol{u}}\{\bar{A}(m-1) \cap A(m)^c \cap D(m-1)\} \\ &\le 2\,\mathbb{P}_{\boldsymbol{u}}\{A(m)^c \mid \bar{A}(m-1) \cap D(m-1)\},\end{aligned} \tag{3.180}$$

where we used (3.179) to get the second inequality. By the Markov property, the conditional probability $\mathbb{P}_{\boldsymbol{u}}\{A(m)^c \mid \mathcal{F}_{m-1}\}$ depends only on $\boldsymbol{u}_{m-1}$. The construction of the coupling operators and Lemma 3.2.6 imply that

$$\mathbb{P}_{\boldsymbol{u}}\{A(m)^c \mid \mathcal{F}_{m-1}\} \le C_5(N)\left|S(u_{m-1}) - S(u'_{m-1})\right|_2.$$

It follows from Theorem 2.1.30 and Remark 2.1.31 that, for $N \gg 1$ and $\mathbb{P}_{\boldsymbol{u}}$-a.e. $\omega \in \bar{A}(m-1) \cap D(m-1)$,

$$\left|S(u_{m-1}) - S(u'_{m-1})\right|_2 \le 2d\left(C\alpha_N^{-1/2}\right)^m e^{C(M+1)m} \le C_6 d\, e^{-2m}.$$

Combining these two inequalities with (3.180), we see that

$$\mathbb{P}_{\boldsymbol{u}}(B(m)) \le C_7(N) d\, e^{-2m}.$$

Choosing $d \ll 1$ and evoking (3.177) and (3.178), we get the required inequality (3.174) with $k = m$. The proof of Proposition 3.4.4 is complete.

3.7 Relevance of the results for physics

The uniqueness of a stationary distribution and the convergence to this distribution are very important for statistical hydrodynamics. In the physical literature, these two properties are usually postulated. For instance, in the classical book of G. K. Batchelor [Bat82], it is taken for granted that *dynamical systems with a large number of degrees of freedom, and with coupling between these degrees of freedom, approach a statistical state which is independent (partially, if not wholly) of the initial condition*; see there pp. 6–7. The results of this chapter rigorously prove this postulate for periodic 2D turbulent flows (with zero space average) and for flows in bounded 2D domains, driven by a non-degenerate random force.

Namely, we show that if in Eq. (1) the random force $f(t, x)$ is a kick force, or a white-in-time force, or a compound Poisson process (see the Introduction and Section 2.2) and if it is non-degenerate, then Eq. (1) has a unique stationary distribution μ, which is a measure in the space of divergence-free vector fields $\{u(x)\}$. The distribution of every solution $u(t, x)$ for (1) at time t (which is a measure in the same function space) converges to μ exponentially fast when $t \to \infty$. The measure μ is one-smoother than the force $f(t, x)$, regarded as a function of the space variables, and any functional $g(u)$ is integrable with respect to $\mu(du)$, unless $g(u)$ grows at infinity very fast (i.e., faster than $\exp(\delta|u|_2^2)$ with some fixed positive δ). So for practically any observable $g(u)$ and for any solution $u(t)$ of (1) our results imply that the averaged-in-ensemble observable $\mathbb{E}\, g(u(t))$ approaches $\int g(u)\,\mu(du)$ exponentially fast when the time t goes to infinity, provided that the random force f is as above and is sufficiently smooth in the x-variable. In particular, this is true if the observable $g(u)$ is the energy of a flow u, or its enstrophy, or the correlation tensor $u^i(x)u^j(y)$ (where i and j are 1 or 2, and x, y are any fixed points).

In the case of periodic boundary conditions when the flows u have zero space average, the stationary measure μ is space-homogeneous if so is the force f. In this situation, we have

$$\int g(u)\,\mu(du) = \int \langle g\rangle(u)\mu(du),$$

where $\langle g\rangle$ is the space-averaged observable defined by

$$\langle g\rangle(u) = (2\pi)^{-2}\int_{\mathbb{T}^2} g(u_y(\cdot))\,dy, \quad u_y(x) = u(x+y)$$

(we assumed that both periods equal 2π). In particular, if $g(u)$ is a linear functional, then $\langle g\rangle(u) = g\langle u\rangle = 0$, so $\int g(u)\,\mu(du) = 0$. For example, choosing $g(u) = u^j(x)$ we get

$$\int u^j(x)\,\mu(du) = 0 \quad \text{for } j = 1, 2 \text{ and for any } x.$$

That is, for homogeneous turbulence the mean value of the velocity, evaluated at any fixed point, vanishes.

Notes and comments

The investigation of the uniqueness of a stationary distribution for the Markov process generated by the randomly forced 2D Navier–Stokes equation began in 1995 by the pioneering article of Flandoli and Maslowski [FM95]. The main result of their paper is Theorem 3.5.1 of this section on the uniqueness of a stationary distribution and the mixing property in total variation norm. The proof in [FM95] is obtained by adjusting Doob's argument [Doo48] to the infinite-dimensional system defined by the Navier–Stokes equations (3.84), under the crucial assumption that the noise is a rough function of x (i.e., the coefficients b_j decay to zero very slowly). The Flandoli–Maslowski theorem was generalised in various directions. Ferrario [Fer97; Fer99] gave a simpler proof of Theorem 3.5.1, allowing for more regular noises. Bricmont, Kupiainen, and Lefevere [BKL01] established the uniqueness of a stationary measure and exponential mixing for the Navier–Stokes system perturbed by a rough kick force. Goldys and Maslowski [GM05] proved a similar result in the case of white-noise force with low spatial regularity. Da Prato, Debussche, and Tubaro [DDT05], Barbu and Da Prato [BD07], and Da Prato, Röckner, Rozovskii, and Wang [DRRW06] studied the Burgers, magneto-hydrodynamics, and porous media equations in analogous settings. Eckmann and Hairer [EH01] and Hairer [Hai02a] considered a one-dimensional Ginzburg–Landau equation with a mildly degenerate rough forcing and proved the exponential convergence to a unique stationary measure in the total variation norm.

From the physical point of view, the roughness condition imposed on the noise is not very natural, and much effort was spent to remove it. In the context of randomly forced PDEs, a first result in this direction was obtained by Sinai [Sin91]. He studied the Burgers equation on the real line, perturbed by a white-in-time space-periodic random force. Sinai proved that there exists a space-periodic stationary measure which attracts distributions of a large class of solutions (these solutions are not space-periodic, and the uniqueness of a space-periodic stationary measure was not established). A few years later Mattingly [Mat99], then a PhD student of Sinai, established that the 2D Navier–Stokes system considered on the torus $\mathbb{T}^2$, with a large viscosity and spatially regular white force, has a unique stationary measure, which attracts distributions of all solutions.[9]

A first result on the uniqueness of a stationary measure for Navier–Stokes equations with any positive viscosity and a smooth-in-x random force was obtained by Kuksin and Shirikyan [KS00] (see also [KS01b; KS02b] for some further developments). They proved Theorem 3.5.11, applicable to a large class of dissipative PDEs perturbed by a smooth random kick force which is non-zero in sufficiently many modes. The proof is based on two key ingredients: a Lyapunov–Schmidt reduction and a version of the Ruelle–Perron–Frobenius theorem. The first of them is a well-known tool for studying dissipative PDEs and was used, for instance, in the theory of inertial manifolds. It is a consequence

[9] The large viscosity case is much simpler since, in this situation, inequality (2.130) implies that any two solutions with non-random initial data converge exponentially fast almost surely; cf. Exercise 2.5.9.

of the Foiaş–Prodi estimates and enables one to reduce the Navier–Stokes equations to a finite-dimensional system with memory (which may be regarded as an abstract Gibbs system in the sense of Bowen [Bow75]). The second ingredient provides a sufficient condition for the uniqueness of a stationary measure for Markov semigroups. Various versions of this result were known earlier (see [Rue68; LY94; Sza97]), however, the uniform Feller property introduced in [KS00] was a crucial point and turned out to be useful in other situations; e.g., see [LS06]. The simple proof of the Ruelle–Perron–Frobenius theorem given in Section 3.5.6 seems to be new. Next, Weinan E, Mattingly, and Sinai [EMS01] and Bricmont, Kupiainen, and Lefevere [BKL02] studied the stochastic Navier–Stokes system in the case when the space variables x belong to the torus and the right-hand side is white noise in time. They showed that there is at most one stationary measure. Moreover, it was established in [BKL02] that the distributions of solutions converge to the stationary measure exponentially fast. These two works are similar to [KS00]: they also are based on the Lyapunov–Schmidt reduction, followed by a study of the resulting finite-dimensional system with memory. The papers [EMS01; BKL02] treat the case when the random force has a large, but finite number of excited modes (so it is a random trigonometric polynomial in x). An important contribution of [EMS01] to this field of research is the introduction of Girsanov's theorem as a tool for studying ergodicity of stochastic PDEs with spatially regular white noise.

A different approach for investigating the mixing behaviour of randomly forced PDEs was developed by Kuksin and Shirikyan [KS01a], Mattingly [Mat02b], Masmoudi and Young [MY02], and Hairer [Hai02b]. It is based on the concept of coupling and enables one to obtain stronger results with shorter proofs. Namely, it was established in [KS01a; KPS02; Kuk02b] that the Navier–Stokes system perturbed by a bounded kick force is exponentially mixing. The approach presented in Sections 3.1.1 and 3.2 is taken from these articles. In particular, the method of the Kantorovich functional as a tool to study mixing for stochastic Navier–Stokes systems was introduced in [KPS02; Kuk02b], while the dual-Lipschitz distance as a suitable metric for convergence to a stationary distribution in the context of randomly forced PDEs was suggested in [KS01a]. These two tools are now commonly used in work in this field. A different, but closely related method was developed independently by Masmoudi and Young [MY02]. Mattingly [Mat02b] combined the coupling with a stopping-time technique to establish the exponential mixing for the Navier–Stokes system with periodic boundary conditions and a random force which is white noise in time and a trigonometric polynomial in x of large finite degree. Similar techniques were used by Hairer [Hai02b] to study some parabolic PDEs with nonlinear dissipation. Various versions of the coupling argument were applied by Kuksin and Shirikyan [KS02a] to prove exponential mixing for the Navier–Stokes system in a bounded domain with a spatially regular white noise of infinite dimension, by Shirikyan [Shi04] to study the Navier-Stokes system perturbed by an unbounded kick force, and by Debussche and Odasso [DO05] to investigate a 1D Schrödinger equation with linear damping. Further results on the problem of mixing for various types of randomly forced PDEs can be found in [Shi05a; Shi06b; Ner08; Oda08; Shi08]. The presentation of Sections 3.1.2 and 3.1.3 follows essentially [Shi04; Shi05a; Shi08], and the proof of the stability property of Section 3.3.3 is based on some ideas taken from [EMS01; Oda08]. We refer the reader to the review papers [ES00; Kuk02a; Bri02; Mat03; Hai05; Shi05b] for further discussions of the results on uniqueness and mixing for stochastic PDEs with sufficiently non-degenerate smooth noise.

All the above results concern the case when all determining modes of the problem in question are perturbed by a random force. In particular, when these results are applied to the Navier–Stokes system, in order to have a unique stationary measure, one has to assume that the dimension of the random perturbation goes to infinity when the viscosity goes to zero; e.g., every mode of the random force is excited. An important question is whether one can prove the uniqueness and mixing for finite-dimensional noise for any positive value of the viscosity. Progress in this direction was achieved by Hairer and Mattingly in 2006. Combining the method of the Kantorovich functional with an infinite-dimensional Malliavin calculus (developed in [MP06]), they proved in [HM06; HM08; HM11] that if the problem is studied on the standard torus and the deterministic part of the noise is zero, then the uniqueness of a stationary measure is true for a four-dimensional white-noise force and the exponential convergence of other solutions to the stationary measure holds in the dual-Lipschitz metric; see Theorem 3.5.4. Two key questions remain open in this context. Namely, does a similar result hold when: (a) the deterministic part of the external force is non-zero; (b) the random perturbation is not white in time. Note that the answers to both questions are positive for different types of degenerate noise. Namely, it was proved recently by Shirikyan [Shi11a] that the Navier–Stokes equations with a random perturbation supported by a given subregion of the space domain possess the exponential mixing property under some mild non-degeneracy assumptions. We also mention the work [AKSS07], in which it is proved that, for a large class of low-dimensional random forces, the support of the law $\mathcal{D}(u(t))$ of a solution of the 2D Navier–Stokes system is infinite-dimensional for any $t > 0$; see Section 6.3.1 for more details.

4

Ergodicity and limiting theorems

In this chapter, we study limiting theorems for the 2D Navier–Stokes system with random perturbations. To simplify the presentation, we shall confine ourselves to the case of spatially regular white noise; however, all the results remain true for random kick forces. The first section is devoted to the derivation of the strong law of large numbers (SLLN), the law of the iterated logarithm (LIL), and the central limit theorem (CLT). Our approach is based on the reduction of the problem to similar questions for martingales and an application of some general results on SLLN, LIL, and CLT. In Section 4.2, we study the relationship between stationary distributions and random attractors. Roughly speaking, it is proved that the support of the random probability measure obtained by the disintegration of the unique stationary distribution is a random point attractor for the RDS in question. The third section deals with the stationary distributions for the Navier–Stokes system perturbed by a random force depending on a parameter. We first prove that the stationary measures continuously depend on spatially regular white noise. We next consider high-frequency random kicks and show that, under suitable normalisation, the corresponding family of stationary measures converges weakly to the unique stationary distribution corresponding to the white-noise perturbation. Finally, in Section 4.4, we discuss the physical relevance of the results of this chapter.

4.1 Ergodic theorems

4.1.1 Strong law of large numbers

Let us consider the Navier–Stokes system (3.84) in which $h \in H$ is a given function and η is a spatially regular white noise of the form (2.66). As was established in Theorem 3.3.1, if sufficiently many first coefficients b_j are non-zero, then Eq. (3.84) has a unique stationary distribution. Moreover, by Theorem 3.5.2, the Markov process $(u_t, \mathbb{P}_u)$ defined by the equation is exponentially mixing, and by Corollary 3.5.3, for any functional $f \in C^\gamma(H, w_\varkappa)$, where $\gamma \in (0, 1]$ and $w_\varkappa(r) = e^{\varkappa r^2}$ with $0 < \varkappa \ll 1$, and any initial datum $v \in H$ the mean value of the corresponding trajectory $\mathbb{E}_v f(u_t)$ converges to (f, μ) as $t \to \infty$ exponentially fast (see inequality (3.133)). The following theorem shows that the mean value with respect to the probability ensemble can be replaced by a time average. In this case, however, the convergence to the mean value holds only at an algebraical rate.

Theorem 4.1.1 *Under the hypotheses of Theorem 3.5.2, there is a constant $\varkappa > 0$ such that for any $\varepsilon > 0$, $\gamma \in (0, 1]$, $v \in H$, and $f \in C^\gamma(H, w_\varkappa)$, with $\mathbb{P}_v$-probability* 1, *we have*

$$\lim_{t\to\infty} t^{\frac{1}{2}-\varepsilon}\left(t^{-1}\int_0^t f(u_s)\,ds - (f, \mu)\right) = 0.$$

Proof We first outline the main idea. Without loss of generality, we can assume that $(f, \mu) = 0$. We need to prove that, for any $v \in H$,

$$t^{-\frac{1}{2}-\varepsilon}\int_0^t f(u_s)\,ds \to 0 \quad \text{with } \mathbb{P}_v\text{-probability } 1. \tag{4.1}$$

Let us denote by $\mathcal{F}_t$ the filtration corresponding to the Markov process $(u_t, \mathbb{P}_v)$ (see Section 1.3.1) and consider Gordin's *martingale approximation*:

$$M_t = \int_0^\infty \big(\mathbb{E}_v(f(u_s)\,|\,\mathcal{F}_t) - \mathbb{E}_v(f(u_s)\,|\,\mathcal{F}_0)\big)\,ds, \qquad v \in H, \quad t \ge 0. \tag{4.2}$$

By Proposition A.13.3, the process M_t is a martingale with respect to the filtration $\{\mathcal{F}_t\}$ and the probability $\mathbb{P}_v$. Furthermore, relation (A.70) implies that

$$\int_0^t f(u_s)\,ds = \int_0^{\hat{t}} f(u_s)\,ds + \int_t^{\hat{t}} f(u_s)\,ds = M_{\hat{t}} - g(u_{\hat{t}}) + g(u_0) + d_t, \tag{4.3}$$

where $\hat{t}$ stands for the integer part of $t \ge 0$,

$$d_t = \int_{\hat{t}}^t f(u_s)\,ds, \quad g(u) = \int_0^\infty \mathfrak{P}_s f(u)\,ds. \tag{4.4}$$

Using representation (4.3) and a priori estimates for solutions of the Navier–Stokes system, we show that

$$\lim_{t\to\infty}\left|t^{-\frac{1}{2}}\int_0^t f(u_s)\,ds - \hat{t}^{-\frac{1}{2}}M_{\hat{t}}\right| = 0 \quad \mathbb{P}_v\text{-almost surely.} \tag{4.5}$$

Hence, to prove (4.1), it suffices to establish the strong law of large number (SLLN) for the discrete-time martingale M_k. This will be done with the help of Theorem A.12.1. We now turn to the accurate proof divided into two steps.

Step 1. We first prove (4.5). In view of (4.3), it suffices to show that

$$\mathbb{P}_v\Big\{\lim_{k\to\infty} k^{-\frac{1}{2}}\Big(\sup_{k\le t\le k+1}|d_t|+|g(u_k)|\Big)=0\Big\}=1. \tag{4.6}$$

In what follows, we denote by C_i positive constants that may depend on f, but not on u. Let us set

$$U_k=\sup_{k\le t\le k+1}|u_t|_2^2, \quad D_k=\sup_{k\le t\le k+1}|d_t|.$$

By inequality (2.130), we can find positive constants C_1 and γ such that

$$\mathbb{P}_v\{U_0\ge\rho\}\le C_1 e^{\gamma|v|_2^2-\gamma\rho} \quad \text{for } v\in H,\ \rho\ge 0.$$

Combining this with the Markov property and inequality (2.125), we obtain

$$\mathbb{P}_v\{U_k\ge\rho\}=\mathbb{E}_v\mathbb{P}_{u_k}\{U_0\ge\rho\}\le C_1e^{-\gamma\rho}\mathbb{E}_v e^{\gamma|u_k|_2^2}\le C_2\big(e^{\gamma|v|_2^2}+1\big)e^{-\gamma\rho}. \tag{4.7}$$

On the other hand, in view of Corollary 3.5.3, we have

$$|g(v)|\le\int_0^\infty|\mathfrak{P}_s f(v)|\,ds\le C_3e^{\varkappa|v|_2^2}, \quad v\in H. \tag{4.8}$$

Furthermore, the definition of d_t implies that

$$D_k\le C_4e^{\varkappa U_k} \quad \text{for } k\ge 0. \tag{4.9}$$

Combining (4.7)–(4.9), for $\varkappa<\frac{\gamma}{4}$ we derive

$$\begin{aligned}\sum_{k=1}^\infty\mathbb{P}_v\big\{D_k+|g(u_k)|\ge k^{1/4}\big\}&\le\sum_{k=1}^\infty\mathbb{P}_v\big\{C_4e^{\varkappa U_k}+C_3e^{\varkappa|u_k|^2}\ge k^{1/4}\big\}\\&\le\sum_{k=1}^\infty\mathbb{P}_v\big\{C_5e^{\varkappa U_k}\ge k^{1/4}\big\}\\&=\sum_{k=1}^\infty\mathbb{P}_v\big\{U_k\ge\tfrac{1}{4\varkappa}\ln k-C_6\big\}\\&\le C_7\big(e^{\gamma|v|_2^2}+1\big)\sum_{k=1}^\infty k^{-\gamma/4\varkappa}<\infty.\end{aligned}$$

Hence, by the Borel–Cantelli lemma, there is a $\mathbb{P}_v$-almost surely finite random integer $k_0\ge 1$ such that

$$\sup_{k\le t\le k+1}|d_t|+|g(u_k)|\le k^{1/4} \quad \text{for } k\ge k_0.$$

This implies the required relation (4.6).

Step 2. We now prove that

$$\mathbb{P}_v\Big\{\lim_{k\to\infty} k^{-\frac{1}{2}-\varepsilon} M_k = 0\Big\} = 1 \quad \text{for any } v \in H.$$

In view of Theorem A.12.1, this claim will be established if we check that the martingale differences $X_k = M_k - M_{k-1}$ satisfies the condition

$$\sum_{k=1}^{\infty} k^{-1-\delta}\,\mathbb{E}_v X_k^2 < \infty \quad \text{for any } v \in H, \delta > 0. \tag{4.10}$$

It follows from (4.3) that

$$X_k = g(u_k) - g(u_{k-1}) + d_k^-, \quad d_k^- = \int_{k-1}^{k} f(u_s)ds. \tag{4.11}$$

By the Markov property, we have

$$\mathbb{E}_v X_k^2 = \mathbb{E}_v \mathbb{E}_v(X_k^2 \mid \mathcal{F}_{k-1}) = \mathbb{E}_v \varphi(u_{k-1}) = (\mathfrak{P}_{k-1}\varphi)(v), \tag{4.12}$$

where $\varphi(v) = \mathbb{E}_v X_1^2$. We see that

$$\sum_{k=1}^{\infty} k^{-1-\delta}\,\mathbb{E}_v X_k^2 = \sum_{k=1}^{\infty} k^{-1-\delta}\,(\mathfrak{P}_{k-1}\varphi)(v). \tag{4.13}$$

To estimate the right-hand side of this relation, let us derive an explicit formula for φ. It follows from (4.11) that

$$\begin{aligned}\varphi(v) = {} & \mathbb{E}_v g^2(u_1) + \mathbb{E}_v g^2(u_0) + \mathbb{E}_v (d_1^-)^2 \\ & - 2\,\mathbb{E}_v\big(g(u_1)g(u_0)\big) + 2\,\mathbb{E}_v\big(d_1^- g(u_1)\big) - 2\mathbb{E}_v\big(d_1^- g(u_0)\big).\end{aligned} \tag{4.14}$$

Now note that

$$\mathbb{E}_v g^2(u_0) = g^2(v), \qquad \mathbb{E}_v g^2(u_1) = \mathfrak{P}_1 g^2(v),$$

$$\mathbb{E}_v\big(g(u_0)g(u_1)\big) = g(v)\mathfrak{P}_1 g(v), \qquad \mathbb{E}_v\big(d_1^- g(u_0)\big) = g(v)\int_0^1 \mathfrak{P}_s f(v)\,ds.$$

Furthermore, using the Markov property, we write

$$\begin{aligned}\mathbb{E}_v (d_1^-)^2 &= \mathbb{E}_v\left(\int_0^1 f(u_s)\,ds\right)^2 = \int_0^1\!\!\int_0^1 \mathbb{E}_v\big(f(u_s)f(u_t)\big)\,dsdt \\ &= 2\int_0^1\!\!\int_0^t \mathfrak{P}_s\big(f\mathfrak{P}_{t-s}f\big)(v)\,dsdt,\end{aligned}$$

$$\mathbb{E}_v\big(d_1^- g(u_1)\big) = \mathbb{E}_v\left(g(u_1)\int_0^1 f(u_s)\,ds\right) = \int_0^1 \mathfrak{P}_s\big(f\mathfrak{P}_{1-s}g\big)(v)\,ds.$$

We now need the following lemma, whose proof is given at the end of this subsection.

Lemma 4.1.2 *Under the hypotheses of Theorem 4.1.1, for any $\gamma \in (0, 1]$ and sufficiently small $\varkappa > 0$ there exist positive constants $\alpha \le \gamma$ and ρ such that the function g defined by (4.4) belongs to $C^\alpha(H, w_\rho)$ for any $f \in C^\gamma(H, w_\varkappa)$ satisfying the condition $(f, \mu) = 0$, and the linear operator taking f to g is bounded in the corresponding spaces. Moreover, the constants α and ρ can be chosen in such a way that $\rho \to 0$ as $\varkappa \to 0$.*

Lemmas 3.5.6 and 4.1.2 show[1] that all the functions on the right-hand side of (4.14) belong to the space $C^\alpha(H, w_\rho)$ with suitable constants α and ρ, and their norms are bounded by $C|f|_{w_\varkappa,\gamma}$. Therefore we have $\varphi \in C^\alpha(H, w_\rho)$ and

$$|\varphi|_{w_\rho,\alpha} \le C_7 |f|_{w_\varkappa,\gamma}. \tag{4.15}$$

Furthermore, since $\rho \to 0$ as $\varkappa \to 0$, Corollary 3.5.3 implies that if $\varkappa > 0$ is sufficiently small, then

$$\big|(\mathfrak{P}_k\varphi)(v) - (\varphi, \mu)\big| \le C_8(|v|_2)\, e^{-\beta k}, \quad k \ge 0.$$

Combining this with (4.13), we see that condition (4.10) is satisfied. This completes the proof of the theorem. □

Proof of Lemma 4.1.2 Since g satisfies (4.8), to prove that $g \in C^\alpha(H, w_\rho)$, it suffices to show that if $|f|_{\gamma,w_\varkappa} \le 1$ and $(f, \mu) = 0$, then

$$\big|g(u) - g(v)\big| \le C_2|u - v|^\alpha\, e^{\rho(|u|_2^2 + |v|_2^2)}, \quad u, v \in H. \tag{4.16}$$

By Corollary 3.5.3, for any such function $f \in C^\gamma(H, w_\varkappa)$, we have

$$|\mathfrak{P}_t f(u)| \le C \exp(-\beta t + \varkappa |u|_2^2), \quad t \ge 0, \quad u \in H. \tag{4.17}$$

Combining this with (3.140), for any $u, v \in H$ and $T > 0$, we derive

$$\begin{aligned}\big|g(u) - g(v)\big| &\le \int_0^T \big|\mathfrak{P}_t f(u) - \mathfrak{P}_t f(v)\big|\, dt + \int_T^\infty \big(|\mathfrak{P}_t f(u)| + |\mathfrak{P}_t f(v)|\big)\, dt \\ &\le C_3\, |u - v|^\alpha\, e^{bT + \rho(|u|_2^2 + |v|_2^2)} + C_4 e^{-\beta T}\big(e^{\varkappa|u|_2^2} + e^{\varkappa|v|_2^2}\big).\end{aligned}$$

Choosing $T = \varepsilon\varkappa\big(|u|_2^2 + |v|_2^2 + \ln(1 \vee |u - v|_2^{-1})\big)$, where $0 < \varepsilon \ll 1$, we arrive at inequality (4.16) in which α and ρ are replaced with $(\alpha - \varepsilon\varkappa) \wedge (\beta\varepsilon\varkappa)$ and $(\rho + b\varepsilon\varkappa) \vee \varkappa$, respectively.

The fact that the linear operator taking f to g is continuous follows from inequalities (4.8) and (4.16), which are true for any $f \in C^\gamma(H, w_\varkappa)$ such that $|f|_{w_\varkappa,\gamma} \le 1$ and $(f, \mu) = 0$. Finally, the explicit form of the constant ρ implies that it goes to zero with $\varkappa$. □

Exercise 4.1.3 Under the hypotheses of Theorem 4.1.1, prove that, for any function $f \in C^\gamma(H, w_\varkappa)$ with a sufficiently small $\varkappa > 0$ and any measure λ

[1] Lemma 3.5.6 was proved for the Navier–Stokes system on the torus. However, when $k = 0$, the assertions remain true in the case of a bounded domain.

satisfying the condition $(w_\varkappa(|u|_2), \lambda) < \infty$, we have

$$\mathbb{P}_\lambda\left\{k^{-1}\sum_{l=0}^{k-1} f(u_l) \to (f,\mu) \text{ as } k\to\infty\right\} = 1.$$

Hint: Use the same scheme as for continuous time. In this case, the martingale approximation is defined by

$$M_k = \sum_{l=0}^{\infty}\big(\mathbb{E}_\lambda(f(u_l)\,|\,\mathcal{F}_k) - \mathbb{E}_\lambda(f(u_l)\,|\,\mathcal{F}_0)\big).$$

4.1.2 Law of the iterated logarithm

The law of large numbers established in the previous subsection can be strengthened to the law of the iterated logarithm (LIL). To simplify the presentation, we shall confine ourselves to the case of stationary solutions. The general case is briefly outlined in Exercise 4.1.6.

As before, we consider the Navier–Stokes system (3.84), in which η is spatially regular white noise of the form (2.66). Let $\{\mathcal{F}_t, t \ge 0\}$ be the filtration associated with the corresponding Markov process $(u_t, \mathbb{P}_v)$. In what follows, we assume that the hypotheses of Theorem 3.5.2 are satisfied and denote by $\mu \in \mathcal{P}(H)$ the unique stationary measure. Recall that, given a function $f \in C^\gamma(H, w_\varkappa)$ with $\varkappa \ll 1$ such that $(f,\mu) = 0$, we defined $g(u) = \int_0^\infty \mathfrak{P}_s f(u)\,ds$; see (4.4). Before formulating the LIL, we establish the following auxiliary result.

Proposition 4.1.4 *Under the above hypotheses, we have*

$$\lim_{t\to\infty} \mathbb{E}_\mu\left(\frac{1}{\sqrt{t}}\int_0^t f(u_s)\,ds\right)^2 = 2(fg,\mu) =: \sigma_f^2, \quad \sigma_f \ge 0. \tag{4.18}$$

Moreover, $\sigma_f > 0$ *if* $f \not\equiv 0$ *and the constants* b_j *entering* (2.66) *are all nonzero.*

Proof Using the Markov property and the stationarity of μ, we write

$$\begin{aligned}
\mathbb{E}_\mu\left(\int_0^t f(u_s)\,ds\right)^2 &= \mathbb{E}_\mu \int_0^t\int_0^t f(u_r)f(u_s)\,dr\,ds \\
&= 2\int_0^t dr\int_r^t \mathbb{E}_\mu\big(f(u_r)\mathbb{E}_\mu(f(u_s)\,|\,\mathcal{F}_r)\big)\,ds \\
&= 2\int_0^t dr\int_r^t \mathbb{E}_\mu\big(f(u_r)(\mathfrak{P}_{s-r}f)(u_r)\big)\,ds \\
&= 2\int_0^t dr\int_r^t (f\mathfrak{P}_{s-r}f,\mu)\,ds = 2\int_0^t (t-s)(f\mathfrak{P}_s f,\mu)\,ds.
\end{aligned}$$

Dividing both sides of this relation by t and using (4.17) together with Lebesgue's theorem to pass to the limit as $t \to \infty$, we arrive at (4.18).

We now assume that $b_j \neq 0$ for all $j \geq 1$ and prove that $\sigma_f > 0$, provided that $f \not\equiv 0$. Suppose, for a contradiction, that $\sigma_f = 0$. We first show that

$$\mathbb{P}_\mu\{M_k = 0 \text{ for all integers } k \geq 0\} = 1. \tag{4.19}$$

Indeed, it follows from relation (4.3), which is true with $\mathbb{P}_\mu$-probability 1, that

$$\int_0^k f(u_s)\,ds = M_k - g(u_k) + g(u_0), \quad k \geq 0. \tag{4.20}$$

In view of the Markov property, we have

$$\begin{aligned}\mathbb{E}_\mu M_k^2 &\leq 3\,\mathbb{E}_\mu\biggl(\int_0^k f(u_s)\,ds\biggr)^2 + 3\,\mathbb{E}_\mu|g(u_k)|^2 + 3\,\mathbb{E}_\mu|g(u_0)|^2\\ &= 3\,\mathbb{E}_\mu\biggl(\int_0^k f(u_s)\,ds\biggr)^2 + 3\int_H \bigl(\mathfrak{P}_k g^2(u) + g^2(u)\bigr)\mu(du).\end{aligned}$$

Combining this with Corollary 3.5.3 and Lemma 4.1.2, we see that

$$\lim_{k\to\infty} k^{-1}\mathbb{E}_\mu M_k^2 \leq 3\,\mathbb{E}_\mu\biggl(\int_0^k f(u_s)\,ds\biggr)^2 = 0. \tag{4.21}$$

On the other hand, since M_k is a martingale with stationary differences X_k, we have

$$\mathbb{E}_\mu M_k^2 = \mathbb{E}_\mu\biggl(\sum_{l=1}^k X_l\biggr)^2 = \mathbb{E}_\mu \sum_{l=1}^k X_l^2 = k\,\mathbb{E}_\mu X_1^2.$$

Substituting this into (4.21), we conclude that $\mathbb{E}_\mu X_k^2 = 0$ for any $k \geq 0$. It follows that $M_k = 0$ almost surely for any $k \geq 0$, and therefore (4.19) holds.

We now prove that relation (4.20) with $M_k = 0$ cannot hold for all $k \geq 0$, unless $f \equiv 0$. Indeed, suppose there is a non-degenerate ball $B \subset H$ and a constant $\varepsilon > 0$ such that $f(u) \geq \varepsilon$ for $u \in B$. Using the fact that $b_j \neq 0$ for all $j \geq 1$, it is not difficult to show that (cf. proof of Lemma 3.3.11)

$$\mathbb{P}_\mu\bigl\{u_t \in B \text{ for } 0 \leq t \leq k\bigr\} > 0 \quad \text{for any } k \geq 1.$$

It follows that

$$\mathbb{P}_\mu\biggl\{\int_0^k f(u_s)\,ds \geq k\varepsilon\biggr\} > 0 \quad \text{for any } k \geq 1. \tag{4.22}$$

On the other hand, since g is bounded on bounded subsets of H, we have $g(v) \leq C$ for $v \in B$. This implies that if $u_0, u_k \in B$, then $|g(u_k) - g(u_0)| \leq 2C$. Choosing $k \geq 1$ so large that $k\varepsilon > 2C$ and recalling (4.22), we see that (4.20) cannot hold with probability 1. The contradiction obtained completes the proof of the proposition. □

Theorem 4.1.5 *Under the hypotheses of Theorem 3.5.2, assume that*[2] $\sigma_f > 0$. *Then for any* $\gamma \in (0, 1]$, *sufficiently small* $\varkappa > 0$, *and any function* $f \in C^\gamma(H, w_\varkappa)$ *such that* $(f, \mu) = 0$ *we have*

$$\mathbb{P}_\mu\left\{\limsup_{t\to\infty} \frac{1}{\omega(t)} \int_0^t f(u_s)\,ds = \sigma_f\right\} = 1, \tag{4.23}$$

$$\mathbb{P}_\mu\left\{\liminf_{t\to\infty} \frac{1}{\omega(t)} \int_0^t f(u_s)\,ds = -\sigma_f\right\} = 1, \tag{4.24}$$

where $\omega(t) = (2t \ln\ln t)^{1/2}$.

Proof In view of (4.5), it suffices to establish the LIL for M_k. To this end, we shall apply Theorem A.12.2. Let us define the conditional variance V_k^2 by relation (A.58) in which $\mathbb{E}$ is replaced by $\mathbb{E}_\mu$. The required result will be established if we show that

$$\mathbb{P}_\mu\{k^{-1} V_k^2 \to \sigma_f^2 \text{ as } k \to \infty\} = 1. \tag{4.25}$$

Repeating the calculations used in Step 2 of the proof of Theorem 4.1.1, we see that

$$k^{-1} V_k^2 = k^{-1} \sum_{l=1}^{k} \varphi(u_{l-1}),$$

where $\varphi(v) = \mathbb{E}_v M_1^2$. As was mentioned in the proof of Theorem 4.1.1 (see inequality (4.15)), the function φ belongs to $C^\alpha(H, w_\rho)$, where $\rho \to 0$ as $\varkappa \to 0$. Therefore, by Exercise 4.1.3 with $\lambda = \mu$ and $f = \varphi$, the required relation (4.25) will be established if we show that $(\varphi, \mu) = \sigma_f^2$.

In view of relation (4.14) and the formulas for the terms on its right-hand side, we have

$$\begin{aligned}(\varphi, \mu) = (\mathfrak{P}_1 g^2, \mu) + (g^2, \mu) + 2\int_0^1\int_0^t \big(\mathfrak{P}_s(f\mathfrak{P}_{t-s} f), \mu\big)\,ds dt \\ - 2(g\mathfrak{P}_1 g, \mu) + 2\int_0^1 \big(\mathfrak{P}_s(f\mathfrak{P}_{1-s} g), \mu\big)\,ds - 2\int_0^1 (g\mathfrak{P}_s f, \mu)\,ds.\end{aligned} \tag{4.26}$$

Let us denote by A_m the mth term on the right-hand side of this relation. Then, using the stationarity of μ, we derive

$$A_1 = (g^2, \mu), \quad A_3 = 2\int_0^1 (1-s)(f\mathfrak{P}_s f, \mu)\,ds \quad A_5 = 2\int_0^1 (f\mathfrak{P}_s g, \mu)\,ds.$$

Substituting these expressions into (4.26) and using the simple relation

$$\mathfrak{P}_s g = \int_0^\infty \mathfrak{P}_{s+t} f\,dt = g - \int_0^s \mathfrak{P}_t f\,dt,$$

we obtain the required result. The proof is complete. □

[2] We recall that, by Proposition 4.1.4, $\sigma_f > 0$ if $b_j \neq 0$ for all $j \geq 1$.

Exercise 4.1.6 Prove that the assertions of Theorem 4.1.5 remain valid for all solutions. *Hint:* There are two possibilities: (a) one can apply Theorem 4.7 of [HH80] to prove that the martingale M_k satisfies the LIL; (b) using the existence of a mixing extension, one can couple a given solution with a stationary one. See the paper [Kuk02a] in which the latter idea is applied to prove a central limit theorem.

Exercise 4.1.7 Under the hypotheses of Theorem 3.5.2, assume that $\sigma_f = 0$. Show that, for any $\gamma \in (0, 1]$, sufficiently small $\varkappa > 0$, and any $f \in C^\gamma(H, w_\varkappa)$ such that $(f, \mu) = 0$, we have

$$\mathbb{P}_\mu \left\{ \lim_{t\to\infty} \frac{1}{\sqrt{t}} \int_0^t f(u_s)\, ds = 0 \right\} = 1. \tag{4.27}$$

Hint: Use (4.19) and (4.5).

4.1.3 Central limit theorem

Let us denote by $N(0, \sigma)$ the centred normal law with variance $\sigma^2 \geq 0$ and by Φ_σ its distribution function (see Section A.4). Recall that the constant $\sigma_f \geq 0$ is defined by (4.18). The following theorem establishes the central limit theorem (CLT) for solutions of (3.84), (2.66).

Theorem 4.1.8 *Under the hypotheses of Theorem 3.5.2, the following convergence holds for an arbitrary $\gamma \in (0, 1]$, sufficiently small $\varkappa > 0$, any function $f \in C^\gamma(H, w_\varkappa)$ with $(f, \mu) = 0$, and any $v \in H$:*

$$\mathcal{D}_v \left(\frac{1}{\sqrt{t}} \int_0^t f(u_s)\, ds \right) \to N(0, \sigma_f) \quad \text{as } t \to \infty, \tag{4.28}$$

where $\mathcal{D}_v$ denote the distribution of a random variable under the law $\mathbb{P}_v$.

Note that, in view of Lemma 1.2.16, the weak convergence (4.28) is equivalent to

$$\mathbb{P}_v \left\{ \frac{1}{\sqrt{t}} \int_0^t f(u_s)\, ds \leq x \right\} \to \Phi_{\sigma_f}(x) \quad \text{as } t \to \infty \text{ for any } x \in \mathbb{R}. \tag{4.29}$$

A more detailed analysis enables one to prove that convergence (4.29) holds with the rate $t^{-1/4-\varepsilon}$ for any $\varepsilon > 0$; see [Shi06c]. This rate is the best possible in the case of martingales (see [Bol82]), but it is not known if it is optimal in our context.

Proof of Theorem 4.1.8 We confine ourselves to the more complicated case $\sigma_f > 0$. The main idea of the proof is the same as in the previous two subsections: we use the martingale approximation to reduce the problem to a similar question for martingales and apply a general result on CLT for martingales.

Let us note that, for any $v \in H$, we have the relation (cf. (4.5))

$$\lim_{t\to\infty} \mathbb{E}_v \left| t^{-1/2} \int_0^t f(u_s)\,ds - \hat{t}^{-1/2} M_{\hat{t}} \right|^2 = 0,$$

which follows from (4.3) and the inequality

$$\mathbb{E}_v \left| d_t - g(u_{\hat{t}}) + g(u_0) \right|^2 \le C_1 \quad \text{for all } t \ge 0,$$

with a constant $C_1 > 0$ depending only on $|v|_2$. Thus, it suffices to prove that

$$\mathcal{D}_v(k^{-1/2} M_k) \to N(0, \sigma_f) \quad \text{as } k \to \infty.$$

By Theorem A.12.3, this convergence will be established if we check that the martingale M_k satisfies Lindeberg's condition (A.64). To this end, it suffices to show that the martingale differences X_k satisfy the inequality

$$\mathbb{E}_v X_k^4 \le C \exp(c\,|v|_2^2) \quad \text{for all } k \ge 0; \tag{4.30}$$

see the remark preceding Theorem A.12.3. To prove (4.30), using the Markov property, we write (cf. (4.12))

$$\mathbb{E}_v X_k^4 = \mathbb{E}_v \mathbb{E}_v \big(X_k^4 \mid \mathcal{F}_{k-1} \big) = \mathbb{E}_v \mathbb{E}_{u_{k-1}} X_1^4 = \mathbb{E}_v \psi(u_{k-1}),$$

where we set $\psi(v) = \mathbb{E}_v X_1^4$. Since

$$\psi(v) \le C_3 \left(\mathbb{E}_v g^4(u_1) + g(v) + \int_0^1 \mathbb{E}_v f^4(u_s)\,ds \right),$$

$f \in C^\gamma(H, w_\varkappa)$, and $g \in C^\alpha(H, w_\rho)$, where $\rho \to 0$ as $\varkappa \to 0$, we see from Lemmas 3.5.6 and 4.1.2 that $\psi(v) \le C_4 \exp(c\,|v|_2^2)$ for $v \in H$ and some small $c > 0$. Applying again Lemma 3.5.6, we arrive at the required estimate (4.30). This completes the proof of Theorem 4.1.8. □

Exercise 4.1.9 Prove analogues of Theorems 4.1.1, 4.1.5, and 4.1.8 for the Navier–Stokes system perturbed by a kick force of the form (2.66). *Hint:* The martingale approximation described in Section A.13 remains true for discrete-time RDS, and the same schemes apply. Note that, in this case the constant σ_f entering the LIL and CLT is defined by the relation $\sigma_f^2 = 2(fg, \mu) - (f^2, \mu)$, where $g(u) = \sum_{k=0}^{\infty} \mathfrak{P}_k f(u)$.

Exercise 4.1.10 Prove the SLLN, LIL, and CLT for Hölder-continuous functionals defined on higher Sobolev spaces. *Hint:* Combine Theorem 3.5.5 with the scheme used above.

4.2 Random attractors and stationary distributions

In this section, we study the relationship between stationary distributions for Markov RDS and random attractors. We first establish a sufficient condition ensuring the existence of a minimal random point attractor in the sense of almost sure convergence. We next prove the Ledrappier–Le Jan–Crauel theorem which shows that there is a one-to-one correspondence between stationary distributions and Markov invariant measures. Finally, we establish the main result of this section claiming that the support of the disintegration of a Markov invariant measure is a minimal random point attractor in the sense of convergence in probability. We shall confine ourselves to the case of continuous-time RDS, which is technically more complicated.

4.2.1 Random point attractors

Let X be a Polish space with a metric dist_X. For any subset $K \subset H$, we define the function

$$\text{dist}_X(u, K) = \inf_{v \in K} \text{dist}_X(u, v),$$

where the infimum over an empty set is equal to $+\infty$. Let $(\Omega, \mathcal{F}, \mathbb{P})$ be a complete probability space and let $\boldsymbol{K} = \{K_\omega, \omega \in \Omega\}$ be a family of closed subsets in X.

Definition 4.2.1 The family $\boldsymbol{K}$ is said to be *measurable* if for any $u \in X$ the real-valued function $\omega \mapsto \text{dist}_X(u, K_\omega)$ is $(\mathcal{F}, \mathcal{B}(\mathbb{R}))$-measurable. In this case, $\boldsymbol{K}$ will be called a *closed random set*. If, in addition, K_ω is compact for almost every ω, then $\boldsymbol{K}$ is called a *compact random set*.

Exercise 4.2.2 Let $\boldsymbol{K}$ be a closed random set in X. Prove the following properties.

(i) If $f : \Omega \to X$ is an $(\mathcal{F}, \mathcal{B}(X))$-measurable function, then the function $\omega \mapsto \text{dist}_X(f(\omega), K_\omega)$ is measurable.
(ii) The function $(\omega, u) \mapsto \text{dist}_X(u, K_\omega)$ is $(\mathcal{F} \otimes \mathcal{B}(X), \mathcal{B}(\mathbb{R}))$-measurable.

Hint: Use approximation by functions taking at most countably many values.

Let us assume that a group of measure-preserving transformations $\boldsymbol{\theta} = \{\theta_t, t \in \mathbb{R}\}$ is defined on Ω and consider an RDS $\boldsymbol{\Phi} = \{\varphi_t^\omega, t \in \mathbb{R}_+\}$ over $\boldsymbol{\theta}$ (see Section 1.3.2). We shall always assume that the trajectories $\varphi_t^\omega(u)$ are continuous in time for any $u \in X$ and $\omega \in \Omega$.

Definition 4.2.3 A compact random set $\boldsymbol{A} = \{A_\omega, \omega \in \Omega\}$ in H is called a *random point attractor for $\boldsymbol{\Phi}$ in the sense of almost sure convergence*[3] if it satisfies the following properties:

(i) *Invariance*. For almost every $\omega \in \Omega$, we have

$$\varphi_t^\omega A_\omega = A_{\theta_t \omega}, \quad t \geq 0. \tag{4.31}$$

[3] In what follows, if there is no confusion, we shall simply say a *random attractor*.

(ii) *Attraction.* For any $u \in X$, we have

$$\mathbb{P}\{\mathrm{dist}_X(\varphi_t^{\theta_{-t}\omega}(u), A_\omega) \to 0 \text{ as } t \to \infty\} = 1. \tag{4.32}$$

Definition 4.2.4 A random point attractor $\boldsymbol{A}$ is said to be minimal if for any other random point attractor $\boldsymbol{A}' = \{A'_\omega, \omega \in \Omega\}$ we have

$$A_\omega \subset A'_\omega \quad \text{for almost every } \omega \in \Omega.$$

The following theorem provides a sufficient condition for the existence of a minimal random point attractor in the sense of almost sure convergence.

Theorem 4.2.5 *Let $\boldsymbol{\Phi} = \{\varphi_t, t \geq 0\}$ be an RDS over $\boldsymbol{\theta}$ in a Polish space X. Suppose that there is a compact random set $\boldsymbol{K} = \{K_\omega\}$ and a subset $\Omega_* \in \mathcal{F}$ of full measure such that $\theta_t(\Omega_*) = \Omega_*$ for all $t \in \mathbb{R}$ and*

$$\mathrm{dist}_X(\varphi_t^{\theta_{-t}\omega}(u), K_\omega) \to 0 \quad \text{as } t \to \infty \text{ for any } \omega \in \Omega_*,\ u \in X. \tag{4.33}$$

Then $\boldsymbol{\Phi}$ possesses a minimal random attractor.

Proof

Step 1. For any $u \in X$ and $\omega \in \Omega_*$, we set

$$A_\omega(u, n) = \overline{\bigcup_{t \geq n} \varphi_t^{\theta_{-t}\omega}(u)}, \tag{4.34}$$

where $\overline{C}$ denotes the closure of C in X. Note that $A_\omega(u, n)$ is the union of the continuous curve $\{\varphi_t^{\theta_{-t}\omega}(u), t \geq n\}$ and of the set of its limit points. The latter is a closed set that belongs to K_ω in view of (4.33). The sequence $\{A_\omega(u, n)\}$ is a nested family of compact sets, and therefore

$$A_\omega(u) = \bigcap_{n \geq 0} A_\omega(u, n)$$

is a non-empty compact subset of X. Furthermore,

$$A_\omega(u) \text{ is the set of all limiting points for the curve } \{\varphi_t^{\theta_{-t}\omega}(u), t \geq 0\}. \tag{4.35}$$

Thus, by (4.33),

$$A_\omega(u) \subset K_\omega \quad \text{for all } u \in X,\ \omega \in \Omega_*. \tag{4.36}$$

We now define a family of subsets $\boldsymbol{A} = \{A_\omega, \omega \in \Omega\}$ by the rule

$$A_\omega = \begin{cases} \overline{\bigcup_{u \in X} A_\omega(u)} & \text{for } \omega \in \Omega_*, \\ \varnothing & \text{for } \omega \in \Omega \setminus \Omega_*. \end{cases} \tag{4.37}$$

Due to (4.36), A_ω is a compact set for each $\omega \in \Omega$. We shall show that $\boldsymbol{A}$ is a minimal random point attractor.

Step 2. Let us show that $\boldsymbol{A}$ is a compact random set. To this end, we need to show that the function $\omega \mapsto \mathrm{dist}_X(v, A_\omega)$ is $(\mathcal{F}, \mathcal{B}(\mathbb{R}))$-measurable for any $v \in X$.

Using the compactness of $A_\omega(u, n)$, it is easy to show that

$$\operatorname{dist}_X(v, A_\omega(u)) = \lim_{n\to\infty} \operatorname{dist}_X(v, A_\omega(u, n)) = \liminf_{k\to\infty} d_X(v, \varphi_t^{\theta_{-t}\omega} u)$$

for any $u, v \in X$ and $\omega \in \Omega_*$. Since the probability space $(\Omega, \mathcal{F}, \mathbb{P})$ is complete and $\mathbb{P}(\Omega_*) = 1$, we conclude that the function

$$d : X \times X \times \Omega \to \mathbb{R}, \quad d(u, v, \omega) = \operatorname{dist}_X(v, A_\omega(u)),$$

is measurable. By Corollary A.3.2 of the projection theorem (see Section A.3), for any $v \in X$ the function

$$\omega \mapsto \inf_{u\in X} d(u, v, \omega) = \operatorname{dist}_X(v, A_\omega)$$

is universally measurable. Using again that $(\Omega, \mathcal{F}, \mathbb{P})$ is complete, we arrive at the required result.

Step 3. Let us prove that $\boldsymbol{A}$ satisfies the invariance property. In what follows, we shall sometimes write $\varphi_t(\omega)$ instead of φ_t^ω to avoid complicated expressions in superscripts.

The invariance property is a consequence of the relation

$$\varphi_t^\omega A_\omega(u) = A_{\theta_t\omega}(u) \quad \text{for any } u \in X, \omega \in \Omega_*. \tag{4.38}$$

Indeed, if (4.38) is proved, then $\varphi_t^\omega A_\omega(u) \subset A_{\theta_t\omega}$, whence it follows that $\varphi_t^\omega A_\omega \subset A_{\theta_t\omega}$. To establish the converse inclusion, note that $\varphi_t^\omega A_\omega$ is a compact set containing $A_{\theta_t\omega}(u)$ for any $u \in X$. Therefore, it must contain the closure of $\cup_u A_{\theta_t\omega}(u)$, which coincides with $A_{\theta_t\omega}$.

To prove (4.38), note that, in view of (4.35), a point $v \in X$ belongs to $A_\omega(u)$ if and only if there is a sequence $t_j \to \infty$ such that

$$\varphi_{t_j}(\theta_{-t_j}\omega)u \to v \quad \text{as } j \to \infty.$$

Applying the continuous operator φ_t^ω and using the cocycle property, we see that

$$\varphi_{t+t_j}(\theta_{-(t+t_j)}(\theta_t\omega))u \to \varphi_t^\omega v \quad \text{as } j \to \infty.$$

This implies that $\varphi_t^\omega v \in A_{\theta_t\omega}(u)$ for any $v \in A_\omega(u)$, that is, $\varphi_t^\omega A_\omega(u) \subset A_{\theta_t\omega}(u)$. Conversely, suppose that $w \in A_{\theta_t\omega}(u)$ and choose a sequence $t_j \to \infty$ such that

$$\varphi_{t_j}(\theta_{-t_j}(\theta_t\omega))u \to w \quad \text{as } j \to \infty. \tag{4.39}$$

Now note that

$$\varphi_{t_j}(\theta_{-t_j}(\theta_t\omega)) = \varphi_t^\omega \circ \varphi_{t_j-k}(\theta_{-(t_j-k)}\omega) \quad \text{for } t_j \geq k. \tag{4.40}$$

It follows from (4.35) that the sequence $\varphi_{t_j-k}(\theta_{-(t_j-k)}\omega)u$ has a limit point v, which belongs to $A_\omega(u)$. Since φ_t^ω is continuous, we conclude from (4.40) that

$$\varphi_{t_j}(\theta_{-t_j}(\theta_t\omega))u \to \varphi_t^\omega v \quad \text{as } j \to \infty.$$

Comparing this with (4.39), we obtain $\varphi_t^\omega v = w$. Since $w \in A_{\theta_t\omega}$ was arbitrary, we have proved that $\varphi_t^\omega A_\omega(u) \supset A_{\theta_t\omega}(u)$.

Step 4. We now prove that $\boldsymbol{A}$ possesses the attraction property. Let $\omega \in \Omega_*$ and let $u \in X$. If $\varphi_t^{\theta_{-t}\omega} u$ does not tend to A_ω, then there is a constant $\varepsilon > 0$ and a sequence $t_j \to \infty$ such that

$$\operatorname{dist}_X(\varphi_{t_j}(\theta_{-t_j}\omega)u, A_\omega) \geq \varepsilon \quad \text{for all } j \geq 1. \tag{4.41}$$

In view of (4.33) and the compactness of K_ω, there is no loss of generality in assuming that the sequence $\{\varphi_{t_j}(\theta_{-t_j}\omega)u\}$ converges to a point in X. This point belongs to A_ω, which contradicts (4.41).

Step 5. It remains to show that $\boldsymbol{A}$ is a minimal random attractor. We claim that if $\boldsymbol{A}' = \{A'_\omega\}$ is another random attractor and Ω'_* is the set of full measure on which the attraction property holds for $\boldsymbol{A}'$, then

$$A_\omega \subset A'_\omega \quad \text{for } \omega \in \Omega_* \cap \Omega'_*. \tag{4.42}$$

Indeed, for any $\omega \in \Omega_* \cap \Omega'_*$ and $u \in X$ the set A'_ω must contain all the limit points of the curve $\{\varphi_t(\theta_{-t}\omega)u\}$. It follows that

$$A_\omega(u) \subset A'_\omega \quad \text{for any } u \in X.$$

Since A'_ω is closed, we conclude that (4.42) holds. To complete the proof of Theorem 4.34, it remains to note that $\Omega_* \cap \Omega'_*$ is a set of full measure. □

In what follows, we shall need another concept of random attractors for which the attraction property holds in a weaker sense.

Definition 4.2.6 A compact random set $\boldsymbol{A} = \{A_\omega, \omega \in \Omega\}$ in X is called a *random point attractor for $\boldsymbol{\Phi}$ in the sense of convergence in probability* (or a *weak random attractor*) if it possesses the following properties (cf. Definition 4.2.3):

(i) *Invariance*. For any $t \geq 0$ and almost every $\omega \in \Omega$, we have $\varphi_t^\omega A_\omega = A_{\theta_t\omega}$.
(ii) *Attraction*. For any $u \in X$, the function $\omega \mapsto \operatorname{dist}_X(\varphi_t^{\theta_{-t}\omega} u, A_\omega)$ converges to zero in probability as $k \to +\infty$, that is, for any $\varepsilon > 0$, we have

$$\mathbb{P}\big\{\operatorname{dist}_X(\varphi_t(\theta_{-t}\omega)u, A_\omega) \geq \varepsilon\big\} \to 0 \quad \text{as } t \to \infty. \tag{4.43}$$

Note that convergence "from the past" (4.43) is equivalent to convergence "in the future"

$$\mathbb{P}\big\{\operatorname{dist}_X(\varphi_t^\omega u, A_{\theta_t\omega}) \geq \varepsilon\big\} \to 0 \quad \text{as } t \to \infty, \tag{4.44}$$

because θ_t preserves $\mathbb{P}$, and for discrete-time RDS, the invariance properties of Definitions 4.2.6 and 4.2.3 are equivalent. Since almost sure convergence implies convergence in probability, any random attractor (in the sense of Definition 4.2.3) is also a weak random attractor. However, it does not need to be minimal (cf. Definition 4.2.4).

A natural question is whether (weak) random attractors contain "all relevant information" on the system as time goes to $+\infty$. In particular, do they support disintegrations of invariant measures? In general, the answer is negative (see [Bax89; Cra01]). There is, however, an important class of invariant measures for which this property is true. They are called *Markov invariant measures* and are studied in the next section.

4.2.2 The Ledrappier–Le Jan–Crauel theorem

Let $(\Omega, \mathcal{F}, \mathbb{P})$ be a complete probability space, let X be a Polish space, and let $\boldsymbol{\Phi} = \{\varphi_t^\omega, t \geq 0\}$ be a Markov RDS over a measure-preserving group of transformations $\boldsymbol{\theta} = \{\theta_t, t \in \mathbb{R}\}$. Recall that the concepts of an invariant measure for $\boldsymbol{\Phi}$ and of its disintegration, as well as the class $\mathcal{P}(\Omega \times X, \mathbb{P})$, were introduced in Section 1.3.4.

Definition 4.2.7 An invariant measure $\mathfrak{M} \in \mathcal{P}(\Omega \times X, \mathbb{P})$ for $\boldsymbol{\Phi}$ is said to be *Markov* if its disintegration $\{\mu_\omega\}$ is $\mathcal{F}_0$-measurable, that is, for any $\Gamma \in \mathcal{B}(X)$ the function $\omega \mapsto \mu_\omega(\Gamma)$ is $(\mathcal{F}_0, \mathcal{B}(\mathbb{R}))$-measurable.

The following proposition shows that the time $t = 0$ does not play any particular role for a Markov invariant measure.

Proposition 4.2.8 *Let $\mathfrak{M} \in \mathcal{P}(\Omega \times H, \mathbb{P})$ be an invariant measure for a Markov RDS $\boldsymbol{\Phi}$. Then the following properties are equivalent:*

(a) *$\mathfrak{M}$ is Markov.*
(b) *μ_ω is $\mathcal{F}_t$-measurable for some $t \in \mathbb{R}$.*
(c) *μ_ω is $\mathcal{F}_t$-measurable for any $t \in \mathbb{R}$.*

Proof We shall only show that (b) implies (a), because the proof of the implication (a) $\Rightarrow$ (c) is similar, and (c) trivially implies (b).

Suppose that μ_ω is $\mathcal{F}_t$-measurable for some $t > 0$. Then the random measure $\mu_{\theta_{-t}\omega}$ is $\mathcal{F}_0$-measurable and the mapping $\omega \mapsto \varphi_t^{\theta_{-t}\omega} u$ acting from Ω to X is $(\mathcal{F}_0, \mathcal{B}(X))$-measurable. We claim that the random measure $(\varphi_t^{\theta_{-t}\omega})_* \mu_{\theta_{-t}\omega}$ is $\mathcal{F}_0$-measurable. If this is proved, then the $\mathcal{F}_0$-measurability of μ_ω follows from relation (1.72) and the fact that $\mathcal{F}_0$ contains all subsets of $\mathcal{F}$ of zero measure.

We need to show that the real-valued function $f(\omega) = \mu_{\theta_{-t}\omega}\big((\varphi_t^{\theta_{-t}\omega})^{-1}(\Gamma)\big)$ is $\mathcal{F}_0$-measurable for any $\Gamma \in \mathcal{B}(X)$. We have

$$f(\omega) = \int_X h(\omega, u) \mu_{\theta_{-t}\omega}(du), \tag{4.45}$$

where $h(\omega, u) = \mathbb{I}_\Gamma(\varphi_t^{\theta_{-t}\omega} u)$. The function h is $(\mathcal{F}_0 \otimes \mathcal{B}(X), \mathcal{B}(\mathbb{R}))$-measurable. If it was the product of two functions $h_1(\omega)$ and $h_2(u)$, then the integral (4.45) could be written in the form

$$h_1(\omega) \int_X h_2(u) \mu_{\theta_{-t}\omega}(du),$$

and the required measurability would follow immediately. The $\mathcal{F}_0$-measurability of (4.45) in the general case can be obtained with the help of the monotone class technique. We leave the details to the reader as an exercise. □

We now turn to the main result of this section, which shows that there is a one-to-one correspondence between the Markov invariant measures and the stationary distributions for $\boldsymbol{\Phi}$.

Theorem 4.2.9 *Let $\boldsymbol{\Phi} = \{\varphi_t^\omega, t \geq 0\}$ be a Markov RDS in a Polish space X. Then the following assertions hold.*

(i) *Let $\mathfrak{M} \in \mathcal{P}(\Omega \times X, \mathbb{P})$ be a Markov invariant measure and let $\{\mu_\omega\}$ be its disintegration. Then $\mathbb{E}\,\mu_\cdot$ is a stationary measure for $\boldsymbol{\Phi}$. Moreover, if $\mathfrak{M}' \in \mathcal{P}(\Omega \times X, \mathbb{P})$ is another Markov invariant measure with disintegration $\{\mu'_\omega\}$ such that $\mathbb{E}\,\mu'_\cdot = \mathbb{E}\,\mu_\cdot$, then $\mathfrak{M}' = \mathfrak{M}$.*

(ii) *Let $\mu \in \mathcal{P}(X)$ be a stationary measure for $\boldsymbol{\Phi}$. Then for any sequence $t_k + \infty$ there is a set $\widetilde{\Omega} \in \mathcal{F}$ of full measure such that there is a weak* limit*

$$\mu_\omega = \lim_{k\to\infty} \varphi_{t_k}(\theta_{-t_k}\omega)_*(\mu) \tag{4.46}$$

for any $\omega \in \widetilde{\Omega}$. Moreover, if $\{t'_k\}$ is another sequence going to $+\infty$, then the corresponding family of measures $\{\mu'_\omega\}$ coincides with $\{\mu_\omega\}$ almost surely. Finally, the measure $\mathfrak{M} \in \mathcal{P}(\Omega \times X, \mathbb{P})$ defined by its disintegration[4] $\{\mu_\omega\}$ is a Markov invariant measure for $\boldsymbol{\Phi}$, and $\mathbb{E}\,\mu_\cdot = \mu$.

Thus, there is a one-to-one correspondence between the stationary measures and Markov invariant measures for $\boldsymbol{\Phi}$.

Proof (i) In view of Proposition 1.3.27, we have

$$(f(\varphi_t u), \mu_\omega) = (f, \mu_{\theta_t\omega}) \quad \mathbb{P}\text{-almost surely}, \tag{4.47}$$

where $f : X \to \mathbb{R}$ is an arbitrary bounded measurable function. We wish to take the mean value. Using the fact that μ_ω is $\mathcal{F}_0$-measurable and $\varphi_t u$ is $\mathcal{F}_t^+$-measurable and recalling that $\mathcal{F}_0$ and $\mathcal{F}_t^+$ are independent, we can write

$$\mathbb{E}\big(f(\varphi_t u), \mu_\omega\big) = \big(\mathbb{E}\, f(\varphi_t u), \mathbb{E}\,\mu_\omega\big) = (\mathfrak{P}_t f, \mu), \tag{4.48}$$

where we set $\mu = \mathbb{E}\,\mu_\cdot$. On the other hand, since $\mathbb{P}$ is invariant with respect to $\boldsymbol{\theta}$, we have

$$\mathbb{E}\,(f, \mu_{\theta_t\omega}) = (f, \mu). \tag{4.49}$$

Combining (4.47)–(4.49), we obtain

$$(f, \mathfrak{P}_t^*\mu) = (f, \mu) \quad \text{for any bounded measurable function } f,$$

whence it follows that μ is a stationary measure for $\boldsymbol{\Phi}$.

[4] We set $\mu_\omega = \gamma$ for $\omega \in \Omega \setminus \widetilde{\Omega}$, where $\gamma \in \mathcal{P}(X)$ is an arbitrary measure.

Exercise 4.2.10 Justify the first equality in (4.48) and relation (4.49). *Hint:* Use the monotone class technique.

We now prove the second part of (i). Let us fix any increasing sequence $\{t_k\}$ going to $+\infty$. In view of Proposition 1.3.27, with probability 1 we have

$$\varphi_{t_k}(\theta_{-t_k}\omega)\mu_{\theta_{-t_k}\omega} = \mu_\omega \quad \text{for all } k \geq 1. \tag{4.50}$$

It follows that

$$\int_X f\big(\varphi_{t_k}(\theta_{-t_k}\omega)u\big)\mu_{\theta_{-t_k}\omega}(du) = \int_X f(u)\mu_\omega(du) \quad \text{for any } k \geq 1,\ f \in C_b(X). \tag{4.51}$$

Let us set $\mathcal{G}_k = \mathcal{F}_{[-t_k,0]}$ for $k \geq 1$ and note that $f(\varphi_{t_k}(\theta_{-t_k}\omega)u)$ is $\mathcal{G}_k$-measurable, while $\mu_{\theta_{-t_k}\omega}$ is independent of $\mathcal{G}_k$. Therefore, taking the conditional expectation of both sides of (4.51) given $\mathcal{G}_k$, with probability 1 we derive

$$\int_X f\big(\varphi_{t_k}(\theta_{-t_k}\omega)u\big)\mu(du) = \mathbb{E}\big((f,\mu_\omega)\,|\,\mathcal{G}_k\big). \tag{4.52}$$

Since μ_ω is $\mathcal{F}_0$-measurable and $\mathcal{F}_0 = \sigma\{\mathcal{G}_k, k \geq 1\}$, we see that the right-hand side of (4.52) is a right-closable martingale with respect to the filtration $\{\mathcal{G}_k\}$. Hence, by Doob's theorem on convergence of right-closed martingale sequences, we have

$$\lim_{k\to\infty}\big(f, \varphi_{t_k}(\theta_{-t_k}\omega)\mu\big) = \lim_{k\to\infty}\int_X f\big(\varphi_{t_k}(\theta_{-t_k}\omega)u\big)\mu(du) = (f,\mu_\omega)$$

for any $f \in C_b(X)$ and almost every $\omega \in \Omega$. Theorem A.5.2 now implies that there is a set of full measure $\widetilde{\Omega} \in \mathcal{F}$ such that (4.46) holds, where the convergence of measures is understood in the weak* topology. It follows that μ_ω is uniquely defined by μ on the set $\widetilde{\Omega}$, and therefore $\mathfrak{M}' = \mathfrak{M}$ for any Markov invariant measure $\mathfrak{M}'$ for which the associated stationary measure coincides with μ.

(ii) Let us take an arbitrary sequence $\{t_k\}$ going to $+\infty$. Without loss of generality, we can assume that $\{t_k\}$ is increasing. Given a bounded continuous function $f : X \to \mathbb{R}$, consider the sequence $\xi_k(\omega) = (f, \varphi_{t_k}(\theta_{-t_k}\omega)\mu)$. Suppose we have shown that $\{\xi_k\}$ is a martingale with respect to the filtration $\mathcal{G}_k = \mathcal{F}_{[-t_k,0]}$. Since $\{\xi_k\}$ is bounded uniformly in k and ω, Doob's martingale convergence theorem implies that $\xi_k(\omega)$ converges almost surely. In particular, for any $f \in C_b(X)$ the sequence (f, μ_ω^k) converges for almost every $\omega \in \Omega$. Therefore, by Theorem A.5.2, there is a random probability measure $\{\mu_\omega\} \subset \mathcal{P}(X)$ and a set of full measure $\widetilde{\Omega} \in \mathcal{F}$ such that (4.46) holds.

We now prove that $\{\xi_k\}$ is a martingale. Setting $g_k(\omega, v) = f(\varphi_{t_k}(\theta_{-t_k}\omega)v)$, we can write

$$\xi_k = \big(g_k(\omega,\cdot),\mu\big) = \big(g_k(\omega,v),\mu(dv)\big).$$

The cocycle and Markov properties imply that

$$\begin{aligned}\mathbb{E}\,\{\xi_{k+1}\mid\mathcal{G}_k\} &= \mathbb{E}\left\{(g_{k+1}(\omega,\cdot),\mu)\mid\mathcal{G}_k\right\}\\ &= \mathbb{E}\left\{(g_k(\omega,\varphi_{s_k}(\theta_{-t_{k+1}}\omega)v),\mu(dv))\mid\mathcal{G}_k\right\}\\ &= \mathbb{E}_{\omega'}\left\{(g_k(\omega,\varphi_{s_k}(\theta_{-t_{k+1}}\omega')v),\mu(dv))\right\}\\ &= \mathbb{E}_{\omega'}\left\{(g_k(\omega,\varphi_{s_k}(\omega')v),\mu(dv))\right\}\\ &= \left(\mathfrak{P}_{s_k}g_k(\omega,\cdot),\mu\right) = \left(g_k(\omega,\cdot),\mathfrak{P}^*_{s_k}\mu\right) = \xi_k,\end{aligned}$$

where $s_k = t_{k+1} - t_k$, and we used that μ is a stationary measure for $\boldsymbol{\Phi}$.

Let us show that the limit in (4.46) does not depend on $\{t_k\}$. Indeed, let $\{t_k'\}$ be another sequence going to $+\infty$ and let μ_ω' be the corresponding limit. Let us consider the altered sequence $\{s_k\} = \{t_1, t_1', t_2, t_2', \dots\}$. Since the limit (4.46) with $t_k = s_k$ also exists almost surely, we conclude that $\mu_\omega = \mu_\omega'$ for a.a. $\omega \in \Omega$.

We now set

$$\mathfrak{M}(d\omega, du) = \mu_\omega(du)\,\mathbb{P}(d\omega) \tag{4.53}$$

and prove that is a Markov invariant measure for $\boldsymbol{\Phi}$. Indeed, the definition of $\mathfrak{M}$ implies that its disintegration is $\mathcal{F}_0$-measurable. To prove that $\mathfrak{M}$ is invariant, it suffices to show that

$$\varphi_t^\omega\mu_\omega = \mu_{\theta_t\omega} \quad \text{for any } t>0 \text{ and a.e. } \omega\in\Omega. \tag{4.54}$$

Let us set $t_k = kt$ and denote by $\widetilde{\Omega}$ the set of convergence in (4.46). Choose any $\omega \in \widetilde{\Omega}$ such that $\theta_t\omega \in \widetilde{\Omega}$. Then convergence (4.46) and the cocycle property imply that

$$\begin{aligned}(f,\varphi_t^\omega\mu_\omega) &= \int_X f(\varphi_t^\omega u)\mu_\omega(du) = \lim_{k\to\infty}\int_X f(\varphi_t^\omega u)\big(\varphi_{t_k}(\theta_{-t_k}\omega)\mu\big)(du)\\ &= \lim_{k\to\infty}\int_X f\big(\varphi_t^\omega\circ\varphi_{t_k}(\theta_{-t_k}\omega)u\big)\,\mu(du)\\ &= \lim_{k\to\infty}\int_X f\big(\varphi_{t_{k+1}}(\theta_{-t_{k+1}}(\theta_t\omega))u\big)\,\mu(du) = (f,\mu_{\theta_t\omega}).\end{aligned}$$

Since $f \in C_b(X)$ is arbitrary, it remains to note that the set $\{\omega\in\widetilde{\Omega} : \theta_t\omega\in\widetilde{\Omega}\}$ is of full measure.

It remains to show that $\mathbb{E}\,\mu_\cdot = \mu$. To this end, take any function $f \in C_b(X)$ and note that, in view of (4.46), we have

$$(f,\mu_\omega) = \lim_{k\to\infty}(f,\varphi_k(\theta_{-k}\omega)\mu) \quad \text{for almost every } \omega\in\Omega.$$

Taking the mean value of both sides and using the Lebesgue theorem on dominated convergence and the stationarity of μ, we obtain

$$\begin{aligned}\mathbb{E}\,(f,\mu_\cdot) &= \lim_{k\to\infty}\mathbb{E}\,\big(f,\varphi_k(\theta_{-k}\cdot)\mu\big)\\ &= \lim_{k\to\infty}\mathbb{E}\int_X f(\varphi_k u)\,\mu(du) = \lim_{k\to\infty}(\mathfrak{P}_k f,\mu) = (f,\mu).\end{aligned}$$

The proof of Theorem 4.2.9 is complete. □

Exercise 4.2.11 In the setting of Theorem 4.2.9, prove that, for any $\Gamma \in \mathcal{B}(X)$ and any sequence $\{t_k\}$ going to $+\infty$, we have

$$\big(\varphi_{t_k}(\theta_{-t_k}\omega)\mu\big)(\Gamma) \to \mu_\omega(\Gamma) \quad \text{as } k \to \infty \text{ for almost every } \omega \in \Omega.$$

Hint: The fact that $\{\xi_k\}$ is a martingale is true for any bounded measurable function f.

We now study the relationship between the support of the disintegration for a Markov invariant measure and weak random attractors.

Proposition 4.2.12 *Let $\boldsymbol{\Phi}$ be a Markov RDS, let $\mathfrak{M} \in \mathcal{P}(\Omega \times \mathbb{P}, \mathbb{P})$ be a Markov invariant measure for $\boldsymbol{\Phi}$ with disintegration $\{\mu_\omega\}$, and let $\mathbf{A} = \{A_\omega\}$ be a weak random attractor for $\boldsymbol{\Phi}$. Then*

$$\operatorname{supp}\mu_\omega \subset A_\omega \quad \textit{for almost all } \omega \in \Omega. \tag{4.55}$$

Proof

Step 1. We first show that if $f(\omega, u)$ is a real-valued bounded measurable function on $\Omega \times H$ that is continuous in u and $\mu \in \mathcal{P}(H)$ is the stationary measure associated with $\mathfrak{M}$ (see Theorem 4.2.9), then

$$\iint\limits_{\Omega\times H} f(\omega, \varphi_k^{\theta_{-k}\omega}u)\,\mu(du)\mathbb{P}(d\omega) \to \iint\limits_{\Omega\times H} f(\omega, u)\,\mu_\omega(du)\mathbb{P}(d\omega) \tag{4.56}$$

as $k \to \infty$. Indeed, by (4.46), for almost every $\omega \in \Omega$, we have

$$\int_H f(\omega, \varphi_k^{\theta_{-k}\omega}u)\,\mu(du) = \int_H f(\omega, u)\,\big(\varphi_k^{\theta_{-k}\omega}\mu\big)(du) \to \int_H f(\omega, u)\,\mu_\omega(du)$$

as $k \to \infty$. Taking the mean value with respect to ω and using the Lebesgue theorem on dominated convergence, we obtain (4.56).

Step 2. We now fix any integer $n \geq 1$ and set

$$\begin{aligned} f_n(\omega, u) &= 1 - \big(n\,\operatorname{dist}_X(u, A_\omega)\big) \wedge 1, \\ A_\omega^n &= \{u \in H : \operatorname{dist}_X(u, A_\omega) \leq 1/n\}. \end{aligned}$$

Applying (4.56) to f_n, we obtain

$$\begin{aligned} \lim_{k\to\infty} \iint\limits_{\Omega\times H} f_n(\omega, \varphi_k^{\theta_{-k}\omega}u)\,\mu(du)\mathbb{P}(d\omega) &= \iint\limits_{\Omega\times H} f_n(\omega, u)\,\mu_\omega(du)\mathbb{P}(d\omega) \\ &\leq \int_\Omega \mu_\omega(A_\omega^n)\mathbb{P}(d\omega). \end{aligned} \tag{4.57}$$

On the other hand, it follows from (4.43) that

$$\lim_{k\to\infty} \int_\Omega f_n(\omega, \varphi_k^{\theta_{-k}\omega}u)\mathbb{P}(d\omega) = 1 \quad \text{for any } u \in H.$$

Integrating this relation with respect to μ and using again the Lebesgue theorem, we derive

$$\lim_{k\to\infty} \iint_{\Omega\times H} f_n(\omega, \varphi_k^{\theta_{-k}\omega} u)\mu(du)\mathbb{P}(d\omega) = 1. \tag{4.58}$$

Comparing (4.57) and (4.58), we see that

$$\int_\Omega \mu_\omega(A_\omega^n)\,\mathbb{P}(d\omega) = 1.$$

It follows that

$$\mathbb{P}\big\{\omega \in \Omega : \mu_\omega(A_\omega^n) = 1\big\} = 1 \quad \text{for any integer } n \geq 1.$$

Since $\cap_n A_\omega^n = A_\omega$, we obtain the relation

$$\mathbb{P}\{\mu_\omega(A_\omega) = 1\} = 1,$$

which is equivalent to (4.55). □

4.2.3 Ergodic RDS and minimal attractors

We now study the connection between invariant measures and random attractors for ergodic RDS. As before, we denote by $\boldsymbol{\Phi} = \{\varphi_t^\omega, t \geq 0\}$ an RDS in a Polish space X over a group of measure-preserving transformations $\boldsymbol{\theta} = \{\theta_t, t \in \mathbb{R}\}$. We shall assume that $\boldsymbol{\Phi}$ satisfies the two hypotheses below.

Condition 4.2.13 *Compactness:* There is a compact random set $\boldsymbol{K} = \{K_\omega\}$ attracting trajectories of φ_t^ω (in the sense specified in Theorem 4.2.5). Moreover, for any $\varepsilon > 0$ there is an $\Omega_\varepsilon \in \mathcal{F}$ and a compact set $\mathcal{C}_\varepsilon \subset X$ such that $\mathbb{P}(\Omega_\varepsilon) \geq 1 - \varepsilon$ and

$$\varphi_t^{\theta_{-t}\omega} u \in \mathcal{C}_\varepsilon \quad \text{for } \omega \in \Omega_\varepsilon,\ u \in X,\ t \geq t_\varepsilon(u), \tag{4.59}$$

where the time $t_\varepsilon(u) > 0$ depends only on u and ε.

Condition 4.2.14 *Mixing:* The RDS $\boldsymbol{\Phi}$ is of mixing type in the following sense: it has a unique stationary measure μ, and for any $f \in L_b(X)$ and any initial point $u \in X$ we have

$$\mathfrak{P}_t f(u) = \mathbb{E}\, f\big(\varphi_t u\big) \to (f, \mu) = \int_X f(u)\mu(du) \quad \text{as } t \to \infty. \tag{4.60}$$

Exercise 4.2.15 Prove that if an RDS $\boldsymbol{\Phi}$ satisfies Condition 4.2.13, then there is a compact random set $\boldsymbol{A} = \{A_\omega\}$ that satisfies the invariance property of Definition 4.2.6 and the attraction property of Definition 4.2.3. In particular, $\boldsymbol{A}$ is a weak random attractor for $\boldsymbol{\Phi}$. *Hint:* Repeat the scheme used in the proof of Theorem 4.2.5.

Recall that the concept of the semigroup associated with an RDS was introduced in Section 1.3.4. The following theorem shows that if a Markov RDS $\boldsymbol{\Phi}$

satisfies Conditions 4.2.13 and 4.2.14, then the associated semigroup $\mathbf{\Theta}_t$ also possesses a mixing property. For any sub-σ-algebra $\mathcal{G} \subset \mathcal{F}$, denote by $\mathbb{L}(X, \mathcal{G})$ the set of measurable functions $F(\omega, u)\colon \Omega \times X \to \mathbb{R}$ that are $\mathcal{G}$-measurable in ω for any fixed $u \in X$ and satisfy the condition

$$\operatorname*{ess\,sup}_{\omega\in\Omega} \|F(\omega, \cdot)\|_L < \infty. \tag{4.61}$$

Theorem 4.2.16 *Let us assume that* $\boldsymbol{\Phi}$ *satisfies Conditions 4.2.13 and 4.2.14. Let* $\mu \in \mathcal{P}(X)$ *be the unique stationary measure for* $\boldsymbol{\Phi}$ *and let* $\mathfrak{M} \in \mathcal{P}(\Omega \times X, \mathbb{P})$ *be the corresponding Markov invariant measure. Then for any* $F \in \mathbb{L}(X, \mathcal{F}^-)$ *we have*

$$\mathbb{E}\,F\big(\Theta_t(\cdot, u)\big) \to (F, \mathfrak{M}) = \int_\Omega \int_X F(\omega, u)\mu_\omega(du)\mathbb{P}(d\omega) \quad \text{as } t \to \infty, \tag{4.62}$$

where $u \in X$ *is an arbitrary initial point.*

Let us denote by μ_ω the disintegration of $\mathfrak{M}$ and set

$$A_\omega = \begin{cases} \operatorname{supp} \mu_\omega, & \omega \in \widetilde{\Omega}, \\ \varnothing, & \omega \notin \widetilde{\Omega}, \end{cases} \tag{4.63}$$

where $\widetilde{\Omega} \in \mathcal{F}$ is a set of full measure on which the limit (4.46) exists. By corollary 1.6.5 in [Arn98], A_ω is a closed random set. Moreover, it follows from (4.46) that A_ω is measurable with respect to $\mathcal{F}^-$ in the sense that the function $\omega \mapsto \operatorname{dist}_X(u, A_\omega)$ is $(\mathcal{F}^-, \mathcal{B}(\mathbb{R}))$-measurable for any $u \in X$.

Theorem 4.2.17 *Let us assume that* $\boldsymbol{\Phi}$ *satisfies Conditions 4.2.13 and 4.2.14. Then the random set* $\{A_\omega\}$ *defined by* (4.63) *is almost surely compact and forms a minimal weak random attractor for* $\boldsymbol{\Phi}$.

Proof of Theorem 4.2.16

Step 1. We first assume that $F(\omega, u) \in \mathbb{L}(X, \mathcal{F}_{[-\ell,0]})$ for some $\ell \geq 0$. Since θ_t preserves $\mathbb{P}$, for any $t \geq s \geq 0$ we have

$$p_t(u) := \mathbb{E}\,F\big(\theta_t\omega, \varphi_t^\omega u\big) = \mathbb{E}\,F\big(\omega, \varphi_t^{\theta_{-t}\omega} u\big) = \mathbb{E}\,\mathbb{E}\big\{F\big(\omega, \varphi_t^{\theta_{-t}\omega} u\big) \,\big|\, \mathcal{F}_{[-s,0]}\big\}.$$

By the cocycle property (see (1.62)),

$$\varphi_t^{\theta_{-t}\omega} = \varphi_s^{\theta_{-s}\omega} \circ \varphi_{t-s}^{\theta_{-t}\omega}, \quad s \leq t.$$

Hence, setting $F_s(\omega, u) = F(\omega, \varphi_s^{\theta_{-s}\omega} u)$, for any $s \leq t$ we derive

$$p_t(u) = \mathbb{E}\,\mathbb{E}\big\{F_s\big(\omega, \varphi_{t-s}^{\theta_{-t}\omega} u\big) \,\big|\, \mathcal{F}_{[-s,0]}\big\}. \tag{4.64}$$

We now note that $F_s \in \mathbb{L}(X, \mathcal{F}_{[-s,0]})$ if $s \geq \ell$. Since $\varphi_{t-s}^{\theta_{-t}\omega} u$ is measurable with respect to $\mathcal{F}_{[-t,-s]}$ and since the σ-algebras $\mathcal{F}_{[-s,0]}$ and $\mathcal{F}_{[-t,-s]}$ are independent, it follows from (4.64) that

$$p_t(u) = \mathbb{E}\,\mathbb{E}'\big\{F_s\big(\omega, \varphi_{t-s}^{\theta_{-t}\omega'} u\big) = \mathbb{E}\,(\mathfrak{P}_{t-s} F_s)(\omega, u), \tag{4.65}$$

where $\ell \le s \le t$ and $\mathbb{E}'$ denotes the expectation with respect to ω'. In view of Condition 4.2.14 and the Lebesgue theorem, for any $s \ge \ell$, the right-hand side of (4.65) tends to $\mathbb{E}\,(F_s(\omega,\cdot),\mu)$ as $t \to +\infty$. Recalling the definition of F_s, we see that

$$\big(F_s(\omega,\cdot),\mu\big) = \big(F(\omega,\cdot),\varphi_s^{\theta_{-s}\omega}\mu\big) \to \big(F(\omega,\cdot),\mu_\omega\big) \quad \text{as } s \to \infty,$$

where we used Proposition 1.3.27. What has been said implies the relation

$$\lim_{t\to+\infty} p_t(u) = \mathbb{E}\,\big(F(\omega,\cdot),\mu_\omega\big),$$

which coincides with (4.62).

Step 2. We now show that (4.62) holds for functions of the form $F(\omega,u) = f(u)g(\omega)$, where $f \in L_b(X)$ and g is a bounded $\mathcal{F}^-$-measurable function. To this end, we use a version of the monotone class theorem.

Let us fix $f \in L_b(X)$ and denote by $\mathcal{H}$ the set of bounded $\mathcal{F}^-$-measurable functions g for which convergence (4.62) with $F = fg$ holds. It is clear that $\mathcal{H}$ is a vector space containing the constant functions. Moreover, as was shown in Step 1, it contains all bounded functions measurable with respect to $\mathcal{F}_{[-\ell,0]}$ for some $\ell \ge 0$. Since the union of $\mathcal{F}_{[-\ell,0]}$, $\ell \ge 0$, generates $\mathcal{F}^-$, the required assertion will be proved as soon as we establish the following property: if $g_n \in \mathcal{H}$ is an increasing sequence of non-negative functions such that $g = \sup g_n$ is bounded, then $g \in \mathcal{H}$.

Suppose that a sequence $\{g_n\} \subset \mathcal{H}$ satisfies the above conditions. Without loss of generality, we shall assume that $0 \le g, g_n \le 1$. By Egorov's theorem, for any $\varepsilon > 0$ there is $\Omega_\varepsilon \in \mathcal{F}$ such that $\mathbb{P}(\Omega_\varepsilon) \ge 1-\varepsilon$ and

$$\lim_{n\to+\infty} \sup_{\omega\in\Omega_\varepsilon} \big|g_n(\omega) - g(\omega)\big| = 0.$$

It follows that for any $\varepsilon > 0$ there is an integer $n_\varepsilon \ge 1$ such that $n_\varepsilon \to +\infty$ as $\varepsilon \to 0$ and

$$g_{n_\varepsilon}(\omega) \le g(\omega) \le g_{n_\varepsilon}(\omega) + \varepsilon + \mathbb{I}_{\Omega_\varepsilon^c}(\omega) \quad \text{for all} \quad \omega \in \Omega.$$

Multiplying this inequality by $f(\varphi_t^{\theta_{-t}\omega}u)$, taking the expectation, passing to the limit as $t \to +\infty$, and using the estimate $\mathbb{P}(\Omega_\varepsilon^c) \le \varepsilon$, we derive

$$\begin{aligned}\mathbb{E}\big\{(f,\mu_\omega)g_{n_\varepsilon}(\omega)\big\} &\le \liminf_{k\to+\infty} \mathbb{E}\big\{f(\varphi_t^{\theta_{-t}\omega}u)g(\omega)\big\} \\ &\le \limsup_{k\to+\infty} \mathbb{E}\big\{f(\varphi_t^{\theta_{-t}\omega}u)g(\omega)\big\} \le \mathbb{E}\big\{(f,\mu_\omega)g_{n_\varepsilon}(\omega)\big\} + 2\varepsilon.\end{aligned}$$

Since $\varepsilon > 0$ is arbitrary and $\mathbb{E}\big\{(f,\mu_\omega)g_{n_\varepsilon}(\omega)\big\} \to \mathbb{E}\big\{(f,\mu_\omega)g(\omega)\big\}$ as $\varepsilon \to 0$ (by the monotone convergence theorem), we conclude that

$$\mathbb{E}\big\{f(\varphi_t^\omega u)g(\theta_t\omega)\big\} = \mathbb{E}\big\{f(\varphi_t^{\theta_{-t}\omega}u)g(\omega)\big\} \overset{t\to+\infty}{\longrightarrow} \mathbb{E}\big\{(f,\mu_\omega)g(\omega)\big\},$$

which means that $g \in \mathcal{H}$. This completes the proof of (4.62) in the case when $F(\omega, u) = f(u)g(\omega)$.

Step 3. Now we consider the general case. Let $F \in \mathbb{L}(X, \mathcal{F}^-)$ be an arbitrary function such that $\|F(\omega, \cdot)\|_{L(X)} \le 1$ for a.e. $\omega \in \Omega$. For any $u \in X$ and $\varepsilon > 0$, we choose $t_\varepsilon(u) \ge 1$, $\Omega_\varepsilon \in \mathcal{F}$, and $\mathcal{C}_\varepsilon \Subset X$ for which (4.59) holds. By the Arzelà–Ascoli theorem, the unit ball $B_\varepsilon = \{f \in L_b(\mathcal{C}_\varepsilon) : \|f\|_L \le 1\}$ is compact in the space $C_b(\mathcal{C}_\varepsilon)$, and therefore there is a finite set $\{h_j\} \subset B_\varepsilon$ whose ε-neighbourhood contains B_ε. It follows that B_ε can be covered by non-intersecting Borel sets $U_j \ni h_j$, $j = 1, \dots, N$, whose diameters do not exceed 2ε. Let us denote by $f_j \in L(X)$ arbitrary extensions of h_j to X such that $\|f_j\|_L \le 2$. For instance, we can take

$$f_j(u) = \inf_{v \in \mathcal{C}_\varepsilon} \big(h_j(v) + \mathrm{dist}_X(u, v) \wedge 1\big).$$

Let us consider the following approximation of F:

$$G_\varepsilon(\omega, u) = \sum_{j=1}^{N} f_j(u) g_j(\omega), \qquad g_j(\omega) = \mathbb{I}_{U_j}\big(F_{\mathcal{C}_\varepsilon}(\omega, \cdot)\big),$$

where $F_{\mathcal{C}_\varepsilon}(\omega, u)$ is the restriction of F to $\Omega \times \mathcal{C}_\varepsilon$. Since only one of the functions g_j can be non-zero, we have $\|G_\varepsilon(\omega, \cdot)\|_\infty \le 2$. Therefore, for any $u \in X$ and a.e. $\omega \in \Omega$, we derive

$$\begin{aligned}\big|G_\varepsilon(\omega, u) - F(\omega, u)\big| &\le 2\varepsilon + \mathbb{I}_{\mathcal{C}_\varepsilon^c}(u)\big(\|G_\varepsilon(\omega, \cdot)\|_\infty + \|F(\omega, \cdot)\|_\infty\big) \\ &\le 2\varepsilon + 3\mathbb{I}_{\mathcal{C}_\varepsilon^c}(u),\end{aligned} \tag{4.66}$$

where we used the inequality $\|F(\omega, \cdot)\|_\infty \le 1$. Let us set

$$p_t(u) = \mathbb{E}\, F(\theta_t\omega, \varphi_t^\omega u), \qquad p_t(u, \varepsilon) = \mathbb{E}\, G_\varepsilon(\theta_t\omega, \varphi_t^\omega u).$$

It is clear that

$$\begin{aligned}\big|p_t(u) - (F, \mathfrak{M})\big| \le \big|p_t(u) - p_t(u, \varepsilon)\big| + \big|p_t(u, \varepsilon) \\ - (G_\varepsilon, \mathfrak{M})\big| + \big|(G_\varepsilon - F, \mathfrak{M})\big|.\end{aligned} \tag{4.67}$$

Let us estimate each term on the right-hand side of (4.67). Combining (4.59) and (4.66), for $t \ge t_\varepsilon(u)$ we derive

$$\begin{aligned}\big|p_t(u) - p_t(u, \varepsilon)\big| &\le \big|\mathbb{E}\big\{F(\omega, \varphi_t^{\theta_{-t}\omega} u) - G_\varepsilon(\omega, \varphi_t^{\theta_{-t}\omega} u)\big\}\big| \\ &\le 2\varepsilon + 3\mathbb{P}\big\{\varphi_t^{\theta_{-t}\omega} u) \notin \mathcal{C}_\varepsilon\big\} \le 2\varepsilon + 3\mathbb{P}(\Omega_\varepsilon^c) \le 5\varepsilon.\end{aligned} \tag{4.68}$$

Furthermore, the functions g_j are $\mathcal{F}^-$-measurable, and hence, by Step 2, for any fixed $\varepsilon > 0$,

$$p_t(u, \varepsilon) \to (G_\varepsilon, \mathfrak{M}) \quad \text{as } t \to +\infty. \tag{4.69}$$

Finally, inequality (4.66) implies that

$$\big|(G_\varepsilon - F, \mathfrak{M})\big| \le 2\varepsilon + 3(\mathbb{I}_{\mathcal{C}_\varepsilon^c}, \mathfrak{M}) = 2\varepsilon + 3\mu(\mathcal{C}_\varepsilon^c). \tag{4.70}$$

Since $\varepsilon > 0$ is arbitrary, it follows from (4.67)–(4.70) that the required convergence (4.62) will be established if we show that $\mu(\mathcal{C}_\varepsilon^c) \to 0$ as $\varepsilon \to 0$.

To this end, we note that

$$\mu(\mathcal{C}_\varepsilon^c) = \int_X \mathbb{P}\{\varphi_t^\omega u \notin \mathcal{C}_\varepsilon\}\mu(du). \tag{4.71}$$

It follows from Condition 4.2.13 that, for any fixed $u \in X$,

$$\limsup_{t\to+\infty} \mathbb{P}\{\varphi_t^\omega u \notin \mathcal{C}_\varepsilon\} = \limsup_{t\to+\infty} \mathbb{P}\{\varphi_t^{\theta_{-t}\omega} u \notin \mathcal{C}_\varepsilon\} \le \varepsilon.$$

Passing to the limit $t \to +\infty$ in (4.71), we conclude that $\mu(\mathcal{C}_\varepsilon^c) \le \varepsilon$ for any $\varepsilon > 0$. This completes the proof of Theorem 4.2.16. □

Proof of Theorem 4.2.17 We first show that A_ω is compact for almost every $\omega \in \Omega$. To this end, note that, by Exercise 4.2.15, the RDS possesses a (compact) weak random attractor $\{A'_\omega\}$. Furthermore, by Proposition 4.2.12, we have $A_\omega \subset A'_\omega$ for almost every ω. Since A_ω is closed, we obtain the required result.

We now prove that the compact random set $\{A_\omega\}$ is a random attractor. Let us fix $\delta \in (0, 1)$ and consider the function

$$F(\omega, u) = 1 - \tfrac{\mathrm{dist}_X(u, A_\omega)}{\delta} \wedge 1, \quad u \in X, \quad \omega \in \Omega.$$

We claim that $F \in \mathbb{L}(X, \mathcal{F}^-)$. Indeed, the definition implies that $F(\omega, u)$ is a bounded measurable function and that

$$\big|F(\omega, u) - F(\omega, v)\big| \le \tfrac{\mathrm{dist}_X(u,v)}{\delta} \quad \text{for all } u, v \in X, \omega \in \Omega.$$

Thus, F satisfies (4.61). Since A_ω is a compact random set measurable with respect to $\mathcal{F}^-$, we conclude that so is the function F. This proves the required properties of F.

Since $F(\omega, u) = 1$ for $u \in A_\omega$, we see that $(F, \mathfrak{M}) = 1$. So, applying Theorem 4.2.16, we get

$$\mathbb{E}\, F(\theta_t\omega, \varphi_t^\omega u) = 1 - \mathbb{E}\left(\frac{\mathrm{dist}_X(\varphi_t^\omega u, A_{\theta_t\omega})}{\delta} \wedge 1\right) \to (F, \mathfrak{M}) = 1.$$

That is,

$$p_t(u) := \mathbb{E}\left(\frac{\mathrm{dist}_X(\varphi_t^\omega u, A_{\theta_t\omega})}{\delta} \wedge 1\right) \to 0. \tag{4.72}$$

We now note that, by Chebyshev's inequality,

$$\mathbb{P}\big\{\mathrm{dist}_X(\varphi_t^\omega u, A_{\theta_t\omega}) > \delta\big\} \le \frac{p_t(u)}{\delta}.$$

In view of (4.72), the right-hand side of this inequality goes to zero as $t \to +\infty$. This completes the proof of the fact that A_ω is a random attractor.

To show that A_ω is a minimal random attractor, it suffices to note that, by Proposition 4.2.12, the Markov invariant measure $\mathfrak{M}$ is supported by any random attractor $\{A'_\omega\}$, and therefore $\operatorname{supp}\mu_\omega \subset A'_\omega$ for a.e. $\omega \in \Omega$. □

4.2.4 Application to the Navier–Stokes system

In this subsection, we show that Theorems 4.2.16 and 4.2.17 can be applied to the Navier–Stokes system (3.84) perturbed by a white-noise force (2.66). To this end, it suffices to check that Conditions 4.2.13 and 4.2.14 are fulfilled. As was proved in Theorem 3.3.1, the second condition holds if the coefficients b_j are all positive and satisfy (2.67). Thus, we shall concentrate on Condition 4.2.13 and prove that it holds for any sequence b_j satisfying the hypothesis

$$\mathfrak{B}_1 = \sum_{j=1}^{\infty} \alpha_j b_j^2 < \infty. \tag{4.73}$$

Recall that a Markov RDS associated with the Navier–Stokes system perturbed by a spatially regular white noise was constructed in Section 2.4.4. We denote by $(\Omega, \mathcal{F}, \mathcal{F}_t, \mathbb{P})$ the corresponding filtered probability space and by $\theta_t : \Omega \to \Omega$ the group of shift operators on Ω. We shall show that the required properties are implied by the following proposition.

Proposition 4.2.18 *There is an increasing sequence of subsets $\Omega_k \in \mathcal{F}$ satisfying the following properties.*

(i) *The union $\Omega_* = \cup_k \Omega_k$ is invariant under the group θ_t and has the full measure.*
(ii) *For any $k \geq 1$ there is an almost surely finite random constant R_k such that*

$$\left\|\varphi_t^{\theta_{-t}\omega}(w)\right\|_1 \leq R_k(\omega) \quad \text{for } \omega \in \Omega_k,\ w \in H,\ t \geq T_k(w), \tag{4.74}$$

where $T_k(w) \geq 0$ is a deterministic constant.

Taking this result for granted, let us check Condition 4.2.13. We first prove the existence of a compact random set attracting the trajectories. Let us set Ω_* to be the set defined in Proposition 4.2.18 and let

$$K_\omega = \begin{cases} B_V\big(R_k(\omega)\big) & \text{for } \omega \in \Omega_k \setminus \Omega_{k-1}, \\ \varnothing, & \text{for } \omega \notin \Omega_*, \end{cases}$$

where $\Omega_0 = \varnothing$. Then $\boldsymbol{K} = \{K_\omega\}$ is a compact random set in H. We claim that it is an absorbing set. Namely, there is a measurable function $T : H \times \Omega_* \to \mathbb{R}_+$

such that

$$\varphi_t^{\theta_{-t}\omega}(w) \in K_\omega \quad \text{for any } w \in H, \omega \in \Omega_*, t \geq T(w, \omega). \tag{4.75}$$

Indeed, if we set $T(w, \omega) = T_k(w)$ for $\omega \in \Omega_k \setminus \Omega_{k-1}$, then (4.75) follows immediately from (4.74). Furthermore, since T_k is almost surely finite, so is T.

We now prove (4.59). Let us fix $\varepsilon > 0$ and choose $k_\varepsilon \geq 1$ so large that $\mathbb{P}(\Omega_{k_\varepsilon}) \geq 1 - \frac{\varepsilon}{2}$. We next find $\rho_\varepsilon > 0$ such that $\mathbb{P}\{R_{k_\varepsilon} > \rho_\varepsilon\} \leq \frac{\varepsilon}{2}$. It is straightforward to see that (4.59) holds with $\mathcal{C}_\varepsilon = B_{\rho_\varepsilon}(V)$ and $t_\varepsilon(u) = T_{k_\varepsilon}(u)$. Thus, it suffices to establish Proposition 4.2.18. Before doing so, we summarise the properties resulting from the application of Theorems 4.2.16 and 4.2.17 to the Navier–Stokes system (3.84).

Theorem 4.2.19 *Let us assume that the random part η of the external force in* (3.84) *is such that $b_j \neq 0$ for all $j \geq 1$ and $\mathfrak{B}_1 < \infty$. Let μ be the corresponding unique stationary measure and let $\{\mu_\omega\}$ be the Markov disintegration of the invariant measure associated with μ. Then there is a full measure event $\widetilde{\Omega} \subset \Omega$ such that the random set A_ω defined by* (4.63) *is a minimal weak random attractor for* (3.84).

Let us mention that similar results hold for the Navier–Stokes system perturbed by a random kick force. Furthermore, the weak random attractor described in Theorem 4.2.19 is contained in the *global random attractor*, which may be even larger than the one constructed in Theorem 4.2.5. A minimal global random attractor $\{\mathcal{A}_\omega\}$ is uniquely defined (in the sense described in Definition 4.2.4), and for many stochastic PDEs (including the 2D Navier–Stokes system), it is known that the Hausdorff dimension of almost every set $\mathcal{A}_\omega$ can be estimated by a deterministic constant; see [Deb97; Deb98; LR06].

Proof of Proposition 4.2.18 We shall need the following auxiliary result established at the end of this subsection.

Lemma 4.2.20 *Suppose that* (4.73) *holds and define a process $\zeta(t)$ as in* (2.98). *Then for any $\alpha > 0$, the stochastic integral*

$$z_\alpha(t) = \int_{-\infty}^{t} e^{-\alpha(t-s)L} d\zeta(s), \quad t \in \mathbb{R}, \tag{4.76}$$

defines an H-valued stationary process possessing the following properties.

(i) *The process z_α is adapted to the filtration $\{\mathcal{F}_t, t \in \mathbb{R}\}$, and its trajectories belong to the space $C(\mathbb{R}; V) \cap L^2_{\text{loc}}(\mathbb{R}; V^2)$ with probability* 1 *and satisfy the equation*

$$\dot{z} + \alpha L z = \partial_t \zeta(t), \quad t \in \mathbb{R}. \tag{4.77}$$

(ii) *For any $\gamma > 0$ and almost every $\omega \in \Omega$, we have*

$$\sup_{-1 \le t \le 0} \|z_\alpha(t)\|_1 < \infty, \tag{4.78}$$

$$\lim_{t \to -\infty} |t|^{-\gamma} |z_\alpha(t)|_2^2 = 0, \tag{4.79}$$

$$\lim_{t \to -\infty} \frac{1}{|t|} \int_t^0 |\nabla z_\alpha(s)|_2^2 ds = \frac{\mathfrak{B}}{2\alpha}, \tag{4.80}$$

$$\lim_{t \to -\infty} \frac{1}{|t|} \int_t^0 |L z_\alpha(s)|_2^2 ds = \frac{\mathfrak{B}_1}{2\alpha}. \tag{4.81}$$

We wish to construct an increasing sequence $\{\Omega_k\} \subset \mathcal{F}$ satisfying (i) and random constants $R_k \ge 0$ such that the solution u of Eq. (3.84) issued from w at time $t = t_0$ satisfies the inequality

$$\|u(0)\|_1 \le R_k(\omega) \quad \text{for } \omega \in \Omega_k,\ w \in H,\ t_0 \le -T_k(w). \tag{4.82}$$

This will imply the required inequality (4.74).

Let $u(t)$ be a solution (3.84) satisfying the initial condition

$$u(t_0) = w.$$

We can write it in the form $u = z_\alpha + v$, where the (large) constant $\alpha > 0$ will be chosen later. Then v must satisfy the equations

$$\dot v + \nu L v + B(v + z_\alpha) = h + (\alpha - \nu) L z_\alpha, \qquad v(t_0) = w - z_\alpha(t_0). \tag{4.83}$$

For fixed $M > 0$ and $k \in \mathbb{Z}_+$, let Ω_k be the set of those $\omega \in \Omega$ for which

$$\frac{1}{|t|} \int_t^0 |L z_\alpha(s)|_2^2 ds \le \frac{M}{\alpha}, \qquad e^{\nu \alpha_1 t} |z_\alpha(t)|_2^2 \le 1 \quad \text{for} \quad t \le -k. \tag{4.84}$$

By (4.81), for a sufficiently large $M > 0$ we have $\mathbb{P}(\Omega_k) \to 1$ as $k \to \infty$, whence it follows that the union $\Omega_* = \cup_k \Omega_k$ is a set of full measure. Moreover, it is easy to see that it is invariant under the shifts. The proof of (4.82) is divided into two steps.

Step 1. We first show that

$$\sup_{-1 \le t \le 0} |v(t)|_2 \le r_k(\omega) \quad \text{for } \omega \in \Omega_k,\ w \in H,\ t_0 \le -T_k(w), \tag{4.85}$$

where r_k is an almost surely finite random constant and T_k depends only on w. Taking the scalar product of (4.83) with v and carrying out some transformations based on Schwarz's inequality and standard estimates of the nonlinear term, we derive

$$\partial_t |v|_2^2 + \nu \|v\|_1^2 \le C_1 g(t) + C_1 |\nabla z_\alpha|_2^2 |v|_2^2, \tag{4.86}$$

where $g = |h|_2^2 + \alpha^2 |\nabla z_\alpha|_2^2 + |z_\alpha|_2 |L z_\alpha|_2$. Since $\|v\|_1 \ge \alpha_1 |v|_2^2$, applying Gronwall's inequality, we obtain

$$|v(t)|_2^2 \le e^{-A(t,t_0)} |w - z_\alpha(t_0)|_2^2 + C_1 \int_{t_0}^t e^{-A(t,s)} g(s)\, ds, \tag{4.87}$$

where we set

$$A(t,s)=\int_s^t\big(\nu\alpha_1-C_1|\nabla z_\alpha(r)|_2^2\big)\,dr.$$

Choosing $\alpha>0$ sufficiently large, we see from the first inequality in (4.84) that

$$A(t,s)\ge\nu\alpha_1(t-s-1)\quad\text{for } -1\le t\le 0,\ s\le -k,\ \omega\in\Omega_k.$$

Substituting this into (4.87) and using the second relation in (4.84), for $t_0\le -k$, $-1\le t\le 0$, and $\omega\in\Omega_k$ we obtain

$$\begin{aligned}|v(t)|_2^2&\le C_2\big(1+e^{\nu\alpha_1 t_0}|w|_2^2\big)\\&\quad+C_2\bigg(\int_{-\infty}^{-k}e^{-\nu\alpha_1(t-s-1)}g(s)\,ds+\int_{-k}^t e^{-A(t,s)}g(s)\,ds\bigg).\end{aligned}$$

This implies the required inequality (4.85), in which

$$r_k(\omega)=C_2\bigg(2+\int_{-\infty}^{-k}e^{\nu\alpha_1(s-2)}g(s)\,ds+\sup_{-1\le t\le 0}\int_{-k}^t e^{-A(t,s)}g(s)\,ds\bigg).$$

Step 2. We now use the regularising property of the Navier–Stokes dynamics to prove that

$$\|v(0)\|_1\le\tilde r_k(\omega)\quad\text{for }\omega\in\Omega_k,\ w\in H,\ t_0\le -T_k(w),\tag{4.88}$$

where $\tilde r_k$ are random constants. Once this inequality is established, the required estimate (4.82) with $R_k(\omega)=\|z_\alpha(0)\|_1+\tilde r_k(\omega)$ will follow from the representation $u(0)=z_\alpha(0)+v(0)$.

Integrating (4.86) in $t\in(-1,0)$ and using (4.85) and the first inequality in (4.84), we get

$$\begin{aligned}\nu\int_0^1\|v(t)\|_1^2dt&\le|v(-1)|_2^2+C_1\int_{-1}^0 g(t)\,dt+C_1\int_{-1}^0|\nabla z_\alpha|_2^2|v|_2^2dt\\&\le r_k(\omega)\bigg(1+C_1\int_{-k}^0|\nabla z_\alpha|_2^2dt\bigg)+C_3\int_{-k}^0\big(1+|Lz_\alpha|_2^2\big)\,dt\\&\le C_4k\,r_k(\omega).\end{aligned}\tag{4.89}$$

We now take the scalar product in H of (4.83) with $2(t+1)Lv$. After some transformations, we obtain

$$\begin{aligned}\partial_t\big((t+1)\,|L^{1/2}v|_2^2\big)+\nu(t+1)\,|Lv|_2^2&\le\|v\|_1^2+C_5(t+1)\big(|h|_2^2+\|z_\alpha\|_2^2\big)\\&\quad+C_5(t+1)\,\big(|v+z_\alpha|_2|L(v+z_\alpha)|_2\big)^{1/2}\|v+z_\alpha\|_1|Lv|_2.\end{aligned}$$

It follows from (4.85) and (4.78) that the last term on the right-hand side of this inequality can be estimated by

$$C_6(\omega)(t+1)\big(|Lv|_2^{3/2}+|Lz_\alpha|_2^{3/2}\big)\big(|L^{1/2}v|_2+1\big),$$

where C_6 is a random constant. Therefore, the function $\psi(t) = (t + 1)|L^{1/2}v(t)|_2$ satisfies the differential inequality

$$\psi'(t) \le C_7(\omega)\psi(t) + C_7(\omega)\bigl(|Lz_\alpha|_2^2 + \|v\|_1^2 + 1\bigr), \quad -1 \le t \le 0.$$

By Gronwall's inequality, we obtain

$$\begin{aligned}\|v(0)\|_1^2 &\le e^{C_7(\omega)} \int_{-1}^{0} \bigl(|Lz_\alpha|_2^2 + \|v\|_1^2 + 1\bigr)\, dt \\ &\le e^{C_7(\omega)}\biggl(\int_{-k}^{0} |Lz_\alpha|_2^2 dt + \int_{-1}^{0} \|v\|_1^2 dt + 1\biggr).\end{aligned}$$

Recalling (4.84) and (4.89), we arrive at the required inequality (4.88). This completes the proof of the proposition. □

Proof of Lemma 4.2.20 The fact that z_α is a well-defined H-valued stationary process possessing properties mentioned in (i) follows easily by finite-dimensional approximations (cf. the proof of Proposition 2.4.2). Therefore we shall confine ourselves to the proof of assertion (ii).

An analogue of Proposition 2.4.10 is true for the process z_α. By stationarity, it follows that

$$\mathbb{E}\Bigl(\sup_{T \le t \le T+1} \exp\bigl(\sigma\nu\|z_\alpha\|_1^2\bigr)\Bigr) = C_1 < \infty \quad \text{for all } T \in \mathbb{R}, \tag{4.90}$$

where $\sigma > 0$ does not depend on ν and T. Inequality (4.90) implies, in particular, that (4.78) holds. Let us introduce the events

$$\Gamma_k = \Bigl\{\sup_{-k \le t \le 1-k} |z_\alpha(t)|_2 > k^{-\gamma/4}\Bigr\}, \quad k \ge 1.$$

Then (4.90) implies that $\sum_k \mathbb{P}(\Gamma_k) < \infty$, whence it follows that $|z_\alpha(t)|_2 \le t^{-\gamma/4}$ for $t \le -t_0$, where $t_0 > 0$ is an almost surely finite random constant. We thus obtain (4.79).

To prove (4.80), we apply Itô's formula to the process $|z_\alpha(t)|_2^2$. This results in

$$|z_\alpha(0)|_2^2 + 2\alpha \int_t^0 |\nabla z_\alpha(s)|_2^2 ds = |z_\alpha(t)|_2^2 + \mathfrak{B}|t| + 2\int_t^0 \langle z_\alpha(s), d\zeta(s)\rangle.$$

Dividing this relation by $2\alpha|t|$, we derive

$$\frac{1}{|t|}\int_t^0 |\nabla z_\alpha(s)|_2^2 ds = \frac{\mathfrak{B}}{2\alpha} + \frac{1}{2\alpha|t|}\biggl(|z_\alpha(t)|_2^2 - |z_\alpha(0)|_2^2 + 2\int_t^0 \langle z_\alpha, d\zeta\rangle\biggr). \tag{4.91}$$

As was proved above,

$$\frac{1}{|t|}\bigl(|z_\alpha(0)|_2^2 + |z_\alpha(t)|_2^2\bigr) \to 0 \quad \text{as } t \to -\infty \text{ almost surely.}$$

On the other hand, using Theorem A.12.1 and inequality (4.90), one can show

$$\frac{1}{|t|}\left|\int_t^0 \langle z_\alpha, d\zeta\rangle\right| \to 0 \quad \text{as } t \to -\infty \text{ almost surely.}$$

Combining this with (4.91), we obtain (4.80). Finally, the proof of (4.81) is based on the application of Itô's formula to the functional $\|z\|_1^2$ and can be carried out by a similar argument. □

4.3 Dependence of a stationary measure on the random force

In this section, we discuss some continuity properties of stationary measures with respect to parameters. We shall study only the dependence of a stationary measure on the random component of an external force. However, similar results remain true when one varies the deterministic component of an external force or the viscosity. We begin with the case in which the random force η entering Eq. (3.84) depends continuously on a parameter a. It is shown that if the hypotheses ensuring the uniqueness of a stationary measure are fulfilled, then the stationary measure continuously depends on a. This implies, in particular, that a solution of the Cauchy problem converges in distribution to a limiting solution *uniformly* in time. We next turn to the case of high-frequency random kicks that converge to a spatially regular white noise and establish a similar result on convergence of stationary measures.

4.3.1 Regular dependence on parameters

Let us consider the Navier–Stokes equations (3.84), in which $\nu > 0$ and $h \in H$ are fixed, and η is a spatially regular white noise depending on a parameter a that belongs to a metric space X:

$$\eta(t,x) = \eta_a(t,x) = \frac{\partial}{\partial t}\zeta_a(t,x), \qquad \zeta_a(t,x) = \sum_{j=1}^{\infty} b_j(a)\beta_j(t)e_j(x). \tag{4.92}$$

Here β_j and e_j are the same as in (2.98), and $b_j : X \to \mathbb{R}_+$ are some continuous functions. We shall always assume that

$$\mathfrak{B} := \sup_{a\in X}\sum_{j=1}^{\infty} b_j^2(a) < \infty. \tag{4.93}$$

We denote by $P_t^a(u,\Gamma)$ the transition function associated with η_a, and by $\mathfrak{P}_t(a)$ and $\mathfrak{P}_t^*(a)$ the corresponding Markov semigroups. The following theorem shows that if the stationary measure is unique for some $a = \hat{a}$, then it is the only accumulation point of stationary distributions in the sense of weak convergence as $a \to \hat{a}$.

Theorem 4.3.1 *Under the above hypotheses, let $b_j(\hat{a}) > 0$ for all $j \geq 1$, where $\hat{a} \in X$ is a given point, and let $\{\mu_a, a \in X\}$ be any family of stationary measures for Eq.* (3.84) *with $\eta = \eta_a$. Then $\mu_a \to \mu_{\hat{a}}$ in the weak topology as $a \to \hat{a}$.*

Proof We first outline the scheme of the proof. We wish to prove that

$$\|\mu_a - \mu_{\hat{a}}\|_L^* \to 0 \quad \text{as } \mathrm{dist}_X(a, \hat{a}) \to 0. \tag{4.94}$$

To this end, we note that $\mathfrak{P}_t^*(a)\mu_a = \mu_a$ for all $t \geq 0$, whence it follows that

$$\mu_a - \mu_{\hat{a}} = \big(\mathfrak{P}_t^*(a)\mu_a - \mathfrak{P}_t^*(\hat{a})\mu_a\big) + \big(\mathfrak{P}_t^*(\hat{a})\mu_a - \mu_{\hat{a}}\big). \tag{4.95}$$

It suffices to show that both terms on the right-hand side of (4.95) go to zero in the dual-Lipschitz metric as $\mathrm{dist}_X(a, \hat{a}) \to 0$. Suppose for any $T > 0$ and $R > 0$ we have proved that

$$\sup_{0\leq t\leq T}\sup_{|v|_2\leq R} \|P_t^a(v,\cdot) - P_t^{\hat{a}}(v,\cdot)\|_L^* \to 0 \quad \text{as} \quad \mathrm{dist}_X(a, \hat{a}) \to 0. \tag{4.96}$$

In this case, using Prokhorov's theorem, one can show that, for any $T > 0$ and any tight family $\Lambda \subset \mathcal{P}(H)$, we have

$$\sup_{0\leq t\leq T}\sup_{\lambda\in\Lambda} \|\mathfrak{P}_t^*(a)\lambda - \mathfrak{P}_t^*(\hat{a})\lambda\|_L^* \to 0 \quad \text{as} \quad \mathrm{dist}_X(a, \hat{a}) \to 0. \tag{4.97}$$

Furthermore, by Theorem 3.3.1,

$$\sup_{\lambda\in\Lambda} \|\mathfrak{P}_t^*(\hat{a})\lambda - \mu_{\hat{a}}\|_L^* \to 0 \quad \text{as} \quad t \to \infty. \tag{4.98}$$

Combining (4.95), (4.97), and (4.98) and noting that the family $\{\mu_a\}$ is tight in H, we arrive at the required result. The accurate proof is divided into four steps.

Step 1. Let us prove (4.96), where $T > 0$ is any fixed number. To this end, we first establish an estimate for the difference between two solutions for Eq. (3.84) with different right-hand sides. Let $u^a(t, x)$ be a solution of problem (3.84), (3.85) in which $\eta = \eta_a$ and $u_0 = v$. Then the difference $u(t) = u^a(t) - u^{\hat{a}}(t)$ vanishes at $t = 0$ and satisfies the equation

$$\dot{u} + \nu L u + B(u^a(t), u) + B(u, u^{\hat{a}}(t)) = \xi(t, x),$$

where we set

$$\xi(t, x) = \frac{\partial}{\partial t}\sum_{j=1}^{\infty}\big(b_j(a) - b_j(\hat{a})\big)\beta_j(t)e_j(x).$$

Let us denote by $z(t) = z(t; a, \hat{a})$ a solution of the linear problem

$$\dot{z} + \nu L z = \xi(t, x), \quad z(0) = 0.$$

Then we can write $u = z + w$, where w vanishes at $t = 0$ and satisfies the equation

$$\dot{w} + \nu L w + B(u^a, z + w) + B(z + w, u^{\hat{a}}) = 0. \tag{4.99}$$

For any curve $\xi \in C(0, T; H) \cap L^2(0, T; H^1)$, let

$$\mathcal{E}_\xi(t) = |\xi(t)|_2^2 + \nu \int_0^t |\nabla \xi(s)|_2^2 ds.$$

The argument used in the proof of Proposition 2.4.2 (see the derivation of (2.109)) enables one to show that

$$\mathbb{E}\Big(\sup_{0\le t\le T} \mathcal{E}_z(t)\Big) \le C_1 \sum_{j=1}^{\infty} |b_j(a) - b_j(\hat{a})|^2.$$

Recalling that b_j are continuous functions that satisfy (4.93), we conclude that

$$\mathbb{E}\Big(\sup_{0\le t\le T} \mathcal{E}_z(t)\Big) \to 0 \quad \text{as} \quad \mathrm{dist}_X(a, \hat{a}) \to 0. \tag{4.100}$$

Furthermore, Corollary 2.4.11 implies that if $|v|_2 \le R$, then

$$\sup_{a\in X} \mathbb{E}\Big(\sup_{0\le t\le T} \mathcal{E}_{u^a}(t)\Big) \le C_2 = C_2(R, T). \tag{4.101}$$

Taking the scalar product of (4.99) with $2w$ and carrying out some standard transformations (cf. the proof of Proposition 2.1.25), we obtain

$$\begin{aligned}\partial_t |w|_2^2 + 2\nu \|w\|_1^2 &\le C_3 \|w\|_1 |w|_2 \|u^{\hat{a}}\|_1 \\ &\quad + C_3\big(\|u^a\|_1 |u^a|_2 + \|u^{\hat{a}}\|_1 |u^{\hat{a}}|_2\big)^{1/2} \big(\|z\|_1 |z|_2\big)^{1/2} \|w\|_1,\end{aligned}$$

whence it follows that

$$\partial_t |w|_2^2 + \nu \|w\|_1^2 \le C_4 \|u^{\hat{a}}\|_1^2 |w|_2^2 + C_4\big(\|u^a\|_1 |u^a|_2 + \|u^{\hat{a}}\|_1 |u^{\hat{a}}|_2\big) \|z\|_1 |z|_2,$$

where $C_4 > 0$ depends on ν. Application of Gronwall's inequality results in

$$\sup_{0\le t\le T} \mathcal{E}_w(t) \le C_5(\nu) \exp\Big(C_5(\nu) \int_0^T \|u^{\hat{a}}(t)\|_1^2 dt\Big) K(a, \hat{a}), \tag{4.102}$$

where we set

$$K(a, \hat{a}) = \int_0^T \big(\|u^a\|_1 |u^a|_2 + \|u^{\hat{a}}\|_1 |u^{\hat{a}}|_2\big) \|z\|_1 |z|_2 dt.$$

Suppose we have shown that, for any positive constants δ and R,

$$\sup_{|v|_2 \le R} \mathbb{P}\Big\{\sup_{0\le t\le T} \mathcal{E}_u(t) > \delta\Big\} \to 0 \quad \text{as } a \to \hat{a}. \tag{4.103}$$

Let us take any function $f : H \to \mathbb{R}$ such that $\|f\|_L \le 1$ and fix an initial datum $v \in B_H(R)$. Denoting by $\Omega_{\delta,a}$ the event on the left-hand side of (4.103), for $0 \le t \le T$ we write

$$\begin{aligned}\left|\mathbb{E}\big(f(u^a(t)) - f(u^{\hat{a}}(t))\big)\right| &\le \mathbb{E}\left|\mathbb{I}_{\Omega_{\delta,a}^c}\big(f(u^a(t)) - f(u^{\hat{a}}(t))\big)\right| + 2\,\mathbb{P}(\Omega_{\delta,a})\\ &\le \sqrt{\delta} + 2\,\mathbb{P}(\Omega_{\delta,a}).\end{aligned}$$

Since f and v were arbitrary, we arrive at the required convergence (4.96).

Step 2. We now prove (4.103). Since $\mathcal{E}_u(t) \le 2(\mathcal{E}_z(t) + \mathcal{E}_w(t))$ for all $t \ge 0$, in view of (4.100), it suffices to show that

$$\sup_{|v|_2 \le R} \mathbb{P}\{\mathcal{E}(w) > \delta\} \to 0 \quad \text{as } a \to \hat{a}, \tag{4.104}$$

where for a function $g(t,x)$ we write

$$\mathcal{E}(g) = \sup_{0 \le t \le T} \mathcal{E}_g(t).$$

For any $M > 0$, we have

$$\begin{aligned}\mathbb{P}\{\mathcal{E}(w) > \delta\} &\le \mathbb{P}\{\mathcal{E}(u^a) > M\} + \mathbb{P}\{\mathcal{E}(u^{\hat{a}}) > M\}\\ &\quad + \mathbb{P}\{\mathcal{E}(w) > \delta,\, \mathcal{E}(u^a) \le M,\, \mathcal{E}(u^{\hat{a}}) \le M\}.\end{aligned}$$

In view of (4.101), the first two terms in the right-hand side of this inequality can be made arbitrarily small, uniformly in $v \in B_H(R)$, by choosing a large M. Therefore, convergence (4.104) will be established if we show that

$$\sup_{|v|_2 \le R} \mathbb{P}\{\mathcal{E}(w) > \delta,\, \mathcal{E}(u^a) \le M,\, \mathcal{E}(u^{\hat{a}}) \le M\} \to 0 \quad \text{as } a \to \hat{a}, \tag{4.105}$$

for any positive δ and M. To this end, note that, in view of (4.102), on the set $\{\mathcal{E}(u^a) \le M, \mathcal{E}(u^{\hat{a}}) \le M\}$, we have

$$\mathcal{E}(w) \le C_6 K(a, \hat{a}) \le C_7 \mathcal{E}(z).$$

Combining this with (4.100) and Chebyshev's inequality, we easily prove (4.105).

Step 3. Let us prove (4.97). Let $\Lambda \subset \mathcal{P}(H)$ be a tight family in H. Then, for any $\varepsilon > 0$ and a sufficiently large $R_\varepsilon > 0$, we have

$$\sup_{\lambda \in \Lambda} \lambda\big(B_H^c(R_\varepsilon)\big) \le \varepsilon. \tag{4.106}$$

On the other hand, the obvious relation

$$\mathfrak{P}_t^*(a)\lambda - \mathfrak{P}_t^*(\hat{a})\lambda = \int_H \big(P_t^a(z,\cdot) - P_t^a(z,\cdot)\big)\lambda(dz)$$

implies that, for any $R > 0$,

$$\|\mathfrak{P}_t^*(a)\lambda - \mathfrak{P}_t^*(\hat{a})\lambda\|_L^* \le \sup_{|v|_2 \le R} \|P_t^a(v,\cdot) - P_t^a(v,\cdot)\|_L^* + \lambda\big(B_H^c(R)\big).$$

Combining this inequality with (4.106) and (4.96) (where $R = R_\varepsilon$), we get (4.97).

Step 4. To complete the proof of (4.94), it remains to show that the family $\{\mu_a\} \subset \mathcal{P}(H)$ is tight in H. To see this, let us note that, in view of (2.157), we have

$$\int_H \|u\|_1^2 \mu_a(du) \le C_8(\nu) \quad \text{for all } a \in X.$$

Chebyshev's inequality and the compactness of the embedding $V \subset H$ prove that the family $\{\mu_a\}$ is tight in H. The proof of Theorem 4.3.1 is complete. □

As a consequence of Theorem 4.3.1, we have the following result on uniform convergence of distributions of solutions for Cauchy's problem.

Corollary 4.3.2 *Under the hypotheses of Theorem 4.3.1, for any compact subset $\Lambda \subset \mathcal{P}(H)$ there is a continuous function $A_\Lambda(\rho) > 0$ going to zero with ρ such that*

$$\sup_{t\ge 0} \big\|\mathfrak{P}_t^*(a)\lambda_1 - \mathfrak{P}_t^*(\hat{a})\lambda_2\big\|_L^* \le A_\Lambda\big(\|\lambda_1 - \lambda_2\|_L^* + \mathrm{dist}_X(a,\hat{a})\big), \tag{4.107}$$

where $\lambda_1, \lambda_2 \in \Lambda$ and $a \in X$.

In particular, if $u^a(t,x)$ is a solution of (3.84), (3.85) with $\eta = \eta_a$, then

$$\sup_{t\ge 0} \big\|\mathcal{D}(u^a(t)) - \mathcal{D}(u^{\hat{a}}(t))\big\|_L^* \to 0 \quad \text{as } a \to \hat{a}.$$

Proof of Corollary 4.3.2 Let us fix a compact subset $\Lambda \subset \mathcal{P}(H)$ and define the function

$$\Delta(t,a,\lambda_1,\lambda_2) = \big\|\mathfrak{P}_t^*(a)\lambda_1 - \mathfrak{P}_t^*(\hat{a})\lambda_2\big\|_L^*.$$

To establish (4.107), it suffices to prove that for any $\varepsilon > 0$ there are positive constants T and δ such that

$$\sup_{t\ge T} \Delta(t,a,\lambda_1,\lambda_2) \le \varepsilon \quad \text{for } \lambda_1,\lambda_2 \in \Lambda,\ \mathrm{dist}_X(a,\hat{a}) \le \delta. \tag{4.108}$$

Indeed, an argument similar to that used in Step 1 of the proof of Theorem 4.3.1 enables one to show that

$$\sup_{0\le t\le T} \Delta(t,\hat{a},\lambda_1,\lambda_2) \to 0 \quad \text{as } \lambda_1,\lambda_2 \in \Lambda,\ \|\lambda_1 - \lambda_2\|_L^* \to 0.$$

Combining this with (4.97), we conclude that

$$\sup_{0\le t\le T} \Delta(t,a,\lambda_1,\lambda_2) \to 0 \quad \text{as } \lambda_1,\lambda_2 \in \Lambda,\ \|\lambda_1 - \lambda_2\|_L^* \to 0,\ \mathrm{dist}_X(a,\hat{a}) \to 0.$$

This convergence and inequality (4.108) imply the required result.

We now prove (4.108). Let μ_a be a stationary measure for (3.84) with $\eta = \eta_a$. By Exercise 3.3.8, such a measure is unique for $\operatorname{dist}_X(a, \hat{a}) \leq \delta \ll 1$ and satisfies the inequality

$$\sup_{a\in B_X(\hat{a},\delta)} \sup_{\lambda\in\Lambda} \|\mathfrak{P}_t^*(a)\lambda - \mu_a\|_L^* \to 0 \quad \text{as } t \to \infty. \tag{4.109}$$

We write

$$\Delta(t, a, \lambda_1, \lambda_2) \leq \left\|\mathfrak{P}_t^*(a)\lambda_1 - \mu_a\right\|_L^* + \left\|\mathfrak{P}_t^*(\hat{a})\lambda_2 - \mu_{\hat{a}}\right\|_L^* + \|\mu_a - \mu_{\hat{a}}\|_L^*.$$

In view of (4.109), the first two terms on the right-hand side of this inequality go to zero uniformly in $\lambda_1, \lambda_2 \in \Lambda$ as $t \to \infty$. Furthermore, by Theorem 4.3.1, the third term vanishes as $a \to \hat{a}$. These two observations immediately imply (4.108). □

4.3.2 Universality of white-noise perturbations

In this section, we consider the Navier–Stokes system

$$\dot{u} + \nu L u + B(u) = \eta, \tag{4.110}$$

where the space variable belongs to a bounded domain and the right-hand side $\eta = \eta_\varepsilon$ is a high-frequency kick force of the form (2.69), i.e.,

$$\eta_\varepsilon(t, x) = \sqrt{\varepsilon} \sum_{k=1}^{\infty} \eta_k(x)\delta(t - k\varepsilon).$$

We assume that the random variables η_k satisfy the hypotheses of Theorem 3.2.9. In this case, if $b_j \neq 0$ for all $j \geq 1$, then for any $\varepsilon > 0$ the RDS associated with the equation

$$u_k = S_\varepsilon(u_{k-1}) + \sqrt{\varepsilon}\,\eta_k, \quad k \geq 1, \tag{4.111}$$

has a unique stationary distribution μ_ε. On the other hand, if η is a spatially regular white noise of the form (2.66), then the Navier–Stokes system (4.110) also has a unique stationary distribution μ. The following theorem shows that, under a mild regularity assumption, the family $\{\mu_\varepsilon\}$ converges to μ in the weak topology.

Theorem 4.3.3 *Let us assume that the above hypotheses are fulfilled and the coefficients b_j are such that $\mathfrak{B}_1 = \sum_j \alpha_j b_j^2 < \infty$. Then $\mu_\varepsilon \to \mu$ as $\varepsilon \to 0^+$ in the weak topology of $\mathcal{P}(H)$.*

Proof We essentially repeat the scheme used in the proof of Theorem 4.3.1. Let $P_t^\varepsilon(u, \Gamma)$ (where $t \in \varepsilon\mathbb{Z}_+$, $u \in H$, and $\Gamma \in \mathcal{B}(H)$) be the transition function for the Markov family defined by (4.111) and let $\mathfrak{P}_t(\varepsilon)$ and $\mathfrak{P}_t^*(\varepsilon)$ be the corresponding Markov operators; cf. Section 2.4.3. In this case, we can write (cf. (4.95))

$$\mu_\varepsilon - \mu = \big(\mathfrak{P}_t^*(\varepsilon)\mu_\varepsilon - \mathfrak{P}_t^*\mu_\varepsilon\big) + \big(\mathfrak{P}_t^*\mu_\varepsilon - \mu\big), \quad t \in \varepsilon\mathbb{Z}_+, \tag{4.112}$$

where $\mathfrak{P}_t$ and $\mathfrak{P}_t^*$ stand for the Markov semigroups associated with Eq. (4.110) in which η is given by (2.66). We wish to show that the right-hand side of (4.112) goes to zero in the dual-Lipschitz distance as $t \to +\infty$ and $\varepsilon \to 0^+$. By Exercise 2.4.16, for any $T > 0$ and any subset $\Lambda \subset \mathcal{P}(H)$ satisfying the condition

$$\sup_{\lambda\in\Lambda} \int_H |u|_2^2 \lambda(du) < \infty, \tag{4.113}$$

we have (cf. (4.97))

$$\sup_{0\le t\le T} \sup_{\lambda\in\Lambda} \big\|\mathfrak{P}_t^*(\varepsilon)\lambda - \mathfrak{P}_t^*\lambda\big\|_L^* \to 0 \quad \text{as} \quad \varepsilon \to 0^+, \tag{4.114}$$

where the Markov semigroup $\mathfrak{P}_t^*(\varepsilon)$ is extended to $\mathbb{R}_+$ in a natural way; cf. Section 2.4.3. Furthermore, by Theorem 3.5.2, for any family $\Lambda \subset \mathcal{P}(H)$ satisfying (4.113), we have

$$\sup_{\lambda\in\Lambda} \big\|\mathfrak{P}_t^*\lambda - \mu\big\|_L^* \to 0 \quad \text{as} \quad t \to \infty. \tag{4.115}$$

If we prove that the family of measures $\Lambda = \{\mu_\varepsilon\}$ satisfies (4.113), then combining (4.112) with relations (4.114) and (4.115), we see that $\|\mu - \mu_\varepsilon\|_L^* \to 0$ as $\varepsilon \to 0^+$. Thus, it remains to show that (4.113) holds for the family $\Lambda = \{\mu_\varepsilon\}$.

In view of (2.51), we have $|S_\varepsilon(u)|_2 \le e^{-\varepsilon}|u|_2$ for $u \in H$. Since η_k and u_k are independent, it follows that

$$\mathbb{E}\,|u_k|_2^2 = \mathbb{E}\,\big|S_\varepsilon(u_{k-1})\big|_2^2 + \varepsilon\,\mathbb{E}\,|\eta_k|_2^2 \le e^{-2\varepsilon}\mathbb{E}\,|u_{k-1}|_2^2 + \varepsilon\mathfrak{B},$$

where we used the relations

$$\mathbb{E}\,\eta_k = 0, \quad \mathbb{E}\,|\eta_k|_2^2 = \sum_{j=1}^\infty b_j^2\,\mathbb{E}\,\xi_{jk}^2 = \sum_{j=1}^\infty b_j^2 = \mathfrak{B}.$$

Iterating the above inequality, we derive

$$\mathbb{E}\,|u_k|_2^2 \le e^{-2\varepsilon k}|u_0|_2^2 + \frac{\varepsilon\mathfrak{B}}{1-e^{-2\varepsilon}} \le e^{-2\varepsilon k}|u_0|_2^2 + \frac{3\mathfrak{B}}{2} \quad \text{for all} \quad k \ge 0.$$

The argument used in the proof of Theorem 2.5.3 now implies the required inequality (4.113). □

In contrast to Section 4.3.1, we cannot derive from Theorem 2.5.3 a result similar to Corollary 4.3.2 since we lack an analogue of (4.109). In other words, we cannot prove that the convergence $\mathcal{D}(u_\varepsilon(t; v)) \to \mu_\varepsilon$ is uniform with respect to the parameter $\varepsilon > 0$.

4.4 Relevance of the results for physics

The strong law of large numbers proved in this chapter shows that, for 2D turbulence driven by a non-degenerate random force, *the time average of an observables equals its ensemble average*. This equality is postulated in the theory of turbulence; e.g., see [Bat82], p. 17, or [Fri95], p. 58. It is important for numerical and experimental simulations, since it allows us to calculate various averaged characteristics of a turbulent flow by running one experiment for a long time, rather than by making a large number of costly independent experiments in the stationary regime. The law of the iterated logarithm shows that the rate of convergence to the time-average is essentially the same as for the sum of independent random variables.

The central limit theorem, in particular, justifies the widely accepted belief that, in the turbulent regime, *in large time-scales the velocity of the fluid has an approximately normal distribution*; see [Bat82], p. 174. For instance, an experiment carried out by A. A. Townsend in the middle of the twentieth century showed that the probability density function of the velocity of a turbulent fluid at an arbitrary point is very close to the Gaussian density; see [Bat82], p. 169. We believe that this observation is explained by the CLT, since any mechanical device measures not the instantaneous velocity, but its average over some time interval.

The random attractors and their deterministic counterparts are well-established tools for studying 2D Navier–Stokes equations. The fact that they have finite dimension for any positive value of the viscosity implies that large-time asymptotics of fluid motion, even in the turbulent regime, can be described by finite-dimensional dynamical systems. Theorem 4.2.19 shows that the random attractor has an additional natural structure – the probability measures μ_ω, which form a disintegration of the unique invariant measure.

The results of Section 4.3.1 indicate that, for statistical hydrodynamics, in striking difference with its deterministic analogue, the distribution of a velocity field $u(t,\cdot)$ satisfying (1) continuously depends on the parameters of the random force f *uniformly in time* t.

Finally, our results in Section 4.3.2 establish that, in a certain sense, *white forces are universal* for the randomly forced 2D Navier–Stokes system (1). Namely, there we prove that if the system is perturbed by the force

$$f_\varepsilon(t,x) = \sqrt{\varepsilon}\,\eta(\varepsilon^{-1}t,x), \quad 0<\varepsilon\ll 1, \tag{4.116}$$

where η is a non-degenerate kick process with zero mean value, then its unique stationary measure converges, as $\varepsilon\to 0$, to the unique stationary measure of a white-forced system. Certainly this is a general phenomenon which holds true for many other rapidly oscillating random forces of the form (4.116), provided that the random field $\eta(t, x)$ satisfies assumptions (a) and (b) from the Introduction.

Notes and comments

The LLN, LIL, and CLT are well understood for independent random variables and for processes with strong mixing properties. Roughly speaking, these results hold as long as the strong mixing coefficient decays to zero sufficiently fast; see the books [Has80; JS87; MT93; Rio00]. In the context of randomly forced PDEs, this condition can be satisfied only if the noise is rough with respect to the space variables; cf. example 1.3 in [Shi06c]. In the context of a randomly forced Navier–Stokes system, the SLNN and CLT were established by Kuksin [Kuk02a], using a coupling argument and some general limit theorems for stationary processes. The martingale approximation introduced by Gordin [Gor69] is a powerful tool for studying limit theorems for stochastic processes. It was used by Shirikyan [Shi06c] to derive the rates of convergence in the SLLN and CLT. The LIL for Navier–Stokes equations was announced by Denisov [Den04], however, a proof has never appeared. The SLLN and CLT discussed in this monograph can be derived from more general results of the book [DDL$^+$07], while the LIL can be obtained as a consequence of the invariance principles established in the recent papers [DM10] and [BMS11].

A systematic study of random dynamical systems can be found in the book of L. Arnold [Arn98]. Random attractors are a natural extension of attractors for non-autonomous equations (cf. [CV02]). They were constructed in the papers of Crauel, Debussche, and Flandoli [CF94; CDF97], and their finite-dimensionality for various stochastic PDEs was proved by Debussche [Deb97; Deb98], Langa and Robinson [LR06], and many others. Theorem 4.2.9 was established at various levels of generality by Ledrappier [Led86], Le Jan [Le 87], and Crauel [Cra91]. Theorems 4.2.16, 4.2.17, and 4.2.19 were proved by Kuksin and Shirikyan [KS04a]

The dependence of stationary measures on various parameters in the context of randomly forced PDEs was studied by Kuksin and Shirikyan [KS03], Chueshov and Kuksin [CK08a], and Hairer and Mattingly [HM08]. The presentation here follows the paper [KS03].

5

Inviscid limit

In this chapter, we consider the Navier–Stokes system on the 2D torus $\mathbb{T}^2$, perturbed by spatially regular white noise (with trivial deterministic part) whose amplitude is proportional to the square root of the viscosity:

$$\partial_t u + \nu L u + B(u) = \sqrt{\nu}\,\eta(t,x), \tag{5.1}$$

$$\eta(t,x) = \frac{\partial}{\partial t}\zeta(t,x), \quad \zeta(t,x) = \sum_{s\in\mathbb{Z}_0^2} b_s \beta_s(t) e_s(x). \tag{5.2}$$

Here $\{e_s\}$ is the basis defined in (2.29), $\{\beta_s\}$ is a sequence of independent standard Brownian motions, and $b_s \geq 0$ are some constants decaying to zero sufficiently fast. According to Theorem 3.3.1, if all the constants b_s are non-zero, then Eq. (5.1) has a unique stationary distribution μ_ν. In this chapter, we show that the family $\{\mu_\nu\}$ is tight and investigate properties of the limiting measures. In particular, it will be proved that any such measure is the law of a stationary process concentrated on solutions of the homogeneous Euler equation. Some a priori estimates and a non-degeneracy property will also be obtained. The results of this chapter (with properly modified constants) remain true for the Navier–Stokes system on a non-standard torus $\mathbb{R}^2/(a\mathbb{Z}\oplus b\mathbb{Z})$, $a, b > 0$.

5.1 Balance relations

5.1.1 Energy and enstrophy

As before, for $k \in \mathbb{Z}_+$ we denote

$$\mathfrak{B}_k = \sum_{s\in\mathbb{Z}_0} \alpha_s^k b_s^2 \leq \infty,$$

where $\alpha_s = |s|^2$ stands for the eigenvalue associated with e_s. Note that $\mathfrak{B}_0 = \mathfrak{B}$.

Let us recall some results established in Section 2.5.2. Theorem 2.5.5 implies that, if $\mathfrak{B}_0 < \infty$ and $u_\nu(t,x)$ is a stationary solution of problem (5.1), (5.2), then

$$\mathbb{E}\,\langle Lu_\nu, u_\nu\rangle = \frac{1}{2}\mathfrak{B}_0, \tag{5.3}$$

$$\mathbb{E}\exp\bigl(\varkappa_0|u_\nu|_2^2\bigr) \le C_0, \tag{5.4}$$

where $\varkappa_0$ and C_0 are positive constants not depending on $\nu > 0$. Furthermore, by Exercise 2.5.7, if $\mathfrak{B}_1 < \infty$, then

$$\mathbb{E}\,|Lu_\nu|_2^2 = \frac{1}{2}\mathfrak{B}_1, \tag{5.5}$$

$$\mathbb{E}\exp\bigl(\varkappa_1\|u_\nu\|_1^2\bigr) \le C_1, \tag{5.6}$$

where the positive constants $\varkappa_1$ and C_1 are also independent of $\nu > 0$. Note that relations (5.3) and (5.4) are true for any bounded domain with smooth boundary, whereas (5.5) and (5.6) are valid only for periodic boundary conditions, because their proof uses the crucial relation $\langle B(u), u\rangle = 0$.

A simple corollary of the above estimates is the inequality

$$\frac{\mathfrak{B}_0^2}{2\mathfrak{B}_1} \le \mathbb{E}\int_{\mathbb{T}^2} |u_\nu(t,x)|^2 dx \le \frac{\mathfrak{B}_0}{2} \quad \text{for all } \nu > 0. \tag{5.7}$$

Indeed, the right-hand estimate is a straightforward consequence of (5.3) and the obvious inequality $|v|_2^2 \le \langle Lv, v\rangle$. To prove the left-hand estimate, note that, by the interpolation inequality for Sobolev spaces (see Property 1.1.4), we have

$$\langle Lv, v\rangle \le |v|_2 |Lv|_2 \quad \text{for any } v \in V^2.$$

It follows that

$$\mathbb{E}\,\langle Lu_\nu, u_\nu\rangle \le \bigl(\mathbb{E}\,|u_\nu|_2^2\bigr)^{1/2}\bigl(\mathbb{E}\,|Lu_\nu|_2^2\bigr)^{1/2}.$$

Recalling relations (5.3) and (5.5), we arrive at the left-hand inequality in (5.7).

5.1.2 Balance relations

Equalities (5.3) and (5.5) are related to the fact that the energy and enstrophy are integrals of motion for the 2D Euler equations on the torus. It is well known that the integral over $\mathbb{T}^2$ of any function of vorticity also is preserved under the dynamics of the Euler system. It turns out that if the right-hand side of (5.1) is sufficiently smooth in x, then to each function of vorticity there corresponds a *balance relation* similar to (5.3) and (5.5). To formulate the corresponding result, we first introduce some notation.

Let us assume that $\mathfrak{B}_2 < \infty$. Then, by Exercise 2.5.8 with $k = 2$, almost every trajectory of a stationary solution $u_\nu(t,x)$ for (5.1) belongs to the space $L^2_{\rm loc}(\mathbb{R}_+; V^3) \cap C(\mathbb{R}_+; V^2)$. Applying the operator curl to (5.1) and using the relation $\operatorname{curl}(\nabla q) = 0$, we obtain the following equation[1] for the stationary process $v_\nu = \operatorname{curl} u_\nu$ whose almost every trajectory belongs to the space $L^2_{\rm loc}(\mathbb{R}_+; H^2) \cap C(\mathbb{R}_+; H^1)$:

$$\partial_t v - \nu\Delta v + \langle u, \nabla\rangle v = \sqrt{\nu}\,\xi. \tag{5.8}$$

Here ξ is an H^1-smooth white noise given by the relation

$$\xi(t,x) = \operatorname{curl}\eta(t,x) = \frac{\partial}{\partial t}\sum_{s\in\mathbb{Z}^2_0} b_s\beta_s(t)\varphi_s(x), \tag{5.9}$$

where we set

$$\varphi_s(x) = \frac{|s|}{\sqrt{2}\pi}\begin{cases} \cos\langle s, x\rangle, & s \in \mathbb{Z}^2_+, \\ -\sin\langle s, x\rangle, & s \in \mathbb{Z}^2_-. \end{cases} \tag{5.10}$$

Theorem 5.1.1 *Let $\mathfrak{B}_2 < \infty$ and let $g(r)$ be a continuous function satisfying the inequality*

$$|g(r)| \le C(1+|r|)^l \quad \textit{for } r \in \mathbb{R}, \tag{5.11}$$

where C and l are some constants. Then for all $\nu > 0$ and $t \ge 0$ we have

$$\mathbb{E}\int_{\mathbb{T}^2} g(v_\nu)|\nabla v_\nu|^2 dx = \frac{1}{2}\sum_{s\in\mathbb{Z}^2_0} b_s^2\,\mathbb{E}\int_{\mathbb{T}^2} g(v_\nu)\varphi_s^2 dx. \tag{5.12}$$

The meaning of the left-hand side of inequality (5.12) needs to be clarified, because the trajectories of v_ν are continuous only as functions of time with range in H^1, and the integral in x does not necessarily converge. However, it follows from Exercise 2.5.8 that $\mathbb{E}\,\|v_\nu(t)\|_2^2 < \infty$ for almost every t, and the mean value of the integral makes sense for almost every t. Since v_ν is stationary, the left-hand side of (5.12) is in fact independent of time. Readers willing to avoid this type of complication may assume that $\mathfrak{B}_3 < \infty$, in which case all the terms are well defined without taking the mean value.

Proof of Theorem 5.1.1 Let G be the second primitive of g vanishing at zero together with its first derivative, that is, $G''(r) = g(r)$ and $G(0) = G'(0) = 0$. Consider a functional $F : H^1 \to \mathbb{R}$ defined by

$$F(w) = \int_{\mathbb{T}^2} G(w(x))\,dx. \tag{5.13}$$

It is straightforward to see that F satisfies the hypotheses of Theorem A.7.5 for the triple $(V, H, V^*) = (H^2, H^1, L^2)$, and therefore we can apply Itô's formula (A.24) to the process $F(v_\nu(t))$. Omitting the subscript ν for simplicity

[1] Note that Eq. (5.8) is valid for any solution whose trajectories possess the above-mentioned regularity.

of notation, we get

$$F\big(v(t\wedge\tau_n)\big)=F\big(v(0)\big)+\int_0^{t\wedge\tau_n}\Big(A(\theta)\,d\theta+\sum_{s\in\mathbb{Z}_0^2}B_s(\theta)\,d\beta_s\Big),\qquad(5.14)$$

where $\tau_n=\inf\{t\geq 0:\|v(t)\|_1>n\}$,

$$A(t)=\int_{\mathbb{T}^2}\Big(G'(v)\big(\nu\Delta v-\langle u,\nabla\rangle v\big)+\frac{\nu}{2}\sum_{s\in\mathbb{Z}_0^2}b_s^2 g(v)\varphi_s^2\Big)\,dx,$$

$$B_s(t)=\sqrt{\nu}\,b_s\int_{\mathbb{T}^2}G'(v)\varphi_s\,dx.$$

Suppose we have shown that condition (A.36) is satisfied for the process $F(v)$. Then, by Corollary A.7.6, relation (5.14) implies that

$$F\big(v(t)\big)=F\big(v(0)\big)+\int_0^t A(\theta)\,d\theta+M_t,\qquad(5.15)$$

where M_t stands for the corresponding stochastic integral. Taking the mean value of both sides of (5.15), using the stationarity of v, and recalling the definition of A, we get $\mathbb{E}A(t)=0$ for any $t\geq 0$. Noting that

$$\int_{\mathbb{T}^2}G'(v)\big(\nu\Delta v-\langle u,\nabla\rangle v\big)\,dx=-\nu\int_{\mathbb{T}^2}g(v)|\nabla v|^2\,dx$$

for any functions $v\in H^2$ and $u\in V^2$, we arrive at the required relation (5.12).

To complete the proof of the theorem, it remains to prove that we can apply Corollary A.7.6. That is, we need to check that

$$\sum_{s\in\mathbb{Z}_0^2}\mathbb{E}\int_0^t|B_s(\theta)|^2d\theta<\infty\quad\text{for any } t>0.\qquad(5.16)$$

In view of (5.10), (5.11), and (5.6), for any $t>0$ we have

$$\sum_{s\in\mathbb{Z}_0^2}\mathbb{E}\int_0^t|B_s(\theta)|^2d\theta=\nu\sum_{s\in\mathbb{Z}_0^2}b_s^2\mathbb{E}\int_0^t\left|\int_{\mathbb{T}^2}G'(v)\varphi_s\right|^2d\theta$$

$$\leq C_1\nu\sum_{s\in\mathbb{Z}_0^2}b_s^2|s|^2\int_0^t\mathbb{E}\,\big|v(\theta)\big|_{2(l+1)}^{2(l+1)}d\theta\leq C_2\mathfrak{B}_1 t\nu,$$

where C_1 and C_2 are some constants not depending on ν and $\{b_s\}$. The proof is complete. □

Exercise 5.1.2 Calculate the first and second derivatives of the functional F defined by (5.13) and verify that the hypotheses of Theorem A.7.5 are satisfied for F. *Hint:* Use the fact that the Sobolev space H^1 is continuously embedded in L^p for any finite $p\geq 1$.

We now consider a particular case in which the stationary measure is unique, and the law of the random perturbation is invariant under translations and reflections. Namely, as was proved in Corollary 3.3.4, if the coefficients b_s are all non-zero and $b_s \equiv b_{-s}$, then the unique stationary measure μ_ν is invariant under translations and reflections. This property enables one to draw some further conclusions on the stationary distributions. Namely, we have the following result.

Proposition 5.1.3 *In addition to the hypotheses of Theorem 5.1.1, suppose that* $\mathfrak{B}_4 < \infty$ *and*

$$b_s = b_{-s} \neq 0 \quad \text{for all } s \in \mathbb{Z}_0^2.$$

Let μ_ν be the unique stationary measure of (5.1)*, let u_ν be a corresponding stationary solution, and let $v_\nu = \operatorname{curl} u_\nu$. Then for all $\nu > 0$, $t \geq 0$, and $x \in \mathbb{T}^2$ we have*[2]

$$\mathbb{E}\big(g(v_\nu(t,x))|\nabla v_\nu(t,x)|^2\big) = \frac{1}{2}(2\pi)^{-2}\mathfrak{B}_1\,\mathbb{E}\big(g(v_\nu(t,x))\big). \tag{5.17}$$

Equalities (5.17) (with various functions g) are called *balance relations* for a stationary solution v_ν.

Proof of Proposition 5.1.3 Since $b_s = b_{-s}$, the right-hand side of (5.12) is equal to

$$\frac{1}{4}\mathbb{E}\sum_{s\in\mathbb{Z}_0^2} b_s^2 \int_{\mathbb{T}^2} g(v_\nu)(\varphi_s^2 + \varphi_{-s}^2)\,dx = \frac{1}{2}(2\pi)^{-2}\mathfrak{B}_1\mathbb{E}\int_{\mathbb{T}^2} g(v_\nu)\,dx,$$

where we used the relation $\varphi_s^2 + \varphi_{-s}^2 = \frac{|s|^2}{2\pi^2}$. Substituting this into (5.12), we derive

$$\int_{\mathbb{T}^2} \mathbb{E}\Big(g(v_\nu)\big(|\nabla v_\nu|^2 - \tfrac{1}{2}(2\pi)^{-2}\mathfrak{B}_1\big)\Big)\,dx = 0.$$

Since the law of v_ν is invariant under translations in x, the integrand is independent of x and vanishes. The proof is complete. □

Relation (5.3) and the space homogeneity of the process $v_\nu(t,x)$ imply that $\mathbb{E}|\nabla v_\nu(t,x)|^2 = \frac{1}{2}\mathfrak{B}_0(2\pi)^{-2}$. So (5.17) means that the random variables $g(v_\nu(t,x))$ and $|\nabla v_\nu(t,x)|^2$ are uncorrelated: the expectation of their product equals the product of their expectations. Jointly the balance relations (5.17) (i.e., the assertion of Proposition 5.1.3) may be reformulated in terms of the conditional expectation with respect to the σ-algebra $\mathcal{F}_{v_\nu(t,x)}$ generated by $v_\nu(t,x)$.

Corollary 5.1.4 *Under the hypotheses of Proposition 5.1.3, for all $\nu > 0$, $t \geq 0$, and $x \in \mathbb{T}^2$ we have*

$$\mathbb{E}\big(|\nabla v_\nu(t,x)|^2 \mid \mathcal{F}_{v_\nu(t,x)}\big) = \frac{1}{2}(2\pi)^{-2}\mathfrak{B}_1.$$

[2] Note that the left- and right-hand sides of (5.17) are well defined, because almost every trajectory of v_ν is a continuous function of time with range in H^3, and the corresponding norm has finite moments; see Exercise 2.5.8.

Proof We first note that, in view of Exercise 2.5.8, almost all trajectories of v_ν are continuous in time with range in H^3, whence it follows that, with probability 1, the expression $|\nabla v_\nu(t,x)|$ is well defined for any $t \geq 0$ and $x \in \mathbb{T}^2$. The required assertion is now a straightforward consequence of (5.17) and the definition of the conditional expectation. □

The other way round, the assertion of the corollary implies that

$$\mathbb{E}\big(g(v_\nu(t,x))|\nabla v_\nu(t,x)|^2\big) = \mathbb{E}\big(g(v_\nu(t,x))|\nabla v_\nu(t,x)|^2 \,|\, \mathcal{F}_{v_\nu(t,x)}\big)$$
$$= \frac{1}{2}(2\pi)^{-2}\mathfrak{B}_1\,\mathbb{E}\big(g(v_\nu(t,x))\big).$$

Thus, Corollary 5.1.4 is equivalent to the balance relations.

The following exercise gives yet another form of the balance relations (5.17).

Exercise 5.1.5 In addition to the hypotheses of Proposition 5.1.3, assume that $\mathfrak{B}_6 < \infty$. Prove that the following properties hold for any fixed $t \geq 0$.

(i) For $\tau \in \mathbb{R}$ and $\omega \in \Omega$, define $\Gamma_t(\tau,\omega) = \{x \in \mathbb{T}^2 : v_\nu(t,x) = \tau\}$. Then for almost all $\omega \in \Omega$ and $\tau \in \mathbb{R}$, the set $\Gamma_t(\tau,\omega)$ is a C^3-smooth curve in $\mathbb{T}^2$.

(ii) Let $d\ell$ be the length element on $\Gamma_t(\tau,\omega)$ (whenever the latter is a smooth curve). Then for almost every $\tau \in \mathbb{R}$, we have the following *co-area form of the balance relations*:

$$\mathbb{E}\int_{\Gamma_t(\tau,\omega)} |\nabla v_\nu|\,d\ell = \frac{1}{2}(2\pi)^{-1}\mathfrak{B}_1\mathbb{E}\int_{\Gamma_t(\tau,\omega)} |\nabla v_\nu|^{-1}d\ell.$$

Hint: Integrate (5.17) over dx and perform the co-area change of variables $dx = |\nabla v_\nu|^{-1}d\tau d\ell$ in the integral. (We refer the reader to [Kuk06b] for a full proof of this result.)

5.1.3 Pointwise exponential estimates

Inequality (5.6), which is true for any stationary solution u_ν of (5.1), implies that

$$\mathbb{P}\{\|u_\nu(t)\|_1 \geq R\} \leq C_1 e^{-\varkappa_1 R^2} \quad \text{for all } R > 0.$$

Thus, large values of the enstrophy are very unlikely in a stationary regime. The aim of this section is to show that if the law of the random perturbation η is invariant under translations, then some similar estimates hold pointwise for u_ν and ∇u_ν. Namely, we have the following result.

Theorem 5.1.6 *Under the hypotheses of Proposition 5.1.3, there are positive constants σ and K, depending only on $\mathfrak{B}_1$, such that, for all $t \geq 0$, $x \in \mathbb{T}^2$, $\nu > 0$, we have*

$$\mathbb{E}\,e^{\sigma|v_\nu(t,x)|} + \mathbb{E}\,e^{\sigma|u_\nu(t,x)|} + \mathbb{E}\,e^{\sigma|\nabla u_\nu(t,x)|^{1/2}} \leq K. \tag{5.18}$$

Proof It suffices to estimate each term on the left-hand side of (5.18) by a constant depending only on $\mathfrak{B}_1$. We begin with the first term.

Integrating relation (5.17) over $x \in \mathbb{T}^2$ in which $g(r) = |r|^{2p}$ with $p > 0$ and applying Hölder's inequality, we derive

$$\mathbb{E}\int_{\mathbb{T}^2} |v_\nu|^{2p}|\nabla v_\nu|^2 dx = C_1\mathfrak{B}_1\mathbb{E}\int_{\mathbb{T}^2} |v_\nu|^{2p} dx$$
$$\leq C_1(2\pi)^{2/(p+1)}\left(\mathbb{E}\int_{\mathbb{T}^2} |v_\nu|^{2(p+1)}dx\right)^{\frac{p}{p+1}}. \tag{5.19}$$

On the other hand, in view of the generalised Poincaré's inequality (A.75), we have

$$\int_{\mathbb{T}^2} |v_\nu|^{2p}|\nabla v_\nu|^2 dx \geq C_2(p+1)^{-2}\int_{\mathbb{T}^2} |v_\nu|^{2(p+1)}dx.$$

Taking the mean value and substituting the resulting inequality into (5.19), we obtain

$$\mathbb{E}\int_{\mathbb{T}^2} |v_\nu|^{2(p+1)}dx \leq (C_3\mathfrak{B}_1)^{p+1}(p+1)^{2(p+1)} \quad \text{for any } p > 0. \tag{5.20}$$

Since the law of $v_\nu(t,x)$ is invariant under translations, we conclude that

$$\mathbb{E}\,|v_\nu(t,x)|^m \leq (C_4\mathfrak{B}_1)^{m/2}m^m \quad \text{for all integers } m \geq 0,$$

where we set $0^0 = 1$. It follows that if $0 < \sigma \leq (e\sqrt{C_4\mathfrak{B}_1})^{-1}$, then

$$\mathbb{E}\, e^{\sigma|v_\nu(t,x)|} = \sum_{m=0}^{\infty}\frac{\sigma^m}{m!}\,\mathbb{E}\,|v_\nu(t,x)|^m \leq \sum_{m=0}^{\infty}\frac{\sigma^m}{m!}(C_4\mathfrak{B}_1)^{m/2}m^m$$
$$\leq \left(1 - \sigma e\sqrt{C_4\mathfrak{B}_1}\,\right)^{-1}, \tag{5.21}$$

where we used the well-known inequality $m! \geq (m/e)^m$. We have thus proved the required estimate for the first term on the left-hand side of (5.18).

The derivation of an upper bound for the second and third terms is based on similar ideas, and we shall confine ourselves to the proof for the more difficult third term. Let us recall that the velocity field u_ν can be recovered from the vorticity v_ν by the relation $u_\nu = \operatorname{curl}(\Delta^{-1}v_\nu)$, where Δ^{-1} stands for the inverse of the Laplace operator on the space of L^2 functions on $\mathbb{T}^2$ with zero mean value. It follows that the components of the matrix ∇u_ν can be written in the form $Rv_\nu = \gamma(\partial_i\partial_j\Delta^{-1})v_\nu$, where $\gamma = \pm 1$. Now note that R is a singular integral operator on $\mathbb{T}^2$, or more precisely, a composition of two Riesz projections; see chapters II and III in [Ste70] for definitions. Therefore, by the Calderón–Zygmund theorem (see sections 2.2 and 6.2 in chapter II of [Ste70]), we have

$$|Rv_\nu|_q \leq Cq\,|v_\nu|_q \quad \text{for any } q \in [2,\infty).$$

Combining this estimate with $q = 2(p+1)$ and inequality (5.20), we obtain

$$\mathbb{E}\int_{\mathbb{T}^2} |\nabla u_\nu|^{2(p+1)}dx \leq (C_5\mathfrak{B}_1)^{p+1}(p+1)^{4(p+1)} \quad \text{for any } p > 0.$$

Since the law of $\nabla u_\nu(t,x)$ is invariant under translations, it follows that

$$\mathbb{E}\,|v_\nu(t,x)|^{m/2} \le (C_6\mathfrak{B}_1)^{m/4} m^m \quad \text{for all } m \ge 0.$$

The required estimate can now be obtained by exactly the same argument as above; see (5.21). □

Exercise 5.1.7 Prove that the middle term in (5.18) can be bounded by the right-hand side. *Hint:* Repeat the scheme of the proof of Theorem 5.1.6 using the fact that $|u_\nu|_p \le C\,|v_\nu|_p$ for any $p \ge 2$, where $C > 0$ does no depend on p; the latter follows from the elliptic regularity for the Laplacian and an interpolation theorem for Sobolev spaces.

5.2 Limiting measures

5.2.1 Existence of accumulation points

Throughout this subsection, we shall assume that

$$\mathfrak{B}_1 < \infty.$$

Let $\{u_\nu, \nu > 0\}$ be a family of stationary solutions for (5.1) and let μ_ν be the law of $u_\nu(t,\cdot)$. In view of the results described in Section 5.1.1, we have

$$\mathbb{E}\,\|u_\nu(t,\cdot)\|_2^2 = \int_H \|v\|_2^2\,\mu_\nu(dv) \le C \quad \text{for all } \nu > 0. \tag{5.22}$$

Since the embedding $H^2 \subset H^{2-\varepsilon}$ is compact for any $\varepsilon > 0$, Prokhorov's theorem implies that the family $\{\mu_\nu\}$ is tight in $H^{2-\varepsilon}$, and therefore any countable subsequence $\{\mu_{\nu_j}, \nu_j \to 0\}$ has a limiting point in the sense of weak convergence of measures on $H^{2-\varepsilon}$. Furthermore, if μ is an accumulation point of the sequence $\{\mu_{\nu_j}\}$, then μ is concentrated on H^2. Indeed, for any $R > 0$ the ball $B_{H^2}(R)$ is closed in $H^{2-\varepsilon}$, and $\mu_{\nu_j}(B_{H^2}(R)) \ge 1 - CR^{-2}$ by (5.22) and Chebyshev's inequality. Therefore, by the portmanteau theorem, we have $\mu(B_{H^2}(R)) \ge 1 - CR^{-2}$ for any $R > 0$, so $\mu(H^2) = 1$. Theorem 5.2.2 established below shows that the laws of the solutions u_{ν_j} (regarded as elements of an appropriate space of trajectories) also converge to a limiting measure. We shall need some notation.

Let us denote by $\boldsymbol{\mu}_\nu$ the law of a stationary solution u_ν. Since the trajectories of solutions of (5.1) almost surely belong to the space $L^2_{\mathrm{loc}}(\mathbb{R}_+; V)$, we see that $\boldsymbol{\mu}_\nu$ is a probability measure on it.

Exercise 5.2.1 Show that if $\mathfrak{B}_1 < \infty$, then the measure $\boldsymbol{\mu}_\nu$ is concentrated on the space $C(\mathbb{R}_+; V) \cap L^2_{\mathrm{loc}}(\mathbb{R}_+; V^2)$.

Given a Banach space X, a finite interval $J \subset \mathbb{R}$, and constants $\alpha \in (0, 1)$ and $p \geq 1$, we define $W^{\alpha,p}(J; X)$ as the space of curves $f \in L^p(J; X)$ such that

$$\|f\|_{W^{\alpha,p}(J;X)} = \|f\|^p_{L^p(J;X)} + \int_J \int_J \frac{\|f(t) - f(s)\|^p_X}{|t - s|^{1+\alpha p}} \, ds \, dt < \infty.$$

The following theorem proves the tightness of the family $\{\boldsymbol{\mu}_\nu\}$ in an appropriate functional space and describes some properties of the limiting measures.

Theorem 5.2.2 *For any $\varepsilon > 0$ the family of measures $\{\boldsymbol{\mu}_\nu\}$ is tight*[3] *in the space $C(\mathbb{R}_+; H^{1-\varepsilon}) \cap L^2_{\rm loc}(\mathbb{R}_+; H^{2-\varepsilon})$. Moreover, if $\boldsymbol{\mu}$ is a limiting point for $\{\boldsymbol{\mu}_\nu\}$ and $u(t, x)$ is a random process whose law coincides with $\boldsymbol{\mu}$, then the following properties hold.*

(i) *Almost every realisation of $u(t, x)$ belongs to the space*

$$L^2_{\rm loc}(\mathbb{R}_+; V^2) \cap W^{1,1}_{\rm loc}(\mathbb{R}_+; V) \cap W^{1,\infty}_{\rm loc}(\mathbb{R}_+; L^p) \tag{5.23}$$

for any $p \in [1, 2)$ and satisfies the Euler equation

$$\dot{u} + B(u) = 0. \tag{5.24}$$

(ii) *The process $u(t, x)$ is stationary in time, and the functions $|u(t)|_2$ and $|\nabla u(t)|_2$ are time-independent random constants.*

Before proving this result, we establish the following corollary.

Corollary 5.2.3 *Let $\boldsymbol{\mu}$ be a limiting point of the family $\{\boldsymbol{\mu}_\nu\}$ in the space $C(\mathbb{R}_+; H^{1-\varepsilon})$ as $\nu \to 0^+$, where $\varepsilon \in (0, 1)$, and let μ be the restriction of $\boldsymbol{\mu}$ at time $t = 0$. Then*

$$\int_H |u|_2^2 \mu(du) \geq \frac{\mathfrak{B}_0^2}{2\mathfrak{B}_1}, \tag{5.25}$$

$$\int_H \|u\|_1^2 \mu(du) = \frac{\mathfrak{B}_0}{2}, \tag{5.26}$$

$$\int_H |Lu|_2^2 \mu(du) \leq \frac{\mathfrak{B}_1}{2}, \tag{5.27}$$

$$\int_H \exp\big(\varkappa \|u\|_1^2\big) \mu(du) \leq C, \tag{5.28}$$

where $\varkappa$ and C are some positive constants. If, in addition, the hypotheses of Proposition 5.1.3 are fulfilled, then for all $x \in \mathbb{T}^2$ we have

$$\int_H \big(e^{\sigma |u(x)|} + e^{\sigma |\nabla u(x)|^{1/2}}\big) \mu(du) \leq K \tag{5.29}$$

[3] Recall that the space $C(\mathbb{R}_+; V)$ is endowed with the topology of uniform convergence on bounded intervals of $\mathbb{R}_+$; see Section 1.1.2. Similarly, the space $L^2_{\rm loc}(\mathbb{R}_+; H^s)$ is endowed with the topology generated by the norms of $L^2(0, n; H^s)$, where $n \geq 1$ is any integer.

with suitable positive constants σ and K. Moreover, in this case the measure $\boldsymbol{\mu}$ is invariant under translations and reflections of $\mathbb{R}^2$.

Proof Inequality (5.25) is a straightforward consequence of (5.26), (5.27), and the interpolation inequality for Sobolev spaces; cf. the proof of (5.7). The derivation of (5.26)–(5.29) is based on passing to the limit in similar estimates for stationary solutions of the Navier–Stokes system. The last assertion of the theorem follows from similar properties of the measures $\boldsymbol{\mu}_\nu$. Therefore, we shall confine ourselves to the proof of (5.27), (5.28), and (5.26).

Since $\boldsymbol{\mu}$ is a limiting point of the family $\{\boldsymbol{\mu}_\nu\}$, there is a sequence $\nu_j \to 0^+$ such that $\boldsymbol{\mu}_{\nu_j} \to \boldsymbol{\mu}$ in the sense of weak convergence of measures on $C(\mathbb{R}_+; H^{1-\varepsilon})$. It follows that $\mu_{\nu_j} \to \mu$ as $j \to \infty$, where μ_{ν_j} stands for the restriction of $\boldsymbol{\mu}_{\nu_j}$ at time $t = 0$, and the convergence holds weakly in the space of Borel measures on $H^{1-\varepsilon}$. Now (5.27) easily follows from Lemma 1.2.17, where π_n is the orthogonal projection $\mathsf{P}_n : H \to H$ to the vector span of the first n eigenfunctions of the Stokes operator L; see Section 2.1.5.

Let us prove (5.28). It follows from (5.27) that the family $\{\mu_{\nu_j}\}$ is tight in $H^{2-\delta}$ for any $\delta > 0$, and by Prokhorov's theorem, there is no loss of generality in assuming that $\mu_{\nu_j} \to \mu$ in the weak topology of $H^{2-\delta}$. Using (5.6) and applying again Lemma 1.2.17, we derive inequality (5.28).

We now prove (5.26). It follows from (5.3) that, for any $R > 0$, we have

$$\int_{B_V(R)} \|u\|_1^2 \mu_{\nu_j}(du) + \int_{B_V^c(R)} \|u\|_1^2 \mu_{\nu_j}(du) = \frac{\mathfrak{B}_0}{2}.$$

Using (2.157), for any $\delta > 0$ we can choose $R = R_\delta > 0$ so large that the second term on the left-hand side is smaller than δ uniformly in j. On the other hand, for any fixed $R > 0$, the first term converges to a similar integral for μ. These two observations combined with Fatou's lemma immediately imply that

$$\frac{\mathfrak{B}_0}{2} - \varepsilon \le \int_{B_V(R_\varepsilon)} \|u\|_1^2 \mu_{\nu_j}(du) \le \int_H \|u\|_1^2 \mu_{\nu_j}(du) \le \frac{\mathfrak{B}_0}{2}.$$

Since $\varepsilon > 0$ is arbitrary, we obtain the required inequality. □

Exercise 5.2.4 Prove inequality (5.29). *Hint:* Use the fact that the functionals taking $u \in V$ to $|u(x)|$ and $|\nabla u(x)|$ are measurable on V, and apply Fatou's lemma.

Proof of Theorem 5.2.2 We first prove the tightness of the family $\{\boldsymbol{\mu}_\nu\}$. To this end, it suffices to show that its restriction to any finite interval $J = [0, n]$ is tight in $C(J; H^{1-\varepsilon}) \cap L^2(J; H^{2-\varepsilon})$. To simplify the formulas, we assume that $J = [0, 1]$, and with a slight abuse of notation, denote the restriction of measures to J by the same symbol.

We fix some constants $\varepsilon > 0$ and $\alpha \in (1/4, 1/2)$ and introduce the functional spaces[4]

$$\mathcal{X} = L^2(J; V^2) \cap \big(H^1(J; H) + W^{\alpha,4}(J; V)\big) = \mathcal{X}_1 + \mathcal{X}_2,$$
$$\mathcal{Y} = L^2(J; H^{2-\varepsilon}) \cap C(J; H^{1-\varepsilon})$$

where we set

$$\mathcal{X}_1 = L^2(J; V^2) \cap H^1(J; H), \quad \mathcal{X}_2 = L^2(J; V^2) \cap W^{\alpha,4}(J; V).$$

Let us note that we have a compact embedding $\mathcal{X} \subset \mathcal{Y}$. Indeed, it follows from theorem 5.1 and 5.2 in [Lio69, chapter I] that $\mathcal{X}_1$ and $\mathcal{X}_2$ are compactly embedded in $L^2(J; H^{2-\varepsilon})$. Furthermore, in view of theorem 3.1 in [LM72], we have a continuous embedding $\mathcal{X}_1 \subset C(J; V)$, whence it follows that the embedding $\mathcal{X}_1 \subset C(J, H^{1-\varepsilon})$ is compact. Finally, by lemma II.2.4 of [Kry02], we have a continuous embedding $W^{\alpha,4}(J; V) \subset C^{\alpha-\frac{1}{4}}(J; V)$, whence it follows that $\mathcal{X}_2$ is also compactly embedded in $C(J; V)$.

Let us denote by $u_\nu(t, x)$ a stationary solution for (5.1) on J whose law coincides with $\boldsymbol{\mu}_\nu$. The tightness of $\{\boldsymbol{\mu}_\nu\}$ in $\mathcal{Y}$ will be established if we prove that

$$\mathbb{E}\, \|u_\nu\|_{\mathcal{X}}^2 \le C_1, \tag{5.30}$$

where $C_1 > 0$ is a constant not depending on ν. To this end, we first note that, in view of (5.5), we have

$$\mathbb{E} \int_0^1 |L u_\nu(t)|_2^2 = \frac{\mathfrak{B}_1}{2}. \tag{5.31}$$

Equation (5.1) implies that

$$u_\nu(t) = u_\nu^1(t) + u_\nu^2(t) + \zeta_\nu(t),$$

where we set $\zeta_\nu = \sqrt{\nu}\,\zeta$,

$$u_\nu^{(1)}(t) = u_\nu(0) - \int_0^t L u_\nu(s)\, ds, \quad u_\nu^{(2)}(t) = -\int_0^t B(u_\nu(s))\, ds.$$

It follows from (5.31) that

$$\mathbb{E}\, \big\|u_\nu^{(1)}\big\|_{H^1(J;H)}^2 \le C_2. \tag{5.32}$$

We now need the following properties of the nonlinear term $B(u)$.

[4] Recall that if B_1 and B_2 are Banach spaces embedded into a topological vector space L, then their sum $B_1 + B_2$ is defined as the vector space of those elements $u \in L$ that are representable in the form $u = u_1 + u_2$, where $u_i \in B_i$. It is a Banach space with respect to the norm $\|u\| = \inf(\|u_1\|_{B_1} + \|u_2\|_{B_2})$, where the infimum is taken over all pairs (u_1, u_2) whose sum is equal to u.

Exercise 5.2.5 For any $p \in [1, 2)$, the bilinear map $(u, v) \mapsto B(u, v)$ satisfies the following inequalities:

$$|B(u,v)|_2 \le C\left(|u|_2\|u\|_2\right)^{1/2}\|v\|_1 \qquad \text{for } u \in H^2,\ v \in H^1, \tag{5.33}$$

$$\|B(u,v)\|_1 \le C\,\|u\|_2\|v\|_2 \qquad \text{for } u \in H^2,\ v \in H^2, \tag{5.34}$$

$$|B(u,v)|_p \le C_p\,\|u\|_1\|v\|_1 \qquad \text{for } u \in H^1,\ v \in H^1. \tag{5.35}$$

Combining (5.33), (5.31), and (5.6), we see that

$$\mathbb{E}\left\|u_\nu^{(2)}\right\|^2_{H^1(J;H)} \le C_3\,\mathbb{E}\int_0^1 \left\|u_\nu^{(2)}\right\|_1^3\left\|u_\nu^{(2)}\right\|_2 dt \le C_4. \tag{5.36}$$

Furthermore, since for a centred Gaussian random variable ξ with variance σ^2 we have $\mathbb{E}\,\xi^4 = 3\sigma^4$, then

$$\mathbb{E}\Big(\sum_{s\in\mathbb{Z}_0^2} c_s\beta_s^2(t)\Big)^2 = \sum_{s,s'\in\mathbb{Z}_0^2} c_s c_{s'}\,\mathbb{E}\big(\beta_s^2(t)\beta_{s'}^2(t)\big)$$
$$\le \sum_{s,s'\in\mathbb{Z}_0^2} c_s c_{s'}\,\mathbb{E}\,\beta_s^4(t) = 3t^2\Big(\sum_{s\in\mathbb{Z}_0^2} c_s\Big)^2,$$

$$\mathbb{E}\Big(\sum_{s\in\mathbb{Z}_0^2} c_s|\beta_s(t)-\beta_s(r)|^2\Big)^2 \le 3(t-r)^2\Big(\sum_{s\in\mathbb{Z}_0^2} c_s\Big)^2,$$

where $c_s \in \mathbb{R}$ are arbitrary constants. It follows that

$$\mathbb{E}\int_0^1 \|\zeta\|_1^4 dt = \mathbb{E}\int_0^1\Big(\sum_{s\in\mathbb{Z}_0^2}\alpha_s b_s^2\beta_s^2(t)\Big)^2 dt \le C_5\mathfrak{B}_1^2,$$

$$E\int_0^1\int_0^1 \frac{\|\zeta(t)-\zeta(r)\|_1^4}{|t-r|^{1+4\alpha}}\,dr\,dt \le C_6\mathfrak{B}_1^2.$$

Combining these inequalities with (5.32)–(5.36), we arrive at (5.30).

We now prove assertions (i) and (ii). As before, it suffices to consider the restrictions of the measures and of the corresponding processes to the interval $J = [0, 1]$. It is easy[5] to see that all the required properties of the process $u(t)$ can be reformulated in terms of its law $\boldsymbol{\mu}$. For instance, the energy and enstrophy of u are random constants if and only if the image of $\boldsymbol{\mu}$ under the mapping

$$w \mapsto \big(|w|_2^2, |\nabla w|_2^2\big), \quad \mathcal{Y} \to C(J;\mathbb{R}^2),$$

is concentrated on the space of constant vector functions. Thus, it suffices to establish assertions (i) and (ii) for any particular choice of the process u.

Now let $\boldsymbol{\mu}$ be a limiting measure for the family $\{\boldsymbol{\mu}_\nu\}$ as $\nu \to 0^+$ in the sense of weak convergence in the space $\mathcal{Y}$. Relation (5.30) and Lemma 1.2.17

[5] It is not straightforward that the space defined by (5.23) is a Borel subset of $C(\mathbb{R}_+; H^{1-\varepsilon}) \cap L^2(\mathbb{R}_+; H^{2-\varepsilon})$. The verification of this property is the content of Exercise 5.2.6.

with $f(u(\cdot)) = \int_J \|u\|_2^2\, dt$ and $\pi_n(u(\cdot)) = \mathsf{P}_n$ (where P_n is the projection from Section 2.1.5) imply that $\boldsymbol{\mu}(L^2(J; V^2)) = 1$.

By Skorokhod's embedding theorem (see theorem 11.7.2 in [Dud02]), there is a sequence of stationary processes $\tilde{u}_{\nu_j}(t, x)$ with $\nu_j \to 0^+$ and a $\mathcal{Y}$-valued random variable $u(t, x)$ that are defined on the same probability space $(\Omega, \mathcal{F}, \mathbb{P})$ and satisfy the properties

$$\mathcal{D}(\tilde{u}_{\nu_j}) = \boldsymbol{\mu}_{\nu_j}, \quad \mathcal{D}(u) = \boldsymbol{\mu}, \tag{5.37}$$

$$\mathbb{P}\big\{\tilde{u}_{\nu_j} \to u \text{ in } \mathcal{Y} \text{ as } j \to \infty\big\} = 1. \tag{5.38}$$

As $\boldsymbol{\mu}(L^2(J; V^2)) = 1$, then $u \in L^2(J; V^2)$ almost surely. Since $\tilde{u}_{\nu_j}$ is a stationary process, so is u.

Let us prove that almost every trajectory of u satisfies the Euler equation (5.24). Indeed, since $\boldsymbol{\mu}_{\nu_j}$ is the law of a stationary solution for (5.1), the process u_{ν_j} can be written as

$$u_{\nu_j}(t) = u_{\nu_j}(0) - \int_0^t \big(\nu_j L u_{\nu_j} + B(u_{\nu_j})\big)\, ds + \sqrt{\nu_j}\, \zeta_j(t), \tag{5.39}$$

where ζ_j is a V-valued spatially regular white noise distributed as ζ. Since the random sequence $\{\sqrt{\nu_j}\, \zeta_j\} \subset C(J; H)$ goes to zero in probability, passing to a subsequence, we can assume that

$$\mathbb{P}\big\{\sqrt{\nu_j}\, \zeta_j \to 0 \text{ in } C(J; H) \text{ as } j \to \infty\big\} = 1. \tag{5.40}$$

Finally, since $\mathbb{E}\|u_{\nu_j}\|^2_{L^2(J,H^2)} \le C$ (see (5.31)), passing again to a subsequence, we see that

$$\mathbb{P}\big\{\nu_j L u_{\nu_j} \to 0 \text{ in } L^2(J; H) \text{ as } j \to \infty\big\} = 1. \tag{5.41}$$

Let us denote by Ω_* the intersection of the events in the left-hand sides of (5.38), (5.40), and (5.41). Then $\mathbb{P}(\Omega_*) = 1$, and for any $\omega \in \Omega_*$, we can pass to the limit in relation (5.39) as $j \to \infty$, regarding it as an equality in $L^1(J \times \mathbb{T}^2)$. This results in the relation

$$u(t) = u(0) - \int_0^t B(u)\, ds, \quad t \in J,$$

which is equivalent to (5.24).

We now prove some further regularity properties for u. It follows from (5.34) and (5.35) that

$$\int_0^1 \|B(u)\|_1 \le C_8 \int_0^1 \|u\|_2^2 dt, \quad \operatorname*{ess\,sup}_{t \in J} |B(u)|_p \le C_9 \operatorname*{ess\,sup}_{t \in J} \|u\|_1^2.$$

Since $u \in L^2(J; V^2) \cap C(J; V)$ with probability 1, we conclude that the right-hand sides in the above inequalities are almost surely finite. Combining this with Eq. (5.24), we see that $u \in W^{1,1}(J; V) \cap W^{1,\infty}(J; L^p)$.

It remains to prove that the energy and enstrophy are random constants. Let us fix any $\omega \in \Omega$ for which $u \in L^2(J; V^2) \cap W^{1,1}(J, V)$. Then the function

$t \mapsto |\nabla u(t)|^2$ is absolutely continuous, and in view of (5.24) and Lemma 2.1.16, for almost all $t \in J$ we have

$$\frac{d}{dt}|\nabla u(t)|^2 = 2\,\langle \nabla \dot{u}(t), \nabla u(t)\rangle = -2\,\langle B(u(t)), \Delta u(t)\rangle = 0.$$

This implies that $|\nabla u(t)|_2^2$ does not depend on t. A similar argument shows that $|u(t)|^2$ is also independent of time. The proof of the theorem is complete. □

Exercise 5.2.6 Prove that the space defined by (5.23) is a Borel subset of the spaces $C(\mathbb{R}_+; H^{s-1})$ and $L^2_{\mathrm{loc}}(\mathbb{R}_+; H^s)$ with $s \le 2$. *Hint:* It suffices to consider the restrictions to finite intervals and to show that the closed balls are Borel subsets. This can be done by describing balls with the help of some countable system of inequalities for functionals.

In Theorem 5.2.2, we showed that the energy and enstrophy of a limiting process are random constants. The proof of this fact was based on some well-known properties of the flow for the homogeneous Euler equations. It turns out that the regularity we obtained for the limiting measure is sufficient to derive an entire family of integrals of motions.

Exercise 5.2.7 Under the hypotheses of Theorem 5.2.2, show that if $g : \mathbb{R} \to \mathbb{R}$ is a continuous function with at most polynomial growth at infinity, then the expression

$$\big\langle g(\operatorname{curl} u(t))\big\rangle = (2\pi)^{-2}\int_{\mathbb{T}^2} g(\operatorname{curl} u(t,x))\,dx \tag{5.42}$$

defines a random constant not depending on time. *Hint:* In the case of regular solutions of the Euler equations, quantity (5.42) is preserved. This fact follows immediately from the observation that the vorticity satisfies a transport equation defined by the vector field $u(t,x)$. Prove that a similar argument applies in our case, on condition that g is a C^1-function with a bounded derivative, and then use a standard regularisation procedure to treat the general situation.

We now construct a topological space $X \subset V$ in which the Euler equation is well posed and any limit point of stationary distributions $\{\mu_\nu\}$ is an invariant measure for the corresponding flow. Let us note that if μ is a limiting measure obtained in Corollary 5.2.3, then for μ-almost every initial function $u_0 \in V$ the Euler equation (5.24) has a global solution belonging to space (5.23). The following exercise shows that the constructed solution is unique.

Exercise 5.2.8 Let u_1 and u_2 be two solutions of (5.24) that belong to the space $L^2(J; V^2) \cap C(J; V)$, where $J = [0, T]$ is a finite interval. Assume that $u_1(\tau) = u_2(\tau)$ for some $\tau \in J$. Show that u_1 and u_2 coincide. *Hint:* The proof can be carried out with the help of an argument due to Yudovich [Jud63] (see also [Che98]).

Let us introduce the Fréchet space $\mathcal{K} = L^2_{\mathrm{loc}}(\mathbb{R}; V^2) \cap W^{1,1}_{\mathrm{loc}}(\mathbb{R}; V)$ and denote by $\mathcal{K}_E$ the set of curves $u \in \mathcal{K}$ that satisfy the Euler equation (5.24) on the real line $\mathbb{R}$. We note that $\mathcal{K}_E$ is a closed subset of $\mathcal{K}$ and provide it with the distance induced from $\mathcal{K}$. Then $\mathcal{K}$ becomes a Polish space. Clearly, the restriction of a function $u \in \mathcal{K}$ to any interval $J \subset \mathbb{R}$ belongs to the space $L^2(J; V^2) \cap C(J; V)$. Therefore, by Exercise 5.2.8, if $u_1, u_2 \in \mathcal{K}_E$ are such that $u_1(0) = u_2(0)$, then $u_1 = u_2$.

Let $\pi : \mathcal{K} \to V$ be the linear continuous operator that takes u to $u(0)$ and let π_E be its restriction to $\mathcal{K}_E$. What has been said above implies that π_E is an injective mapping. Let us denote by X its image and endow it with the induced distance.[6] Thus, X is a complete metric space embedded in V and homeomorphic to $\mathcal{K}_E$. The very definition of X implies that for any $u_0 \in X$ the function $\pi_E^{-1}(u_0) \in \mathcal{K}_E$ is the unique solution of the Euler equation (5.24) supplemented with the initial condition

$$u(0) = u_0. \tag{5.43}$$

Moreover, this solution continuously depends on u_0. Thus, we can define a group of homeomorphisms $S_t : X \to X$ such that $\{S_t u_0, t \in \mathbb{R}\}$ is a unique global solution of (5.24), (5.43). The mapping π_E conjugates any S_t with the time shift $\theta_t : \mathcal{K}_E \ni u(\cdot) \mapsto u(\cdot + t)$. The following theorem establishes some further properties of X and S_t.

Theorem 5.2.9

(i) *The space V^3 is continuously embedded in X.*

(ii) *For any $u_0 \in X$, the curve $\{S_t u_0, t \in \mathbb{R}\} \subset X$ is continuous.*

(iii) *For any $u \in \mathcal{K}_E$ and any bounded continuous function $g : \mathbb{R} \to \mathbb{R}$, the integral* (5.42) *is independent of time, as is the L^2-norm $|u(t)|_2$.*

(iv) *Under the hypotheses of Theorem 5.2.2, any limit point of the family $\{\mu_\nu\}$ as $\nu \to 0^+$ is concentrated on X and is an invariant measure for S_t.*

Proof **(i)** It is well known that the 2D Euler equations are well posed in the space V^3, and the corresponding solutions belong to $C(\mathbb{R}; V^3)$; e.g., see chapter 17 in [Tay97]. Using the equation and properties of the nonlinear term, it is straightforward to see that any solution u with an initial condition from V^3 satisfies the inclusion $\dot{u} \in C(\mathbb{R}; V^2)$, whence we conclude that $u \in \mathcal{K}_E$. This means that $V^3 \subset X$.

(ii) This follows from the fact that the shifts θ_t are continuous homeomorphisms of $\mathcal{K}_E$.

(iii) We confine ourselves to a sketch of the proof, leaving the details to the reader as an exercise. It suffices to show that, for any time-dependent vector filed $u \in \mathcal{K}_E$, the Cauchy problem

$$\dot{y} = u(t, y), \quad y(s) = y_0, \quad s \in \mathbb{R}, \quad y_0 \in \mathbb{T}^2, \tag{5.44}$$

[6] In other words, $\mathrm{dist}_X(\pi_E(u), \pi_E(v)) = \mathrm{dist}_{\mathcal{K}_E}(u, v)$ for all $u, v \in \mathcal{K}_E$.

is well posed, and the corresponding flow $U(t, s) : \mathbb{T}^2 \to \mathbb{T}^2$ that takes y_0 to $y(t)$ preserves the Lebesgue measure on $\mathbb{T}^2$. Indeed, if these properties are established, then the classical argument of the theory of first-order linear PDEs shows that $(\operatorname{curl} u)(t, x) = (\operatorname{curl} u)(0, U(0, t)x)$. Combining this with the measure-preserving property of U, for any function $g \in C_b(\mathbb{R})$ we obtain

$$\int_{\mathbb{T}^2} g\big(\operatorname{curl} u(t, x)\big)\, dx = \int_{\mathbb{T}^2} g\big(\operatorname{curl} u(0, U(0, t)x\big)\, dx = \int_{\mathbb{T}^2} g\big(\operatorname{curl} u(0, x)\big)\, dx.$$

We now prove the above-mentioned result on problem (5.44). It is well known that any function from the unit ball in H^2 possesses the modulus of continuity $\sigma(r) = r|\log r|$. It follows that any time-dependent vector field $u \in \mathcal{K}$ has a modulus of continuity in x that has the form $\varphi(t)\sigma(r)$, where $\varphi \in L^2_{\mathrm{loc}}(\mathbb{R})$. Thus, by Osgood's criterion (see corollary 6.2 in chapter III of [Har64]), problem (5.44) is well posed. The fact that the corresponding flow preserves the Lebesgue measure would follow from the relation $\operatorname{div}_x u \equiv 0$ if the vector field u was C^1-smooth. In our situation, it suffices to approximate it by a regular divergence-free vector field and to pass to the limit.

(iv) Let $\mu \in \mathcal{P}(H)$ be a limit point of the family $\{\mu_\nu\}$. Then there is a subsequence $\nu_j \to 0^+$ such that $\mu_{\nu_j} \to \mu$ as $j \to \infty$. Let $\{u_j\}$ be a sequence of stationary solutions of the Navier–Stokes system (5.1) with $\nu = \nu_j$ that are defined throughout the real line and distributed at any fixed time as μ_{ν_j} and let $\boldsymbol{\mu}_j$ be the law of the process u_j. Analogues of Theorem 5.2.2 and Corollary 5.2.3 for the real line imply that $\boldsymbol{\mu}_j$ converges weakly to the law of a stationary process $u(t)$ whose almost every trajectory belongs to the set $\mathcal{K}_E$. It follows from what has been said that $\mathcal{D}\big(u(t)\big) = \mu$ for any $t \in \mathbb{R}$ and $\mu(X) = 1$. Furthermore, with probability 1, we have

$$S_t\big(u(0)\big) = u(t) \quad \text{for all } t \in \mathbb{R},$$

whence we conclude that μ is an invariant measure for S_t. The proof of the theorem is complete. □

Remark 5.2.10 It is an interesting open question to decide whether X is a *vector* subspace of V. We believe that it is not the case, though we cannot prove this fact.

We proved that the group of homeomorphisms $S_t : X \to X$ possesses an invariant measure μ, which was obtained as an inviscid limit of stationary distributions for the Navier–Stokes system. Let us define the space $Y = \mathcal{P}(\mathbb{R}) \times \mathbb{R}_+$ with the natural metric on the product space (where $\mathcal{P}(\mathbb{R})$ is endowed with the dual-Lipschitz distance) and consider the mapping

$$\Psi : X \to Y, \quad u \mapsto \big((\operatorname{curl} u)_*(\mathrm{d}x), |u|_2\big),$$

where $\mathrm{d}x = (2\pi)^{-2}dx$ is the normalised Lebesgue measure on the torus $\mathbb{T}^2$. Using the continuity of the embedding $X \subset V$, it is straightforward to see that Ψ is continuous. Furthermore, assertion (iii) of Theorem 5.2.9 implies that $\Psi(S_t u_0)$ is constant for any $u_0 \in X$. In other words, we have $\Psi \circ S_t = \Psi$ for any $t \in \mathbb{R}$. Thus, denoting by $X_{\boldsymbol{b}}$ the pre-image[7] of $\boldsymbol{b} \in Y$ under Ψ and endowing it with the topology induced from X, we see that $S_t : X_{\boldsymbol{b}} \to X_{\boldsymbol{b}}$ is a homeomorphism for all $t \in \mathbb{R}$. Let us set $\lambda = \Psi_*(\mu)$. The following result is an immediate consequence of a disintegration theorem for probability measures; see sections III.70–73 in [DM78].

Theorem 5.2.11 *There is a random probability measure $\{\lambda_{\boldsymbol{b}}, \boldsymbol{b} \in Y\}$ on the space X such that the following properties hold.*

(i) *We have $\lambda_{\boldsymbol{b}}(X_{\boldsymbol{b}}) = 1$ for all $\boldsymbol{b} \in Y$ and*

$$\mu\big(A \cap \Psi^{-1}(B)\big) = \int_B \lambda_{\boldsymbol{b}}(A)\lambda(d\boldsymbol{b}) \quad \textit{for any } A \in \mathcal{B}(X),\ B \in \mathcal{B}(Y).$$

(ii) *For λ-almost every $\boldsymbol{b} \in Y$, the measure $\lambda_{\boldsymbol{b}}$ is invariant for the dynamical system $S_t : X_{\boldsymbol{b}} \to X_{\boldsymbol{b}}$.*

Proof Assertion (i) is just a reformulation of the disintegration theorem, and we confine ourselves to the proof of (ii). The relation $\Psi \circ S_t = \Psi$ implies that $S_t(\Psi^{-1}(B)) = \Psi^{-1}(B)$ for any Borel subset $B \subset Y$. Since $(S_t)_*\mu = \mu$, we have

$$\int_B \lambda_{\boldsymbol{b}}\big(S_t(A)\big)\lambda(d\boldsymbol{b}) = \int_B \lambda_{\boldsymbol{b}}(A)\lambda(d\boldsymbol{b}) \quad \text{for any } A \in \mathcal{B}(X),\ B \in \mathcal{B}(Y).$$

It follows that $(S_t)_*\lambda_{\boldsymbol{b}} = \lambda_{\boldsymbol{b}}$ for any $t \in \mathbb{R}$ and λ-almost every $\boldsymbol{b} \in Y$. Combining this with the continuity of S_t in time, we arrive at the required result. □

5.2.2 Estimates for the densities of the energy and enstrophy

The balance relations established in Theorem 5.1.1 and Proposition 5.1.3 give infinitely many identities for stationary measures of the Navier–Stokes system (5.1). Unfortunately, the convergence established in Theorem 5.2.2 is not strong enough to pass to the limit in those relations and to obtain similar information for the accumulation points of the family of stationary distributions, even though both sides of relation (5.12) are well defined for a limit process u and any bounded continuous function g. Instead, in this section we prove that the concept of local time, combined with relation (5.5), enables one to get some qualitative properties of the energy and enstrophy functionals in the limiting stationary regimes. Namely, we establish the following result.

Theorem 5.2.12 *Let $\mathfrak{B}_1 < \infty$ and let $\{\mu_\nu, \nu > 0\}$ be a family of stationary measures for* (5.1). *Then the following properties hold for any limit point μ_0 of the family $\{\mu_\nu\}$ in the sense of weak convergence of measures on H.*

[7] Note that $X_{\boldsymbol{b}}$ is a closed subset of X, since it is the pre-image of a closed set under a continuous mapping.

(i) *Let* $b_s \neq 0$ *for at least two indices in* $\mathbb{Z}_0^2$. *Then* μ_0 *has no atom at zero. Moreover, there is a constant* $C > 0$ *not depending on* $\{b_s\}$ *such that*

$$\mu_0\big(B_H(\delta)\big) \le C\,\gamma^{-1}\sqrt{\mathfrak{B}_1}\,\delta \quad \text{for all } \delta > 0, \tag{5.45}$$

where we set $\gamma = \inf\{\mathfrak{B}_0 - b_s^2,\, s \in \mathbb{Z}_0^2\}$.

(ii) *Let* $b_s \neq 0$ *for all* $s \in \mathbb{Z}_0^2$. *Then there is an increasing continuous function* $p(r)$, *vanishing at* $r = 0$ *and depending only on the sequence* $\{b_s\}$, *such that*

$$\mu_0\big(\{u \in H : |u|_2 \in \Gamma\}\big) + \mu_0\big(\{u \in H : |\nabla u|_2 \in \Gamma\}\big) \le p\big(\ell(\Gamma)\big) \tag{5.46}$$

for any Borel subset $\Gamma \subset \mathbb{R}$, *where* ℓ *stands for the Lebesgue measure on* $\mathbb{R}$.

Before presenting a proof of this result, we outline its main idea. We claim that it suffices to establish inequalities (5.45) and (5.46) for any measure of the family $\{\mu_\nu\}$ with some universal constant C and function p. Indeed, consider, for instance, the case of the second inequality. It follows from (1.29) that inequality (5.46) remains true for any limiting measure μ_0 and any open subset $\Gamma \subset \mathbb{R}$. Now recall that the Lebesgue measure is regular:

$$\ell(\Gamma) = \inf\{\ell(G) : G \supset \Gamma,\, G \text{ is open}\}.$$

Combining this relation with the continuity of p, we conclude that (5.46) is true for any Borel set $\Gamma \subset \mathbb{R}$.

The proof of inequalities (5.45) and (5.46) for the measures of the family $\{\mu_\nu\}$ is based on an application of some properties of the local times for Itô processes. If we were just interested in absolute continuity of the law for the components $\langle u, e_s\rangle$ under a measure μ_ν with a fixed $\nu > 0$, then it would be sufficient to remark that the corresponding stationary process $\langle u_\nu(t), e_s\rangle$ is an Itô process with a non-zero constant diffusion, and the required assertion would follow immediately from a well-known technique involving the concept of local time (see Step 2 in the proof below) or from Krylov's estimate (see Theorem A.9.1). In our situation, the estimates must be uniform in $\nu > 0$, and diffusion terms in the equations satisfied by $|u_\nu(t)|_2^2$ and $|\nabla u_\nu(t)|_2^2$ are not separated from zero. This difficulty will be overcome with the help of a thorough study of the local times for the processes in question.

Proof of Theorem 5.2.12 As was mentioned above, it suffices to find a constant $C > 0$ and a function p with the properties stated in the theorem such that (5.45) and (5.46) hold for any measure μ' of the family $\{\mu_\nu\}$. This will be done in several steps. In what follows, we denote by $u(t, x)$ a stationary solution for (5.1) whose law coincides with μ' and write $u_s(t) = \langle u(t), e_s\rangle$ for its components in the Hilbert basis $\{e_s,\, s \in \mathbb{Z}_0^2\}$.

Step 1. We first show that, for any real-valued function $g \in C^2(\mathbb{R})$ whose second derivative has at most polynomial growth at infinity and for any Borel subset $\Gamma \subset \mathbb{R}$, we have

$$\mathbb{E}\int_\Gamma \mathbb{I}_{(a,\infty)}\big(g(|u|_2^2)\big)\Big(g'(|u|_2^2)\big(\tfrac{\mathfrak{B}_0}{2} - |\nabla u|_2^2\big) + g''(|u|_2^2)\sum_{s\in\mathbb{Z}_0^2} b_s^2 u_s^2\Big)\,da$$
$$+\sum_{s\in\mathbb{Z}_0^2} b_s^2\,\mathbb{E}\Big(\mathbb{I}_\Gamma\big(g(|u|_2^2)\big)\big(g'(|u|_2^2)\,u_s\big)^2\Big) = 0. \quad (5.47)$$

Indeed, let us fix any function $g \in C^2(\mathbb{R})$ and consider the process $f(t) = g(|u(t)|_2^2)$. By Itô's formula, we have

$$f(t) = f(0) + \nu\int_0^t A(s)\,ds + 2\sqrt{\nu}\sum_{s\in\mathbb{Z}_0^2} b_s \int_0^t g'(|u|_2^2)u_s\,d\beta_s,$$

where we set

$$A(t) = 2\Big(g'(|u|_2^2)\big(\tfrac{\mathfrak{B}_0}{2} - |\nabla u|_2^2\big) + g''(|u|_2^2)\sum_{s\in\mathbb{Z}_0^2} b_s^2 u_s^2\Big).$$

Let $\Lambda_t(a)$ be the local time for f. Then, in view of relation (A.42) with $h = \mathbb{I}_\Gamma$, we have

$$2\int_\Gamma \Lambda_t(a)\,da = 4\nu\sum_{s\in\mathbb{Z}_0^2} b_s^2\int_0^t \mathbb{I}_\Gamma(f(r))\big(g'(|u|_2^2)u_s\big)^2\,dr.$$

Taking the mean value and using the stationarity of u, we derive

$$\int_\Gamma \big(\mathbb{E}\Lambda_t(a)\big)\,da = 2\nu t\sum_{s\in\mathbb{Z}_0^2} b_s^2\,\mathbb{E}\Big(\mathbb{I}_\Gamma(f)\big(g'(|u|_2^2)u_s\big)^2\Big).$$

On the other hand, taking the mean value in (A.45) and using again the stationarity of u, we obtain

$$\mathbb{E}\Lambda_t(a) = -\nu t\,\mathbb{E}\big(\mathbb{I}_{(a,\infty)}(f(0))\,A(0)\big).$$

The last two equalities and the definition of the drift A imply the required relation (5.47).

Step 2. We now prove that (cf. (5.45))

$$\mu'\big(\{u \in H : 0 < |u|_2 \le \delta\}\big) \le C\gamma^{-1}\sqrt{\mathfrak{B}_1}\,\delta \quad \text{for any } \delta > 0. \quad (5.48)$$

Let us apply relation (5.47) in which $\Gamma = [\alpha, \beta]$ with $\alpha > 0$ and $g \in C^2(\mathbb{R})$ is a function that coincides with $\sqrt{x}$ for $x \ge \alpha$ and vanishes for $x \le 0$. This results in

$$\mathbb{E}\int_{\alpha}^{\beta} \mathbb{I}_{(a,\infty)}(|u|_2)\left(\frac{\mathfrak{B}_0 - 2|\nabla u|_2^2}{4|u|_2} - \frac{1}{4|u|_2^3}\sum_{s\in\mathbb{Z}_0^2} b_s^2 u_s^2\right) da$$
$$+\frac{1}{4}\sum_{s\in\mathbb{Z}_0^2} b_s^2\,\mathbb{E}\Big(\mathbb{I}_{[\alpha,\beta]}(|u|_2)|u|_2^{-2}u_s^2\Big) = 0.$$

It follows that

$$\mathbb{E}\int_{\alpha}^{\beta} \frac{\mathbb{I}_{(a,\infty)}(|u|_2)}{|u|_2^3}\left(\mathfrak{B}_0|u|_2^2 - \sum_{s\in\mathbb{Z}_0^2} b_s^2 u_s^2\right) da \le 2(\beta-\alpha)\,\mathbb{E}\left(\frac{|\nabla u|_2^2}{|u|_2}\right). \tag{5.49}$$

Now note that

$$\mathfrak{B}_0|u|_2^2 - \sum_{s\in\mathbb{Z}_0^2} b_s^2 u_s^2 = \sum_{s\in\mathbb{Z}_0^2}(\mathfrak{B}_0 - b_s^2)u_s^2 \ge \gamma|u|_2^2,$$
$$\mathbb{E}\left(\frac{|\nabla u|_2^2}{|u|_2}\right) \le C_1\mathbb{E}\,|Lu|_2 \le C_1\sqrt{\mathfrak{B}_1},$$

where we used interpolation and inequality (5.27). Substituting these estimates into (5.49) and passing to the limit as $\alpha \to 0^+$, we derive

$$\mathbb{E}\int_0^{\beta} \mathbb{I}_{(a,\infty)}(|u|_2)|u|_2^{-1}da \le C_1\gamma^{-1}\sqrt{\mathfrak{B}_1}\,\beta. \tag{5.50}$$

We now fix a constant $\delta > 0$ and note that the left-hand side of (5.50) can be minorised by

$$\mathbb{E}\int_0^{\beta} \mathbb{I}_{(a,\delta]}(|u|_2)|u|_2^{-1}da \ge \delta^{-1}\mathbb{E}\int_0^{\beta} \mathbb{I}_{(a,\delta]}(|u|_2)\,da$$
$$= \delta^{-1}\int_0^{\beta} \mathbb{P}\big(\{a < |u|_2 \le \delta\}\big)\,da.$$

Substituting this inequality into (5.50), we obtain

$$\frac{1}{\beta}\int_0^{\beta} \mathbb{P}\big(\{a < |u|_2 \le \delta\}\big)\,da \le C_1\gamma^{-1}\sqrt{\mathfrak{B}_1}\,\delta.$$

Passing to the limit as $\beta \to 0^+$, we arrive at the required inequality (5.48).

Step 3. In view of (5.48), inequality (5.45) will be established if we prove that μ' has no atom at zero. To this end, we could apply the local time technique used in Step 1 to a one-dimensional projection of u. However, the following proof based on Krylov's estimate for Itô processes is simpler.

Let us fix an index $s \in \mathbb{Z}_0^2$ such that $b_s \ne 0$. The stationary process $u_s(t)$ satisfies the equation

$$u_s(t) = u_s(0) + \int_0^t v(r)\,dr + \sqrt{\nu}\,b_s\beta_s(t),$$

where we set $v(t) = \langle -\nu Lu - B(u), e_s\rangle$. Since $|u|_2 \ge |u_s|$, it suffices to show that $P = \mathbb{P}\{u_s(0) = 0\} = 0$. By Theorem A.9.1 with $d = 1$ and $f = \mathbb{I}_{(-\varepsilon,\varepsilon)}$, we have

$$\sqrt{\nu}\, b_s \mathbb{P}\{|u_s(0)| < \varepsilon\} \le C\varepsilon \quad \text{for any } \varepsilon > 0,$$

whence it follows that $P = 0$.

Step 4. It remains to establish (5.46). We shall confine ourselves to the proof of the fact that the first term on the left-hand side can be bounded by $p(\ell(\Gamma)$. A similar argument shows that the same is true for the second term.

Applying relation (5.47) with $g(x) = x$, we obtain

$$\mathbb{E}\Big(\mathbb{I}_\Gamma\big(|u|_2^2\big)\sum_{s\in\mathbb{Z}_0^2} b_s^2 u_s^2\Big) \le \int_\Gamma \mathbb{E}\Big(\mathbb{I}_{(a,\infty)}\big(|u|_2^2\big)|\nabla u|_2^2\Big)\,da \le \tfrac{1}{2}\mathfrak{B}_0\ell(\Gamma), \tag{5.51}$$

where we used (5.3) to get the second inequality. We wish to estimate the left-hand side of this inequality from below. To this end, note that if $|u|_2 \ge \delta$ and $|Lu|_2 \le \delta^{-1/2}$, then for any integer $N \ge 1$ we have

$$\begin{aligned}\sum_{s\in\mathbb{Z}_0^2} b_s^2 u_s^2 &\ge \underline{b}_N^2 \sum_{0<|s|\le N} u_s^2 = \underline{b}_N^2\Big(|u|_2^2 - \sum_{|s|>N} u_s^2\Big)\\ &\ge \underline{b}_N^2\Big(|u|_2^2 - N^{-4}|Lu|_2^2\Big) \ge \underline{b}_N^2(\delta^2 - N^{-4}\delta^{-1}),\end{aligned}$$

where $\underline{b}_N = \min\{b_s, |s| \le N\}$. Choosing $N = N(\delta)$ sufficiently large, we find an increasing function $\varepsilon(\delta) > 0$ going to zero with δ such that

$$\sum_{s\in\mathbb{Z}_0^2} b_s^2 u_s^2 \ge \varepsilon(\delta) \quad \text{for } |u|_2 \ge \delta,\ \|u\|_2 \le \delta^{-1/2}. \tag{5.52}$$

Define now the event $G_\delta = \{|u(0)|_2 \le \delta \text{ or } |Lu(0)|_2 \ge \sqrt{\delta}\}$ and note that, in view of (5.5), (5.45), and Chebyshev's inequality, we have

$$\mathbb{P}(G_\delta) \le \mathbb{P}\{|u|_2 \le \delta\} + \mathbb{P}\{|Lu|_2 \ge \sqrt{\delta}\} \le C_5\delta.$$

Combining this with (5.51) and (5.52), we write

$$\begin{aligned}\mathbb{P}\{|u|_2^2 \in \Gamma\} &= \mathbb{P}\big(\{|u|_2^2 \in \Gamma\} \cap G_\delta\big) + \mathbb{P}\big(\{|u|_2^2 \in \Gamma\} \cap G_\delta^c\big)\\ &\le C_5\delta + \varepsilon(\delta)^{-1}\mathbb{E}\Big(\mathbb{I}_\Gamma\big(|u|_2^2\big)\sum_{s\in\mathbb{Z}_0^2} b_s^2 u_s^2\Big)\\ &\le C_5\delta + C_6\varepsilon(\delta)^{-1}\ell(\Gamma).\end{aligned}$$

This inequality immediately implies the required result. □

Exercise 5.2.13 Prove that the second term on the left-hand side of (5.46) can be bounded by $p(\ell(\Gamma))$. *Hint:* A relation similar to (5.47) holds for the process $g(|\nabla u|_2^2)$. To see this, it suffices to repeat the arguments used in Step 1 of the above proof.

5.2.3 Further properties of the limiting measures

In the previous subsection, we have shown that the limiting measures of the family of stationary distributions for (5.1) possess some non-degeneracy properties. Indeed, Theorem 5.2.12 implies, in particular, that the image of such a measure under the energy and enstrophy functionals has no atoms. In this section, we combine balance relations with Krylov's estimate for Itô diffusions to show that a similar assertion is true for a large class of multidimensional functionals.

Let $f_1, \dots, f_d : \mathbb{R} \to \mathbb{R}$ be real-analytic functions whose second derivatives have at most polynomial growth at infinity and are bounded from below:

$$f_k''(z) \geq -C \quad \text{for } z \in \mathbb{R}, k = 1, \dots, d. \tag{5.53}$$

These hypotheses are satisfied, for instance, if the functions f_k are trigonometric polynomials or polynomials of even degree with positive leading coefficients. Since $H^1(\mathbb{T}^2)$ is continuous embedded in $L^p(\mathbb{T}^2)$ for any $p < \infty$, we can consider the map $F : V^2 \to \mathbb{R}^d$ defined by

$$F(u) = \big(F_1(u), \dots, F_d(u)\big), \quad F_k(u) = \int_{\mathbb{T}^2} f_k(\operatorname{curl} u(x))\, dx.$$

Assume that the functions f_k' are independent modulo constants, that is, if $c_1 f_1' + \cdots + c_d f_d' \equiv \text{const}$, then all c_k vanish. The following theorem gives some information on the distribution of F under any limiting point of the family of stationary measures for (5.1).

Theorem 5.2.14 *Let $b_s = b_{-s} \neq 0$ for all $s \in \mathbb{Z}_0^2$, let $\mathfrak{B}_2 < \infty$, and let the functions $f_1, \dots, f_d$ be as above. Then there is an increasing continuous function $p(r)$ vanishing at $r = 0$ such that for any limiting point μ of the family $\{\mu_\nu\}$, as $\nu \to 0$, in the sense of weak convergence on H and any Borel subset $\Gamma \in \mathbb{R}^d$, we have*[8]

$$F_*(\mu)(\Gamma) \leq p\big(\ell(\Gamma)\big). \tag{5.54}$$

Before proving this result, we establish a corollary showing that the limiting measures are not concentrated on subsets of low dimension. Its formulation uses the concept of Hausdorff dimension (e. g., see section X.1 of the book [BV92]).

Corollary 5.2.15 *Under the hypotheses of Proposition 5.1.3, let $X \subset H$ be a closed subset whose intersection with any compact set $\mathcal{K} \subset V^2$ has finite Hausdorff dimension with respect to the metric of V^2 and let μ be a limiting point of $\{\mu_\nu\}$, as $\nu \to 0$, in the sense of weak convergence on H. Then $\mu(X) = 0$.*

[8] Recall that, by (5.27), any limiting point μ is concentrated on the space V^2.

Proof We shall only outline the proof, leaving it to the reader to fill in the details. We know that μ is concentrated on V^2. By Ulam's theorem (see Section 1.2.1), any Borel measure on a Polish space is regular, and therefore we can find an increasing sequence of compact subsets $\mathcal{K}_n \subset V^2$ such that $\mu(\mathcal{K}_n) \to 1$ as $n \to \infty$. Hence, it suffices to show that $\mu(\mathcal{K}_n \cap X) = 0$ for any $n \geq 1$. Thus, we can assume from the very beginning that X is a compact subset in V^2 of finite Hausdorff dimension.

Let us take an integer $d \geq 1$ larger than Hausdorff's dimension of X and consider any map $F : V^2 \to \mathbb{R}^d$ satisfying the hypotheses of Theorem 5.2.14. It is easy to see that F is uniformly Lipschitz continuous on any compact subset of V^2. Combining this with the definition of the Hausdorff dimension, we see that the compact set $F(X) \subset \mathbb{R}^d$ has zero Lebesgue measure. Inequality (5.54) with $\Gamma = F(X)$ now implies that $\mu(X) = 0$. □

Proof of Theorem 5.2.14 It suffices to prove that inequality (5.54) holds for any measure μ_ν with a function p not depending on ν. Let us fix a stationary solution $u(t)$ distributed as μ_ν and define the stochastic process $y(t) = F(u(t))$. The first and second derivatives of the mapping $F : V^2 \to \mathbb{R}^d$ have the form

$$F'(u;h) = \left(\int_{\mathbb{T}^2} f_k'\big(\operatorname{curl} u(x)\big) \operatorname{curl} h(x)\, dx,\ k = 1, \dots, d\right),$$

$$F''(u;h,h) = \left(\int_{\mathbb{T}^2} f_k''\big(\operatorname{curl} u(x)\big)\big(\operatorname{curl} h(x)\big)^2 dx,\ k = 1, \dots, d\right).$$

Combining these relations with estimates (2.162), Corollary A.7.6, and Eq. (5.8) for the vorticity $v = \operatorname{curl} u$, we prove that $y(t)$ is an Itô process whose components are representable in the form (cf. the proof of Proposition 5.1.3)

$$y_k(t) = y_k(0) + \int_0^t A_k(r)\, dr + \sum_{s \in \mathbb{Z}_0^2} \int_0^t B_{ks}(r)\, d\beta_s, \quad k = 1, \dots, d, \tag{5.55}$$

where we set

$$\begin{aligned} A_k(t) &= \int_{\mathbb{T}^2} f_k'(v(t,x))\Delta v(t,x)\, dx + \frac{1}{2}\sum_{s \in \mathbb{Z}_0^2} b_s^2 \int_{\mathbb{T}^2} f_k''(v(t,x))\varphi_s^2(x)\, dx \\ &= \int_{\mathbb{T}^2} f_k''(v(t,x))\big(-|\nabla v|^2 + \tfrac{\mathfrak{B}_1}{8\pi^2}\big)\, dx, \\ B_{ks}(t) &= b_s \int_{\mathbb{T}^2} f_k'(v(t,x))\varphi_s(x)\, dx. \end{aligned}$$

We now apply Krylov's estimate (A.46) with $f = \mathbb{I}_\Gamma$ to the process y. In view of the stationarity of y, A_k, and B_{ks}, this results in

$$\mathbb{E}\left((\det \sigma_0)^{1/d}\, \mathbb{I}_\Gamma(y(0))\right) \leq C_d \ell(\Gamma)^{1/d}\, \mathbb{E}\,|A(0)|, \tag{5.56}$$

where $A = (A_1, \dots, A_d)$, and σ_t stands for the $d \times d$ diffusion matrix BB^* (that is, $\sigma_t^{kl} = \sum_s B_{ks}(t) B_{ls}(t)$). Next, we estimate $\mathbb{E}\,|A(0)|$ and $\det \sigma_0$ from above and below, respectively.

It follows from (5.53) that $|f_k''(z)| \le f_k''(z) + 2C$ for all $z \in \mathbb{R}$. Therefore,

$$\mathbb{E}\,|A_k(0)| \le \mathbb{E}\int_{\mathbb{T}^2} |f_k''(v(t,x))| \cdot \big|-|\nabla v|^2 + \tfrac{\mathfrak{B}_1}{8\pi^2}\big|\,dx$$
$$\le \mathbb{E}\int_{\mathbb{T}^2} \big(f_k''(v(t,x)) + 2C\big)\big(|\nabla v|^2 + \tfrac{\mathfrak{B}_1}{8\pi^2}\big)\,dx.$$

The balance relation (5.17) with $g = f_k''$ implies that

$$\mathbb{E}\int_{\mathbb{T}^2} f_k''\big(v(t,x)\big)|\nabla v(t,x)|^2 dx = C\,\mathbb{E}\int_{\mathbb{T}^2} f_k''\big(v(t,x)\big)\,dx.$$

Combining this with the above inequality, we derive

$$\mathbb{E}\,|A_k(0)| \le C_1\mathbb{E}\int_{\mathbb{T}^2} \big(|f_k''(v(t,x))| + |\nabla v(t,x)|^2 + 1\big)\,dx.$$

Recalling that f_k'' has at most polynomial growth at infinity and invoking estimates (5.5) and (5.6), we obtain

$$\mathbb{E}\,|A_k(0)| \le C_2 \quad \text{for } k = 1, \dots, d, \tag{5.57}$$

where $C_2 > 0$ is a constant depending only on the sequence $\{b_s\}$.

To estimate from below the quantity $\det\sigma_0$, we need an auxiliary lemma established at the end of this section. Recall that $\dot H^m(\mathbb{T}^2)$ stands for the space of functions in $H^m(\mathbb{T}^2)$ with zero mean value. We abbreviate $f'(w) = (f_1'(w), \dots, f_d'(w))$ and define a function $D : \dot H^1(\mathbb{T}^2) \to \mathbb{R}$ by the relation $D\big(w(\cdot)\big) = \det\sigma(f' \circ w)$. Here for a vector function $g : \mathbb{T}^2 \to \mathbb{R}^d$ we denote by $\sigma(g)$ a $d \times d$ matrix with the entries

$$\sigma^{kl}(g) = \sum_{s\in\mathbb{Z}_0^2} b_s^2 \int_{\mathbb{T}^2} g_k(x)\varphi_s(x)\,dx \int_{\mathbb{T}^2} g_l(x)\varphi_s(x)\,dx;$$

cf. the definition of the diffusion matrix σ_t.

Lemma 5.2.16 *For any functions $f_1, \dots, f_d$ satisfying the hypotheses of Theorem 5.2.14, there is an $r < 1$ such that $D(w)$ admits a continuous extension to the space $\dot H^r = \dot H^r(\mathbb{T}^2)$ and does not vanish there outside the origin.*

Let us note that $\det\sigma_0 = D(v(0))$. For any $\delta > 0$, we introduce the event $Q_\delta = \{v(0) \in \mathcal{O}_\delta\}$, where $\mathcal{O}_\delta = \{w \in \dot H^1(\mathbb{T}^2) : \|w\|_1 \le \delta^{-1}, \|w\|_r \ge \delta\}$, and r denotes the constant constructed in the above lemma. Since the embedding $\dot H^1(\mathbb{T}^2) \subset \dot H^r(\mathbb{T}^2)$ is compact, $\mathcal{O}_\delta$ is a compact subset of $\dot H^r(\mathbb{T}^2)$ not containing zero. Lemma 5.2.16 implies that there $D(w) \ge c_3(\delta) > 0$. It follows that

$$\det\sigma_0 \ge c_3(\delta) \quad \text{for } \omega \in Q_\delta^c.$$

Combining this with (5.56) and (5.57), we obtain

$$\begin{aligned}\mathbb{P}\{y(0)\in\Gamma\} &\le \mathbb{E}\big(\mathbb{I}_{Q_\delta}\mathbb{I}_\Gamma(y(0))\big)+\mathbb{P}(Q_\delta^c)\\ &\le \big(c_3(\delta)\big)^{-1/d}\,\mathbb{E}\left((\det\sigma_0)^{1/d}\,\mathbb{I}_\Gamma(y(0))\right)+\mathbb{P}(Q_\delta^c)\\ &\le C_4(\delta)\,\ell(\Gamma)^{1/d}+\mathbb{P}(Q_\delta^c). \end{aligned} \tag{5.58}$$

Let us show that the second term in the right-hand side goes to zero with δ. Indeed, we have

$$\begin{aligned}\mathbb{P}(Q_\delta^c) &\le \mathbb{P}\{\|v(0)\|_1\ge\delta^{-1}\}+\mathbb{P}\{\|v(0)\|_r\ge\delta\})\\ &\le \mathbb{P}\{\|u(0)\|_2\ge\delta^{-1}\}+\mathbb{P}\{|u(0)|_2\ge\delta\},\end{aligned}$$

where we used the inequalities

$$\|\operatorname{curl}u\|_1\le\|u\|_2,\qquad \|\operatorname{curl}u\|_r\ge|\operatorname{curl}u|_2\ge|u|_2.$$

It follows from (5.5) and (5.45) that $\mathbb{P}(Q_\delta^c)\to 0$ as $\delta\to 0$. Combining this with (5.58), we arrive at the required inequality (5.54). □

Proof of Lemma 5.2.16 The conditions imposed on f_k imply that

$$|f_k'(z)|+|f_k''(z)|\le C_1(1+|z|)^q\quad\text{for all } z\in\mathbb{R},$$

where $q\ge 1$ is a constant. Let $r<1$ be such that $\dot H^r$ is continuously embedded in $L^{2q}(\mathbb{T}^2)$. By the Cauchy–Schwarz inequality,

$$|\sigma^{kl}(g)|\le\Big(\sum_{s\in\mathbb{Z}_0^2}b_s^2|\varphi_s|_2^2\Big)\,|g_k|_2|g_l|_2\le C_5\mathfrak{B}_1|g|_2^2.$$

Therefore D is continuous as the composition of the continuous maps

$$\begin{aligned} &w\mapsto g=f'(w(x)), &&\dot H^r(\mathbb{T}^2)\to L^2(\mathbb{T}^2;\mathbb{R}^d),\\ &g\mapsto\det\sigma(g), &&L^2(\mathbb{T}^2;\mathbb{R}^d)\to\mathbb{R}.\end{aligned}$$

We now prove that $D(w)\ne 0$ for $w\in\dot H^r\setminus\{0\}$. It is an elementary exercise in linear algebra to show that if $\mathcal{H}$ is a Hilbert space and $B:\mathcal{H}\to\mathbb{R}^d$ is a surjective continuous operator, then the application $BB^*:\mathbb{R}^d\to\mathbb{R}^d$ is an isomorphism. Let us denote by $\dot L^2(\mathbb{T}^2)$ the space of L^2 functions with zero mean value and define a family of continuous operators $B(w):\dot L^2(\mathbb{T}^2)\to\mathbb{R}^d$ by the relation (cf. the relation for the first derivative of F)

$$B(w)\xi=\Big(\sum_{s\in\mathbb{Z}_0^2}|s|^{-1}b_s\langle\xi,\varphi_s\rangle\int_{\mathbb{T}^2}f_k'\big(w(x)\big)\varphi_s(x)\,dx,\ k=1,\dots,d\Big).$$

It is straightforward to check that

$$B_{ks}(w)=b_s\int_{\mathbb{T}^2}f_k'(w(x))\varphi_s(x)\,dx,\quad 1\le k\le d,\quad s\in\mathbb{Z}_0^2,$$

are the matrix entries of the operator $B(w)$ with respect to the orthonormal basis $\{|s|^{-1}\varphi_s, s \in \mathbb{Z}_0^2\}$. Therefore $\sigma(f'(w)) = B(w)B^*(w)$. It follows that the application $\sigma(w) : \mathbb{R}^d \to \mathbb{R}^d$ is an isomorphism and its matrix is nonsingular for those $w \in \dot{H}^r$ for which $B(w)$ is surjective. Hence, the required assertion will be established if we show that $B(w)$ is surjective for any non-zero function $w \in \dot{H}^r$.

Suppose that $B(w)$ is not surjective. Then there exists a non-zero vector $c = (c_1, \dots, c_d) \in \mathbb{R}^d$ such that

$$\langle c, B(w)\xi\rangle = \sum_{k=1}^{d} c_k \sum_{s\in\mathbb{Z}_0^2} |s|^{-1} b_s \langle \xi, \varphi_s\rangle \int_{\mathbb{T}^2} f_k'\big(w(x)\big)\varphi_s(x)\,dx = 0$$

for any $\xi \in \dot{L}^2(\mathbb{T}^2)$. That is,

$$\sum_{s\in\mathbb{Z}_0^2} |s|^{-1} b_s \langle \xi, \varphi_s\rangle \int_{\mathbb{T}^2} h\big(w(x)\big)\varphi_s(x)\,dx = 0 \quad \text{for any } \xi \in \dot{L}^2(\mathbb{T}^2),$$

where $h(w) = c_1 f_1'(w) + \cdots + c_d f_d'(w)$. Choosing $\xi = \varphi_s$ with s varying in $\mathbb{Z}_0^2$ and recalling that the numbers b_s are non-zero, we get

$$\int_{\mathbb{T}^2} h\big(w(x)\big)\xi(x)\,dx = 0 \quad \text{for any } \xi \in C^\infty(\mathbb{T}^2) \text{ with zero mean value.}$$

Therefore there exists a constant $C = C_w$ such that $h(w(x)) = C$ almost everywhere, whence we conclude that

$$w(x) \in N_C \quad \text{for almost every } x \in \mathbb{T}^2, \tag{5.59}$$

where $N_C = \{z \in \mathbb{R} : h(z) = C\}$. Recalling that $f_1', \dots, f_d'$ are analytic functions linearly independent modulo constants, we see that N_C is a discrete[9] subset of $\mathbb{R}$. Combining (5.59) with Lemma A.15.1, we see that the function w must be a constant almost everywhere. Since the mean value of w is zero, we conclude that $w \equiv 0$. □

5.2.4 Other scalings

We have studied so far the inviscid limit of stationary measures for the Navier–Stokes equations in the case when the random perturbation is white in time and has an amplitude proportional to $\sqrt{\nu}$. A natural question is to find out what happens for white-noise perturbations of different size. In this section, we prove that the scaling we studied is the only one for which a non-trivial limit exists.

Let us consider the equation (cf. (5.1))

$$\partial_t u + \nu L u + B(u) = \nu^a \eta(t, x), \tag{5.60}$$

where $a \in \mathbb{R}$ is a constant, and η is a regular white noise.

[9] Recall that a subset of $\mathbb{R}$ is said to be *discrete* if it has no finite accumulation points.

Theorem 5.2.17 *Let $\{\mu_\nu, \nu > 0\}$ be a family of stationary measures for* (5.60). *Then the following assertions hold.*

(i) *Let $a > 1/2$. Then $\{\mu_\nu\}$ converges weakly on V^2 to the Dirac measure δ_0 as $\nu \to 0^+$.*

(ii) *Let $a < 1/2$. Then $\{\mu_\nu\}$ has no accumulation points in the sense of weak convergence on H as $\nu \to 0^+$.*

Proof We first assume that $a > 1/2$. By (2.161), we have

$$\int_H |Lu|_2^2 \mu_\nu(du) = \frac{\nu^{2a-1}\mathfrak{B}_1}{2}. \tag{5.61}$$

Let $f \in L_b(V^2)$ be an arbitrary function such that $\|f\|_L \le 1$. Then, applying the Cauchy–Schwarz inequality and using (5.61), we derive

$$\left|\int_H f(u)\mu_\nu(du) - f(0)\right| \le \int_H \|u\|_2 \mu_\nu(du) \le C_1 \nu^{a-1/2}.$$

Recalling assertion (ii) of Theorem 1.2.15, we conclude that $\mu_\nu \to \delta_0$ weakly on V.

We now assume that $a < 1/2$. Let $u_\nu(t)$ be a stationary solution for (5.60) whose law coincides with μ_ν. We fix a constant $b > 0$ and consider the random process $v_\nu(t) = \nu^b u_\nu(\nu^b t)$. It is a matter of direct verification to show that v_ν is a stationary solution of the equation

$$\partial_t v + \nu^{1+b} L v + B(v) = \nu^{a+\frac{3b}{2}}\, \partial_t \tilde{\zeta}, \tag{5.62}$$

where $\tilde{\zeta}(t) = \nu^{-\frac{b}{2}} \zeta(\nu^b t)$. Using the scaling properties of the Brownian motion, it is easy to see that $\tilde{\zeta}$ is a process with the same distribution as ζ. Choosing $b = \frac{1}{2} - a$, we conclude from (5.62) that v_ν satisfies Eq. (5.1), in which ζ and ν are replaced by $\tilde{\zeta}$ and $\tilde{\nu} = \nu^{1+b}$, respectively.

To prove that $\{\mu_\nu\}$ has no accumulation points as $\nu \to 0^+$, we argue by contradiction: suppose that $\{\mu_{\nu_j}\}$ converges weakly on H for some sequence $\nu_j \to 0^+$. Then, by Prokhorov's theorem, for any $\varepsilon > 0$ there is an $R > 0$ such that

$$\mu_{\nu_j}\big(B_H(R)^c\big) = \mathbb{P}\{|u_{\nu_j}(0)|_2 > R\} \le \varepsilon \quad \text{for all } j \ge 1.$$

Recalling the definition of v_ν, we see that

$$\mathbb{P}\{|v_{\nu_j}(0)|_2 > \nu_j^b R\} = \mathbb{P}\{|u_{\nu_j}(0)|_2 > R\} \le \varepsilon \quad \text{for all } j \ge 1,$$

whence it follows that the law of $v_{\nu_j}(0)$ converges weakly to the Dirac measure δ_0 as $j \to \infty$. On the other hand, by Theorem 5.2.2 and Corollary 5.2.3, the accumulation points of the family of stationary measures for (5.1) are non-degenerate in the sense that they are not concentrated at zero (see (5.25)). Since the law of $v_{\nu_j}(0)$ is a stationary measure for (5.1) with $\nu = \nu_j^{1+b}$, we arrive at a contradiction. The proof of the theorem is complete. □

Exercise 5.2.18 Prove that the process v_ν defined in the above proof satisfies (5.62). *Hint:* Rewrite Eq. (5.1) in the form

$$u(t) = u(0) - \int_0^t \big(\nu L u + B(u)\big)\, ds + \sqrt{\nu}\,\zeta(t), \quad t \geq 0,$$

and make a suitable change of the time variable.

Remark 5.2.19 If in Eq. (5.1) we replace the viscous term $\nu L u$ with a hyperviscous one of the form $\nu L^m u$, $m > 1$, then the equation still has a unique stationary measure μ_ν^m, and there exist limits along sequences $\mu_0^m = \lim_{\nu_j \to 0} \mu_{\nu_j}^m$. For the same reason as in Section 5.2.1, we have $\mathbb{E}_{\mu_0^m} \|u\|_m^2 = \frac{1}{2}\mathfrak{B}_0$ and $\mathbb{E}_{\mu_0^m} \|u\|_{m+1}^2 \leq \frac{1}{2}\mathfrak{B}_1$. As $\mathbb{E}_{\mu_0} \|u\|_1^2 = \frac{1}{2}\mathfrak{B}_0$, then the non-degeneracy property of Corollary 5.2.15 implies that $\mathbb{E}_{\mu_0} \|u\|_m^2 > \frac{1}{2} B_0$. So the measure μ_0^m differs from μ_0 (it is smoother). That is, introducing in Eq. (5.1) a hyperviscosity, we change the limiting, as $\nu \to 0$, distributions of velocity and vorticity.

5.2.5 Kicked Navier–Stokes system

We now discuss the existence of an inviscid limit for the kicked Navier–Stokes system on $\mathbb{T}^2$ with a random force proportional to $\sqrt{\nu}$ (cf. (3.74)):

$$\dot{u} + \nu L u + B(u) = \sqrt{\nu} \sum_{k=1}^{\infty} \eta_k \delta(t-k). \tag{5.63}$$

Here $\{\eta_k\}$ is a sequence of i.i.d. random variables satisfying the hypotheses imposed in Section 3.2.4. We assume, in addition, that

$$\mathbb{E}\xi_{sk} = 0 \text{ for all } s \in \mathbb{Z}_0^2, \qquad \mathfrak{B}_1 = \sum_{s \in \mathbb{Z}_0^2} |s|^2 b_s^2 < \infty,$$

and introduce the quantities

$$D_0 = \sum_{s \in \mathbb{Z}_0^2} b_s^2\, \mathbb{E}\, \xi_{sk}^2, \quad D_1 = \sum_{s \in \mathbb{Z}_0^2} b_s^2 |s|^2\, \mathbb{E}\, \xi_{sk}^2.$$

Exercise 5.2.20 Let $u(t)$, $t \geq 0$, be a stationary solution for the Markov semigroup associated with (5.63). Prove that

$$\int_{k-1}^{k} |\nabla u(t)|_2^2 dt = \frac{D_0}{2}, \quad \int_{k-1}^{k} |Lu(t)|_2^2 dt = \frac{D_1}{2} \quad \text{for any } k \geq 1.$$

Let $C(\mathbb{R}_+ \setminus \mathbb{Z}_+; V)$ be the space of V-valued functions on $\mathbb{R}_+$ that are continuous on each interval $I_k = [k-1, k)$ with $k \geq 1$ and have a limit as $t \to k^-$. This space is endowed with the family of seminorms

$$\|u\|_{C(I_k;V)} = \sup_{t \in I_k} \|u(t)\|_1, \quad k \geq 1.$$

It is easily seen that $C(\mathbb{R}_+ \setminus \mathbb{Z}_+; V)$ is a Fréchet space. The following result can be proved with the help of the argument used to establish Theorem 5.2.2; we leave its proof to the reader as an exercise.

Theorem 5.2.21 *Under the above hypotheses, let* $\{\mu_\nu, \nu > 0\}$ *be any family of stationary measures for the discrete-time Markov semigroup associated with* (5.63) *and let* $\{\boldsymbol{\mu}_\nu\}$ *be the corresponding family of measures on the space of trajectories. Then for any* $\varepsilon > 0$ *the family* $\{\boldsymbol{\mu}_\nu\}$ *is tight in* $C(\mathbb{R}_+ \setminus \mathbb{Z}_+; V) \cap L^2_{\mathrm{loc}}(\mathbb{R}_+; H^{2-\varepsilon})$*. Moreover, if* $\boldsymbol{\mu}$ *is a limiting point for* $\{\boldsymbol{\mu}_\nu\}$ *and* $u(t,x)$ *is a random process whose law coincides with* $\boldsymbol{\mu}$*, then the following properties hold.*

(i) *Almost every realisation of* $u(t,x)$ *belongs to space*

$$L^2_{\mathrm{loc}}(\mathbb{R}_+; V^2) \cap W^{1,1}_{\mathrm{loc}}(\mathbb{R}_+; V) \cap W^{1,\infty}_{\mathrm{loc}}(\mathbb{R}_+; L^p),$$

for any $p \in [1,2)$*, and satisfies the Euler equation* (5.24)*.*

(ii) *The process* $u(t,x)$ *is* 1*-periodic in time. Moreover, the functions* $|u(t)|_2$ *and* $|\nabla u(t)|_2$*, as well as quantities* (5.42)*, are time-independent random constants.*

(iii) *We have the relations*

$$\mathbb{E}\,|\nabla u(t)|_2^2 = \frac{D_0}{2} \quad \text{for all } t \geq 0,$$

$$\mathbb{E}\,|\nabla u(t)|_2^m \leq C_m \quad \text{for all } t \geq 0,\ m \geq 1,$$

$$\int_{k-1}^{k} \|u(t)\|_2^2 \leq \frac{D_1}{2} \quad \text{for all } k \geq 1,$$

where $C_m > 0$ *are some constants not depending on the limiting measure.*

The investigation of the inviscid limit was based essentially on the existence of two "good" integrals of motion for the limiting equation. In the case of Navier–Stokes equations, they are given by the energy and enstrophy. The additional set of infinitely many integrals given by the functions of vorticity enabled us to study further properties of the limiting measures, such as the existence of density for finite-dimensional functionals. A similar analysis can be carried out for other problems possessing those properties, e.g., the damped-driven Korteweg–de Vries equation. Moreover, due to complete integrability of the latter, it is possible to prove uniqueness of the limiting measure and to describe it as a stationary measure for a nonlocal stochastic PDE; see [KP08; Kuk10a; Kuk12].

5.2.6 Inviscid limit for the complex Ginzburg–Landau equation

The complex Ginzburg–Landau equation is an example for which we know the existence of only two integrals of motion, and this turns out to be sufficient to draw some non-trivial conclusions. They are, however, less complete than those obtained in the case of the Navier–Stokes system. We now briefly discuss the corresponding results.

Let us consider the following problem in a bounded domain $Q \subset \mathbb{R}^d$, $d \leq 4$, with a smooth boundary ∂Q:

$$\dot{u} - (\nu + i)\Delta u + i\lambda|u|^2 u = \sqrt{\nu}\,\eta(t, x), \quad u\big|_{\partial Q} = 0. \tag{5.64}$$

Here ν and λ are positive parameters and η is a white-noise force of the form (2.66), where $\{e_j\}$ stands for a complete set of normalised real eigenfunctions for the Dirichlet Laplacian with eigenvalues α_j, $\{\beta_j = \beta_j^+ + i\beta_j^-\}$ is a sequence of independent standard complex-valued Brownian motions, and $\{b_j\} \subset \mathbb{R}$ is a sequence. In this case, as was mentioned in Section 3.5.5, the Cauchy problem for (5.64) is well posed in the space $H_0^1(Q; \mathbb{C})$, and almost every trajectory of solutions belongs to the space

$$L^2_{\text{loc}}(\mathbb{R}_+; H^2) \cap \big(W^{1,\frac{4}{3}}_{\text{loc}}(\mathbb{R}_+; L^{\frac{4}{3}}) + W^{\alpha,4}_{\text{loc}}(\mathbb{R}_+; H^1)\big),$$

where $\alpha \in (\frac{1}{4}, \frac{1}{2})$ is an arbitrary constant. Moreover, using the Bogolyubov–Krylov argument, one can prove that Eq. (5.64) has a stationary measure μ_ν for any $\nu > 0$; cf. Sections 2.5. Let $\{\boldsymbol{\mu}_\nu\}$ be the corresponding family of measures on the space of trajectories.

Using essentially the same argument as in Sections 5.2.1 and 5.2.2, one can prove the following assertions.

Tightness: The family $\{\mu_\nu\}$ is tight in $H^{2-\varepsilon}$ and $\{\boldsymbol{\mu}_\nu\}$ is tight in the space

$$L^2_{\text{loc}}(\mathbb{R}_+; H^{2-\varepsilon}) \cap C(\mathbb{R}_+; H^{-\frac{d}{4}-\varepsilon}).$$

Limiting process: Any limiting measure $\boldsymbol{\mu}$ for the family $\{\boldsymbol{\mu}_\nu\}$ is concentrated on the space $L^2_{\text{loc}}(\mathbb{R}_+; H^2) \cap W^{\frac{4}{3}}_{\text{loc}}(\mathbb{R}_+; L^{\frac{4}{3}})$, and $\boldsymbol{\mu}$-almost every curve $v(t)$ of that space satisfies the nonlinear Schrödinger equation

$$\dot{v} - i\Delta v + i\lambda|v|^2 v = 0.$$

Moreover, the quantities

$$\mathcal{H}_0(u) = \frac{1}{2}\int_D |u(x)|^2 dx, \quad \mathcal{H}_1(u) = \int_D \Big(\frac{1}{2}\,|\nabla u(x)|^2 + \frac{\lambda}{4}\,|u(x)|^4\Big)\,dx$$

are preserved along any such trajectory, and their mean values admit some explicit bounds in terms of the coefficients b_j.

Non-degeneracy: Suppose that all numbers b_j are non-zero. Then the laws of $\mathcal{H}_0(u)$ and $\mathcal{H}_1(u)$ with respect to any limit point of the family $\{\mu_\nu\}$ possess densities against the Lebesgue measure on the real line.

We refer the reader to the papers [KS04b; Shi11b] for an exact statement of the above results and their proofs and to [Kuk12] for some related results.

5.3 Relevance of the results for physics

The results of this chapter form the beginning of a rigorous mathematical theory of space-periodic 2D turbulence in a fluid stirred by a random force. Namely, here we study periodic flows of a 2D fluid with high Reynolds numbers and finite energy. We are concerned with their stationary measures (i.e., the statistical equilibria). Denoting by μ_ν the unique stationary measure for a fluid with viscosity ν, we show that if μ_ν has a non-trivial limit as $\nu \to 0$, then the force must be proportional to $\sqrt{\nu}$. Accordingly, we take the force of the form[10] "*$\sqrt{\nu}$ times a random force $\eta(t,x)$ independent of ν*":

$$\partial_t u + \nu L u + B(u) = \sqrt{\nu}\,\eta(t,x), \qquad \eta = \sum_{s\in\mathbb{Z}^2\setminus 0} b_s \dot\beta_s(t) e_s(x). \tag{5.65}$$

Here $\{e_s(x)\}$ is the usual trigonometric basis of the L_2-space of periodic solenoidal vector fields with zero mean value (see (2.29)), $\dot\beta_s(t)$ are standard independent white noises, and b_s are real numbers such that

$$\mathcal{B}_0 = \sum b_s^2 < \infty, \quad \mathcal{B}_1 = \sum |s|^2 b_s^2 < \infty.$$

In this case, when $\nu \to 0$, the energy of the fluid remains of order one, while its Reynolds number grows like ν^{-1}.

Equation (5.65) has a unique stationary measure μ_ν. The latter has limits as ν goes to zero along sequences, $\mu_0 = \lim_{\nu_j \to 0} \mu_{\nu_j}$. We do not know if the limiting measure μ_0 is unique, but each μ_0 has a number of universal properties.

(i) The measure μ_0 is supported by the space H^2, $\mu_0(H^2) = 1$, and

$$\mathbb{E}_{\mu_0} \|u\|_2^2 \le \tfrac{1}{2}\mathcal{B}_1 < \infty. \tag{5.66}$$

It is space homogeneous if so is the force.

(ii) The μ_0-averaged enstrophy $\frac{1}{2}\mathbb{E}_{\mu_0}\int |\operatorname{curl} u|^2\,dx$ equals $\frac{1}{4}\mathcal{B}_0$, while the μ_0-averaged energy $\mathbb{E}_{\mu_0}\int |u|^2 dx$ has explicit lower and upper bounds. The random variables $\exp(\kappa|u(x)|)$ and $\exp(\kappa|\operatorname{curl} u(x)|^{1/2})$ are μ_0-integrable for any x, where the constant $\kappa > 0$ does not depend on μ_0.

(iii) μ_0 is an invariant measure for the free 2D Euler equation.

(iv) It is genuinely infinite-dimensional. Namely, $\mu_0(K) = 0$ if K is a finite-dimensional set.

(v) It is μ_0-unlikely that the energy $E = \frac{1}{2}\int |u|^2 dx$ of the fluid is very small or very large. That is,

$$\mu_0\{E(u) \le \delta\} \le C\sqrt{\delta}, \quad \mu_0\{E(u) \ge \delta^{-1}\} \le Ce^{-c\delta^{-1}} \quad \text{for all } \delta > 0,$$

where c and C are independent of μ_0.

[10] Note that the Navier–Stokes equations with viscosity ν and a random force $\nu^a \eta(t,x)$ with any $a \in \mathbb{R}$ reduce to (5.65) by suitable scaling of ν, u, and t; see Section 5.2.4.

(vi) The stationary measures μ_ν, $\nu > 0$, satisfy infinitely many explicit algebraical relations. These relations are independent of ν and depend only on two scalar characteristics $\mathcal{B}_0$ and $\mathcal{B}_1$ of the random force. We cannot pass to the limit in these relations to show that they remain true for μ_0, but we use them to study μ_0; e.g., to derive the second assertion in (ii).

In Remark 5.2.19, we explained that the limiting measure(s) μ_0 will change if in Eq. (5.65) we replace the viscosity $\nu L u$ with a hyperviscosity $\nu L^m u$, $m > 1$.

Note that (iii) agrees with the belief (which goes back at least to Onsager [Ons49]) that the 2D Euler equation describes certain classes of 2D turbulence. Unfortunately, the equation does this in an implicit way since Euler-invariant measures form a large class and we do not know how to single out from it the measure μ_0. Also note that (vi) indicates a certain universality of the space-periodic 2D turbulence. Various universalities of 2D turbulence are often suggested in the physical literature.

For $k \geq 1$ denote by E_k the energy density at wavenumber k. That is, for $u(x) = \sum_{s\in\mathbb{Z}_0^2} u_s e_s(x)$ we define

$$E_k = (2C)^{-1} \sum_{k-C\leq|s|<k+C} \tfrac{1}{2}\, \mathbb{E}_{\mu_0} |u_s|^2,$$

where $C \geq 1$ is of order one and we are most interested in large k. Then

$$\mathbb{E}_{\mu_0} \|u\|_m^2 \sim \int_1^\infty E_k k^{2m}\, dk \qquad \forall m. \tag{5.67}$$

The function $k \mapsto E_k$ is called the *energy spectrum of the turbulence*. Making the bold assumption that $E_k \sim k^{-r}$ for some r, we get from (i) that $r \geq 5$. We neglect the weak logarithmic divergence of the integral in (5.67) which occurs when $r = 5$ (to remove the divergence one may correct the anzats for E_k by a suitable logarithmic factor). The exponent $r = 5$ is distinguished due to relation (5.66), which follows from the conservation of enstrophy in the 2D Euler equation. We conjecture that $\mu_0(H^{2+\delta}) = 0$ for positive δ and that

$$\mathbb{E}_{\mu_0} \|u\|_{2+\delta}^2 = \infty \quad \forall\, \delta > 0.$$

Then

$$E_k \sim k^{-5} \quad \text{for} \quad k \gg 1, \tag{5.68}$$

modulo a possible logarithmic correction. This relation implies that periodic 2D turbulence exhibits a *direct cascade of energy*: even though the force $\eta(t, x)$

in (5.65) is concentrated in low modes (and has an insignificant smooth component in high modes to guarantee the non-degeneracy), the energy spectrum decays only as k^{-5}.

It is interesting to compare our results and predictions with those of the heuristic theory of 2D turbulence originated by Batchelor [Bat82] (pp. 186–187), [Bat69] and Kraichnan [Kra67], and called below the BK theory. That theory is based on the assumption that the rate of dissipation of enstrophy $\nu \mathbb{E}_{\mu_\nu} \| \operatorname{curl} u \|_1^2$ converges to a finite non-zero limit ϵ as $\nu \to 0$. Since $\| \operatorname{curl} u \|_1 = \| u \|_2$ for divergence-free vector fields, then for the space-periodic 2D turbulence we have $\epsilon = 0 \cdot \mathbb{E}_{\mu_\nu} \| u \|_2^2$ which vanishes by (5.66). So the assumption fails. The explanation for this is simple: for the BK theory to hold, the 2D flow of the fluid should be not 2π-periodic, but $2\pi L$-periodic, where $\mathbb{N} \ni L \to \infty$ (it is not quite clear what relation should be imposed on ν^{-1} and L when both of them go to infinity). The force $\eta(t, x)$ remains essentially the same: most of its energy is supported by modes $e_s(x)$, where s is an integer of order one, while other modes $e_{s'}(x)$, $s' \in (\mathbb{Z}/L)^2$, carry a tiny proportion of the energy, just to make the force non-degenerate with respect to the period $2\pi L$. In this large-volume setup the force η enters the right-hand side of the Navier–Stokes equations without any scaling factor, so the equations take the form

$$\partial_t u + \nu L u + B(u) = \eta(t, x). \tag{5.69}$$

The BK theory predicts that for $\nu \ll 1$ and $L \gg 1$ the energy spectrum E_k behaves as follows:

$$E_k \sim k^{-3}, \qquad 1 \ll k \ll k_+, \tag{5.70}$$

$$E_k \sim k^{-5/3}, \qquad k_- \ll k \ll 1, \tag{5.71}$$

where $k_+ \to \infty$ and $k_- \to 0$ as $\nu \to 0$ and $L \to \infty$. The theory also specifies ϵ-dependent factors in front of k^{-3} and $k^{-5/3}$ which we do not discuss. The asymptotic (5.70) differs from what we have in the space-periodic 2D turbulence, where the assumption $E_k \sim k^{-r}$ implies that $r \geq 5$ and we predict that $r = 5$. Presumably, this happens since the 2π-periodic boundary conditions, imposed on the fluid, regularise it and make the direct cascade of energy weaker than the BK cascade (5.70).[11]

Relation (5.71), known as the *inverse cascade of energy*, has no analogy in space-periodic turbulence, where $k \geq 1$. Some experts in turbulence believe that for the limits (5.70) and (5.71) to exist when $\nu \to 0$ and $L \to \infty$, the right-hand side of Eq. (5.69) has to be modified by the friction term $-\tau u$ (sometimes

[11] When asked if his theory of turbulence applies to 3D periodic flows, A. N. Kolmogorov said that he is not certain, since periodic flows may be "too regular" (we are grateful to A. V. Fursikov and M. I. Vishik for this recollection).

called the *Ekman damping* in honor of Swedish oceanologist V. W. Ekman). The parameter τ has to be sent to zero with ν and L^{-1} (again, the relation between these three quantities is not clear). See [Ber00] for an advanced version of the corresponding argument and see [BV11] for a detailed discussion.

So, in contrast to the BK theory, the theory of space-periodic 2D turbulence, presented in this section, has a rigorous component (i)–(vi). It gives rise to a direct cascade of energy with an exponent ≥ 5, which is predicted to be five, and differs from the BK exponent 3 in (5.70). Further rigorous development of this theory by means of modern mathematical tools (including those presented in this book) seems to us more feasible than rigorous justification of the BK theory.

Notes and comments

The existence of an inviscid limit of stationary measures (see Theorems 5.2.2 and 5.2.21) was proved by Kuksin [Kuk04]. The balance relations described in Section 5.1.2 were established by Kuksin and Penrose [KP05]. The co-area form of the balance relations (see Exercise 5.1.5), pointwise exponential estimates of Section 5.1.3, and the properties of the limiting measures described in Sections 5.2.1–5.2.4 were obtained by Kuksin [Kuk06b; Kuk08; Kuk07]. The inviscid limit for the complex Ginzburg–Landau equation was studied by Kuksin and Shirikyan [KS04b], and the local-time technique used in Section 5.2.2 is taken from [Shi11b]. We refer the reader to the review papers [Kuk07; Kuk10b] for further references and a discussion of the inviscid limit for damped/driven integrable PDEs.

6

Miscellanies

This chapter is mostly devoted to the 3D Navier–Stokes equations with random perturbations. We begin with the problem in thin domains and state a result on the convergence of the unique stationary distribution to a unique measure which is invariant under the flow of the limiting 2D Navier–Stokes system. We next turn to the 3D problem in an arbitrary bounded domain or a torus. We describe two different approaches for constructing Markov processes whose trajectories are concentrated on weak solutions of the Navier–Stokes system and investigate the large-time asymptotics of their trajectories. Finally, we discuss some qualitative properties of solutions in the case of perturbations of low dimension. Almost all the results of this chapter are presented without proofs.

6.1 3D Navier–Stokes system in thin domains

In this section, we present a result that justifies the study of 2D Navier–Stokes equations in the context of hydrodynamical turbulence. Namely, we study the 3D Navier–Stokes system in a thin domain and prove that, roughly speaking, if the domain is sufficiently thin, then the problem in question has a unique stationary measure, which attracts exponentially all solutions in a large ball and converges to a limiting measure invariant under the 2D dynamics. Moreover, when the width of the domain shrinks to zero, the law of a 3D solution converges to that of a 2D solution uniformly in time. The accurate formulation of these results requires some preliminaries from the theory of Navier–Stokes equations in thin domains. They are discussed in the first subsection. We next turn to the large-time asymptotics of solutions and the limiting behaviour of stationary measures and solutions.

6.1.1 Preliminaries on the Cauchy problem

Let $Q_\varepsilon = \mathbb{T}^2 \times (0, \varepsilon) \subset \mathbb{R}^3$, where $\mathbb{T}^2$ stands for the two-dimensional torus[1] and $\varepsilon > 0$ is a small parameter. We denote by $x = (x_1, x_2, x_3)$ the space variable, so that (x_1, x_2) varies in $\mathbb{T}^2$ while x_3 belongs to the interval $(0, \varepsilon)$. Consider the 3D Navier–Stokes system

$$\dot{u} + \langle u, \nabla \rangle u - \nu \Delta u + \nabla p = \eta(t, x), \quad \operatorname{div} u = 0, \quad x \in Q_\varepsilon, \tag{6.1}$$

where $u = (u_1, u_2, u_3)$ and p are the unknown velocity field and pressure, and η denotes a random kick force of the form (2.65). Equations (6.1) are supplemented with the free boundary conditions in the vertical direction x_3:

$$u_3 = \partial_3 u_i = 0 \quad \text{for } x_3 = 0 \text{ and } \varepsilon, \tag{6.2}$$

where $i = 1, 2$. As in the 2D case with periodic or Dirichlet boundary conditions, problem (6.1), (6.2) can be reduced to an evolution equation in an appropriate functional space. Namely, denote by V_ε^2 the space of divergence-free vector fields $u \in H^2(Q_\varepsilon; \mathbb{R}^3)$ that satisfy the boundary conditions (6.2) and whose first and second components have zero mean value on Q_ε. Let H_ε and V_ε be the closures of V_ε^2 in the spaces $L^2(Q_\varepsilon)$ and $H^1(Q_\varepsilon)$, respectively, and let Π_ε be the orthogonal projection in $L^2(Q_\varepsilon)$ to the closed subspace H_ε. The space H_ε is endowed with the usual L^2 norm $|\cdot|_\varepsilon$, and V_ε is a Hilbert space with respect to the norm $\|u\|_\varepsilon = |\nabla u|_\varepsilon$, due to Poincaré's and Friedrichs' inequalities. Applying Π_ε formally to the first relation in (6.1), we get the nonlocal equation

$$\dot{u} + \nu L_\varepsilon u + B_\varepsilon(u) = \Pi_\varepsilon \eta(t, x). \tag{6.3}$$

Here L_ε is the Stokes operator in H_ε given on its domain V_ε^2 by $L_\varepsilon u = -\Pi_\varepsilon \Delta u$ and $B_\varepsilon(u) = B_\varepsilon(u, u)$, where $B_\varepsilon : V_\varepsilon \to V_\varepsilon^*$ is a quadratic mapping defined for $u, v \in V_\varepsilon^2$ by the relation $B_\varepsilon(u, v) = \Pi_\varepsilon\big(\langle u, \nabla \rangle v\big)$. Introduce the following orthogonal projections in $L^2(Q_\varepsilon; \mathbb{R}^3)$:

$$M_\varepsilon u = \left(\varepsilon^{-1} \int_0^\varepsilon u_1(x', y)\, dy, \varepsilon^{-1} \int_0^\varepsilon u_2(x', y)\, dy, 0\right), \quad N_\varepsilon = \mathrm{Id} - M_\varepsilon,$$

where $x' = (x_1, x_2)$. It is straightforward to check that these projections also are continuous and orthogonal in V_ε, and their images are closed subspaces in both cases. Furthermore, recall that the spaces H and V were introduced at the end of Section 2.1.5 and note that they can be identified with the images of M_ε in H_ε and V_ε, respectively, by associating to each element (u_1, u_2) the vector function $(u_1, u_2, 0)$. The following result due to Raugel and Sell [RS93] (see also the paper [TZ96] by Temam and Ziane) enables one to construct

[1] To simplify the presentation, we assume that the sides of the torus are equal. However, all the results remain true in the case of an arbitrary ratio of the sides.

global-in-time solutions of the homogeneous equation for initial data in a large ball, provided that $\varepsilon > 0$ is sufficiently small.

Theorem 6.1.1 *Let $R : (0, 1] \to \mathbb{R}_+$ be an arbitrary function such that*

$$\lim_{\varepsilon\to 0^+} \varepsilon^\theta R(\varepsilon) = 0 \quad \text{for some } \theta \in \big(0, \tfrac{1}{2}\big). \tag{6.4}$$

Then there is an $\varepsilon_0 > 0$ depending on $R(\cdot)$ such that, for any initial function $u_0^\varepsilon \in V_\varepsilon$ satisfying the inequality $\|u_0^\varepsilon\|_\varepsilon \le R(\varepsilon)$, Eq. (6.3) *has a unique solution*

$$u^\varepsilon \in C(\mathbb{R}_+; V_\varepsilon) \cap L^2_{\text{loc}}(\mathbb{R}_+; V_\varepsilon^2)$$

issued from u_0^ε at time $t = 0$. Moreover, if $M_\varepsilon u_0^\varepsilon \to v_0$ as $\varepsilon \to 0$ weakly in H, then for any $T > 0$ we have

$$\lim_{\varepsilon\to 0} M_\varepsilon u^\varepsilon = v \quad \text{in} \quad C(0, T; H) \cap L^2(0, T; V), \tag{6.5}$$

where $v(t, x)$ is the solution of the 2D Navier–Stokes equation (2.19) *with $f \equiv 0$ issued from v_0.*

6.1.2 Large-time asymptotics of solutions

We first describe the random kick forces we are dealing with. For simplicity, we shall assume that $\eta(t, \cdot)$ belongs to the image of the projection Π_ε for all $t \ge 0$, so that we can omit Π_ε from the right-hand side of (6.3). As was explained in Section 2.3, the Navier–Stokes system (6.1) with the random kick force (2.65) is equivalent to the discrete-time dynamical system

$$u_k = S_T^\varepsilon(u_{k-1}) + \eta_k, \tag{6.6}$$

where $u_k = u(kT)$, S_t^ε denotes the resolving operator for the homogeneous problem (see Theorem 6.1.1), and the solutions are normalised by the condition of right-continuity at the points of the form kT. In the 3D case, there is an additional problem related to the fact that S_t may not be defined for all $t \ge 0$. However, it is possible to construct large subsets of V_ε on which the RDS (6.6) is well defined. Namely, for given positive functions $a(\varepsilon)$ and $b(\varepsilon)$, we set

$$\mathcal{B}_\varepsilon = \{u \in V_\varepsilon : \|M_\varepsilon u\|_\varepsilon \le a(\varepsilon), \|N_\varepsilon u\|_\varepsilon \le b(\varepsilon)\}.$$

The following result is established in [CK08a] (see proposition 4.1).

Proposition 6.1.2 *Let $a(\varepsilon)$ and $b(\varepsilon)$ be two positive functions such that* (6.4) *holds for $R(\varepsilon) = a(\varepsilon) + b(\varepsilon)$ and*

$$\lim_{\varepsilon\to 0} \frac{\sqrt{\varepsilon}\, b^2(\varepsilon)}{a(\varepsilon)} = 0. \tag{6.7}$$

Then for any $T_0 > 0$ there are positive constants ε_0 and c_0 such that if

$$\mathbb{P}\big\{\|M_\varepsilon \eta_k\|_\varepsilon \le c_0 a(\varepsilon), \|N_\varepsilon \eta_k\|_\varepsilon \le c_0 b(\varepsilon), k \ge 1\big\} = 1, \tag{6.8}$$

then the RDS (6.6) *is well defined on the set $\mathcal{B}_\varepsilon$ for any $T \ge T_0$ and $\varepsilon \le \varepsilon_0$.*

This proposition enables one to define a family of Markov chains on $\mathcal{B}_\varepsilon$ associated with (6.6). We denote by $P_k^\varepsilon(u,\Gamma)$ its transition function and by $\mathfrak{P}_k(\varepsilon)$ and $\mathfrak{P}_k^*(\varepsilon)$ the corresponding Markov operators.

We now turn to the question of the large-time behaviour of trajectories for (6.6) under some additional hypotheses on the kicks η_k. To formulate them, let us introduce a special orthogonal basis in the space V_ε. The Stokes operator L_ε is self-adjoint and has a compact resolvent, so the spaces H_ε and V_ε decompose into the direct sum of one-dimensional orthogonal subspaces spanned by the eigenfunctions of L_ε. Moreover, since the Stokes operator preserves the direct decomposition $V_\varepsilon = M_\varepsilon V_\varepsilon \oplus N_\varepsilon V_\varepsilon$, its eigenfunctions must belong to these subspaces, and it is easy to see that those belonging to $M_\varepsilon V_\varepsilon$ are also eigenfunctions for the Stokes operator L on the 2D torus $\mathbb{T}^2$ and are independent of ε. We denote them by e_j and write λ_j for the corresponding eigenvalues. As for the eigenvectors belonging to $N_\varepsilon V_\varepsilon$, they depend on ε and will be denoted by e_j^ε. We normalise the eigenfunctions by the condition

$$\|e_j\|_\varepsilon = \|e_j^\varepsilon\|_\varepsilon = \sqrt{\varepsilon} \quad \text{for all } j \ge 1, \tag{6.9}$$

whence it follows, in particular, that the norm in V of e_j considered as a vector field on $\mathbb{T}^2$ is equal to 1.

We shall assume that the kicks η_k entering the random force η of the form (2.65) can be written as

$$\eta_k(x) = \eta_k^\varepsilon(x) = \sum_{j=1}^{\infty} b_j \xi_{jk} e_j(x) + \sum_{j=1}^{\infty} d_j \zeta_{jk} e_j^\varepsilon(x), \tag{6.10}$$

where b_j and d_j are non-negative numbers such that

$$\mathfrak{B} = \sum_{j=1}^{\infty} b_j^2 < \infty, \quad \mathfrak{D} = \sum_{j=1}^{\infty} d_j^2 < \infty,$$

and $\{\xi_{jk}\}$ and $\{\zeta_{jk}\}$ are two independent sequences of independent random variables whose laws possess densities $p_j(r)$ and $q_j(r)$ with respect to the Lebesgue measure on the real line. Furthermore, the functions p_j and q_j have bounded total variation and their supports are subsets of $[-1,1]$ containing the point $r=0$. Note that, in view of (6.9), the V_ε-norm of the kicks satisfies the inequality

$$\|\eta_k\|_\varepsilon^2 \le \sum_{j=1}^{\infty} \big(b_j^2 \|e_j\|_\varepsilon^2 + d_j^2 \|e_j^\varepsilon\|_\varepsilon^2\big) = \varepsilon(\mathfrak{B}+\mathfrak{D}).$$

Thus, by Proposition 6.1.2, the RDS (6.6) with the above hypotheses on η_k is well defined on $\mathcal{B}_\varepsilon$, provided that the functions a and b are minorised $C\sqrt{\varepsilon}$ with a large constant $C>0$ and satisfy relations (6.4) and (6.7) in which $R = a + b$.

Theorem 6.1.3 *Let the above-mentioned hypotheses on η_k be satisfied and let $a(\varepsilon)$ and $b(\varepsilon)$ be two functions such that* (6.7) *holds and*

$$C_1\sqrt{\varepsilon} \le a(\varepsilon) \le C_2\sqrt{\varepsilon}\left(\ln\tfrac{1}{\varepsilon}\right)^{\sigma}, \quad C_1\sqrt{\varepsilon} \le b(\varepsilon) \le C_2\left(\ln\tfrac{1}{\varepsilon}\right)^{\sigma/2},$$

where $0 \le \sigma < \frac{1}{2}$ and $C_1 > 0$ is sufficiently large. Then there is a constant $\varepsilon_0 > 0$ and an integer $N \ge 1$ not depending on ε such that the following assertions are true for $0 < \varepsilon \le \varepsilon_0$, provided that

$$b_j \ne 0 \quad \textit{for } j = 1, \dots, N: \tag{6.11}$$

Existence and uniqueness: *The RDS* (6.6) *has a unique stationary measure μ_ε.*

Exponential mixing: *There are positive constants α_ε and C_ε such that for any $\lambda \in \mathcal{P}(\mathcal{B}_\varepsilon)$ we have*

$$\|\mathfrak{P}_k^*(\varepsilon)\lambda - \mu_\varepsilon\|_L^* \le C_\varepsilon e^{-\alpha_\varepsilon k} \quad \textit{for } k \ge 0,$$

where $\|\cdot\|_L^*$ stands for the dual-Lipschitz norm on $\mathcal{P}(V_\varepsilon)$.

We refer the reader to section 5 in [CK08a] for a proof of this result. Note also that if the random force η is more regular in x, then the stationary measure is supported by a space of smoother functions and the convergence to it holds in a stronger norm. Furthermore, under some additional assumptions, the constants α_ε and C_ε may be chosen to be independent of ε.

6.1.3 The limit $\varepsilon \to 0$

We now turn to the main result of this section. Recall that the 2D Navier–Stokes system on the torus $\mathbb{T}^2$ perturbed by a random force of the form (2.65) gives rise to a Markov chain in the space H; see Section 2.3. In view of Theorem 3.2.9, if the kicks have the form (6.10) with $d_j = 0$ for all $j \ge 1$ and the non-degeneracy condition (6.11) is fulfilled, then the Markov chain has a unique stationary measure μ. The following theorem established by Chueshov and Kuksin [CK08a] shows that the averaged horizontal component of the unique stationary measure for (6.6) can be approximated by μ.

Theorem 6.1.4 *Under the hypotheses of Theorem 6.1.3, if* (6.11) *holds with sufficiently large N, then*

$$(M_\varepsilon)_*\mu_\varepsilon \to \mu \quad \textit{as } \varepsilon \to 0, \tag{6.12}$$

where the convergence holds in the weak topology of $\mathcal{P}(H)$. If, in addition, $\sum_j \lambda_j^2 b_j^2 < \infty$, then (6.12) *holds in the weak topology of $\mathcal{P}(V^{2-\delta})$ for any $\delta > 0$.*

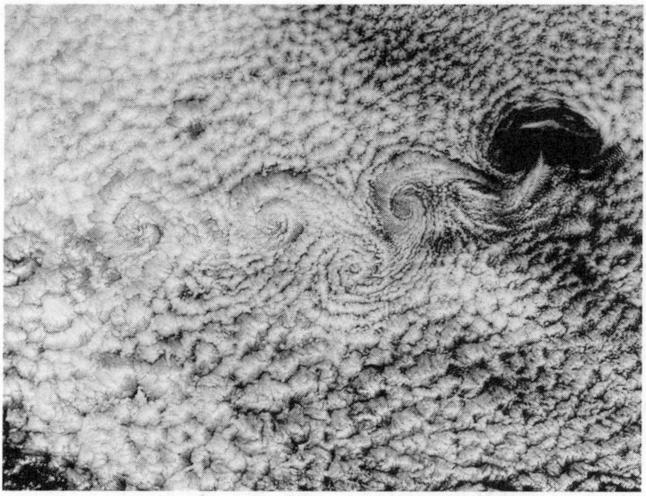

Figure 6.1 Motion of the atmosphere above the island of Madeira. On scales of order the island's size, the motion clearly looks two-dimensional; cf. Figure 1 and Section 6.1. Jeff Schmaltz, MODIS Rapid Response Team, NASA/GSFC. http://visibleearth.nasa.gov/view.php?id=63391.

Theorem 6.1.4 enables one to strengthen convergence (6.5) for solutions of (6.1), showing that it holds uniformly in time for an appropriate topology. Namely, we have the following result (cf. Corollary 4.3.2).

Corollary 6.1.5 *Under the hypotheses of Theorem 6.1.4, let* $\{u_0^\varepsilon\} \in V_\varepsilon$ *be a family of initial functions such that* $\|u_0^\varepsilon\|_\varepsilon \le C\sqrt{\varepsilon}$ *for some constant* $C > 0$ *and* $M_\varepsilon u_0^\varepsilon \to v_0$ *weakly in* H*, where* $v_0 \in V$*. Let* $u^\varepsilon(t)$ *be a solution of* (6.1), (6.2) *issued from* u_0^ε *and let* $v(t)$ *be a solution of Eq.* (2.76) *with* $h \equiv 0$ *on* $\mathbb{T}^2$ *such that* $v(0) = v_0$*. Then*

$$\sup_{t\ge 0} \left\| \mathcal{D}\big(M_\varepsilon u^\varepsilon(t)\big) - \mathcal{D}(v(t)) \right\|_L^* \to 0 \quad \text{as } \varepsilon \to 0, \tag{6.13}$$

where $\|\cdot\|_L^*$ *stands for the dual-Lipschitz distance over* H.

We note that the condition $\|u_0^\varepsilon\|_\varepsilon \le C\sqrt{\varepsilon}$ is not a smallness assumption for u_0^ε since it signifies not that the vector field $u_0^\varepsilon(x)$ is small, but that the volume of the domain Q_ε is of order ε. A proof of the above results can be found in section 5.3 of [CK08a]. Let us also mention that convergence (6.13) holds for the dual-Lipschitz distance over $V^{2-\delta}$ if the coefficients b_j satisfy the additional decay condition of Theorem 6.1.4.

For the 3D stochastic Navier–Stokes system in Q_ε with a white-in-time random force $\eta(t,x)$, the argument of this section does not work, because Proposition 6.1.2 is no longer true, as domains of the form $\mathcal{B}_\varepsilon$ are not invariant

Figure 6.2 Jupiter's Red Spot. This is a huge 2D structure in the Jovian atmosphere (twice as big as the Earth), that has been persistent for more than 180 years. It is believed that the motion inside the spot is driven by 2D turbulence, cf. Figure 2. NASA/JPL/courtesy of nasaimages.org.

for the corresponding RDS. However, if we smooth out the stochastic system (6.1) by replacing there the bilinear term $B(u)$ with $B((I - \alpha\Delta)^{-1}u, u)$, where $\alpha > 0$ is arbitrary, then we get a well-posed stochastic PDE, and the corresponding RDS converges weakly to trajectories of (6.1) as $\alpha \to 0^+$. An analogue of Theorem 6.1.3 is valid for this system with any positive α, and an analogue of convergence (6.12) holds true as $\alpha \to 0^+$ and $\varepsilon \to 0$ (in any order); see [CK08b].

Relevance of the results for physics

The results of this section show that, in the thin domain $\mathbb{T}^2 \times (0, \varepsilon)$, non-isotropic 3D turbulence described by the Navier–Stokes system with a *not too big* vertical component may be well approximated by the 2D turbulence in $\mathbb{T}^2$. The same results hold also for non-isotropic 3D turbulence in the thin spherical layer $\mathbb{S}^2 \times (0, \varepsilon)$.

6.2 Ergodicity and Markov selection

The aim of this section is to discuss some results on 3D stochastic Navier–Stokes equations. Since the corresponding results are rather technical and certainly far from their final form, to make them more transparent, we begin with the case of stochastic differential equations in $\mathbb{R}^d$ (SDE). We use a simple example to explain informally two fundamental results: the martingale problem and Markov selection. We next turn to 3D Navier–Stokes equations and discuss two different approaches for constructing Markov processes concentrated on solutions and proving a mixing property in the total variation distance.

6.2.1 Finite-dimensional stochastic differential equations

Let us consider the SDE

$$dx_t = b(x_t)dt + dw_t, \quad x_t \in \mathbb{R}^d, \tag{6.14}$$

where $b : \mathbb{R}^d \to \mathbb{R}^d$ is a continuous function and w_t is an $\mathbb{R}^d$-valued standard Brownian motion. If b is globally Lipschitz continuous, then for any $x \in \mathbb{R}^d$ Eq. (6.14) has a unique solution defined on the positive half-line and satisfying the initial condition

$$x_0 = x, \tag{6.15}$$

and the family of all solutions forms a Markov process $(x_t, \mathbb{P}_x)$ in $\mathbb{R}^d$ with the transition function $P_t(x, \Gamma) = \mathbb{P}_x\{x_t \in \Gamma\}$. Itô's formula implies that if f is a C^2-smooth function with bounded derivatives, then

$$f(x_t) = f(x_0) + \int_0^t Lf(x_r)\,dr + \int_0^t \langle \nabla f(x_r), dw_r \rangle, \tag{6.16}$$

where we set

$$L = \frac{1}{2}\Delta + \langle b(x), \nabla \rangle.$$

The operator L is called the *generator* corresponding to the SDE (6.14). Let us denote by $\mathcal{F}_t$ the filtration associated with the Markov process and apply the condition expectation $\mathbb{E}_x(\cdot \mid \mathcal{F}_s)$ to (6.16). Recalling that the stochastic integral defines a martingale, we obtain

$$\begin{aligned}
\mathbb{E}_x\Big(f(x_t) - \int_0^t Lf(x_r)\,dr \,\Big|\, \mathcal{F}_s \Big) &= f(x) + \mathbb{E}_x\Big(\int_0^t \langle \nabla f(x_r), dw_r \rangle \,\Big|\, \mathcal{F}_s \Big) \\
&= f(x) + \int_0^s \langle \nabla f(x_r), dw_r \rangle \\
&= f(x_s) - \int_0^s Lf(x_r)\,dr,
\end{aligned}$$

where the equality holds with $\mathbb{P}_x$-probability 1. Hence, we see that the process

$$M_t^f = f(x_t) - \int_0^t Lf(x_r)\,dr, \quad t \geq 0, \tag{6.17}$$

is a martingale with respect to the filtration $\mathcal{F}_t$ and the probability $\mathbb{P}_x$, with an arbitrary $x \in \mathbb{R}^d$. The following proposition shows that the latter condition is sufficient for the existence of a weak solution for (6.14), (6.15). Let Ω be the Fréchet space $C(\mathbb{R}_+; \mathbb{R}^d)$ with its Borel σ-algebra and let $\{x_t(\omega)\}_{t\geq 0}$ be the canonical process.

Proposition 6.2.1 *Let $b : \mathbb{R}^d \to \mathbb{R}^d$ be a bounded continuous function and let $\mathbb{P}$ be a probability measure on Ω such that M_t^f is a continuous martingale with respect to $\mathbb{P}$ for any C^2-function f with bounded derivatives. Then the process $x_t - x_0 - \int_0^t b(x_r)\,dr$ is an $\mathbb{R}^d$-valued standard Brownian motion. In particular, problem* (6.14), (6.15) *has a weak solution defined on $\mathbb{R}_+$.*

We refer the reader to section 5.4.B of the book [KS91] for the proof of a more general result. In what follows, any probability measure $\mathbb{P}$ satisfying the hypotheses of Proposition 6.2.1 is called a *solution of the martingale problem associated with L*. What has been said above implies that problem (6.14), (6.15) has a weak solution associated with an initial measure $\mu \in \mathcal{P}(\mathbb{R}^d)$ if and only if the martingale problem associated with its generator L has a solution $\mathbb{P}$ whose restriction at $t = 0$ coincides with μ. Moreover, the uniqueness of the law for a weak solution is equivalent to the uniqueness of a solution for the martingale problem with a given initial measure. We now study a sufficient condition for the latter property.

Let $\mathbb{P}$ be a solution of the martingale problem associated with L, let x_t be the canonical process, and let w_t be the Brownian motion constructed in Proposition 6.2.1. Let $u(t,x)$ be an arbitrary function having continuous partial derivatives $\partial_t^j \partial_x^\alpha u$ for $2j + |\alpha| \leq 2$ that are bounded in the strip $[0,T] \times \mathbb{R}^d$ for some $T > 0$. Then, applying Itô's formula to the process $y_t = u(T-t, x_t)$, we derive

$$y_t - y_0 = \int_0^t \big(-\partial_t u + Lu\big)(T-r, x_r)\,dr + \int_0^t \langle u(T-r, x_r), dw_r\rangle.$$

Assuming now that u is the solution[2] of the Cauchy problem for *Kolmogorov's backward equation*

$$\partial_t u = Lu, \quad u(0,x) = f(x), \quad x \in \mathbb{R}^d, \quad t > 0, \tag{6.18}$$

where $f \in C_0^\infty(\mathbb{R}^d)$ is a given function, we see that the process y_t is a martingale under $\mathbb{P}$. In particular, taking the expectation with respect to $\mathbb{P}$, we obtain

$$\mathbb{E}\,u(T-t, x_t) = \mathbb{E}\,y_t = \mathbb{E}\,y_0 = \mathbb{E}\,u(T, x_0) \quad \text{for any } t \in [0,T].$$

[2] Solutions with the above-mentioned regularity exist, for instance, if b is Hölder continuous; see chapter 8 of [Kry96].

Setting $t = T$, we see that

$$\mathbb{E}\, f(y_T) = \mathbb{E}\, u(0, x_T) = \mathbb{E}\, f(x_T) = \mathbb{E}\, u(T, x_0).$$

The right-hand side of this relation depends only on the function f and the law of the restriction of $\mathbb{P}$ at time $t = 0$. Since $C_0^\infty(\mathbb{R}^d)$ is a determining family for the Borel measures on $\mathbb{R}^d$, we conclude that the law of x_T under a solution of the martingale problem depends only on the initial law $\mathbb{P}|_{t=0}$. Further analysis shows that the law of the canonical process $\{x_t\}$ under $\mathbb{P}$ also depends only on $\mathbb{P}|_{t=0}$, provided that problem (6.18) has a solution satisfying the above-mentioned boundedness condition for any $T > 0$. More precisely, we have the following result due to Stroock and Varadhan; see section 5.4.E of the book [KS91].

Theorem 6.2.2 *Let $b : \mathbb{R}^d \to \mathbb{R}^d$ be a continuous function such that for any $f \in C_0^\infty(\mathbb{R}^d)$ problem* (6.18) *has a solution $u(t, x)$ whose derivatives $\partial_t^j \partial_x^\alpha u$ with $2j + |\alpha| \le 2$ are bounded in the strip $[0, T] \times \mathbb{R}^d$ for any $T > 0$. Then for every $x \in \mathbb{R}^d$ the martingale problem associated with L has at most one solution $\mathbb{P}$ such that $\mathbb{P}|_{t=0} = \delta_x$.*

It is a "natural" idea to try to apply the above approach to 3D Navier–Stokes equations, for which well-posedness is not known to hold. Unfortunately, the existing methods do not allow one to construct a sufficiently regular solution for the infinite-dimensional analogue of problem (6.18), and the possibility of carrying out the Stroock–Varadhan program for 3D Navier–Stokes equations is unclear. Nevertheless, one can prove that Kolmogorov's equation possesses a solution with weak regularity properties, and this turns out to be sufficient for building a Markov family concentrated on trajectories of Navier–Stokes equations. This approach, together with an investigation of mixing properties, was realised by Da Prato, Debussche, and Odasso in the series of papers [DD03; DO06; Oda07] and is presented briefly in Section 6.2.2.

We now turn to the case in which the uniqueness of the law for the SDE (6.14) is unknown. In this situation, Krylov [Kry73] suggested a general scheme for constructing a *Markov selection* – a family of strong Markov processes concentrated on martingale solutions of (6.14). We now describe his approach following essentially the presentation in [SV79] and confining ourselves to the Markov property. The strong Markov property of the selection is more delicate and will not be discussed here.

As before, we assume that $b(x)$ is a bounded continuous function and denote by L the generator corresponding to (6.14). Given $s \ge 0$ and $x \in \mathbb{R}^d$, we shall say that a probability measure $\mathbb{P}$ on the canonical space Ω is a *solution of the martingale solution for L starting from* (s, x) if $\mathbb{P}\{x_t = x \text{ for } 0 \le t \le s\} = 1$ and the process M_t^f defined by (6.17) is a martingale after time s for any $f \in C_0^\infty(\mathbb{R}^d)$. Let us denote by $\mathcal{C}(s, x)$ the set of all solutions to the martingale problem for L starting from (s, x). Thus, $\mathcal{C}(s, x)$ is a subset in the space $\mathcal{P}(\Omega)$, which is endowed with the dual-Lipschitz distance and the topology of weak convergence. We shall write $\mathcal{C}(x)$ instead of $\mathcal{C}(0, x)$. Since the drift coefficient b does not depend on time, it is clear that $\mathcal{C}(s, x)$ can be obtained from the set $\mathcal{C}(x)$ by "shifting" it to the half-line $[s, +\infty)$ and "gluing" to the resulting measure the Dirac mass at the function equal to x for $0 \le t \le s$. To formulate accurately this and some other properties of $\mathcal{C}(s, x)$, we need the following lemma, whose proof in a more general setting can be found in [SV79] (see section 6.1).

Lemma 6.2.3 *Let $s \ge 0$ and let $Q_\omega : \Omega \to \mathcal{P}(\Omega)$ be a random probability measure such that $Q.(\Gamma)$ is $\mathcal{F}_s$-measurable for any $\Gamma \in \mathcal{B}(\Omega)$ and*

$$Q_\omega\big(\{x_t = \omega(s) \text{ for } 0 \le t \le s\}\big) = 1 \quad \text{for all } \omega \in \Omega.$$

Then for any $P \in \mathcal{P}(\Omega)$ there is a unique measure $P \otimes_s Q. \in \mathcal{P}(\Omega)$ such that

$$(P \otimes_s Q.)(B) = P(B), \tag{6.19}$$

$$(P \otimes_s Q.)(B \cap \Gamma) = \int_B Q_\omega(\Gamma) P(d\omega) \tag{6.20}$$

for any $B \in \mathcal{F}_s$ and $\Gamma \in \mathcal{F}_s^+$, where $\mathcal{F}_s^+$ stands for the σ-algebra generated by the canonical process on $[s, +\infty)$.

In other words, if Q_ω is a random probability measure that is measurable with respect to $\mathcal{F}_s$ and whose restriction to $\mathcal{F}_s$ coincides with the Dirac mass at the function identically equal to $\omega(s)$ on the interval $[0, s]$, then there is a unique measure in $\mathcal{P}(\Omega)$ such that its restriction to $\mathcal{F}_s$ coincides with P (see (6.19)) and the restriction to $\mathcal{F}_s^+$ of its regular conditional probability given $\mathcal{F}_s$ coincides with $Q_\omega|_{\mathcal{F}_s^+}$ (see (6.20)). Note also that the uniqueness is a straightforward consequence of relation (6.20), the monotone class technique, and the fact that $\mathcal{F}$ is generated by the π-system of the sets of the form $B \cap \Gamma$ with $B \in \mathcal{F}_s$ and $\Gamma \in \mathcal{F}_s^+$.

The time-independence of b implies that if $P \in \mathcal{C}(s, x)$ for some $s > 0$, then $(\theta_s)_* P \in \mathcal{C}(x)$, where θ_s stands for the shift operator on Ω. Conversely, if $P \in \mathcal{C}(x)$, then $\delta_x \otimes_s ((\theta_{-s})_* P) \in \mathcal{C}(s, x)$, where δ_x is the Dirac mass at the function identically equal to x and, with a slight abuse of notation, we denote by $(\theta_{-s})_* P$ the measure defined by $((\theta_{-s})_* P)(\Gamma) = P(\theta_s \Gamma)$. That is, the mapping

$$P \mapsto \delta_x \otimes_s ((\theta_{-s})_* P) \in \mathcal{C}(s, x)$$

defines a bijection between $\mathcal{C}(x)$ and $\mathcal{C}(s, x)$.

The proposition below establishes some further properties of the sets $\mathcal{C}(s, x)$. It is the first of the two key ingredients for the construction of a Markov selection for (6.14).

Proposition 6.2.4 *The family $\{\mathcal{C}(s, x), s \ge 0, x \in \mathbb{R}^d\}$ possesses the following properties.*

(a) *The set* $\mathcal{C}(s,x)$ *is compact in* $\mathcal{P}(\Omega)$ *for any* $s \geq 0$ *and* $x \in \mathbb{R}^d$.
(b) *The mapping* $(s,x) \mapsto \mathcal{C}(s,x)$ *is lower-semicontinuous. That is, if* $s_n \to s$, $x_n \to x$, $P_n \in \mathcal{C}(s_n, x_n)$, *and* $P_n \to P$ *in* $\mathcal{P}(\Omega)$, *then* $P \in \mathcal{C}(s,x)$. *In particular, the mapping* $(s,x) \mapsto \mathcal{C}(s,x)$ *defines a random compact set in* $\mathcal{P}(\Omega)$, *with the underlying space* $\mathbb{R}_+ \times \mathbb{R}^d$.
(c) *Let* $P \in \mathcal{C}(x)$ *and let* $P_s(\omega, \Gamma)$ *be the regular conditional probability of* P *given* $\mathcal{F}_s$. *Then there is a* P*-null set* $N \in \mathcal{F}_s$ *such that*[3]

$$\delta_{\omega(s)} \otimes_s P_s(\omega, \cdot) \in \mathcal{C}(s, \omega(s)) \quad \text{for } \omega \notin N. \tag{6.21}$$

(d) *Let* $P \in \mathcal{C}(x)$ *and let* $Q_\cdot : \Omega \to \mathcal{P}(\Omega)$ *be an* $\mathcal{F}_s$*-measurable function such that* $\delta_{\omega(s)} \otimes_s Q_{\omega(s)} \in \mathcal{C}(s, \omega(s))$ *for all* $\omega \in \Omega$. *Then* $P \otimes_s Q_\cdot \in \mathcal{C}(x)$.

We do not give a proof of this proposition, leaving it to the reader as an exercise or referring to [SV79, lemma 12.2.1] for a more general result. Let us, however, clarify informally the meaning of properties (a)–(d) and give some hints for their proof.

The compactness of $\mathcal{C}(s,x)$ is almost obvious because, due to the boundedness of b, any measure in $\mathcal{C}(s,x)$ is concentrated on a set of functions whose C^1-norm is bounded with high probability on any finite interval. The lower-semicontinuity of $\mathcal{C}(s,x)$ means essentially that the limit of a sequence of solutions is again a solution. Since the noise is additive, this property is a straightforward consequence of Skorohod's embedding theorem. Property (c) says, roughly speaking, that if $\{y_t(\omega), t \geq 0\}$ is a weak solution of (6.14) starting from $(0,x)$, then its restriction to the half-line $[s, +\infty)$ is a weak solution of (6.14) starting from $(s, y_s(\omega))$. Finally, property (d) is a formalisation of the claim that if we cut a solution starting from $(0,x)$ at a time s and then continue it as another solution, then the resulting process is again a solution starting from $(0,x)$, provided that the two solutions agree at time s.

The second key ingredient for constructing a Markov selection is the following theorem, which says that one can choose elements in $\mathcal{C}(x)$ that depend on x in a measurable manner and satisfy the Markov propoerty.

Theorem 6.2.5 *Let* $b : \mathbb{R}^d \to \mathbb{R}^d$ *be a bounded continuous function, let* $\mathcal{C}(x)$ *be the set of solutions for martingale problem starting from* $(0,x)$, *and let* $\{x_t\}_{t \geq 0}$ *be the canonical process on* Ω. *Then there exists a measurable mapping* $x \mapsto \mathbb{P}_x$ *from* $\mathbb{R}^d$ *to* $\mathcal{P}(\Omega)$ *such that* $\mathbb{P}_x \in \mathcal{C}(x)$ *for all* x *and* $(x_t, \mathbb{P}_x)$ *is a homogeneous family of Markov processes.*

We refer the reader to Krylov's original paper [Kry73] or to chapter 12 of the book [SV79] for a proof of this theorem and some further results. Note that the Markov process constructed in Theorem 6.2.5 may not satisfy the

[3] Note that the second factor $P_s(\omega, \cdot)$ in the product $\otimes_s$ of (6.21) is considered to be a constant measure.

Feller property. For instance, the one-dimensional equation $\dot{x} = \operatorname{sgn}(x)|x|^{1/2}$ has no Feller selection; see exercise 12.4.2 in [SV79]. Furthermore, not every selection $x \mapsto \mathcal{C}(x)$ will produce a Markov family, as is easily seen by studying the equation $\dot{x} = |x|^{1/2}$.

The above scheme can be carried out in the case of 3D Navier–Stokes equations with additive noise. There is, however, a number of essential difficulties, which result in that one can establish only almost sure Markov properties. This question, as well as the problem of mixing under a non-degeneracy assumption on the noise, was analysed in the papers of Flandoli and Romito [FR08; Rom08] and is discussed in Section 6.2.3.

6.2.2 The Da Prato–Debussche–Odasso theorem

Let us consider the 3D Navier–Stokes equations with a white-noise perturbation. To simplify the presentation, we assume that the space variable x belongs to the 3D torus, even though most of the results in this and the next subsections remain valid for any bounded domain with smooth boundary. Projecting to the space H, we get the equation

$$\dot{u} + \nu L u + B(u) = h(x) + \partial_t \zeta(t,x), \quad \zeta(t,x) = \sum_{s\in\mathbb{Z}_0^3} b_s \beta_s(t) e_s(x), \tag{6.22}$$

where $h \in H$ is a deterministic function, $b_s \in \mathbb{R}$ are some constants such that

$$\mathfrak{B} = \sum_{s\in\mathbb{Z}_0^3} b_s^2 < \infty, \tag{6.23}$$

$\{\beta_s\}$ is a family of independent standard Brownian motions, and $\{e_s\}$ is the normalised trigonometric basis in H (in the 3D case); cf. Sections 2.1.4, 2.1.5, and 2.4. Using the methods of the deterministic theory and treating the equation pathwise, one can prove the existence of a global weak solution for (6.22) satisfying the initial condition

$$u(0,x) = u_0(x), \tag{6.24}$$

where $u_0 \in H$ is a given function. Indeed, denoting by $z(t)$ a solution of the stochastic Stokes equation (2.100) issued from zero and writing $u = v + z$, we see that v must be a solution of problem (2.111), (2.112). The regularity of z guaranteed by Proposition 2.4.2 turns out to be sufficient to prove the existence of a weak solution v with probability 1. The resulting process u is a weak solution of the original problem, and further analysis shows that an energy inequality holds for it. We thus obtain the following result established in [BT73; VF88]; see also [CC92; FG95] for the case of multiplicative noise.

Proposition 6.2.6 *Under the above hypotheses on h and η, for any $u_0 \in H$ there is a random process $u : \mathbb{R}_+ \to H$ whose almost every trajectory belongs* *to the space*

$$L^{\infty}_{\text{loc}}(\mathbb{R}_+; H) \cap L^2_{\text{loc}}(\mathbb{R}_+; V) \cap C(\mathbb{R}_+, V^{-r}) \tag{6.25}$$

with any $r > 0$ and satisfies Eq. (2.99), where the equality holds in $C(\mathbb{R}_+; V^*)$. Moreover,

$$\mathbb{E}\biggl(|u(t)|_2^2 + \nu \int_0^t |\nabla u(s)|_2^2 ds\biggr) \le |u_0|_2^2 + \bigl(\mathfrak{B} + C|h|_2^2\bigr)\, t, \quad t \ge 0, \tag{6.26}$$

where $C > 0$ is a constant depending only on ν.

The question of uniqueness of a weak solution remains open. However, under strong non-degeneracy hypotheses on the noise, it is possible to construct a Markov process concentrated on weak solutions of (6.22) and to prove for it the property of exponential mixing. To state the corresponding results, we first reformulate the above proposition on the existence of a solution in different terms.

For technical reasons, we consider solutions of weaker regularity than those constructed in Proposition 6.2.6. Let us fix a small constant $r > 0$ and introduce the Fréchet space (cf. (6.25))

$$\Omega = L^2_{\text{loc}}(\mathbb{R}_+; V^{1-r}) \cap C(\mathbb{R}_+, V^{-r}), \tag{6.27}$$

endowed with a natural metric. Let $\mathcal{F}$ be the Borel σ-algebra on Ω and let $\mathcal{F}_t$ be the natural filtration on Ω. The law $\mathbb{P}$ of the weak solution for (6.22), (6.24) constructed in Proposition 6.2.6 is a probability measure on $(\Omega, \mathcal{F})$ possessing the following properties:

(i) $\mathbb{P}\bigl(\{\omega \in \Omega : \omega(0) = u_0\}\bigr) = 1$.
(ii) The $\mathbb{P}$-law of the process[4]

$$M_t = \omega(t) - \omega(0) + \int_0^t \bigl(\nu L\omega + B(\omega)\bigr)\, ds - ht, \quad t \ge 0, \tag{6.28}$$

coincides with that of ζ.

In what follows, any probability measure $\mathbb{P}$ on $(\Omega, \mathcal{F})$ satisfying these properties is called a *martingale solution* for problem (6.22), (6.24). Similarly, a probability measure $\mathbb{P}$ on $(\Omega, \mathcal{F})$ is called a *stationary martingale solution* for Eq. (6.22) if the canonical process is stationary under $\mathbb{P}$ and property (ii) mentioned above holds for it.

Let us assume that the coefficients of the random force ζ entering Eq. (6.22) satisfy the inequality

$$c\,|s|^{-3+\delta} \le |b_s| \le C\,|s|^{-\frac{5}{2}-\delta} \quad \text{for all } s \in \mathbb{Z}_0^3, \tag{6.29}$$

[4] It is easy to see that the trajectories of M_t belong to $C(\mathbb{R}_+; V^{-2})$ with $\mathbb{P}$-probability 1. Thus, the law of M is a probability measure on that space.

where C, c, and δ are positive constants. This means, roughly speaking, that the trajectories of ζ have regularity higher than $H^{5/4}$, but lower than $H^{3/2}$. The following result shows that it is possible to construct a family of martingale solutions which satisfy a Markov property.

Theorem 6.2.7 *Let $h \in V$ and let ζ be a random force for which (6.29) holds. Then there is a measurable mapping $u_0 \mapsto \mathbb{P}_{u_0}$ from V^2 to the space of probability measures on $(\Omega, \mathcal{F})$ with the weak topology such that the following properties hold.*

(i) *For any $u_0 \in V^2$, the measure $\mathbb{P}_{u_0}$ is a martingale solution for (6.22), (6.24) such that $P_t(u_0, V^2) = 1$, where $P_t(u_0, \cdot)$ stands for the law of $\omega(t)$ under $\mathbb{P}_{u_0}$:*

$$P_t(u_0, \Gamma) = \mathbb{P}_{u_0}\{\omega(t) \in \Gamma\}, \quad \Gamma \in \mathcal{B}(V^{-r}). \tag{6.30}$$

(ii) *For any $u_0 \in V^2$ and $t > 0$, with $\mathbb{P}_{u_0}$-probability 1 we have*

$$\mathbb{P}_{u_0}\{\omega(t + \cdot) \in \Gamma \mid \mathcal{F}_t\} = \mathbb{P}_{\omega(t)}(\Gamma), \tag{6.31}$$

where $\Gamma \in \mathcal{F}$ is an arbitrary cylindrical set,

$$\Gamma = \{\omega \in \Omega : \omega(t_i) \in B_i \text{ for } i = 1, \dots, n\},$$

with $0 \le t_1 < \cdots < t_n$ and $B_i \in \mathcal{B}(V^2)$.

(iii) *The function $t \mapsto P_t(u_0, \cdot)$ from V^2 to $\mathcal{P}(V^2)$ is continuous in the weak topology.*

We refer the reader to the original work [DO06] for a proof, confining ourselves to some comments on this result. Its formulation in [DO06] is slightly different: weak solutions are constructed there on an abstract probability space depending on the initial condition. Nevertheless, it is easy to see in the case of an additive noise that the law of a weak solution for (6.22) considered as a probability measure on the canonical space (6.27) is a martingale solution. Thus, we consider a (measurable) mapping taking an initial condition u_0 to its law $\mathbb{P}_{u_0}$. Property (i) means that, at any time $t \ge 0$, the solution is H^2 smooth. Note, however, that this property is not sufficient to ensure that the trajectories lie in H^2. Assertion (ii) is a weak form of the Markov property: in relation (6.31), one can take only cylindrical subsets Γ that are defined with the help of some Borel sets in V^2.[5] Finally, the last property means that the Markov semigroup $\mathfrak{P}_t$ associated with the transition function $P_t(u_0, \Gamma)$ is stochastically continuous.

Once there is a Markov semigroup associated with the 3D Navier–Stokes equations, a natural question is whether it is ergodic. The following theorem gives a positive answer to it.

[5] The set $G = G_\Gamma \subset \Omega$ on which (6.31) holds is a full-measure event depending a priori on Γ, and we cannot rule out that the intersection of all G_Γ, with Γ varying in $\mathcal{F}$, is empty.

Theorem 6.2.8 *Under the hypotheses of Theorem 6.2.7, the Markov semigroup $\{\mathfrak{P}_t\}$ constructed there is irreducible and possesses the strong Feller property in V^2. In particular, it has a unique stationary distribution $\mu \in \mathcal{P}(V^2)$, and for any $\lambda \in \mathcal{P}(V^2)$ we have*

$$\|\mathfrak{P}_t^*\lambda - \mu\|_{\mathrm{var}} \to 0 \quad \text{as } t \to \infty, \tag{6.32}$$

where $\|\cdot\|_{\mathrm{var}}$ stands for the total variation norm in V^2. Furthermore, if $\|h\|_1$ is sufficiently small, then convergence (6.32) *is exponentially fast for any measure λ with finite second moment on H.*

The proof of the strong Feller and irreducibility properties can be found in [DO06], while the exponential convergence in the total variation norm is established in [Oda07]. In conclusion, let us note that Theorem 6.2.8 does not imply uniqueness of a stationary distribution for the 3D Navier–Stokes equations, since different Markov semigroups may have different stationary measures.

6.2.3 The Flandoli–Romito theorem

The results discussed in the foregoing subsection show that it is possible to construct martingale solutions for the Navier–Stokes equations that satisfy a weak form of the Markov property, provided that the spatially regular white noise is sufficiently rough. A different approach inspired by Krylov's paper [Kry73] (see the discussion in Section 6.2.1) was taken by Flandoli and Romito, who proved the existence of a (weak) Markov selection without any non-degeneracy assumption on the noise. To formulate their results, we shall need some auxiliary concepts.

Definition 6.2.9 Let X be a separable Banach space and let $\Omega = C(\mathbb{R}_+; X)$ be the canonical space, which is endowed with its Borel σ-algebra $\mathcal{F}$ and the natural filtration $\{\mathcal{F}_t, t \geq 0\}$. We shall say that a family of probability measures $\{\mathbb{P}_u, u \in X\}$ possesses the *a. s. Markov property* if the mapping $u \mapsto \mathbb{P}_u$ is measurable from X to the space $\mathcal{P}(\Omega)$ with the weak topology, and for any $u \in X$ there is a subset $T_u \subset (0, +\infty)$ of zero Lebesgue measure such that, with $\mathbb{P}_u$-probability 1, we have

$$\mathbb{E}_u\big(f(\omega(t)) \,|\, \mathcal{F}_r\big) = \mathbb{E}_{\omega(r)} f(\omega(t - r)) \quad \text{for } r \notin T_u \text{ and all } t \geq r, \tag{6.33}$$

where $f : X \to \mathbb{R}$ is an arbitrary bounded measurable function and $\omega(t)$ stands for the canonical process.

For the purposes of this section, we need the concept of a martingale solution for (6.22) which is slightly different from (and is stronger than) that used in the foregoing subsection. To simplify the presentation, we shall always assume that $h \equiv 0$ and the coefficients b_j satisfy (6.23). Let Ω be the Fréchet space defined by (6.27) and endowed with its natural filtration and let z be a solution of the Stokes equations that vanishes at $t = 0$ (see (2.101)).

Definition 6.2.10 A probability measure $\mathbb{P}$ on Ω is called an *admissible martingale solution* of (6.22) with an initial measure $\mu \in \mathcal{P}(H)$ if it possesses the following properties:

(i) The restriction of $\mathbb{P}$ to $t = 0$ coincides with μ and

$$\mathbb{P}\big(L^\infty_{\mathrm{loc}}(\mathbb{R}_+; H) \cap L^2_{\mathrm{loc}}(\mathbb{R}_+, V)\big) = 1.$$

(ii) The law of process (6.28) with $h \equiv 0$ coincides with that of ζ.

(iii) There is a set $T \subset (0, +\infty)$ of zero Lebesgue measure such that

$$\mathbb{P}\big\{\mathcal{E}_t(v, z) \leq \mathcal{E}_r(v, z)\big\} = 1 \quad \text{for } r \notin T \text{ and all } t \geq r, \tag{6.34}$$

where we set $v(t) = \omega(t) - z(t)$ and

$$\mathcal{E}_t(v) = \frac{1}{2}|v(t)|_2^2 + \int_0^t \big(\nu|\nabla v|_2^2 + \langle B(v + z, z), v\rangle\big)\, ds.$$

In what follows, we call (6.34) the *energy inequality* for solutions of (6.22). The main difference between admissible martingale solutions and the solutions considered in the preceding section is that the former satisfy the energy inequality. This property is crucial for proving the following weak–strong uniqueness result (see [Soh01] for the deterministic problem and [Shi07b] for the stochastic case).

Proposition 6.2.11 *Let $\mathbb{P}$ be an admissible martingale solution for* (6.22) *issued from a point $u_0 \in V$ and let $\{u(t), t \in [0, T]\}$ be a random process with continuous V-valued trajectories that satisfy* (6.22) *and coincide with u_0 at $t = 0$ with probability* 1. *Then the restriction to* $[0, T]$ *of* $\mathbb{P}$ *coincides with the law of* $\{u(t), t \in [0, T]\}$.

We now turn to the problem of a Markov selection among admissible martingale solutions of (6.22). The following result established in [FR08, theorem 4.1] is an analogue of Theorem 6.2.5 in the case of Navier–Stokes equations.

Theorem 6.2.12 *Assume that the coefficients b_s satisfy* (6.23). *Then there is a family $\{\mathbb{P}_{u_0}, u_0 \in H\}$ of measures on Ω possessing the a.s. Markov property such that, for any $u_0 \in H$, $\mathbb{P}_{u_0}$ is an admissible martingale solution for* (6.22) *with the initial measure δ_{u_0}.*

In what follows, any such family will be called a *Markov selection* for (6.22). Theorem 6.2.12 is true without any non-degeneracy hypothesis on the noise. In particular, it holds in the case $\zeta \equiv 0$. If we assume that the noise is very rough, then the Markov property will be true for any time $s \geq 0$, and the

Markov family $(\omega(t), \mathbb{P}_{u_0})$ will be ergodic. More precisely, let us assume that the coefficients b_s satisfy the inequality (cf. (6.29))

$$c|s|^{-\frac{11}{6}-\delta} \le |b_s| \le C|s|^{-\frac{11}{6}-\delta} \quad \text{for all } s \in \mathbb{Z}_0^3, \tag{6.35}$$

where $C > c > 0$ are some constants. The following theorem is established in [FR08; Rom08].

Theorem 6.2.13 *Assume that* (6.35) *holds. Then there is a* $\theta \ge 0$ *depending on* δ *such that the assertions below hold for the Markov selection* $\{\mathbb{P}_{u_0}, u_0 \in H\}$ *constructed in Theorem 6.2.12.*

Markov property: *The family* $(\omega(t), \mathbb{P}_{u_0})$ *is a Markov process. That is, relation* (6.33) *holds for all* $t \ge r \ge 0$ *and any bounded measurable function* $f : H \to \mathbb{R}$.

Strong Feller property: *The family* $(\omega(t), \mathbb{P}_{u_0})$ *is strong Feller in the topology of* $V^{1+\theta}$. *That is, the function* $u_0 \mapsto P_t(u_0, \Gamma)$ *is continuous on* $V^{1+\theta}$ *for any* $\Gamma \in \mathcal{B}(H)$.

Stationary distribution: *There exists a unique stationary distribution* μ *for* $(\omega(t), \mathbb{P}_{u_0})$.

Exponential mixing: *Convergence* (6.32) *holds exponentially fast for any initial measure* $\lambda \in \mathcal{P}(H)$ *with finite second moment.*

A striking consequence of the existence of a Markov selection is that the local existence of a strong solution on an arbitrary small interval implies global existence, provided that the selection possesses the Feller property on $V^{1+\theta}$ with $\theta > 0$ and is irreducible. In particular, the result is true under the hypotheses of Theorem 6.2.13.

Proposition 6.2.14 *Let* $\{\mathbb{P}_u, u \in H\}$ *be an arbitrary Markov selection for the 3D Navier–Stokes equations* (6.22) *which possesses the following properties for some* $\theta \ge 0$.

Feller property: *The Markov semigroup associated with* $\mathbb{P}_u$ *is Feller on* $V^{1+\theta}$. *That is,* $\mathfrak{P}_t f \in C_b(V^{1+\theta})$ *for any* $f \in C_b(V^{1+\theta})$.

Irreducibility: *For any* $t > 0$ *and* $u \in V^{1+\theta}$, *we have* $P_t(u, G) > 0$ *for all open subsets* $G \subset V^{1+\theta}$.

Let us assume that Eq. (6.22) *possesses a martingale solution* $\mathbb{P}$ *on an interval* $[0, \tau]$ *such that* $\mathbb{P}\big(C(0, \tau; V^{1+\theta})\big) = 1$. *Then,* $\mathbb{P}_u\big(C(\mathbb{R}_+; V^{1+\theta})\big) = 1$ *for any* $u \in V^{1+\theta}$. *In particular, the stochastic Navier–Stokes equations* (6.22) *are well-posed in* $V^{1+\theta}$.

Let us emphasise that no initial condition is imposed on $\mathbb{P}$. So $\mathbb{P}|_{t=0}$ may be any measure on $V^{1+\theta}$, e.g., a Dirac mass at zero.

Sketch of the proof We follow essentially the argument used by Flandoli and Romito [FR08, section 6.4]. To simplify formulas, we shall assume that $\theta = 0$. Let μ be the projection of $\mathbb{P}$ to time $t = 0$:

$$\mu(\Gamma) = \mathbb{P}\big(\{\omega \in \Omega : \omega(0) \in \Gamma\}\big), \quad \Gamma \in \mathcal{B}(H).$$

We define $\mathbb{P}_\mu = \int_H \mathbb{P}_u \mu(du)$ and note that $\mathbb{P}_\mu$ is an admissible martingale solution for (6.22) with the initial measure μ. By Proposition 6.2.11, we see that the restriction of $\mathbb{P}_\mu$ to $[0, \tau]$ coincides with $\mathbb{P}$. The Markov property of $\mathbb{P}_u$ now implies that[6]

$$\begin{aligned} 1 = \mathbb{P}_\mu\big(C(0, \tau; V)\big) &= \mathbb{E}_\mu \mathbb{P}_\mu\big\{C(0, \tau; V) \,\big|\, \mathcal{F}_\theta\big\} \\ &= \mathbb{E}_\mu\big\{\mathbb{I}_{C(0,\theta;V)} \mathbb{P}_{\omega(\theta)}\big(C(0, \tau - \theta; V)\big)\big\}, \end{aligned}$$

where we set $\theta \in (0, \tau)$ as an arbitrary point for which the Markov property holds. Since the non-negative function under the second expectation above does not exceed 1, we conclude that

$$\mathbb{P}_{\omega(\theta)}\big(C(0, \tau - \theta; V)\big) = 1 \quad \text{for } \mathbb{P}_\mu\text{-almost every } \omega \in \Omega.$$

The irreducibility property implies that there is a dense subset $A \subset V$ such that $\mathbb{P}_v\big(C(0, \tau - \theta; V)\big) = 1$ for any $v \in A$. We claim that this relation is true for any $v \in V$. If this is established, then the Markov property will imply by iteration that $\mathbb{P}_u\big(C(\mathbb{R}_+; V)\big) = 1$ for any $u \in V$. Since a solution of the Navier–Stokes system is unique in the class $C(\mathbb{R}_+; V)$, we conclude that it is well-posed in V.

We now prove $\mathbb{P}_v\big(C(0, \theta_1; V)\big) = 1$ for any $v \in V$, where $\theta_1 = \tau - \theta$. For any integer $N \ge 1$, denote by $\mathbb{Q}_N(\theta_1)$ the set of all rational numbers $2^{-N} p \in [0, \theta_1]$, where p is integer. Then

$$C(0, \theta_1; V) = \bigcap_{m=1}^{\infty} \bigcup_{k=1}^{\infty} \bigcap_{N=1}^{\infty} G(N, k, m), \tag{6.36}$$

where we set

$$G(N, k, m) = \big\{\omega \in \Omega : \|\omega(t) - \omega(s)\|_V \le \tfrac{1}{m} \text{ for } t, s \in \mathbb{Q}_N(\theta_1), |t - s| \le \tfrac{1}{k}\big\}.$$

Note that each subset $G(N, k, m)$ depends on the values of $\omega \in \Omega$ at finitely many points $t_1, \ldots, t_l \in [0, \theta_1]$, where $l \ge 1$ is an integer, and can be represented in the form

$$G(N, k, m) = \{\omega \in \Omega : (\omega(t_1), \ldots, \omega(t_l)) \in \Gamma(N, k, m)\},$$

where $\Gamma(N, k, m)$ is a closed set in the direct product of l copies of V.

[6] With a slight abuse of notation, we use the same symbol for a measure on Ω and its restriction to the interval $[0, \tau]$.

Now let $v \in V$ and let $\{v_n\} \subset A$ be a sequence converging to v in V. Then $\mathbb{P}_{v_n}\big(C(0, \theta_1; V)\big) = 1$, whence it follows that

$$\mathbb{P}_{v_n}\big(G(N, k, m)\big) \geq 1 - \varepsilon_k \quad \text{for all } m, n, N,$$

where $\varepsilon_k \to 0$ as $k \to \infty$. The Feller and Markov properties and the above-mentioned structure of $G(N, k, m)$ imply that

$$\mathbb{P}_v\big(G(N, k, m)\big) \geq \limsup_{n\to\infty} \mathbb{P}_{v_n}\big(G(N, k, m)\big) \geq 1 - \varepsilon_k \quad \text{for all } m, N.$$

Combining this with (6.36), we arrive at the required result, which completes the proof of the proposition. □

In conclusion, let us note that the irreducibility condition of Proposition 6.2.14 implies that the noise η is non-zero, and therefore it cannot be satisfied for the deterministic problem. On the other hand, the energy inequality (for admissible martingale solutions) and the Feller property make sense also for the deterministic Navier–Stokes system. The energy inequality is true for any solution obtained by a constructive procedure (such as Galerkin approximations or regularisation of the nonlinear term), while the Feller property means, roughly speaking, that any sequence of strong solutions converge to a strong solution, provided that the initial conditions converge in $V^{1+\theta}$. This is a very strong property and would immediately imply global existence of a strong solution in the deterministic case.

6.3 Navier–Stokes equations with very degenerate noise

The understanding of 3D Navier–Stokes dynamics with smooth stochastic forcing is rather poor, and only some partial results showing its mixing character are available. They are based on a detailed study of controllability properties of the Navier–Stokes system. The counterparts of these results in the 2D case are much more complete and have independent interest. In this section, we first discuss some results for 2D Navier–Stokes equations and then turn to their generalisations for the 3D case. We confine ourselves to the case of periodic boundary conditions, even though some of the results remain true for other geometries.

6.3.1 2D Navier–Stokes equations: controllability and mixing properties

Let us consider controlled[7] Navier–Stokes equations on a 2D torus:

$$\dot{u} + \nu L u + B(u) = h + \eta(t). \tag{6.37}$$

[7] The meaning of this term is clarified below.

To simplify the presentation, we shall consider only the case of a square torus $\mathbb{T}^2$; however, all the results are true for a rectangular torus. We assume that $h \in H$ is a given function and $\eta(t)$ is a *control* taking values in a finite-dimensional space $E \subset V^2$. Recall that $S_t(u_0, h + \eta)$ denotes the solution of (6.37) issued from $u_0 \in H$. We shall need the following two concepts of controllability.

Definition 6.3.1 We shall say that Eq. (6.37) is *approximately controllable at a time $T > 0$ by an E-valued control* if for any $u_0, \hat{u} \in H$ and any $\varepsilon > 0$ there is an $\eta \in C^\infty(0, T; E)$ such that

$$|S_T(u_0, h + \eta) - \hat{u}|_2 < \varepsilon. \tag{6.38}$$

We shall say that Eq. (6.37) is *solidly controllable in projections at a time $T > 0$ by an E-valued control* if for any $u_0 \in H$, any $R > 0$, and any finite-dimensional subspace $F \subset H$ there is a compact set $X \subset L^2(0, T; E)$ and a constant $\varepsilon > 0$ such that for any continuous mapping $\Phi : X \to F$ satisfying the inequality $\sup_{\xi \in X} \|\Phi(\xi) - S_T(u_0, \xi)\| \le \varepsilon$ we have

$$\mathsf{P}_F \Phi(X) \supset B_F(R), \tag{6.39}$$

where $\mathsf{P}_F : H \to H$ denotes the orthogonal projection in H onto F.

Thus, when studying the controlled Navier–Stokes system (6.37), we assume that the function η is at our disposal, and we try to make sure that at the final time T the solutions possess some particular properties.

To formulate the main result on controllability, we introduce some notation. Given a finite-dimensional subspace $G \subset V^2$, we denote by $\mathcal{E}(G)$ the largest vector space in V^2 whose elements can be represented in the form

$$\eta_1 = \eta - \sum_{j=1}^{N} B(\zeta^j),$$

where $\eta, \zeta^1, \dots, \zeta^N \in G$ and N is an integer depending on η_1. Let us define an increasing sequence $\{E_k\}$ of subspaces in V^2 by the rule

$$E_0 = E, \qquad E_k = \mathcal{E}(E_{k-1}) \quad \text{for } k \ge 1,$$

and denote by E_∞ the union of $E_k, k \ge 1$. We shall say that E is *saturating* if E_∞ is dense in H. The following result is due to Agrachev and Sarychev [AS05; AS06].

Theorem 6.3.2 *Let $h \in H$ and let $E \subset V^2$ be a saturating subspace. Then for any $T > 0$ Eq. (6.37) is approximately controllable and solidly controllable at time T by an E-valued control.*

A necessary and sufficient condition for a subspace to be saturating can be found in [AS06, theorem 4]. In particular, if E contains the basis vectors $e_{(s_1,s_2)}$ with $|s_1| \vee |s_2| \le 1$, then it is saturating.

We now fixe $T > 0$ and assume that the external force η in (6.37) is a random process on $[0, T]$ whose trajectories belong to $L^2(0, T; E)$ almost surely. In this case, the dynamics of (6.37) is well defined because we can apply the corresponding deterministic result to any realisation of η (see Theorem 2.1.13). We wish to investigate the qualitative properties of solutions for (6.37) at time T. To this end, we impose the following *decomposability* assumption on the law of η. Let λ be the law of the process η regarded as a random variable in $L^2(0, T; E)$.

Definition 6.3.3 We shall say that λ is *decomposable* if there is an orthonormal basis in $L^2(0, T; E)$ such that the measure λ can be written as the direct product of its projections to the straight lines spanned by the basis vectors.

Given a decomposable measure λ, we shall denote by λ_j its projection to the straight line spanned by the jth basis vector. Thus, we can write $\lambda = \otimes_{j=1}^{\infty} \lambda_j$. Combining the controllability properties of the Navier–Stokes system with a result on the image of probability measures under analytic mappings, Agrachev, Kuksin, Sarychev, and Shirikyan [AKSS07] established the following theorem.

Theorem 6.3.4 *Let $T > 0$, let $h \in H$, let $E \subset V^2$ be a finite-dimensional subspace, and let η be a random process whose law λ is a decomposable measure in $L^2(0, T; E)$ such that the support of λ coincides with the entire space, the second moment $\mathfrak{m}_2(\lambda)$ is finite, and the one-dimensional projections λ_j possess continuous densities with respect to the Lebesgue measure. Then for any $u_0 \in H$ the solution $u(t, u_0)$ of (6.37) issued from u_0 possesses the following properties.*

Non-degeneracy: *The support of the law of $u(T; u_0)$ coincides with H.*

Regularity: *For any finite-dimensional subspace $F \subset H$, the law of $\mathsf{P}_F u(T; u_0)$ possesses a density with respect to the Lebesgue measure and continuously depends on $u_0 \in H$ in the total variation norm.*

Note that a similar result is true when the random perturbation η is a spatially regular white noise or a kick force. In this case, the above properties can be reformulated in terms of the transition function $P_t(u, \Gamma)$. Namely, the non-degeneracy is equivalent to the relation $\operatorname{supp} P_t(u, \cdot) = H$, while the regularity means that the finite-dimensional projections of $P_t(u, \cdot)$ are absolutely continuous with respect to the Lebesgue measure and continuously depend on u in the total variation norm. These properties imply, in particular, that if μ is a stationary measure, then its support coincides with H and any finite-dimensional projection is absolutely continuous.

6.3.2 3D Navier–Stokes equations with degenerate noise

The Cauchy problem for the Navier–Stokes system is not known to be well posed in the 3D case. Nevertheless, the control problem still makes sense, and an analogue of Theorem 6.3.2 is true. We shall not give here an exact formulation, referring the reader to the papers [Shi06a; Shi07a]. Theorem 6.3.4 can be extended partially to the 3D case, and we now describe the corresponding result.

To this end, we first introduce the concept of an *admissible weak solution* for the 3D Navier–Stokes equation; cf. Definition 6.2.10. We consider Eq. (6.37) on the 3D torus, assuming that $h \in H$ is a deterministic function and η is a spatially regular white noise defined on a filtered probability space $(\Omega, \mathcal{F}, \mathcal{F}_t, \mathbb{P})$ satisfying the usual hypotheses; see Section 1.2.1.

Definition 6.3.5 We shall say that a random process $u(t)$ is an *admissible solution* for (6.37) if it can be written in the form $u(t) = z(t) + v(t)$, where z is the solution of the stochastic Stokes equation constructed in Proposition 2.4.2 and v is a random process satisfying the following properties.

(i) The process v is adapted to the filtration $\mathcal{F}_t$, and its almost every trajectory belongs to the space $L^\infty_{\rm loc}(\mathbb{R}_+; H) \cap L^2_{\rm loc}(\mathbb{R}_+; V)$ and satisfies Eq. (2.111).

(ii) The following energy inequality holds for all $t \geq 0$ with probability 1:

$$\frac{1}{2}|v(t)|_2^2 + \nu \int_0^t |\nabla v|_2^2 ds + \int_0^t \big(B(v+z, z)\big)\, ds \leq \frac{1}{2}|v(0)|_2^2 + \int_0^t (h, v)\, ds.$$

We shall say that $\tilde{u}$ is an *admissible weak solution* for (6.37) if there is a spatially regular white noise $\tilde{\eta}$ distributed as η such that $\tilde{u}$ is an admissible solution for Eq. (6.37) with η replaced by $\tilde{\eta}$.

A proof of the existence of an admissible weak solution for (6.37) can be found in [VF88; CG94; FG95]. Moreover, it is proved in [VF88; FG95] that one can construct an admissible weak solution with a time-independent law μ satisfying the condition

$$\mathfrak{m}_2(\mu) = \int_H \|u\|_V^2 \mu(du) < \infty.$$

In this case, we call μ a *stationary measure* for (6.37). Given two measures μ_1 and μ_2 on a Polish space X, we write $\mu_1 \leq \mu_2$ if $\mu_1(\Gamma) \leq \mu_2(\Gamma)$ for any $\Gamma \in \mathcal{B}(X)$. The following result is proved by Shirikyan [Shi07b].

Theorem 6.3.6 *Let $h \in H$ and let η be a spatially regular white noise of the form* (2.66). *Then there is an integer $N \geq 1$ not depending on ν such that if $b_j \neq 0$ for $1 \leq j \leq N$, then the following two assertions hold for any stationary measure μ for* (6.37) *that satisfies the inequality $\mathfrak{m}_2(\mu) \leq \mathfrak{m}$, where $\mathfrak{m} > 0$ is a constant.*

Non-degeneracy: *For any ball $Q \subset V$ there is a constant $p = p(Q, \mathfrak{m}) > 0$ such that $\mu(B) \geq p(Q, \mathfrak{m})$.*

Weak regularity: *Let $F \subset H$ be a finite-dimensional subspace and let μ_F be the projection of μ to F. Then there is a function $\rho_F \in C^\infty(F)$ depending only on $\mathfrak{m}$ such that $\mu_F \geq \rho_F \ell_F$, where ℓ_F stands for the Lebesgue measure on F.*

In conclusion, let us mention that the non-degeneracy property remains true in the case of a bounded domain, provided that the noise is non-zero in all Fourier modes. This result was established by Flandoli [Fla97].

Appendix

A.1 Monotone class theorem

Let Ω be a set and let $\mathcal{M}$ be a family of subsets of Ω. We shall say that $\mathcal{M}$ is a *monotone class* if it contains Ω and possesses the following properties:

- if $A, B \in \mathcal{M}$ and $A \cap B = \varnothing$, then $A \cup B \in \mathcal{M}$;
- if $A, B \in \mathcal{M}$ and $A \subset B$, then $B \setminus A \in \mathcal{M}$;
- if $A_i \in \mathcal{M}$ for $i = 1, 2 \ldots$, and $A_1 \subset A_2 \subset \cdots$, then $\bigcup_i A_i \in \mathcal{M}$.

It is clear that any σ-algebra is a monotone class, but not vice versa. The following well-known theorem gives a sufficient condition ensuring that the minimal monotone class containing a family of subsets coincides with the σ-algebra generated by that family. Recall that a family $\mathcal{C}$ of subsets of Ω is called a π*-system* if $A \cap B \in \mathcal{C}$ for any $A, B \in \mathcal{C}$.

Theorem A.1.1 *Let Ω be a set and let $\mathcal{C}$ be a π-system of subsets of Ω. Then the minimal monotone class containing $\mathcal{C}$ coincides with the σ-algebra generated by $\mathcal{C}$.*

Proof It suffices to show that the minimal monotone class $\mathcal{M}$ containing $\mathcal{C}$ is a σ-algebra. This will be established if we prove that the intersection of any two sets in $\mathcal{M}$ is also an element of $\mathcal{M}$.

Let us fix an arbitrary set $A \in \mathcal{C}$ and define the family

$$\mathcal{M}_A = \{B \in \mathcal{M} : A \cap B \in \mathcal{M}\}.$$

It is clear that $\mathcal{M}_A$ is a monotone class and $\mathcal{M}_A \supset \mathcal{C}$. Therefore, by the definition of $\mathcal{M}$, we must have $\mathcal{M}_A \supset \mathcal{M}$. We have thus shown that, for any $A \in \mathcal{C}$ and $B \in \mathcal{M}$, the intersection $A \cap B$ belongs to $\mathcal{M}$.

We now fix $A \in \mathcal{M}$ and consider the family $\mathcal{M}_A$. Repeating literally the above argument, we can show that $\mathcal{M}_A \supset \mathcal{M}$. This completes the proof of the lemma. □

The following two results provide important applications of the monotone class technique.

Corollary A.1.2 *Let $(\Omega, \mathcal{F})$ be a measurable space and let $\mathcal{C} \subset \mathcal{F}$ be a π-system generating the σ-algebra $\mathcal{F}$. Suppose that μ_1 and μ_2 are two probability measures* *on* $(\Omega, \mathcal{F})$ *such that*

$$\mu_1(\Gamma) = \mu_2(\Gamma) \quad \textit{for any } \Gamma \in \mathcal{C}. \tag{A.1}$$

Then $\mu_1 = \mu_2$.

Proof The σ-additivity of measures imply that if (A.1) holds, then μ_1 and μ_2 are equal on the minimal monotone class $\mathcal{M}$ containing $\mathcal{C}$. In view of Theorem A.1.1, the σ-algebra $\mathcal{F}$ coincides with $\mathcal{M}$. □

Corollary A.1.3 *Let X be a Polish space with its Borel σ-algebra $\mathcal{B}(X)$. Then there is a countable family $\{f_j, j \geq 1\}$ of bounded Lipschitz functions such that any two measures $\mu_1, \mu_2 \in \mathcal{P}(X)$ satisfying the following relation are equal:*

$$(f_j, \mu_1) = (f_j, \mu_2) \quad \textit{for any } j \geq 1. \tag{A.2}$$

Proof For any closed subset $F \subset X$, there is a sequence of Lipschitz functions $\{g_k\}$ such that $0 \leq g_k \leq 1$ and $\{g_k\}$ converges pointwise to the indicator function of F. For instance, we can take

$$g_k(u) = \frac{\operatorname{dist}_X(u, G_{1/k})}{\operatorname{dist}_X(u, G_{1/k}) + \operatorname{dist}_X(u, F)},$$

where G_ε stands for the complement of the ε-neighbourhood for F. Now note that if $(g_k, \mu_1) = (g_k, \mu_2)$ for all k, then passing to the limit as $k \to \infty$, we obtain $\mu_1(F) = \mu_2(F)$.

Let $\mathcal{C} \subset \mathcal{F}$ be a countable π-system of closed subsets that generates $\mathcal{F}$. For any $F \in \mathcal{C}$, we can construct a sequence $\{g_k\}$ that satisfies the above properties. Taking the union of all such sequences, we obtain a countable family $\{f_j\}$ such that, for any two measures $\mu_1, \mu_2 \in \mathcal{P}(X)$ satisfying (A.2), we have

$$\mu_1(F) = \mu_2(F) \quad \text{for any } F \in \mathcal{C}.$$

Since $\mathcal{C}$ is a π-system generating $\mathcal{F}$, we see that $\mu_1 = \mu_2$. □

A.2 Standard measurable spaces

Recall that two measurable spaces $(\Omega_i, \mathcal{F}_i)$, $i = 1, 2$, are said to be *isomorphic* if there is a bijective map $f : \Omega_1 \to \Omega_2$ such that $f(\Gamma) \in \mathcal{F}_2$ if and only if $\Gamma \in \mathcal{F}_1$.

Definition A.2.1 Let $(\Omega, \mathcal{F})$ be a measurable space. We shall say that $(\Omega, \mathcal{F})$ is a *standard measurable space* if it is isomorphic to one of the following Polish spaces endowed with their Borel σ-algebras:

- the finite set $\{1, \dots, N\}$ with the discrete topology;
- the set of positive integers $\mathbb{N}$ with the discrete topology;
- the closed interval $[0, 1]$ with its standard metric.

A proof of the following theorem can be found in [Dud02] (see chapter 13).

Theorem A.2.2 *Any Polish space endowed with its Borel σ-algebra is a standard measurable space.*

An important property of a standard measurable space is that its σ-algebra is *countably generated*. Namely, we have the following result.

Proposition A.2.3 *Let $(\Omega, \mathcal{F})$ be a standard measurable space. Then there is a π-system $\mathcal{C} \subset \mathcal{F}$ that is at most countable and generates the σ-algebra $\mathcal{F}$. Moreover, if Ω is a Polish space endowed with its Borel σ-algebra, then there is a π-system $\mathcal{C}$ of closed subsets that satisfies the above properties.*

Proof We shall confine ourselves to the proof of the assertion concerning the Polish spaces, because the first part of the theorem follows from it and the definition of a standard measurable space. Let $\{\omega_j\} \subset \Omega$ be a countable dense subset and let $\mathcal{C}_0$ be the countable family of all closed balls centred at ω_j with rational radii. We claim that the family $\mathcal{C}$ of all finite intersections of the elements in $\mathcal{C}_0$ satisfies the required properties. Indeed, the construction implies that $\mathcal{C}$ is an at most countable π-system. Furthermore, any open set in Ω can be represented as a countable union of the elements of $\mathcal{C}_0$. It follows that the Borel σ-algebra is the minimal one that contains $\mathcal{C}$. □

A.3 Projection theorem

Let $(\Omega, \mathcal{F})$ be a measurable space and let X be a Polish space endowed with its Borel σ-algebra $\mathcal{B}(X)$. Consider the measurable space $(\Omega \times X, \mathcal{F} \otimes \mathcal{B}(X))$, and the natural projection $\Pi_\Omega : \Omega \times X \to \Omega$. It is well known that the image under Π_Ω of a measurable subset of $\Omega \times X$ does not necessarily belong[1] to $\mathcal{F}$. Recall that a subset $A \subset \Omega$ is said to be *universally measurable* if it belongs to the completion of $\mathcal{F}$ with respect to any probability measure on $(\Omega, \mathcal{F})$. A proof of the following result can be found in [CV77] (see theorem III.23).

Theorem A.3.1 (Projection theorem) *For any set $A \in \mathcal{F} \otimes \mathcal{B}(X)$, its projection $\Pi_\Omega(A)$ is universally measurable.*

As an immediate consequence of the above theorem, we establish the following result (see chapter 2 in [Cra02]). Recall that a function $g : \Omega \to \mathbb{R}$ is said to be *universally measurable* if so is the set $\{\omega \in \Omega : g(\omega) > a\}$ for any $a \in \mathbb{R}$.

Corollary A.3.2 *Let $f : \Omega \times X \to \mathbb{R}$ be a measurable function. Then the functions*

$$\overline{f}(\omega) = \sup_{u \in X} f(\omega, u), \qquad \underline{f}(\omega) = \inf_{u \in X} f(\omega, u)$$

are universally measurable.

Proof Note that

$$\{\omega \in \Omega : \overline{f}(\omega) > a\} = \{\omega \in \Omega : f(\omega, u) > a \text{ for some } u \in X\} = \Pi_\Omega(A),$$

where we set

$$A = \{(\omega, u) \in \Omega \times X : f(\omega, u) > a\}.$$

It remains to note that the set A belongs to $\mathcal{F} \otimes \mathcal{B}(X)$, and therefore, by Theorem A.3.1, its projection is universally measurable. The universal measurability of $\underline{f}$ can be established by a similar argument. □

A.4 Gaussian random variables

Let us recall that a one-dimensional normal distribution $N(m, \sigma)$, with mean value $m \in \mathbb{R}$ and variance $\sigma^2 > 0$, is given by its density against the Lebesgue measure:

$$p_{m,\sigma}(x) = \frac{1}{\sqrt{2\pi}\,\sigma} e^{-\frac{(x-m)^2}{2\sigma^2}}, \quad x \in \mathbb{R}.$$

[1] This property is not true even in the case when Ω is the closed interval $[0, 1]$, so that the map Π_Ω is continuous.

In the case $\sigma = 0$, the normal distribution $N(m, 0)$ coincides with the Dirac measure concentrated at $x = m$. The distribution function of $N(0, \sigma)$ is denoted by Φ_σ:

$$\Phi_\sigma(x) = \frac{1}{\sigma\sqrt{2\pi}} \int_{-\infty}^{x} e^{-y^2/2\sigma^2} dy.$$

Recall that the *characteristic function* of a measure $\mu \in \mathcal{P}(\mathbb{R})$ is defined as $\varphi_\mu(t) = \int_{\mathbb{R}} e^{itx}\,\mu(dx)$. A measure μ is a normal distribution $N(m, \sigma)$ if and only if its characteristic function is given by

$$\varphi_{m,\sigma}(t) = \exp(imt - \sigma^2 t^2/2), \quad t \in \mathbb{R}. \tag{A.3}$$

Let X be a separable Banach space and let X^* be its dual space. A probability measure μ on $(X, \mathcal{B}(X))$ is said to be *Gaussian* if $\ell_*(\mu)$ has a normal distribution for any $\ell \in X^*$. A Gaussian measure μ is said to be *centred* if the mean value of $\ell_*(\mu)$ is zero for any $\ell \in X^*$. An X-valued random variable is said to be *Gaussian* if its distribution is a Gaussian probability measure. A comprehensive study of Gaussian measures on topological vector spaces can be found in [Bog98]. The following proposition establishes some simple properties that were used in the main text.

Proposition A.4.1

(i) *Let X be a separable Banach space and let $\{\mu_n\}$ be a sequence of Gaussian measures that converges weakly to $\mu \in \mathcal{P}(X)$. Then μ is a Gaussian measure.*
(ii) *A finite linear combination of independent Gaussian random variables is also Gaussian.*

Proof **(i)** It suffices to consider the one-dimensional case. Let us denote by m_n and σ_n the mean value and the variance for μ_n. Then the characteristic function of μ_n is given by $\varphi_{\mu_n}(t) = \exp(im_n t - \sigma_n^2 t^2/2)$. Furthermore, since μ_n converges weakly to μ, the corresponding characteristic functions must converge pointwise. It follows that the sequences $\{m_n\}$ and $\{\sigma_n\}$ have some limits $m \in \mathbb{R}$ and $\sigma \geq 0$, respectively, and the characteristic function of μ coincides with (A.3). We thus conclude that μ is a normal distribution on $\mathbb{R}$.

(ii) This property is a straightforward consequence of the fact that the characteristic function of the sum of independent random variables is equal to the product of the corresponding characteristic functions. □

We now consider an important example of a Gaussian random variable. Let $\beta(t)$ be a standard Brownian motion. For a constant $\gamma \in \mathbb{R}$, we define the stochastic process

$$z(t) = \int_0^t e^{\gamma(t-s)} d\beta(s), \quad t \in [0, 1].$$

It is straightforward to see that almost every trajectory of z belongs to the space $\dot{C}(0, 1)$ of real-valued continuous functions on $[0, 1]$ vanishing at zero.

Proposition A.4.2 *The process $\{z(t), 0 \leq t \leq 1\}$ regarded as a random variable with range in $\dot{C}(0, 1)$ has a centred Gaussian distribution whose support coincides with the entire space.*

Sketch of the proof To prove that z is a centred Gaussian random variable, we need to show that if $\ell : \dot{C}(0, 1) \to \mathbb{R}$ is a continuous linear functional, then $\ell(z)$ has a normal distribution with zero mean value. By the Riesz theorem (e.g., see theorem 7.4.1

in [Dud02]), there is a function of bounded variation $\varphi : [0, 1] \to \mathbb{R}$ such that

$$\ell(g) = \int_0^1 g(t)d\varphi(t) \quad \text{for any } g \in \dot{C}(0, 1).$$

It follows that

$$\ell(z) = \int_0^1 \left(\int_0^t e^{\gamma(t-s)} d\beta(s) \right) d\varphi(t).$$

We wish to change the order of integration. To this end, we use the following lemma whose proof is left to the reader as an exercise.

Lemma A.4.3 *Let β be a standard Brownian motion, let $f : [0, 1] \times [0, 1] \to \mathbb{R}$ be a continuously differentiable function, and let $\varphi : [0, 1] \to \mathbb{R}$ be a function of bounded variation. Then, with probability* 1, *we have*

$$\int_0^1 \left(\int_0^t f(t, s)d\beta(s) \right) d\varphi(t) = \int_0^1 \left(\int_s^1 f(t, s)d\varphi(t) \right) d\beta(s). \tag{A.4}$$

Relation (A.4) implies that

$$\ell(z) = \int_0^1 h(s)d\beta(s), \quad h(s) = e^{-\gamma s} \int_s^1 e^{\gamma t} d\varphi(t).$$

The mean value of a stochastic integral is always zero. Thus, the required assertion will be proved if we show that the stochastic integral of any square-integrable deterministic function h has a normal distribution.

If h is a piecewise constant function equal to c_k on $[t_{k-1}, t_k)$, $k = 1, \dots, N$, then

$$\int_0^1 h(s)d\beta(s) = \sum_{k=1}^N c_k \big(\beta(t_k) - \beta(t_{k-1})\big).$$

Assertion (ii) of Proposition A.4.1 implies that this random variable has a normal distribution. In the general case, it suffices to approximate h by piecewise constant functions and to use assertion (i) of Proposition A.4.1.

We now prove that the support of the distribution for $\{z(t), 0 \le t \le 1\}$ coincides with the space $\dot{C}(0, 1)$. To this end, it suffices to show that for any function $g \in C^1(0, 1)$ vanishing at zero and any $\varepsilon > 0$ we have

$$\mathbb{P}\Big\{ \sup_{0 \le t \le 1} |z(t) - g(t)| < \varepsilon \Big\} > 0. \tag{A.5}$$

Let $f \in C^1(0, 1)$ be a function vanishing at zero such that

$$g(t) = \int_0^t e^{\gamma(t-s)} f'(s)\, ds.$$

Integrating by parts, we can write

$$|z(t) - g(t)| \le |\beta(t) - f(t)| + |\gamma| \int_0^t e^{\gamma(t-s)} |\beta(s) - f(s)|\, ds.$$

It follows that if

$$\omega \in A_\delta := \Big\{ \sup_{0 \le t \le 1} |\beta(t) - f(t)| < \delta \Big\},$$

then $|z(t) - g(t)| \le \delta(1 + e^{|\gamma|})$ for any $t \in [0, 1]$. Hence, the required inequality (A.5) will be established if we show that $\mathbb{P}(A_\delta) > 0$ for any $\delta > 0$. This property is an immediate consequence of the fact that the support of the law for $\{\beta(t), 0 \le t \le 1\}$ is the space $\dot{C}(0, 1)$. □

Exercise A.4.4 Prove Lemma A.4.3. *Hint:* If φ is continuously differentiable, then (A.4) is obvious. In the general case, construct a sequence of C^1-functions with uniformly bounded variations that converges pointwise to φ and apply Helly's first theorem (see [KF75]).

A.5 Weak convergence of random measures

Measurable isomorphisms between Polish spaces (see Section A.2) need not preserve the topology. In particular, the properties of continuity of functions and of weak convergence of measures are not invariant under such isomorphisms. The following result shows that any Polish space is homeomorphic to a subset of a compact metric space.

Theorem A.5.1 *Let X be a Polish space. Then there is a metric d on X such that the topology defined by d coincides with the original one and (X, d) is relatively compact, that is, the completion $\overline{X}$ of (X, d) is compact. Moreover, $\mathcal{B}(X)$ is formed by intersections with X of the sets from $\mathcal{B}(\overline{X})$.*

Proof For the first assertion, see theorem 2.8.2 in [Dud02]. To prove the second, denote by $\mathcal{C}$ the intersections with X of the sets from $\mathcal{B}(\overline{X})$. This is a σ-algebra. Denote by $i : X \to \overline{X}$ the natural embedding. For any closed set $F \subset X$, we have $i^{-1}\big(\overline{i(F)}\big) = F$. That is, the σ-algebra $\mathcal{C}$ contains all closed sets. So it equals $\mathcal{B}(X)$. □

The well-known *Alexandrov theorem* says that if a sequence of measures $\{\mu_k\}$ on a Polish space X is such that (f, μ_k) converges for any bounded continuous function $f : X \to \mathbb{R}$, then the sequence $\{\mu_k\}$ weakly converges to a limit (see [Ale43] or the corollary of theorem 1 in [GS80, section VI.1])). The following theorem due to Berti, Pratelli, and Rigo [BPR06] establishes a similar result for random probability measures.

Theorem A.5.2 *Let $(\Omega, \mathcal{F}, \mathbb{P})$ be a probability space and let $\{\mu_\omega^k, \omega \in \Omega\}$ be a sequence of random probability measures on a Polish space X, endowed with the Borel σ-algebra, such that the sequence $\{(f, \mu_\omega^k)\}$ converges almost surely*[2] *for any $f \in C_b(X)$. Then there is a random probability measure $\{\mu_\omega\}$ such that*

$$\mu_\omega^k \to \mu_\omega \quad \text{as } k \to \infty \text{ for almost every } \omega \in \Omega. \tag{A.6}$$

Proof

Step 1. In view of Theorem A.5.1, there is no loss of generality in assuming that X is a subset of a compact metric space Y such that $\Gamma \cap X \in \mathcal{B}(X)$ for any $\Gamma \in \mathcal{B}(Y)$. Let us define a sequence of random measures on Y by the relation

$$\hat{\mu}_\omega^k(\Gamma) = \mu_\omega^k(\Gamma \cap X), \qquad \Gamma \in \mathcal{B}(Y).$$

Since Y is compact, for any $\omega \in \Omega$ there is a sequence $k_n \to \infty$ and a measure $\hat{\mu}_\omega \in \mathcal{P}(Y)$ such that $\hat{\mu}_\omega^k \to \hat{\mu}_\omega$. It is well known that the space $C_b(Y)$ is separable (see exercise 1.2.4 or corollary 11.2.5 in [Dud02]). Let $\{\hat{f}_j\}$ be a dense sequence in $C_b(Y)$. Then a sequence of measures $\nu_k \in \mathcal{P}(Y)$ converges weakly to $\nu \in \mathcal{P}(Y)$ if and only if $(\hat{f}_j, \nu_k) \to (\hat{f}_j, \nu)$ as $k \to \infty$ for any $j \ge 1$.

Let us denote by f_j the restriction of $\hat{f}_j$ to X, and let Ω_0 be the set of $\omega \in \Omega$ for which the sequence $\{(f_j, \mu_\omega^k)\}_{k\ge1}$ converges for any $j \ge 1$. By assumption, we have $\mathbb{P}(\Omega_0) = 1$. Since

$$(\hat{f}_j, \hat{\mu}_\omega^k) = (f_j, \mu_\omega^k) \quad \text{for any } j, k \ge 1,$$

we conclude that

$$\hat{\mu}_\omega^k \to \hat{\mu}_\omega \quad \text{as } k \to \infty \text{ for any } \omega \in \Omega_0. \tag{A.7}$$

[2] We emphasise that the set of convergence may depend on f.

Suppose we have constructed $\mathcal{C} \subset X$ and a full-measure set $\Omega_1 \subset \Omega_0$ such that $\mathcal{C} \in \mathcal{B}(Y)$ and

$$\hat{\mu}_\omega(\mathcal{C}) = 1 \quad \text{for } \omega \in \Omega_1. \tag{A.8}$$

In this case, setting

$$\mu_\omega = \begin{cases} \hat{\mu}_\omega(\cdot \cap X) & \text{for } \omega \in \Omega_1, \\ \gamma & \text{otherwise,} \end{cases}$$

where $\gamma \in \mathcal{P}(X)$ is an arbitrary measure, we see that (A.7) implies (A.6).

Step 2. To prove (A.8), note that, by the Lebesgue theorem on dominated convergence, for any $f \in C_b(X)$, the sequence $\mathbb{E}\,(f, \mu_\omega^k)$ converges. By Alexandrov's theorem, it follows that the measures $\mathbb{E}\,\mu_\cdot^k$ converge weakly in $\mathcal{P}(X)$. On the other hand, in view of (A.7), we have $\mathbb{E}\,\hat{\mu}_\cdot^k \to \mathbb{E}\,\mu_\cdot$ as $k \to \infty$.

Let us fix any constant $\varepsilon > 0$ and use Prokhorov's compactness criterion to find a compact set $\mathcal{C}_\varepsilon \subset X$ such that

$$\mathbb{E}\,\mu_\cdot^k(\mathcal{C}_\varepsilon) \geq 1 - \varepsilon \quad \text{for any } k \geq 1.$$

Since $\mathcal{C}_\varepsilon$ is closed in Y, we obtain

$$\mathbb{E}\,\hat{\mu}_\cdot(X) \geq \mathbb{E}\,\hat{\mu}_\cdot(\mathcal{C}_\varepsilon) \geq \limsup_{k\to\infty} \mathbb{E}\,\hat{\mu}_\cdot^k(\mathcal{C}_\varepsilon) = \limsup_{k\to\infty} \mathbb{E}\,\mu_\cdot^k(\mathcal{C}_\varepsilon) \geq 1 - \varepsilon.$$

Recalling that $\varepsilon > 0$ was arbitrary and defining $\mathcal{C} = \cup_n \mathcal{C}_{1/n}$, we conclude that $\mathbb{E}\,\hat{\mu}_\cdot(\mathcal{C}) = 1$, whence follows (A.8). The proof is complete. □

A.6 The Gelfand triple and Yosida approximation

Let H and V be two separable real Hilbert[3] spaces such that V is continuously and densely embedded into H. We denote by $(\cdot,\cdot)_H$ and $(\cdot,\cdot)_V$ the scalar products in H and V, respectively, and by $\|\cdot\|_H$ and $\|\cdot\|_V$ the corresponding norms. The space H can be regarded as a dense subspace of the dual space V^*. Indeed, each element $u \in H$ defines a continuous linear functional $\ell_u : V \to \mathbb{R}$ by the relation

$$\langle \ell_u, v\rangle = (u, v)_H \quad \text{for } v \in V, \tag{A.9}$$

where $\langle\cdot,\cdot\rangle$ stands for the pairing between a Hilbert space and its dual. Since the embedding $V \subset H$ is continuous, we have

$$\|\ell_u\|_{V^*} = \sup_{\|v\|_V \leq 1} \bigl|(u, v)_H\bigr| \leq C\,\|u\|_H.$$

Furthermore, if $u_1 \neq u_2$, then $\ell_{u_1} \neq \ell_{u_2}$, because V is dense in H. Finally, to show that H is dense in V^*, assume that $f : V^* \to \mathbb{R}$ is a continuous linear functional vanishing on H. Then

$$\langle f, \ell_u\rangle = \langle \ell_u, v_f\rangle = (u, v_f)_H = 0 \quad \text{for any } u \in H, \tag{A.10}$$

where $v_f \in V$ is the element corresponding to f in the natural isomorphism between V^{**} and V. Relation (A.10) implies that $v_f = 0$ and, hence, $f = 0$.

Thus, we have dense and continuous embeddings

$$V \subset H \subset V^*. \tag{A.11}$$

In what follows, any separable Hilbert spaces satisfying these conditions are called a *Gelfand triple* and H is called a *rigged Hilbert space*.

[3] The construction below also applies if V is a reflexive Banach space.

We now describe a general construction, called the Yosida approximation, that enables one to approach the elements of H and V^* by those of V in a regular way. By the Riesz representation theorem, there is a unique isometry $L : V \to V^*$ such that

$$\langle Lu, v\rangle = \langle Lv, u\rangle = (u, v)_V \quad \text{for any } u, v \in V. \tag{A.12}$$

Proposition A.6.1 *The operator $I + \varepsilon L$ taking $v \in V$ to $v + \varepsilon Lv \in V^*$ is invertible for any $\varepsilon > 0$, and its inverse is continuous. Moreover, for $X = H$, V, or V^* and $u \in X$, we have*

$$\|(I + \varepsilon L)^{-1}u\|_X \le \|u\|_X \quad \textit{for } \varepsilon > 0, \tag{A.13}$$

$$(I + \varepsilon L)^{-1}u \to u \quad \textit{in } X \textit{ as } \varepsilon \to 0^+. \tag{A.14}$$

Proof Relations (A.9) and (A.12) imply that

$$\langle (I + \varepsilon L)u, u\rangle = \|u\|_H^2 + \varepsilon\, \|u\|_V^2 \quad \text{for any } u \in V, \tag{A.15}$$

and therefore the operator $I + \varepsilon L$ is injective. To prove that it is surjective, we fix $f \in V^*$ and consider the equation $(I + \varepsilon L)u = f$ for $u \in V$. In view of (A.12), it is equivalent to

$$(u, v)_\varepsilon := (u, v)_H + \varepsilon\, (u, v)_V = \langle f, v\rangle \quad \text{for any } v \in V. \tag{A.16}$$

Endowing the space V with the new scalar product $(\cdot, \cdot)_\varepsilon$, we can use the Riesz representation theorem to find, for any $\varepsilon > 0$, a unique element $u_\varepsilon \in V$ satisfying (A.16). Thus, $I + \varepsilon L$ is a bijection of V onto V^*. The Banach inverse mapping theorem implies that $(I + \varepsilon L)^{-1} : V^* \to V$ is continuous.

We now prove (A.13) and (A.14) for $X = H$; the proof in the other two cases is similar. Let $u \in H$ and $u_\varepsilon = (I + \varepsilon L)^{-1}u$. Then $(I + \varepsilon L)u_\varepsilon = u$. Applying both sides of this relation to u_ε and using (A.15), we obtain

$$\|u_\varepsilon\|_H^2 + \varepsilon\, \|u_\varepsilon\|_V^2 = \langle u, u_\varepsilon\rangle = (u, u_\varepsilon)_H \le \|u\|_H\, \|u_\varepsilon\|_H,$$

whence (A.13) follows immediately. To prove convergence (A.14), we first note that it suffices to consider the case in which $u \in V$; the general case can be treated by a standard approximation argument combined with (A.13). Let $u \in V$ and let

$$\delta_\varepsilon := (I + \varepsilon L)^{-1}u - u = -\varepsilon (I + \varepsilon L)^{-1}Lu.$$

Setting $g = Lu$ and $g_\varepsilon = (I + \varepsilon L)^{-1}g$ and again using relation (A.15), we derive

$$\|g_\varepsilon\|_H^2 + \varepsilon\, \|g_\varepsilon\|_V^2 = \langle g, g_\varepsilon\rangle \le \|g\|_{V^*}\, \|g_\varepsilon\|_V \le \frac{1}{4\varepsilon}\|g\|_{V^*}^2 + \varepsilon\, \|g_\varepsilon\|_V^2,$$

whence it follows that

$$\|g_\varepsilon\|_H^2 \le \frac{1}{4\varepsilon}\|g\|_{V^*}^2.$$

Recalling the definition of δ_ε, we obtain

$$\|\delta_\varepsilon\|_H = \varepsilon\, \|g_\varepsilon\|_H \le \frac{\sqrt{\varepsilon}}{2}\|g\|_{V^*} \to 0 \quad \text{as } \varepsilon \to 0^+.$$

This completes the proof of the proposition. □

Exercise A.6.2 Prove (A.13) and (A.14) for $X = V$ and $X = V^*$. *Hint:* To prove (A.13), apply both sides of the relation $(I + \varepsilon L)u_\varepsilon = u$ regarded as elements of V^* to Lu_ε and $L^{-1}u_\varepsilon$. To establish (A.14), modify the argument used in the above proof.

A.7 Itô formula in Hilbert spaces

Let $(\Omega, \mathcal{F}, \mathbb{P})$ be a complete probability space with a right-continuous filtration $\{\mathcal{F}_t, t \geq 0\}$. We assume that $\mathcal{F}_t$ is augmented with respect to $(\mathcal{F}, \mathbb{P})$, that is, the σ-algebra $\mathcal{F}_t$ contains all $\mathbb{P}$-null sets of $\mathcal{F}$. Let H be a separable Hilbert space, let $\{u_t, t \geq 0\}$ be a random process in H, and let $\{\beta_j(t), j \geq 1\}$ be a sequence of independent standard Brownian motions. We assume that all these processes are defined on $(\Omega, \mathcal{F}, \mathbb{P})$ and are $\mathcal{F}_t$-adapted, and that almost all trajectories of u_t are continuous in time.

Definition A.7.1 We shall say that $\{u_t\}$ is an *Itô process in H with constant diffusion* if it can be represented in the form

$$u_t = u_0 + \int_0^t f_s ds + \sum_{j=1}^{\infty} \beta_j(t) g_j, \quad t \geq 0, \tag{A.17}$$

where the equality holds almost surely, $\{g_j\} \subset H$ is a sequence satisfying the condition

$$\sum_{j=1}^{\infty} \|g_j\|_H^2 < \infty, \tag{A.18}$$

and f_t is an $\mathcal{F}_t$-progressively measurable H-valued process such that

$$\mathbb{P}\left\{\int_0^T \|f_t\|_H dt < \infty \text{ for any } T > 0\right\} = 1. \tag{A.19}$$

Before formulating the first result, we make some comments on the above definition. Conditions (A.18) and (A.19) ensure that the right-hand side of (A.17) makes sense. Indeed, the integral is almost surely finite due to (A.19), and the series converges in $L^2(\Omega; H)$ for any fixed $t > 0$, because

$$\mathbb{E}\left\|\sum_{j=m}^{n} \beta_j(t) g_j\right\|_H^2 = \mathbb{E}\sum_{j=m}^{n} \|g_j\|_H^2 \beta_j^2(t) = t \sum_{j=m}^{n} \|g_j\|_H^2 \to 0 \quad \text{as } m, n \to \infty.$$

Moreover, the Doob–Kolmogorov inequality (A.55) implies that

$$\mathbb{P}\left\{\sup_{0\leq t\leq T} \left\|\sum_{j=m}^{\infty} \beta_j(t) g_j\right\|_H > \varepsilon\right\} \leq \varepsilon^{-2} T \sum_{j=m}^{\infty} \|g_j\|_H^2 \to 0 \quad \text{as } m \to \infty,$$

and therefore the series on the right-hand side of (A.17) converges in probability in the space $C(0, T; H)$. It defines an H-valued process whose almost all trajectories are continuous in time.

Theorem A.7.2 *Let $J = [0, T]$, let H be a separable real Hilbert space, let $F : J \times H \to \mathbb{R}$ be a twice continuously differentiable function that is bounded and uniformly continuous together with its derivatives, and let $\{u_t, t \geq 0\}$ be an Itô process in H with constant diffusion. Then*

$$F(t, u_t) = F(0, u_0) + \int_0^t A(s)\, ds + \sum_{j=1}^{\infty} \int_0^t B_j(s) d\beta_j(s), \quad t \in J, \tag{A.20}$$

where we set

$$A(t) = (\partial_t F)(t, u_t) + (\partial_u F)(t, u_t; f_t) + \frac{1}{2} \sum_{j=1}^{\infty} (\partial_u^2 F)(t, u_t; g_j), \tag{A.21}$$

$$B_j(t) = (\partial_u F)(t, u_t; g_j), \tag{A.22}$$

and $(\partial_u F)(u; v)$ and $(\partial_u^2 F)(u; v)$ denote the values of the first and second derivatives of F with respect to u at a point $v \in H$.

This result can be established by literal repetition of the derivation of Itô's formula in the finite-dimensional case. A detailed proof of a more general result can be found in [DZ92].

Exercise A.7.3 Show that all the terms on the right-hand side of (A.20) are almost surely continuous functions of time.

As in the case of finite-dimensional Itô processes, relation (A.20) remains valid for a more general class of functionals. Namely, let us assume that $F(t, u)$ is twice continuously differentiable on $\mathbb{R}_+ \times H$, and F is uniformly continuous, together with its derivatives, on any bounded subset of $\mathbb{R}_+ \times H$. In this case, relation (A.20) holds in the following sense. Let us introduce the stopping times[4]

$$\tau_n = \inf\{t \geq 0 : \|u_t\|_H > n\}, \tag{A.23}$$

where the infimum of an empty set is equal to $+\infty$. Then

$$F(t \wedge \tau_n, u_{t\wedge\tau_n}) = F(0, u_0) + \int_0^{t\wedge\tau_n} A(s)\,ds + \sum_{j=1}^{\infty} \int_0^{t\wedge\tau_n} B_j(s)d\beta_j(s), \quad t \geq 0. \tag{A.24}$$

It is not difficult to see that all the terms on the right-hand side of (A.24) make sense. We refer the reader to [Kry02, chapter III] and [DZ92] for a justification of (A.24) in the finite- and infinite-dimensional cases, respectively.

The above version of Itô's formula is not applicable to study solutions of nonlinear stochastic PDEs. Indeed, these solutions are random processes in a function space and Itô processes in some bigger space, while very often one needs to study the behaviour of functionals defined on the former space. In the literature, there are versions of Itô's formula that apply to some nonlinear PDEs (e.g., see [Roz90]). However, they are not flexible enough to treat all the situations encountered in this book.

Below we suggest a version of Itô's formula sufficient for our purposes. To state the result, we assume that H is a rigged Hilbert space. More precisely, let $V \subset H \subset V^*$ be a Gelfand triple and let $\{u_t, t \geq 0\}$ be an H-valued random process. We assume that it satisfies the following hypothesis.

Condition A.7.4 Almost all trajectories of $\{u_t\}$ are continuous with range in H and locally square integrable with range in V. Moreover, $\{u_t\}$ is an Itô process in V^* with constant diffusion (see Definition A.7.1) such that $g_j \in H$ for any $j \geq 1$, inequality (A.18) holds, and (A.19) is replaced by the stronger condition

$$\mathbb{P}\left\{\int_0^T \|f_t\|_{V^*}^2 dt < \infty\right\} = 1. \tag{A.25}$$

Note that, due to (A.18), the diffusion term in (A.17) is a continuous H-valued function of $t \geq 0$ for almost all $\omega \in \Omega$. Combining this with the continuity of u_t, we see that the integral $\int_0^t f_s ds$ is also continuous with range in H. However, it is not absolutely continuous in general, and the function f_t is locally (square) integrable only with range in V^*.

[4] Note that, since the trajectories of u_t are continuous, the stopping times τ_n go to $+\infty$ for almost all ω.

Theorem A.7.5 *Let $F : \mathbb{R}_+ \times H \to \mathbb{R}$ be a twice continuously differentiable function that is uniformly continuous, together with its derivatives, on any bounded subset of $\mathbb{R}_+ \times H$. Assume also that F satisfies the following two conditions.*

(i) *For any $T > 0$ there is a positive continuous function $K_T(r)$, $r \ge 0$, such that*[5]

$$\big|(\partial_u F)(t, u; v)\big| \le K_T\big(\|u\|_H\big)\, \|u\|_V \|v\|_{V^*} \quad \text{for } t \in [0, T],\ u \in V,\ v \in V^*. \tag{A.26}$$

(ii) *For any sequence $\{w_k\} \subset V$ converging to $w \in V$ in the topology of V and any $t \in \mathbb{R}_+$ and $v \in V^*$, we have*

$$(\partial_u F)(t, w_k; v) \to (\partial_u F)(t, w; v) \quad \text{as } k \to \infty.$$

Then for an arbitrary random process $\{u_t, t \ge 0\}$ satisfying Condition A.7.4 relation (A.24) *holds for almost every $\omega \in \Omega$.*

Before proving this result, let us show that all the terms on the right-hand side of (A.24) make sense and for almost every $\omega \in \Omega$ are continuous in time. Indeed, approximating f_t by piecewise constant functions and using (A.26), we easily show that $(\partial_u F)(t, u_t; f_t)$ is a measurable function of (t, ω). Moreover, it follows from (A.21), (A.25), and (A.26) that

$$|A(t)| \le C_1 + K_T(R_T)\|u_t\|_V \|f_t\|_{V^*} + C_2 \sum_{j=1}^{\infty} \|g_j\|_H^2 \le C_3\big(1 + \|u_t\|_V^2 + \|f_t\|_{V^*}^2\big),$$

where $R_T = \sup\{\|u_t\|_H, 0 \le t \le T\}$ and C_i are almost surely finite random constants. Hence, the function A is locally square integrable in time for almost all $\omega \in \Omega$, and its integral is continuous in time. Furthermore, if $\|u_t\|_H \le n$, then $|B_j(t)| \le C_4(n)\|g_j\|_H$ for any $j \ge 1$, whence it follows that

$$\mathbb{E}\left|\sum_{j=1}^{\infty} \int_0^{t\wedge\tau_n} B_j(s)\, d\beta_j(s)\right|^2 \le C_4^2(n)\, t \sum_{j=1}^{\infty} \|g_j\|_H^2 < \infty.$$

Thus, for any $T > 0$, the series on the right-hand side of (A.24) converges in $L^2(\Omega)$ uniformly with respect to $t \in [0, T]$. Moreover, by the Doob–Kolmogorov inequality (A.55), for any $\varepsilon > 0$ we have

$$\mathbb{P}\left\{\sup_{0\le t\le T}\left|\sum_{j=m}^{\infty} \int_0^{t\wedge\tau_n} B_j(s)\, d\beta_j(s)\right| \ge \varepsilon\right\} \le \varepsilon^{-2} C_4^2(n)\, T \sum_{j=m}^{\infty} \|g_j\|_H^2 \to 0$$

as $m \to \infty$, whence we conclude that the series of (A.24) converges in probability in the space $C(0, T)$ for any $T > 0$, and its limit is almost surely continuous in time.

Proof of Theorem A.7.5 Both the left- and right-hand sides of (A.24) are continuous functions of time for almost all $\omega \in \Omega$. Therefore, if we show that, for any fixed $t \ge 0$, the equality in (A.24) holds almost surely, then a standard argument based on continuity will imply that (A.24) takes place *for all $t \ge 0$* outside a universal $\mathbb{P}$-null set.

Let us fix a constant $\varepsilon > 0$ and define the vectors $g_j^\varepsilon = (I + \varepsilon L)^{-1} g_j$ and the processes $u_t^\varepsilon = (I + \varepsilon L)^{-1} u_t$ and $f_t^\varepsilon = (I + \varepsilon L)^{-1} f_t$, where $L : V \to V^*$ is the operator

[5] Inequality (A.26) implies that the derivative $(\partial_u F)(t, u)$ defined initially on H admits a continuous extension to V^* for any $u \in V$.

satisfying (A.12). It follows from Proposition A.6.1 that $\{u_t^\varepsilon\}$ is an Itô process in H with constant diffusion. So, Itô's formula (A.24) holds for it:

$$F(t \wedge \tau_n^\varepsilon, u_{t\wedge\tau_n^\varepsilon}^\varepsilon) = F(0, u_0^\varepsilon) + \int_0^{t\wedge\tau_n^\varepsilon} A^\varepsilon(s)\,ds + \sum_{j=1}^{\infty} \int_0^{t\wedge\tau_n^\varepsilon} B_j^\varepsilon(s)d\beta_j(s), \quad t \geq 0, \tag{A.27}$$

where $\tau_n^\varepsilon = \inf\{t \geq 0 : \|u_t^\varepsilon\|_H > n\}$, and the processes A^ε and B_j^ε are defined by relations (A.21) and (A.22), where u_t, f_t, and g_j are replaced by u_t^ε, f_t^ε, and g_j^ε, respectively. We wish to pass to the limit in (A.27) as $\varepsilon \to 0^+$.

Step 1. Proposition A.6.1 and the continuity of trajectories of u_t imply that, for almost all $\omega \in \Omega$, we have

$$\tau_n^\varepsilon \geq \tau_n \quad \text{for } \varepsilon > 0, \qquad \tau_n^\varepsilon \to \tau_n \quad \text{as } \varepsilon \to 0^+. \tag{A.28}$$

Since F is a continuous function of its arguments, we conclude that for any $s \geq 0$ and almost all ω, we have

$$F(s \wedge \tau_n^\varepsilon, u_{s\wedge\tau_n^\varepsilon}^\varepsilon) \to F(s \wedge \tau_n, u_{s\wedge\tau_n}) \quad \text{as } \varepsilon \to 0^+. \tag{A.29}$$

It follows that we can pass to the limit on the left-hand side and the first term of the right-hand side of (A.27).

Step 2. We now study the first integral in (A.27). In view of Proposition A.6.1, for almost all ω, we have

$$\sup_{0\leq t\leq T} \big|(\partial_t F)(t, u_t^\varepsilon) - (\partial_t F)(t, u_t)\big| \to 0 \quad \text{as } \varepsilon \to 0^+, \tag{A.30}$$

$$\sup_{0\leq t\leq T} \big|(\partial_u^2 F)(t, u_t^\varepsilon; g_j^\varepsilon) - (\partial_u^2 F)(t, u_t; g_j)\big| \leq C_5(\varepsilon)\, \|g_j\|_H^2 + C_6\|g_j^\varepsilon - g_j\|_H^2, \tag{A.31}$$

where $C_5(\varepsilon)$ is a random constant going to zero as $\varepsilon \to 0^+$. Proposition A.6.1, inequality (A.18), and Lebesgue's theorem on dominated convergence imply that

$$\sum_{j=1}^{\infty} \|g_j^\varepsilon - g_j\|_H^2 \to 0 \quad \text{as } \varepsilon \to 0^+. \tag{A.32}$$

Combining this with (A.31) and (A.28), we see that

$$\sum_{j=1}^{\infty}\left| \int_0^{t\wedge\tau_n^\varepsilon} (\partial_u^2 F)(s, u_s^\varepsilon; g_j^\varepsilon)\,ds - \int_0^{t\wedge\tau_n} (\partial_u^2 F)(s, u_s; g_j)\,ds\right| \to 0 \quad \text{a. s. as } \varepsilon \to 0^+.$$

Furthermore, it follows from inequality (A.26) that

$$\begin{aligned}&\big|(\partial_u F)(t, u_t^\varepsilon; f_t^\varepsilon) - (\partial_u F)(t, u_t; f_t)\big| \\ &\qquad\leq C_7\|u_t^\varepsilon\|_V \|f_t^\varepsilon - f_t\|_{V^*} + \big|(\partial_u F)(t, u_t^\varepsilon; f_t) - (\partial_u F)(t, u_t; f_t)\big|.\end{aligned}$$

In view of (A.25) and Proposition A.6.1, for any $T > 0$, we have

$$\int_0^T \|u_s^\varepsilon\|_V \|f_s^\varepsilon - f_s\|_{V^*} ds \leq \left(\int_0^T \|u_s^\varepsilon\|_V^2 ds\right)^{1/2} \left(\int_0^T \|f_s^\varepsilon - f_s\|_{V^*}^2 ds\right)^{1/2} \to 0$$

as $\varepsilon \to 0^+$. Using the Lebesgue theorem on dominated convergence and property (ii) of the function F, we derive

$$\int_0^{t\wedge\tau_n} \big|(\partial_u F)(t, u_t^\varepsilon; f_t) - (\partial_u F)(t, u_t; f_t)\big|\,ds \to 0 \quad \text{as } \varepsilon \to 0^+.$$

Finally, it follows from (A.28) and Proposition A.6.1 that

$$\int_{t\wedge\tau_n}^{t\wedge\tau_n^\varepsilon} \left|(\partial_u F)(s,u_s^\varepsilon;f_s^\varepsilon)\right| ds \le C_8(n)\int_{t\wedge\tau_n}^{t\wedge\tau_n^\varepsilon} \|u_s\|_V\|f_s\|_{V^*} ds \to 0 \quad \text{as } \varepsilon\to 0^+.$$

Combining all that has been said above, we obtain

$$\int_0^{t\wedge\tau_n^\varepsilon} A^\varepsilon(s)\,ds \to \int_0^{t\wedge\tau_n} A(s)\,ds \quad \text{as } \varepsilon\to 0^+ \text{ for almost all } \omega\in\Omega. \tag{A.33}$$

Step 3. It remains to study the sum on the right-hand side of (A.27). By Itô's isometry, we have

$$\begin{aligned}\mathbb{E}\left|\int_{t\wedge\tau_n}^{t\wedge\tau_n^\varepsilon} B_j^\varepsilon(s)d\beta_j(s)\right|^2 &\le \int_0^t \mathbb{E}\,\mathbb{I}_{[\tau_n,\tau_n^\varepsilon]}(s)\left|B_j^\varepsilon(s)\right|^2 ds\\ &\le C_9(n)\|g_j^\varepsilon\|_H^2\,\mathbb{E}\,|t\wedge\tau_n^\varepsilon - t\wedge\tau_n|.\end{aligned}$$

Combining this with (A.28) and inequality (A.13) with $X=H$, and using the Lebesgue theorem on dominate convergence, we obtain

$$\mathbb{E}\left|\sum_{j=1}^\infty\int_{t\wedge\tau_n}^{t\wedge\tau_n^\varepsilon} B_j^\varepsilon(s)d\beta_j(s)\right|^2 \le C_9(n)\left(\mathbb{E}\,|t\wedge\tau_n^\varepsilon - t\wedge\tau_n|\right)\sum_{j=1}^\infty\|g_j\|_H^2 \to 0 \tag{A.34}$$

as $\varepsilon\to 0^+$. Furthermore, since $\partial_u F$ is uniformly continuous on bounded subsets, we have

$$\begin{aligned}\sup_{0\le s\le t\wedge\tau_n}|B_j^\varepsilon(s)-B_j(s)| &\le \|g_j\|_H \sup_{0\le s\le t\wedge\tau_n}\|(\partial_u F)(s,u_s^\varepsilon)-(\partial_u F)(s,u_s)\|_{\mathcal{L}(H,\mathbb{R})}\\ &\quad + \|g_j^\varepsilon - g_j\|_H \sup_{0\le s\le t\wedge\tau_n}\|(\partial_u F)(s,u_s)\|_{\mathcal{L}(H,\mathbb{R})}\\ &= C_{10}(n,\varepsilon)\|g_j\|_H + C_{11}(n)\|g_j^\varepsilon - g_j\|_H,\end{aligned}$$

where $C_{10}(n,\varepsilon)$ is a random constant going to zero as $\varepsilon\to 0^+$. It follows that

$$\begin{aligned}\mathbb{E}\left|\sum_{j=1}^\infty\int_0^{t\wedge\tau_n}\left(B_j^\varepsilon(s)-B_j(s)\right)d\beta_j\right|^2 &= \sum_{j=1}^\infty\int_0^t \mathbb{E}\,\mathbb{I}_{[0,t\wedge\tau_n]}(s)\left|B_j^\varepsilon(s)-B_j(s)\right|^2 ds\\ &\le 2C_{10}^2(n,\varepsilon)\sum_{j=1}^\infty\|g_j\|_H^2 + 2C_{11}^2(n)\sum_{j=1}^\infty\|g_j^\varepsilon - g_j\|_H^2.\end{aligned}$$

In view of (A.32), the right-hand side of this inequality goes to zero as $\varepsilon\to 0^+$. Combining this with (A.34), we find a sequence $\varepsilon_k\to 0^+$ such that, for almost all $\omega\in\Omega$,

$$\sum_{j=1}^\infty\int_0^{t\wedge\tau_n^\varepsilon} B_j^\varepsilon(s)d\beta_j(s) \to \sum_{j=1}^\infty\int_0^{t\wedge\tau_n} B_j(s)d\beta_j(s) \quad \text{as } \varepsilon=\varepsilon_k\to 0^+. \tag{A.35}$$

Comparing (A.29), (A.33), and (A.35) and passing to the limit as $\varepsilon=\varepsilon_k\to 0^+$ in relation (A.27), we obtain (A.24). The proof of Theorem A.7.5 is complete. □

Corollary A.7.6 *Suppose that the conditions of Theorem A.7.5 are satisfied, and*

$$\sum_{j=1}^\infty\mathbb{E}\int_0^t |B_j(s)|^2 ds < \infty \quad \textit{for any } t\ge 0. \tag{A.36}$$

Then the sum of stochastic integrals

$$M_t = \sum_{j=1}^{\infty} \int_0^t B_j(s)\, d\beta_j(s), \quad t \geq 0,$$

defines a square-integrable martingale with almost surely continuous trajectories, and with probability 1, *we have*

$$F(t, u_t) = F(u_0) + \int_0^t A(s)\, ds + M_t, \quad t \geq 0. \tag{A.37}$$

Proof The fact that (A.36) is a square-integrable martingale with continuous trajectories is a standard assertion in the theory of Itô's integral; e.g., see chapter 3 in [Øks03]. When proving (A.37), as in the case of Theorem A.7.5, it suffices to show that the equality holds almost surely for any fixed $t \geq 0$. We wish to pass to the limit in (A.24) as $n \to \infty$. Since $\tau_n \to \infty$ as $n \to \infty$ and u_t is continuous in time with range in H, for almost every $\omega \in \Omega$ we have

$$F(t \wedge \tau_n, u_{t\wedge\tau_n}) \to F(t, u_t) \quad \text{as } n \to \infty. \tag{A.38}$$

Furthermore, it follows from (A.18) and (A.21) that $A(\cdot)$ is a continuous function of time, and therefore,

$$\int_0^{t\wedge\tau_n} A(s)\, ds \to \int_0^t A(s)\, ds \quad \text{with probability 1 as } n \to \infty. \tag{A.39}$$

Finally, condition (A.36) and Itô's isometry (see [Øks03]) imply that

$$\begin{aligned} \varepsilon_n &= \mathbb{E}\left|\sum_{j=1}^{\infty} \int_0^{t\wedge\tau_n} B_j(s)\, d\beta_j(s) - \sum_{j=1}^{\infty} \int_0^t B_j(s)\, d\beta_j(s)\right|^2 \\ &= \mathbb{E} \int_0^t \sum_{j=1}^{\infty} \left|\mathbb{I}_{[0,t\wedge\tau_n]}(s) - 1\right|^2 |B_j(s)|^2\, ds. \end{aligned}$$

Applying the Lebesgue theorem on dominated convergence, we see that $\varepsilon_n \to 0$ as $n \to \infty$. Combining this with (A.38) and (A.39), we arrive at (A.37). □

A.8 Local time for continuous Itô processes

In this section, we discuss distributions of convex functionals evaluated on one-dimensional Itô processes. Recall that a function $f : \mathbb{R} \to \mathbb{R}$ is said to be *convex* if

$$f(\alpha x + (1-\alpha)y) \leq \alpha f(x) + (1-\alpha) f(y) \quad \text{for any } x, y \in \mathbb{R}, \alpha \in [0, 1].$$

It is well known that a convex function has left- and right-hand derivatives $\partial^- f$ and $\partial^+ f$ at any point $x \in \mathbb{R}$. Furthermore, theses derivatives are non-decreasing functions, they satisfy the inequality $\partial^- f(x) \leq \partial^+ f(x)$ for any $x \in \mathbb{R}$, and the equality breaks on a set which is at most countable. Finally, the second derivative $\partial^2 f(x)$ in the sense of distributions is a positive measure, which satisfies the relation

$$\mu_f\big([a, b)\big) = \partial^- f(b) - \partial^+ f(a) \quad \text{for all } a < b.$$

The following result is a straightforward consequence of theorem 7.1 in [KS91, chapter 3].

Theorem A.8.1 *Let* $(\Omega, \mathcal{F}, \mathbb{P})$ *be a probability space with filtration* $\mathcal{F}_t$, $t \geq 0$, *let* $\{\beta_j\}$ *be a sequence of independent Brownian motions with respect to* $\mathcal{F}_t$, *and let* y_t *be a scalar Itô process of the form*

$$y_t = y_0 + \int_0^t x_s ds + \sum_{j=1}^{\infty} \int_0^t \theta_s^j\, d\beta_j(s), \tag{A.40}$$

where x_t and θ_t^j are $\mathcal{F}_t$-adapted processes such that

$$\mathbb{E}\int_0^t \left(|x_s| + \sum_{j=1}^{\infty}|\theta_s^j|^2\right) ds < \infty \quad \textit{for any } t > 0. \tag{A.41}$$

Then there is a random field $\Lambda_t(a,\omega)$, $t \geq 0$, $a \in \mathbb{R}$, $\omega \in \Omega$, such that the following properties hold.

(i) *The mapping $(t, a, \omega) \mapsto \Lambda_t(a,\omega)$ is measurable, and for any $a \in \mathbb{R}$ the process $t \mapsto \Lambda_t(a,\omega)$ is $\mathcal{F}_t$-adapted, continuous, and non-decreasing. Moreover, for every $t \geq 0$ and almost every $\omega \in \Omega$ the function $a \mapsto \Lambda_t(a,\omega)$ is right-continuous.*

(ii) *For any non-negative Borel-measurable function $g : \mathbb{R} \to \mathbb{R}$, with probability* 1 *we have*

$$\int_0^t g(y_s)\left(\sum_{j=1}^{\infty}|\theta_s^j|^2\right) ds = 2\int_{-\infty}^{\infty} g(a)\Lambda_t(a,\omega)da, \quad t \geq 0. \tag{A.42}$$

(iii) *For any convex function $f : \mathbb{R} \to \mathbb{R}$, with probability* 1 *we have*

$$\begin{aligned} f(y_t) = f(y_0) &+ \sum_{j=1}^{\infty}\int_0^t \partial^- f(y_s)\theta_s^j d\beta_j + \int_0^t \partial^- f(y_s)x_s ds \\ &+ \int_{-\infty}^{\infty}\Lambda_t(a,\omega)\partial^2 f(da), \quad t \geq 0. \end{aligned} \tag{A.43}$$

The random field $\Lambda_t(a,\omega)$ constructed in the above theorem is called a *local time* for y_t, and relation (A.43) is usually referred to as the *change of variable formula*. This is an analogue of Itô's formula (A.20) for one-dimensional processes and non-smooth convex functionals. Let us comment on properties (ii) and (iii). Taking $g(x) = \mathbb{I}_{[\alpha,\beta]}(x)$ in (A.42), we obtain

$$\int_0^t \mathbb{I}_{[\alpha,\beta]}(y_s)\left(\sum_{j=1}^{\infty}|\theta_s^j|^2\right) ds = 2\int_\alpha^\beta \Lambda_t(a,\omega)\,da. \tag{A.44}$$

Assuming that the quadratic variation of the diffusion term is bounded from above and separated from zero, we see that

$$\ell\big(\{s \in [0,t] : \alpha \leq y_s \leq \beta\}\big) \sim \int_\alpha^\beta \Lambda_t(a,\omega)\,da.$$

Thus, one can say roughly that $\Lambda_t(a,\omega)$ measures the fraction of time spent by the process y_s, $0 \leq s \leq t$, in the vicinity of a.

Furthermore, relation (A.43) enables one to find an explicit formula for $\Lambda_t(a,\omega)$. Indeed, taking $f(x) = (x-a)_+ = \max(0, x-a)$, we obtain

$$(y_t - a)_+ = (y_0 - a)_+ + \sum_{j=1}^{\infty}\int_0^t \mathbb{I}_{[a,+\infty)}(y_s)\theta_s^j d\beta_j + \int_0^t \mathbb{I}_{[a,+\infty)}(y_s)x_s ds + \Lambda_t(a,\omega). \tag{A.45}$$

A.9 Krylov's estimate

There is no analogue of the concept of local time for multidimensional Itô processes, so a multidimensional version of formula (A.44) is not available. In many cases, however, it is possible to get rather sharp estimates for a process' law, using Itô's formula and martingales inequalities. This type of result was obtained by Krylov; see [Kry74; Kry86] and the book [Kry80]. In Chapter 5, we use a simplified version of such an estimate. Namely, consider a d-dimensional Itô process written in the form (A.40), where x_t and θ_t^j are adapted d-dimensional processes satisfying condition (A.41), and β_j are independent Brownian motions. We shall denote by θ_t^{jk}, $k = 1\dots, d$, the components of the vector function θ_t^j and by σ_t the diffusion, which is a $d \times d$ symmetric matrix with entries

$$\sigma_t^{kl} = \sum_{j=1}^{\infty} \theta_t^{jk}\theta_t^{jl}, \quad k, l = 1, \dots, d.$$

Note that σ_t is non-negative for any $t \geq 0$. To simplify the presentation, we shall assume that y_t is a stationary process; this is the situation considered in Chapter 5.

Theorem A.9.1 *Under the above hypotheses, there is a constant $C_d > 0$ depending only on the dimension d such that, for any bounded measurable function $f : \mathbb{R}^d \to \mathbb{R}$, we have*

$$\mathbb{E}\int_0^1 \big(\det\sigma_t\big)^{1/d} f(x_t)\,dt \leq C_d |f|_d\,\mathbb{E}\int_0^1 |x_s|\,ds, \tag{A.46}$$

where $|f|_d$ stands for the L^d-norm of f.

Proof A standard argument based on the monotone class theorem and Fatou's lemma shows that it suffices to establish inequality (A.46) for smooth functions $f \geq 0$ with compact support. The general idea of the proof is rather simple: we first apply Itô's formula to the stationary semimartingale $\psi(y_t)$ and derive an identity for it and then, choosing a particular function ψ, we obtain an expression minorised by the left-hand side of (A.46).

Let $\psi : \mathbb{R}^d \to \mathbb{R}$ be a twice continuously differentiable function with bounded derivatives. By Itô's formula, we have

$$\psi(y_t) = \psi(y_0) + \int_0^t \bigg((\nabla\psi(y_s), dy_s) + \frac{1}{2}\sum_{k,l=1}^{d} \sigma_s^{kl}(\partial_k\partial_l\psi)(y_s)\,ds\bigg).$$

Taking the mean value, using the stationarity of y_t, and setting $t = 1$, we get

$$2\,\mathbb{E}\int_0^1 (\nabla\psi(y_s), x_s)\,ds + \mathbb{E}\int_0^1 \sum_{k,l=1}^{d} \sigma_s^{kl}(\partial_k\partial_l\psi)(y_s)\,ds = 0. \tag{A.47}$$

We now need the following proposition, whose proof[6] can be found in [Kry87] (see theorem III.2.3).

Proposition A.9.2 *For any $\lambda > 0$ and any non-negative function $f \in C_0^\infty(\mathbb{R}^d)$ there is a function $\psi \in C^2(\mathbb{R}^d)$ with bounded derivatives such that*

$$\sum_{k,l=1}^{d} a_{kl}(\partial_k\partial_l\psi)(x) - \lambda(\operatorname{Tr} a)\psi(x) + \big(\det a\big)^{1/d} f(x) \leq 0, \tag{A.48}$$

$$|\nabla\psi(x)| \leq \lambda^{1/2}\psi(x), \quad \psi(x) \leq A_d\lambda^{-1/2}|f|_1, \tag{A.49}$$

where $a = (a_{kl})$ is an arbitrary non-negative symmetric $d \times d$ matrix, A_d is a constant depending only on d, and the inequalities hold for all $x \in \mathbb{R}^d$.

[6] Note that Proposition A.9.2 is obvious for $d = 1$. In this case, it suffices to take for ψ the unique bounded solution of the equation $\psi'' - \psi + f = 0$.

Let us take for ψ in (A.47) the function constructed in Proposition A.9.2. Recalling that σ_t is a non-negative matrix for any $t \geq 0$ and taking into account (A.48), we derive

$$\mathbb{E}\int_0^1 \big(\det\sigma_t\big)^{1/d} f(y_s)\,ds \leq 2\,\mathbb{E}\int_0^1 (\nabla\psi(y_s), x_s)\,ds + \lambda\,\mathbb{E}\int_0^1 \psi(y_s)\big(\operatorname{Tr}\sigma_s\big)\,ds. \tag{A.50}$$

Using inequalities (A.49), we see that the right-hand side of (A.50) can be estimated from above by

$$C_1|f|_1\mathbb{E}\int_0^1 |x_s|\,ds + C_2\sqrt{\lambda}|f|_1\mathbb{E}\int_0^1 \operatorname{Tr}\sigma_s ds,$$

where C_1 and C_2 are some constants depending only on d. Substituting this expression into (A.50) and passing to the limit as $\lambda \to 0^+$, we arrive at the required inequality (A.46). □

A.10 Girsanov's theorem

In this section, we deal with processes defined on a complete probability space $(\Omega, \mathcal{F}, \mathbb{P})$ with a right-continuous filtration $\{\mathcal{F}_t, t \geq 0\}$. The following result is a simple consequence of Girsanov's theorem (see [Øks03]).

Theorem A.10.1 *Let y be an $\mathbb{R}^N$-valued process of the form*

$$y(t) = \sum_{j=1}^N b_j\beta_j(t)e_j, \quad t \geq 0,$$

where $\{e_j\}$ is an orthonormal basis in $\mathbb{R}^N$, β_j are independent standard Brownian motions, and $b_j > 0$ are some constants. Let $\tilde{y}$ be another process given by

$$\tilde{y}(t) = y(t) + \int_0^t a(s)\,ds,$$

where $a = (a_1, \dots, a_N)$ is a progressively measurable process such that

$$\mathbb{E}\exp\Big(C\int_0^\infty |a(t)|^2 dt\Big) < \infty \quad \textit{for any } C > 0. \tag{A.51}$$

Then the distributions of the processes y and $\tilde{y}$ regarded as random variables in $C(\mathbb{R}_+, \mathbb{R}^N)$ satisfy the inequality

$$\|\mathcal{D}(y) - \mathcal{D}(\tilde{y})\|_{\text{var}} \leq \frac{1}{2}\bigg(\Big(\mathbb{E}\exp\Big\{6\int_0^\infty \sum_{j=1}^N b_j^{-2}|a_j(t)|^2 dt\Big\}\Big)^{1/2} - 1\bigg)^{1/2}. \tag{A.52}$$

Proof Let us consider an exponential supermartingale M_t corresponding to the stochastic integral $\xi_t = -\int_0^t \sum_{j=1}^N b_j^{-1}a_j(t)d\beta_j(t)$. That is,

$$M_t(\omega) = \exp\Big(-\int_0^t \sum_{j=1}^N b_j^{-1}a_j(t)d\beta_j(t) - \frac{1}{2}\int_0^t \sum_{j=1}^N b_j^{-2}|a_j(t)|^2 dt\Big), \quad t \geq 0.$$

By Girsanov's theorem (see theorem 8.6.5 in [Øks03]), condition (A.51) implies that this is a martingale. So $\mathbb{E}\,M_t = \mathbb{E}\,M_0 = 1$ for each $t \geq 0$. Since $M_t \geq 0$, by the *convergence theorem* of Section A.11, there exists a limit $\lim_{t\to\infty} M_t = M$, and $\mathbb{E}\,M = 1$. Let us consider the probability measure $\mathbb{Q} = M(\omega)\mathbb{P}$ on the space $(\Omega, \mathcal{F})$. Due to Girsanov's theorem (see [Øks03]), the $\mathbb{P}$-law of y coincides with the $\mathbb{Q}$-law of $\tilde{y}$. That is, $\mathcal{D}(y) = y_*(\mathbb{P}) = \tilde{y}_*(\mathbb{Q})$ and $\mathcal{D}(\tilde{y}) = \tilde{y}_*(\mathbb{P})$. By Exercise 1.2.13(i), the left-hand side of (A.52) can be estimated by $\|\mathbb{P} - \mathbb{Q}\|_{\text{var}}$. Proposition 1.2.7 implies that the latter quantity does not exceed

$$\frac{1}{2}\mathbb{E}\,|M-1| \le \frac{1}{2}\big(\mathbb{E}M^2 - 1\big)^{1/2}. \tag{A.53}$$

To estimate the right-hand side, let us note that

$$\mathbb{E}M_t^2 = \mathbb{E}\,e^{2\xi_t - \langle\xi\rangle_t} \le \big(\mathbb{E}\,e^{4\xi_t - 8\langle\xi\rangle_t}\big)^{1/2}\big(\mathbb{E}\,e^{6\langle\xi\rangle_t}\big)^{1/2} \le \big(\mathbb{E}\,e^{6\langle\xi\rangle_t}\big)^{1/2}$$

for any $t \ge 0$, where we used that $\exp(4\xi_t - 8\langle\xi\rangle_t)$ is the exponential supermartingale corresponding to ξ_t. By the above inequality and the monotone convergence theorem, we have $\mathbb{E}\,M^2 \le (\mathbb{E}\,e^{6\langle\xi\rangle_\infty})^{1/2}$. Substituting this inequality into (A.53), we arrive at the required result. □

A.11 Martingales, submartingales, and supermartingales

The results of this section are valid for both discrete- and continuous-time processes. To simplify the presentation, we mostly restrict ourselves to the case of continuous time. We assume that the corresponding processes are continuous in t for a.e. ω since in the main text we do not use discontinuous processes.

Let $(\Omega, \mathcal{F}, \mathcal{F}_t, \mathbb{P})$ be a filtered probability space satisfying the usual hypotheses. Recall that a (continuous) random process $\{\xi_t, t \ge 0\}$ valued in a Hilbert space H is called a *martingale* with respect to $\{\mathcal{F}_t\}$ if it is adapted to $\mathcal{F}_t$, each random variable ξ_t is integrable, and for $t \ge s \ge 0$ we have $\mathbb{E}\,(\xi_t \mid \mathcal{F}_s) = \xi_s$ with probability 1. A martingale $\{\xi_t\}$ is said to be *square integrable* if $\mathbb{E}\|\xi_t\|^2 < \infty$ for any $t \ge 0$. A *quadratic variation* of an H-valued square-integrable martingale $\{\xi_t\}$ is a real-valued random process $\langle\xi\rangle_t$ adapted to $\{\mathcal{F}_t\}$ such that $\langle\xi\rangle_0 = 0$ almost surely, almost every trajectory of $\langle\xi\rangle_t$ is non-decreasing, and the process $\|\xi_t\|_H^2 - \langle\xi\rangle_t$ is a real-valued martingale.

Important examples of continuous martingales are given by stochastic integrals. Namely, consider a Hilbert space K with an orthonormal basis $\{e_j\}$, standard independent $\mathcal{F}_t$-adapted Brownian motions $\{\beta_j(t)\}$, and an $\mathcal{F}_t$-adapted process B_t valued in $\mathcal{L}(K, H)$ such that

$$\mathbb{E}\int_0^T \|B_t\|_{\text{HS}}^2 dt < \infty \quad \text{for any } T > 0.$$

Here $\|C\|_{\text{HS}}$ stands for the Hilbert–Schmidt norm of an operator $C \in \mathcal{L}(K, H)$ (that is, $\|C\|_{\text{HS}}^2 = \sum_j \|Ce_j\|^2$). Then the stochastic integral

$$\xi_t = \int_0^t \sum_{j=1}^{\infty} B_t e_j d\beta_j(t) \tag{A.54}$$

is a continuous square-integrable martingale in H; see section 5.2 in [DZ96]. Moreover, its quadratic variation is given by

$$\langle\xi\rangle_t = \int_0^t \|B_s\|_{\text{HS}}^2 ds.$$

Example A.11.1 The H-valued process $\zeta(t)$ defined in (2.98) can be written in the form (A.54), where $K = H$ and B is an operator in H such that $Be_j = b_j e_j$ for any $j \geq 1$. In this case, $\|B\|_{\mathrm{HS}}^2 = \mathfrak{B}$ and $\langle \zeta \rangle_t = \mathfrak{B}t$.

A continuous random process $\{M_t, t \geq 0\}$ in $\mathbb{R}$ is called a *submartingale* (or *supermartingale*) with respect to $\{\mathcal{F}_t\}$ if it is integrable and adapted to $\mathcal{F}_t$, and for all $t \geq s \geq 0$ we have

$$\mathbb{E}\,(M_t \mid \mathcal{F}_s) \geq M_s \quad (\text{or } \mathbb{E}\,(M_t \mid \mathcal{F}_s) \leq M_s).$$

If ξ_t is a martingale in H and $f : H \to \mathbb{R}$ is a continuous convex function, then $f(\xi_t)$ is a submartingale, while $-f(\xi_t)$ is a supermartingale; see [DZ92].

We now formulate without proof several results on martingales, submartingales, and supermartingales.

Doob–Kolmogorov inequality

Let $\{M_t, 0 \leq t \leq T\}$ be a non-negative submartingale. Then

$$\mathbb{P}\Big\{\sup_{0 \leq t \leq T} M_t > c\Big\} \leq \frac{1}{c}\,\mathbb{E}\,M_T. \tag{A.55}$$

In particular, if ξ_t is an H-valued martingale, then (A.55) is true with $M_t = \|\xi_t\|_H^p$ for any $p \geq 1$.

Doob's moment inequality

Let $\{M_t, t \geq 0\}$ be a non-negative submartingale. Then, for any $p \in (1, \infty)$, we have

$$\mathbb{E}\Big(\sup_{t \geq 0} M_t^p\Big) \leq \Big(\frac{p}{p-1}\Big)^p \lim_{t \to \infty} \mathbb{E}\,M_t^p. \tag{A.56}$$

A similar inequality is true on any finite interval $[0, T]$, in which case the supremum should be taken over $t \in [0, T]$ and the limit of $\mathbb{E}\,M_t^p$ should be replaced by $\mathbb{E}\,M_T^p$.

Exponential supermartingale

Let $\{\xi_t, t \geq 0\}$ be a martingale whose quadratic variation $\langle \xi \rangle_t$ is a.s. finite for all $t \geq 0$ and let $\zeta_t := \xi_t - \frac{1}{2}\langle \xi \rangle_t$. Then the process $\exp(\zeta_t)$ is a supermartingale, and we have the inequality

$$\mathbb{P}\Big\{\sup_{t \geq 0} \zeta_t \geq c\Big\} \leq e^{-c} \quad \text{for any } c > 0. \tag{A.57}$$

A similar result holds on a finite interval $[0, T]$, in which case the supremum should be taken over $t \in [0, T]$.

Doob's optional sampling theorem

Let $\{\xi_t, t \geq 0\}$ be a submartingale and let $\sigma \leq \tau$ be two almost surely finite stopping times for the filtration $\{\mathcal{F}_t\}$. Then $\mathbb{E}\,\xi_\sigma = \mathbb{E}\,\xi_\tau$.

A proof of (A.55) and (A.56) can be found in section III.3 of [Kry02]. The fact that the process $\exp \zeta_t$ is a supermartingale is established in [RY99, section 4.3], and (A.57) is the classical supermartingale inequaity; see [Kry02, theorem III.6.11] or [Mey66, theorem VI.T1]. Finally, Doob's optional sampling theorem is proved in section 1.3.C of [KS91].

The assertion below is a special case of a more general theorem due to Doob. Its proof can be found in [Dud02]; see theorems 10.5.1 and 10.5.4 and corollary 10.5.2.

Convergence theorem

Let $\{M_k, k \geq 0\}$ be a real-valued martingale such that $\sup_{k\geq 0} \mathbb{E}\,|M_k| < \infty$. Then there is an integrable random variable M_∞ such that

$$M_k \to M_\infty \text{ almost surely}, \quad \mathbb{E}\,|M_k - M_\infty| \to 0 \quad \text{as } k \to \infty.$$

A.12 Limit theorems for discrete-time martingales

Let $(\Omega, \mathcal{F}, \mathbb{P})$ be a probability space, let $\{\mathcal{F}_k, k \geq 0\}$ be a discrete-time filtration on it, and let $\{M_k, k \geq 1\}$ be a martingale with respect to $\mathcal{F}_k$. The following theorem establishes a strong law of large numbers. Its proof can be found in [Fel71]; see theorem 3 in section VII.9.

Theorem A.12.1 *Let $\{M_k, k \geq 1\}$ be a zero-mean square-integrable martingale and let $\{c_k\}$ be an increasing sequence going to $+\infty$ such that*

$$\sum_{k=1}^{\infty} c_k^{-2}\,\mathbb{E}\,X_k^2 < \infty,$$

where $X_k = M_k - M_{k-1}$ and $M_0 = 0$. Then

$$\mathbb{P}\{c_k^{-1} M_k \to 0 \text{ as } k \to \infty\} = 1.$$

We now turn to the law of the iterated logarithm (LIL). Given a square-integrable martingale $\{M_k\}$, define its conditional variance by the formula

$$V_k^2 = \sum_{l=1}^{k} \mathbb{E}\left(X_l^2 | \mathcal{F}_{l-1}\right). \tag{A.58}$$

The following result on the LIL for martingales with identically distributed increments is due to Heyde [Hey73].

Theorem A.12.2 *Let a zero-mean martingale $\{M_k\}$ be such that the martingale differences X_k are identically distributed, $\mathbb{E}\,X_1^2 = \sigma^2 > 0$, and*

$$k^{-1} V_k^2 \to \sigma^2 \quad \text{in probability as } k \to \infty. \tag{A.59}$$

Then for almost every $\omega \in \Omega$ we have

$$\limsup_{k\to\infty} \frac{M_k}{\sqrt{2k \ln \ln k}} = \sigma, \quad \liminf_{k\to\infty} \frac{M_k}{\sqrt{2k \ln \ln k}} = -\sigma. \tag{A.60}$$

Finally, let us discuss the central limit theorem (CLT). We define

$$s_k^2 = \mathbb{E}\,M_k^2 = \sum_{l=1}^{k} \mathbb{E}\,X_l^2.$$

In what follows, we assume that

$$\mathbb{E}\,X_k^2 \to \sigma^2 \quad \text{as } k \to \infty, \tag{A.61}$$

$$s_k^{-2} V_k \to \sigma^2 \quad \text{in probability as } k \to \infty, \tag{A.62}$$

where $\sigma \geq 0$ is a constant, and that the martingale differences X_k satisfy *Lindeberg's condition*, that is,

$$s_n^{-2} \sum_{k=1}^{n} \mathbb{E}\left(X_k^2 \mathbb{I}_{\{|X_k|>\varepsilon s_n\}}\right) \to 0 \quad \text{as } n \to \infty, \tag{A.63}$$

where $\varepsilon > 0$ is arbitrary. Note that, if $\sigma > 0$, then $s_k^2 \sim k\sigma^2$ as $k \to \infty$, and Lindeberg's condition holds if and only if, for any $\varepsilon > 0$,

$$n^{-1} \sum_{k=1}^{n} \mathbb{E}\left(X_k^2 \mathbb{I}_{\{|X_k|>\varepsilon\sqrt{n}\}}\right) \to 0 \quad \text{as } n \to \infty. \tag{A.64}$$

This condition is not very restrictive. For instance, it holds if $\mathbb{E}\, X_k^4 \leq C < \infty$ for all $k \geq 1$. Indeed, using Schwarz's and Chebyshev's inequalities, we derive

$$\begin{aligned} \mathbb{E}\big(X_k^2 \mathbb{I}_{|X_k|>\varepsilon\sqrt{n}}\big) &\leq \big(\mathbb{E}\, X_k^4\big)^{1/2} \mathbb{P}_u\{|X_k| > \varepsilon\sqrt{n}\}^{1/2} \\ &\leq \varepsilon^{-2} n^{-1}\, \mathbb{E}\, X_k^4 \leq C\varepsilon^{-2} n^{-1}, \end{aligned}$$

whence it follows that (A.63) holds. The following theorem due to Brown [Bro71] establishes the CLT for martingales.

Theorem A.12.3 *Let $\{M_k\}$ be a zero-mean martingale such that the martingale differences X_k satisfy* (A.61), (A.62) *with some $\sigma \geq 0$ and* (A.63). *Then*

$$\mathcal{D}\big(k^{-1/2} M_k\big) \to N(0, \sigma) \quad \text{as } k \to \infty.$$

That is (*see Lemma 1.2.16*),

$$\lim_{k\to\infty} \mathbb{P}\{k^{-1/2} M_k \leq x\} = \Phi_\sigma(x) \quad \text{for any } x \in \mathbb{R}.$$

A.13 Martingale approximation for Markov processes

Let X be a separable Banach space and let $(u_t, \mathbb{P}_u)$ be a continuous-time Markov process in X defined on a measurable space $(\Omega, \mathcal{F})$ with filtration $\{\mathcal{F}_t, t \geq 0\}$. We denote by $P_t(u, \Gamma)$ the transition function for $(u_t, \mathbb{P}_u)$ and by $\mathfrak{P}_t$ and $\mathfrak{P}_t^*$ the corresponding Markov semigroups (see Section 1.3.3). Given a weight function w (see Section 3.3.1) and a constant $\gamma \in (0, 1]$, we denote by $C^\gamma(X, w)$ the space of continuous functions $f : X \to \mathbb{R}$ for which

$$|f|_{w,\gamma} = \sup_{u\in X} \frac{|f(u)|}{w(\|u\|)} + \sup_{0<\|u-v\|_X\leq 1} \frac{|f(u) - f(v)|}{\|u - v\|^\gamma \big(w(\|u\|_X) + w(\|v\|_X)\big)} < \infty.$$

This is a subspace of the space of Hölder-continuous functions on X.

Definition A.13.1 A Markov process $(u_t, \mathbb{P}_u)$ is said to be *uniformly mixing* for the class $C^\gamma(X, w)$ if it has a unique stationary distribution $\mu \in \mathcal{P}(X)$, and for any $f \in C^\gamma(X, w)$, we have[7]

$$\big|\mathfrak{P}_t f(u) - (f, \mu)\big| \leq \alpha(t)|f|_{w,\gamma} w_1(\|u\|_X), \quad t \geq 0, \tag{A.65}$$

where α is a non-increasing integrable function on $\mathbb{R}_+$, w_1 is a weight function, and both of them do not depend on f.

[7] We assume, in particular, that any function $f \in C^\gamma(X, w)$ is integrable with respect to μ.

The exercise below shows that an analogue of inequality (A.65) is true for random initial conditions.

Exercise A.13.2 Let $(u_t, \mathbb{P}_u)$ be a Markov process in X that is uniformly mixing for the class $C^\gamma(X, w)$ and let $\lambda \in \mathcal{P}(X)$ be a measure such that

$$(w_1, \lambda) := \int_X w_1(\|u\|_X)\lambda(du) < \infty, \tag{A.66}$$

where w_1 is the function from (A.65). Show that, for any $f \in C^\gamma(X, w)$, we have

$$\big|\mathbb{E}_\lambda f(u_t) - (f, \mu)\big| \le \alpha(t)|f|_{w,\gamma}(w_1, \lambda), \quad t \ge 0. \tag{A.67}$$

Hint: Integrate (A.65) with respect to $\lambda(du)$.

The following construction due to Gordin [Gor69] is very useful when studying limit theorems for stochastic processes. In the context of uniformly mixing Markov processes, it enables one to reduce the problem to the case of martingales.

Let us take any function $f \in C^\gamma(X, w)$ with $(f, \mu) = 0$ and a measure $\lambda \in \mathcal{P}(X)$ and set

$$M_t^\lambda = \int_0^\infty \big(\mathbb{E}_\lambda(f(u_s) \,|\, \mathcal{F}_t) - \mathbb{E}_\lambda(f(u_s) \,|\, \mathcal{F}_0)\big)\, ds, \quad t \ge 0, \tag{A.68}$$

where $\mathbb{E}_\lambda$ stands for the mean value with respect to the measure $\mathbb{P}_\lambda$ defined by

$$\mathbb{P}_\lambda(\Gamma) = \int_\Omega \mathbb{P}_u(\Gamma)\lambda(du), \quad \Gamma \in \mathcal{F}.$$

Proposition A.13.3 *Let $(u_t, \mathbb{P}_u)$ be a Markov process in X uniformly mixing for the class $C^\gamma(X, w)$. Assume, in addition, that*

$$\mathbb{E}_u w_1(\|u_t\|_X) \le C\big(1 + w_1(\|u\|_X)\big) \quad \textit{for any } t \ge 0,\ u \in X, \tag{A.69}$$

where $C > 0$ is independent of $t \ge 0$ and $u \in X$. Then, for any measure $\lambda \in \mathcal{P}(X)$ satisfying (A.66) *and any function $f \in C^\gamma(X, w)$ with zero mean value, the process $\{M_t^\lambda, t \ge 0\}$ is well defined and forms a zero-mean martingale with respect to the filtration $\{\mathcal{F}_t\}$ and the probability $\mathbb{P}_\lambda$. Moreover, we have, with $\mathbb{P}_\lambda$-probability* 1,

$$\int_0^t f(u_s)\, ds = M_t^\lambda - g(u_t) + g(u_0), \quad t \ge 0, \tag{A.70}$$

where $g(v) = \int_0^\infty \mathfrak{P}_s f(v)\, ds$.

Proof To simplify notation, we shall write M_t instead of M_t^λ. We first show that

$$\mathbb{E}_\lambda |M_t| \le \big(|f|, \mu\big)t + C_1\big(1 + (w_1, \lambda)\big), \tag{A.71}$$

where $C_1 > 0$ is a constant not depending on λ and $t \ge 0$. Indeed, the Markov property and inequality (A.67) imply that, with $\mathbb{P}_\lambda$-probability 1, we have

$$\big|\mathbb{E}_\lambda(f(u_s) \,|\, \mathcal{F}_t)\big| \le \big|\mathfrak{P}_{s-t} f(u_t)\big| \le \alpha(s-t)|f|_{w,\gamma} w_1(\|u_t\|_X), \quad s \ge t. \tag{A.72}$$

Integrating (A.69) with respect to $\lambda(du)$, we see that

$$\mathbb{E}_\lambda w_1(\|u_t\|_X) \le C\big(1 + (w_1, \lambda)\big) \quad \text{for any } t \ge 0. \tag{A.73}$$

Taking the mean value $\mathbb{E}_\lambda$ of inequality (A.72) and using (A.73), for $s \ge t$ we obtain

$$\mathbb{E}_\lambda \big|\mathbb{E}_\lambda(f(u_s) \,|\, \mathcal{F}_t)\big| \le C\alpha(s-t)|f|_{w,\gamma}\big(1 + (w_1, \lambda)\big). \tag{A.74}$$

On the other hand, since u_s is $\mathcal{F}_t$-measurable for $t \ge s$, inequality (A.67) with f replaced by $|f|$ implies that

$$\mathbb{E}_\lambda\big|\mathbb{E}_\lambda(f(u_s)\,|\,\mathcal{F}_t)\big| = \mathbb{E}_\lambda|f(u_s)| \le \big(|f|,\mu\big) + \alpha(s)|f|_{w,\gamma}(w_1,\lambda), \qquad s \le t.$$

Combining this with (A.74) and using the fact that $\int_0^\infty \alpha(t)\,dt < \infty$, we obtain the required inequality (A.71).

The fact that M_t is a martingale can be checked easily by applying the conditional expectation $\mathbb{E}_\lambda(\,\cdot\,|\,\mathcal{F}_r)$ to M_t with $t \ge r$. Furthermore, it is straightforward to check that $\mathbb{E}_\lambda M_t = 0$ for all $t \ge 0$. Thus, it remains to establish (A.70). To this end, we use the Markov property and write

$$\begin{aligned} M_t &= \int_0^t f(u_s)\,ds + \int_t^\infty \mathbb{E}_\lambda(f(u_s)\,|\,\mathcal{F}_t)\,ds - \int_0^\infty \mathbb{E}_\lambda(f(u_s)\,|\,\mathcal{F}_0)\,ds \\ &= \int_0^t f(u_s)\,ds + \int_t^\infty (\mathfrak{P}_{s-t}f)(u_t)\,ds - \int_0^\infty (\mathfrak{P}_s f)(u_0)\,ds. \end{aligned}$$

This implies the required relation (A.70). □

A.14 Generalised Poincaré inequality

Proposition A.14.1 *For any integer $d \ge 1$ there is a constant $C_d > 0$ such that, for any $p \ge 1$, we have*

$$\int_{\mathbb{T}^d} |u|^{2p}dx \le C_d \int_{\mathbb{T}^d} \big|\nabla\big(|u|^{2p}\big)\big|^2 dx = C_d\, p^2 \int_{\mathbb{T}^d} |u|^{2(p-1)}|\nabla u|^2 dx, \qquad \text{(A.75)}$$

where $u \in H^1(\mathbb{T}^d)$ is an arbitrary function with zero mean value for which the right-hand side is finite.

Proof We argue[8] by contradiction. If (A.75) does not hold, then we can construct a sequence of functions $u_k \in H^1(\mathbb{T}^d)$ with zero mean value and some constants $p_k \ge 1$ such that[9]

$$\int_{\mathbb{T}^d} |u_k|^{2p_k}dx = (2\pi)^d, \quad p_k^2 \int_{\mathbb{T}^d} |u_k|^{2(p_k-1)}|\nabla u_k|^2 dx \le k^{-1}, \quad k \ge 1. \qquad \text{(A.76)}$$

Let us set $v_k = |u_k|^{p_k-1}u_k$. Then inequalities (A.76) imply that

$$\int_{\mathbb{T}^d} |v_k|^2 dx = (2\pi)^d, \quad \int_{\mathbb{T}^d} |\nabla v_k|^2 dx \le k^{-1}, \quad k \ge 1. \qquad \text{(A.77)}$$

Thus, $\{v_k\}$ is a bounded sequence in H^1. Since the embedding $H^1 \subset L^2$ is compact (see (1.5)), there is no loss of generality in assuming that $v_k \to v$ in L^2, where v is a function in H^1. Moreover, it follows from the second inequality in (A.77) that $\nabla v = 0$ almost everywhere, and therefore $v \equiv C$. On the other hand, the first relation in (A.77) implies that $|v|_2^2 = (2\pi)^d$, whence it follows that $C = 1$. We have thus shown that

$$\int_{\mathbb{T}^d} \big||u_k|^{p_k-1}u_k - 1\big|^2 dx \to 0 \quad \text{as } k \to \infty.$$

[8] We thank S. Brandle and M. Struwe for communicating this proof to us.

[9] The case in which the L^{2p_k}-norm of u_k is infinite can easily be treated by truncation. Namely, it suffices to consider the sequence $\tilde{u}_k = \varphi(u_k/N_k)$, where N_k is sufficiently large, and $\varphi : \mathbb{R} \to \mathbb{R}$ is a function such that $\varphi(r) = r$ for $|r| \le 1$ and $\varphi(r) = \operatorname{sgn}(r)$ for $|r| > 1$.

Combining this with the elementary inequality $|y-1| \le 2\big||y|^{p-1}y-1\big|$, valid for any $y \in \mathbb{R}$ and $p \ge 1$, we see that $u_k \to 1$ in L^2. This contradicts the fact that the mean value of u_k is zero. □

A.15 Functions in Sobolev spaces with a discrete essential range

Let $(X, \mathcal{F})$ be a measurable space, let μ be a positive measure on it, let Y be a Polish space endowed with its Borel σ-algebra, and let $f : X \to Y$ be a measurable map. The *essential range* of f is defined as the set of points $y \in Y$ such that $\mu(f^{-1}(\mathcal{O})) > 0$ for any open set $\mathcal{O} \subset Y$ containing y. In other words, the essential range of f is the support of the measure $f_*(\mu)$.

It is an obvious fact that if $D \subset \mathbb{R}^d$ is a connected open set, then a continuous function $f : D \to \mathbb{R}$ with a discrete essential range must be constant. The following lemma shows that a similar result is true for functions in a Sobolev space of order $r > 1/2$. Its stronger versions can be found in [HKL90; BD95; BBM00]. For the reader's convenience, we shall give a proof of the lemma, following the argument in [BD95].

Lemma A.15.1 *If $u \in H^r(\mathbb{T}^2)$ with $r > \frac{1}{2}$ and the essential range of u is a discrete subset of $\mathbb{R}$, then $u = C$ almost everywhere.*

Proof For $i = 1, 2$, denote by $\mathbb{T}_i$ the one-dimensional torus $\mathbb{R}/2\pi\mathbb{Z}$, $\mathbb{T}_i = \{x_i\}$. Then $\mathbb{T}^2 = \mathbb{T}_1 \times \mathbb{T}_2$. It follows immediately from the definition of the Sobolev norm that

$$\int_0^{2\pi} \|u(x_1, \cdot)\|^2_{H^r(\mathbb{T}_2)} dx_1 \le \|u\|_r^2.$$

Hence, there exists a full measure set $A \subset \mathbb{T}_1$ such that $u(x_1, \cdot) \in H^r(\mathbb{T}_2)$ for $x_1 \in A$. Since $H^r(\mathbb{T}_2) \subset C(\mathbb{T}_2)$ and the essential range of u is a discrete subset, Fubini's theorem implies that, for almost every $x_1 \in A$, the essential range of $u(x_1, \cdot)$ is a discrete subset. It follows that, for almost every $x_1 \in \mathbb{T}_1$, the function $u(x_1, \cdot)$ is constant almost everywhere. Similarly, for almost every $x_2 \in \mathbb{T}_2$, the function $u(\cdot, x_2)$ is constant almost everywhere.

For $x = (x_1, x_2)$ and $y = (y_1, y_2)$, we now write

$$|u(x) - u(y)| \le |u(x_1, x_2) - u(x_1, y_2)| + |u(x_1, y_2) - u(y_1, y_2)|.$$

Taking into account what has been said above and applying Fubini's theorem, we see that

$$\iint_{\mathbb{T}^2 \times \mathbb{T}^2} |u(x) - u(y)|\, dx dy = 0.$$

It follows that $u(x) - u(y) = 0$ almost everywhere on $\mathbb{T}^2 \times \mathbb{T}^2$. Let $B \subset \mathbb{T}^2$ be a subset of full measure such that, for $x \in B$, we have $u(x) = u(y)$ almost everywhere in y. Integrating this relation in y, we conclude that $u(x) = \int_{\mathbb{T}^2} u(y)\, dy$ for $x \in B$. This completes the proof of the lemma. □

Solutions to selected exercises

Exercise 1.2.4 (i) Let us assume that X is compact and prove that $C_b(X)$ is separable. Using relation (1.19) and taking into account inequality (1.20), it is easy to approximate each continuous function $f : X \to \mathbb{R}$ by bounded Lipschitz functions. Therefore, it suffices to construct countably many continuous functions whose finite linear combinations are dense in the unit ball of $L_b(X)$ for the norm $\|\cdot\|_\infty$. Let us fix $\varepsilon > 0$ and denote by $\{x_1, \dots, x_n\} \subset X$ an ε-net in X and by $\{\varphi_k\}_{k=1}^n \subset C_b(X)$ a partition of unity subordinate to the covering of X by the balls $B_X(x_k, 2\varepsilon)$. In this case, if f is an element of the unit ball in $L_b(X)$, then the norm of the function $f - \sum_k f(x_k)\varphi_k$ does not exceed 2ε.

Conversely, let us assume that X is not compact and prove that $C_b(X)$ is not separable. Indeed, since X is not compact, there is an $\varepsilon > 0$ and a sequence $\{x_k\} \subset X$ such that $\mathrm{dist}_X(x_k, x_m) \ge 2\varepsilon > 0$. Let $\varphi_k \in C_b(X)$ be such that $\operatorname{supp}\varphi_k \subset B_X(x_k, \varepsilon)$ and $\varphi_k(x_k) = 1$. Then the infinite linear combinations of the form $\sum_k c_k\varphi_k$ with $c_k = \pm 1$ form an uncountable family of functions such that the distance between any two functions is equal to 2.

(ii) Let us denote by $\varphi_\alpha(x)$ a function equal to 0 for $x < \alpha$ and to 1 for $x \ge \alpha$ and define $\psi_\alpha(x) = \int_0^x \varphi dy$. Then the family $\{\psi_\alpha, 0 < \alpha < 1\} \subset L_b(X)$ is uncountable and the distance in $L_b(X)$ between any of its two elements is equal to 1. □

Exercise 1.2.10 Let us denote by $\|\mu_1 - \mu_2\|'_{\mathrm{var}}$ the right-hand side of (1.26). Since $C_b(X) \subset L^\infty(X)$, it follows from (1.13) that $\|\mu_1 - \mu_2\|_{\mathrm{var}} \le \|\mu_1 - \mu_2\|'_{\mathrm{var}}$. To prove the converse inequality, we use a simple approximation argument. Namely, let us fix an arbitrary $\varepsilon > 0$ and choose a function $f \in L^\infty(X)$ with $\|f\|_\infty \le 1$ such that

$$\|\mu_1 - \mu_2\|'_{\mathrm{var}} \le \frac{1}{2}\big((f, \mu_1) - (f, \mu_2)\big) + \varepsilon.$$

We now choose finitely many disjoint subsets $\Gamma_k \in \mathcal{B}(X)$ and real constants c_k such that $|c_k| \le 1$ and $\big|f(x) - \sum_k c_k \mathbb{I}_{\Gamma_k}(x)\big| \le \varepsilon$ for all $x \in X$. Combining this with the above inequality and using (1.22), we derive

$$\begin{aligned}
\|\mu_1 - \mu_2\|'_{\mathrm{var}} &\le \frac{1}{2}\sum_k c_k\big(\mu_1(\Gamma_k) - \mu_2(\Gamma_k)\big) + 3\varepsilon \\
&\le \frac{1}{2}\sum_{k\in\Lambda^+}\big(\mu_1(\Gamma_k) - \mu_2(\Gamma_k)\big) + \frac{1}{2}\sum_{k\in\Lambda^-}\big(\mu_2(\Gamma_k) - \mu_1(\Gamma_k)\big) + 3\varepsilon \\
&\le \frac{1}{2}\big(\mu_1(\Gamma^+) - \mu_2(\Gamma^+)\big) + \frac{1}{2}\big(\mu_2(\Gamma^-) - \mu_1(\Gamma^-)\big) + 3\varepsilon \\
&\le \|\mu_1 - \mu_2\|_{\mathrm{var}} + 3\varepsilon,
\end{aligned}$$

where $\Lambda^\pm$ stands for the set of indices k such that $\pm c_k > 0$ and $\pm(\mu_1(\Gamma_k) - \mu_2(\Gamma_k)) > 0$, and $\Gamma^\pm$ is the union of Γ_k with $k \in \Lambda^\pm$. Since $\varepsilon > 0$ was arbitrary, we arrive at the required inequality. □

Exercise 1.2.12 We first recall that if a sequence of functions $\{f_n\}$ converges in the space $L^1(X, m)$ to a limit f, then there is a subsequence $\{f_{n_k}\}$ that converges to f almost surely. Thus, given a bounded measurable function $f : X \to \mathbb{R}$ and a constant $\varepsilon > 0$, it suffices to construct f_ε such that $\|f - f_\varepsilon\|_{L^1(X,m)} \le \varepsilon$ and $|f(x)| \le \|f\|_\infty$ for all $x \in X$. A simple approximation argument shows that it suffices to consider the case when f is the indicator function of a Borel set Γ. Let $K \subset \Gamma$ be a compact set and let $G \supset \Gamma$ be an open set such that $m(G \setminus K) \le \varepsilon$. Then the function $f_\varepsilon(x) = \frac{\mathrm{dist}_X(x,G)}{\mathrm{dist}_X(x,K)+\mathrm{dist}_X(x,G)}$ is continuous and satisfies the required inequality. □

Exercise 1.2.13 We confine ourselves to the proof of (i). For any $\Gamma \in \mathcal{F}_2$, we have

$$|f_*(\mu_1)(\Gamma) - f_*(\mu_2)(\Gamma)| = |\mu_1(f^{-1}(\Gamma)) - \mu_2(f^{-1}(\Gamma))| \le \|\mu_1 - \mu_2\|_{\mathrm{var}}.$$

Since Γ was arbitrary, we arrive at the required inequality. □

Exercise 1.2.18 (i) Let $\{x_k\} \subset X$ be an arbitrary dense sequence. We claim that for any $\mu \in \mathcal{P}(X)$ and $\varepsilon > 0$ there is a sequence $\{c_k\} \subset \mathbb{R}$ with finitely many non-zero elements such that

$$\Big\|\mu - \sum_k c_k \delta_{x_k}\Big\|_L^* \le \varepsilon, \tag{1}$$

where δ_y stands for the Dirac measure concentrated at y. This property immediately implies the required result.

Let us fix $\mu \in \mathcal{P}(X)$ and $\varepsilon > 0$ and choose a compact subset $\mathcal{C} \subset X$ such that $\mu(X \setminus \mathcal{C}) \le \varepsilon/2$. We next cover $\mathcal{C}$ by finitely many disjoint subsets B_j with non-empty interior and diameters $\le \varepsilon/2$. Choosing arbitrary points $x_{k_j} \in B_j$, for any function $f \in L_b(X)$ with norm $\|f\|_L \le 1$ we obtain

$$\begin{aligned}\Big|(f,\mu) - \sum_j f(x_{k_j})\mu(B_j \cap \mathcal{C})\Big| &\le \sum_j \Big|(f\mathbb{I}_{B_j\cap\mathcal{C}},\mu) - f(x_{k_j})\mu(B_j\cap\mathcal{C})\Big| + |(f\mathbb{I}_{X\setminus\mathcal{C}},\mu)\\ &\le \sum_j \mu(B_j\cap\mathcal{C}) \sup_{x\in B_j\cap\mathcal{C}} |f(x) - f(x_{k_j})| + \varepsilon/2 \le \varepsilon,\end{aligned}$$

where we used the fact f is Lipschitz continuous with a constant ≤ 1. Since f was arbitrary, we arrive at inequality (1) in which $c_{k_j} = \mu(B_j \cap \mathcal{C})$ and $c_k = 0$ for the other indices.

(ii) It is straightforward to see that $\{\delta_y, y \in \mathbb{R}\}$ is an uncountable subset $\mathcal{P}(\mathbb{R})$ such that the distance between any two elements is equal to 1. This implies that the space $\mathcal{P}(\mathbb{R})$ endowed with the total variation distance is not separable. □

Exercise 1.2.19 We need to show that

$$\sup_{\|f\|_L\le 1} \big|\mathbb{E}\,(f(\zeta_m) - f(\zeta))\big| \to 0 \quad \text{as } m \to \infty. \tag{2}$$

Let us take any function $f \in L_b(X)$ with $\|f\|_L \le 1$ and write

$$\begin{aligned}\big|\mathbb{E}\,(f(\zeta_m) - f(\zeta))\big| &\le |\mathbb{E}\,(f(\zeta_m) - f(\zeta_m^n))| + |\mathbb{E}\,(f(\zeta) - f(\zeta^n))| + |\mathbb{E}\,(f(\zeta_m^n) - f(\zeta^n))|\\ &\le \sup_{m\ge1} \mathbb{E}\big(\mathrm{dist}_X(\zeta_m,\zeta_m^n) + \mathrm{dist}_X(\zeta,\zeta^n)\big) + \|\mathcal{D}(\zeta_m^n) - \mathcal{D}(\zeta^n)\|_L^*.\end{aligned}$$

The first term on the left-hand side can be made arbitrarily small by choosing $n \ge 1$ while the second term goes to zero, for any fixed $n \ge 1$, as $m \to \infty$. Since f was arbitrary, we arrive at the required assertion. □

Exercise 1.2.22 Since $\|\mu_1 - \mu_2\|_{\mathrm{var}} = 1$, by Corollary 1.2.11, there is a set $A \in \mathcal{B}(X)$ for which $\mu_1(A) = \mu_2(X \setminus A) = 1$. It follows that if ξ_1 and ξ_2 are independent, then

$$\begin{aligned}\mathbb{P}\{\xi_1 \neq \xi_2\} &= \iint\limits_{\{u_1 \neq u_2\}} \mu_1(du_1)\mu_2(du_2) = \int_A \mu_2(X \setminus \{u_1\})\mu(du_1)\\ &\geq \int_A \mu_2(X \setminus A)\mu(du_1) = 1.\end{aligned}$$

Furthermore, it is straightforward to see that the random variables ξ_1 and ξ_2 conditioned on the event $N = \{\xi_1 \neq \xi_2\}$ are independent.

Conversely, if (ξ_1, ξ_2) is a maximal coupling for (μ_1, μ_2), then $\mathbb{P}(N) = 1$, and for any $\Gamma_1, \Gamma_2 \in \mathcal{B}(X)$ we have

$$\begin{aligned}\mathbb{P}\{\xi_1 \in \Gamma_1, \xi_2 \in \Gamma_2\} &= \mathbb{P}\{\xi_1 \in \Gamma_1, \xi_2 \in \Gamma_2 \mid N\}\\ &= \mathbb{P}\{\xi_1 \in \Gamma_1 \mid N\}\,\mathbb{P}\{\xi_1 \in \Gamma_1 \mid N\}\\ &= \mathbb{P}\{\xi_1 \in \Gamma_1\}\,\mathbb{P}\{\xi_1 \in \Gamma_1\},\end{aligned}$$

whence we conclude that ξ_1 and ξ_2 are independent. □

Exercise 1.2.23 Let us set

$$a = \mathbb{P}(A_1 \mid E) = \mathbb{P}(A_2 \mid E), \quad b_1 = \mathbb{P}(A_1 \mid N), \quad b_2 = \mathbb{P}(A_2 \mid N),$$

where $E = \{\xi_1 = \xi_2\}$ and $A_i = \{\xi_i \in \Gamma\}$. Noting that $\mathbb{P}(E) + \mathbb{P}(N) = 1$, we can write

$$\begin{aligned}\mathbb{P}\{\xi_1 \in \Gamma, \xi_2 \in \Gamma\} &= \mathbb{P}(A_1A_2) = \mathbb{P}(A_1A_2 \mid E)\,\mathbb{P}(E) + \mathbb{P}(A_1A_2 \mid N)\,\mathbb{P}(N)\\ &= \mathbb{P}(A_1 \mid E)\mathbb{P}(E) + \mathbb{P}(A_1 \mid N)\,\mathbb{P}(A_2 \mid N)\,\mathbb{P}(N) = ax + b_1b_2(1-x),\\ \mathbb{P}\{\xi_1 \in \Gamma\}\,\mathbb{P}\{\xi_2 \in \Gamma\} &= \mathbb{P}(A_1)\,\mathbb{P}(A_2) = \big(ax + b_1(1-x)\big)\big(ax + b_2(1-x)\big),\end{aligned}$$

where $x = \mathbb{P}(E)$. We wish to show that the function

$$f(x) := ax + b_1b_2(1-x) - \big(ax + b_1(1-x)\big)\big(ax + b_2(1-x)\big)$$

is non-negative for $x \in [0, 1]$. Note that f is a quadratic function such that $f(0) = 0$ and $f(1) = a - a^2 \geq 0$. Therefore, the required property will be established if we show that the coefficient in front of the linear term is non-negative. Thus, it suffices to prove that

$$a + b_1b_2 - a(b_1 + b_2) \geq 0 \quad \text{for} \quad a, b_1, b_2 \in [0, 1].$$

This inequality can easily be checked. □

Exercise 1.2.27 Since convergence in the dual-Lipschitz metric is equivalent to the weak convergence of measures (see Theorem 1.2.15), the corresponding topology on the space $\mathcal{P}(X)$ is generated by the open sets of the form $\{\mu \in \mathcal{P}(X) : (f, \mu) < a\}$, where $f \in C_b(X)$ and $a \in \mathbb{R}$. It follows that a function $z \mapsto \mu(z, \cdot)$ from Z to $\mathcal{P}(X)$ is measurable if and only if, for any $f \in C_b(X)$, so is the function $z \mapsto (f, \mu(z, \cdot))$ acting from Z to $\mathbb{R}$.

Now let $\mu(z, \cdot)$ be a random probability measure and let $f \in C_b(X)$. Approximating f by finite linear combinations of indicator functions, we see that $(f, \mu(z, \cdot))$ is a pointwise limit of measurable functions of the form $\sum_k c_k\mu(z, \Gamma_k)$. This observation immediately implies that $(f, \mu(z, \cdot))$ is measurable. Conversely, if $(f, \mu(z, \cdot))$ is measurable for any $f \in C_b(X)$ and $F \subset X$ is a closed subset, then approximating $\mathbb{I}_F$ pointwise by continuous functions (see the proof of Corollary A.1.3), we see that $\mu(z, F) = (\mathbb{I}_F, \mu(z, \cdot))$ is measurable. The measurability of $\mu(z, \Gamma)$ for an arbitrary $\Gamma \in \mathcal{B}(X)$ follows by the monotone class technique. □

Exercise 1.2.29 The fact that $\hat{\mu}_i(z,\cdot)$ and $\mu(z,\cdot)$ are random probability measures follows from their definition and the measurability of $\delta(z)$. To prove that δ is measurable, recall that (see Proposition 1.2.7)

$$\delta(z) = \|\mu_1(z,\cdot) - \mu_2(z,\cdot)\|_{\mathrm{var}} = \frac{1}{2}\int_X |\rho_1(z,u) - \rho_2(z,u)|\, m_z(du),$$

where $m_z = \frac{1}{2}(\mu_1(z,\cdot) + \mu_2(z,\cdot))$, and $\rho_i(z,u)$ is the density of μ_i with respect to m_z. By a parameter version of the Radon–Nikodym theorem, the functions ρ_i can be assumed to be measurable in (z,u). Thus, the required property will be established if we show that $(f(z,\cdot), m_z)$ is measurable for any measurable function $f : Z \times X \to \mathbb{R}$ such that $f(z,\cdot) \in L^1(X, m_z)$ for any $z \in Z$. A standard approximation argument shows that it suffices to consider the case when $f = \mathbb{I}_\Gamma$, where $\Gamma \in \mathcal{B}(Z \times X)$. The claim is obvious for sets of the form $\Gamma = \Gamma_1 \times \Gamma_2$ with $\Gamma_1 \in \mathcal{B}(Z)$ and $\Gamma_2 \in \mathcal{B}(X)$. The general case follows by the monotone class technique. □

Exercise 1.2.30 Let us define measures $\tilde{\mu}_i \in \mathcal{P}(X \times Y)$ by the relation

$$\tilde{\mu}_i(A \times B) = \mu_i(A \cap f^{-1}(B)) \quad \text{for } A \in \mathcal{B}(X),\ B \in \mathcal{B}(Y).$$

Then $(P_Y)_*\tilde{\mu}_i = f_*(\mu_i)$, where $P_Y : X \times Y \to Y$ denotes the natural projection, and by the disintegration theorem, there are random probability measures $\mu_i(y, dx)$ on X such that $\tilde{\mu}_i(dx, dy) = f_*(\mu_i)(dy)\mu_i(y, dx)$, $i = 1, 2$. Now let (η_1, η_2) be a maximal coupling for $(f_*(\mu_1), f_*(\mu_2))$ and let $\zeta_i(y,\cdot)$, $i = 1, 2$, be some X-valued random variables independent of (η_1, η_2) such that $\mathcal{D}(\zeta_i(y,\cdot)) = \mu_i(y,\cdot)$ for any $y \in Y$. Then the random variables $\xi_i(\omega) = \zeta_i(\eta_i(\omega), \omega)$, $i = 1, 2$, satisfy the required property. This proves assertion (i). To establish (ii), it suffices to use in the above argument some parameter versions of the corresponding results. □

Exercise 1.3.1 In view of (1.48), for any $f \in L^\infty(X)$ and any $B \in \mathcal{F}_s$, we have

$$\mathbb{E}_u\big(f(u_{t+s})\mathbb{I}_B\big) = \mathbb{E}_u\big((\mathfrak{P}_t f)(u_s)\mathbb{I}_B\big).$$

Integrating this relation with respect to $\mathbb{P}_\lambda(du)$, we obtain

$$\mathbb{E}_\lambda\big(f(u_{t+s})\mathbb{I}_B\big) = \mathbb{E}_\lambda\big((\mathfrak{P}_t f)(u_s)\mathbb{I}_B\big).$$

Since $B \in \mathcal{F}_s$ was arbitrary, we arrive at (1.48).

When proving (1.49), we assume for simplicity that $m = 2$. It suffices to consider the case when $f(u_1, u_2) = \mathbb{I}_{\Gamma_1}(u_1)\mathbb{I}_{\Gamma_2}(u_2)$. Using (1.48), we write

$$\begin{aligned}\mathbb{E}_\lambda\{f(u_{t_1+s}, u_{t_2+s}) \,|\, \mathcal{F}_s\} &= \mathbb{E}_\lambda\big(\mathbb{I}_{\Gamma_1}(u_{t_1+s})\,\mathbb{E}_\lambda\{\mathbb{I}_{\Gamma_2}(u_{t_2+s}) \,|\, \mathcal{F}_{t_1+s}\} \,|\, \mathcal{F}_s\big)\\ &= \mathbb{E}_\lambda\big(\mathbb{I}_{\Gamma_1}(u_{t_1+s})\, P_{t_2-t_1}(u_{t_1+s}, \Gamma_2) \,|\, \mathcal{F}_s\big)\\ &= \mathfrak{P}_{t_1}\big(\mathbb{I}_{\Gamma_1} P_{t_2-t_1}(\cdot, \Gamma_2)\big)(u_s). \end{aligned} \tag{3}$$

On the other hand, using again the Markov property, we derive

$$\begin{aligned}\mathbb{E}_v f(u_{t_1}, u_{t_2}) &= \mathbb{E}_v\big(\mathbb{I}_{\Gamma_1}(u_{t_1})\mathbb{I}_{\Gamma_2}(u_{t_2})\big) = \mathbb{E}_v\big(\mathbb{I}_{\Gamma_1}(u_{t_1})\mathbb{E}_v\{\mathbb{I}_{\Gamma_2}(u_{t_2}) \,|\, \mathcal{F}_{t_1}\}\big)\\ &= \mathbb{E}_v\big(\mathbb{I}_{\Gamma_1}(u_{t_1})P_{t_2-t_1}(u_{t_1}, \Gamma_2)\big) = \mathfrak{P}_{t_1}\big(\mathbb{I}_{\Gamma_1} P_{t_2-t_1}(\cdot, \Gamma_2)\big)(v).\end{aligned}$$

Taking $v = u_s$ in this relation and comparing it with (3), we obtain the required result. □

Exercise 1.3.3 It is clear that $\mathfrak{P}_0$ is the identity operator in $C_b(X)$, and we shall prove the semigroup property. A simple approximation argument shows that it suffices to prove the relation

$$\mathfrak{P}_t(\mathfrak{P}_s\mathbb{I}_\Gamma) = \mathfrak{P}_{t+s}\mathbb{I}_\Gamma \quad \text{for any } \Gamma \in \mathcal{B}(X).$$

However, since $\mathfrak{P}_r\mathbb{I}_\Gamma(u) = P_r(u, \Gamma)$, the above equality coincides with the Kolmogorov–Chapman relation. The proof of the claim concerning $\mathfrak{P}_t^*$ is similar.

We now prove the duality relation. It suffices to consider the case $f = \mathbb{I}_\Gamma$. In this situation, we have

$$(\mathfrak{P}_t\mathbb{I}_\Gamma, \mu) = \int_X P_t(u, \Gamma)\mu(du) = \mathfrak{P}_t^*\mu(\Gamma) = (\mathbb{I}_\Gamma, \mathfrak{P}_t^*\mu). \qquad \square$$

Exercise 1.3.4 By definition, the $\mathbb{P}_v$ law of u_t coincides with $P_t(v, \cdot)$. It follows that

$$\mathbb{P}_\lambda\{u_t \in \Gamma\} = \int_X \mathbb{P}_v\{u_t \in \Gamma\}\lambda(dv) = \int_X P_t(v, \Gamma)\lambda(dv) = \mathfrak{P}_t^*\lambda(\Gamma),$$

where $\Gamma \in \mathcal{B}(X)$ is an arbitrary set. To prove (1.51), we write

$$\mathbb{E}_\lambda f(u_t) = \int_\Omega f(u_t)\,\mathbb{P}_\lambda(d\omega) = \int_X f(v)(\mathfrak{P}_t^*\lambda)(dv) = \int_X \mathfrak{P}_t f(v)\lambda(dv),$$

where we used the duality relation of Exercise 1.3.3. $\square$

Exercise 1.3.9 We need to prove that $\{\tau \le t\} \cap \{\tau \le s\}$ belongs to $\mathcal{F}_t$ for any $t, s \in \mathcal{T}_+$. This fact follows immediately from the definition of a stopping time. $\square$

Exercise 1.3.12 We confine our discussion to the case of discrete time. It suffices to establish relation (1.59) for indicator functions $f = \mathbb{I}_B$ with a Borel set B of the form

$$B = \{v = (v_j, j \in \mathbb{Z}_+) : v_j \in B_j \text{ for } j = 0, \dots, n\},$$

where $B_j \in \mathcal{B}(X)$. The case of an arbitrary Borel set $B \subset X^{\mathbb{Z}_+}$ will follow by the monotone class technique, and an approximation argument will yield relation (1.59) for functions f satisfying the hypotheses of the exercise.

Thus, we wish to prove that

$$\mathbb{E}_u\biggl(\mathbb{I}_{\{\tau<\infty\}}\prod_{j=0}^m \mathbb{I}_{B_j}(u_{\tau+j})\,\Big|\,\mathcal{F}_\tau\biggr) = \mathbb{I}_{\{\tau<\infty\}}\biggl(\mathbb{E}_v\prod_{j=0}^m \mathbb{I}_{B_j}(u_{\tau+j})\biggr)\bigg|_{v=u_\tau}$$

with $\mathbb{P}_u$-probability 1. To this end, it suffices to show that

$$\mathbb{E}_u\biggl(\mathbb{I}_{\Gamma\cap\{\tau<\infty\}}\prod_{j=0}^m \mathbb{I}_{B_j}(u_{\tau+j})\biggr) = \mathbb{E}_u\biggl(\mathbb{I}_{\Gamma\cap\{\tau<\infty\}}\,\mathbb{E}_{u_\tau}\prod_{j=0}^m \mathbb{I}_{B_j}(u_j)\biggr), \tag{4}$$

where $\Gamma \in \mathcal{F}_\tau$ is an arbitrary subset. Using (1.49) and (1.56), we write

$$\begin{aligned}
\mathbb{E}_u\biggl(\mathbb{I}_{\Gamma\cap\{\tau<\infty\}}\prod_{j=0}^m \mathbb{I}_{B_j}(u_{\tau+j})\biggr) &= \sum_{n=0}^\infty \mathbb{E}_u\biggl(\mathbb{I}_{\Gamma\cap\{\tau=n\}}\prod_{j=0}^m \mathbb{I}_{B_j}(u_{n+j})\biggr)\\
&= \sum_{n=0}^\infty \mathbb{E}_u\biggl(\mathbb{I}_{\Gamma\cap\{\tau=n\}}\mathbb{E}_u\biggl\{\prod_{j=0}^m \mathbb{I}_{B_j}(u_{n+j})\,\Big|\,\mathcal{F}_n\biggr\}\biggr)\\
&= \sum_{n=0}^\infty \mathbb{E}_u\biggl(\mathbb{I}_{\Gamma\cap\{\tau=n\}}\mathbb{E}_{u_n}\prod_{j=0}^m \mathbb{I}_{B_j}(u_j)\biggr)\\
&= \mathbb{E}_u\biggl(\mathbb{I}_{\Gamma\cap\{\tau<\infty\}}\mathbb{E}_{u_\tau}\prod_{j=0}^m \mathbb{I}_{B_j}(u_j)\biggr).
\end{aligned}$$

This completes the proof of (1.59) in the case of discrete time. $\square$

Exercise 1.3.16 Since $\theta_s(\omega_j, j \in \mathbb{Z}) = (\omega_{j+s}, j \in \mathbb{Z})$, the definition of $\boldsymbol{\Phi}$ implies that $\varphi_t^{\theta_s\omega} u$ depends only on $\omega_{s+1}, \dots, \omega_{s+t}$. It follows that $\mathcal{F}_{[p,q]}$ is contained in the σ-algebra generated by the cylindrical sets depending on $\omega_{p+1}, \dots, \omega_q$. On the other hand, for any integer k satisfying the inequality $p+1 \le k \le q$ the σ-algebra generated by the random variables $\varphi_1^{\theta_{k-1}\omega} u = S(u) + \omega_k, u \in H$, coincides with the cylindrical sets defined by ω_k. What has been said implies that $\mathcal{F}_{[p,q]}$ coincides with the σ-algebra of the sets of the form (1.65). □

Exercise 1.3.17 To be precise, we assume that p and q are finite. Since the family of those $\Gamma \in \mathcal{F}[p,q]$ for which $\theta_r^{-1}(\Gamma) \in \mathcal{F}_{[p+r,q+r]}$ is a σ-algebra, it suffices to prove this inclusion for any subset of the form

$$\Gamma = \{\omega \in \Omega : \varphi_t^{\theta_s\omega} u \in A\}, \quad A \in \mathcal{B}(X),$$

where p, q, t, s, and u satisfy conditions (1.64). To this end, note that, by the group property for $\boldsymbol{\theta}$, we have

$$\theta_r^{-1}(\Gamma) = \{\omega \in \Omega : \varphi_t^{\theta_s(\theta_r\omega)} u \in A\} = \{\omega \in \Omega : \varphi_t^{\theta_{s+r}\omega} u \in A\},$$

whence we conclude that $\theta_r^{-1}(\Gamma) \in \mathcal{F}_{[p+r,q+r]}$. □

Exercise 1.3.19 It was proved in Exercise 1.3.16 that $\mathcal{F}^-$ and $\mathcal{F}^+$ coincide with the σ-algebras generated by the cylindrical sets depending on ω_j with $j \le 0$ and $j \ge 1$, respectively. The required assertion follows now from the definition of the probability $\mathbb{P}$. □

Exercise 1.3.20 Let us note that, for any $\Gamma^+ \in \mathcal{F}^+$ and $\Gamma^- \in \mathcal{F}^-$, we have

$$\begin{aligned} \mathbb{P}\big(\theta_t^{-1}(\Gamma^+) \cap \theta_t^{-1}(\Gamma^-)\big) &= \mathbb{P}\big(\theta_t^{-1}(\Gamma^+\Gamma^-)\big) = \mathbb{P}(\Gamma^+\Gamma^-) \\ &= \mathbb{P}(\Gamma^+)\,\mathbb{P}(\Gamma^-) = \mathbb{P}\big(\theta_t^{-1}(\Gamma^+)\big)\,\mathbb{P}\big(\theta_t^{-1}(\Gamma^-)\big). \end{aligned}$$

Combining this relation with Exercise 1.3.17, we arrive at the required result. □

Exercise 1.3.23 We need to show that if $f \in C_b(X)$, then $\mathfrak{P}_t f(u)$ is continuous in u. Recall that

$$\mathfrak{P}_t f(u) = \int_\Omega f(\varphi_t^\omega u)\mathbb{P}(d\omega).$$

Since f is bounded and $\varphi_t^\omega u$ is continuous in u, Lebesgue's theorem on dominated convergence implies the required result. □

Exercise 1.3.24 Recall that $P_k(u, \cdot)$ is the law of the trajectory issued from u. Since the mapping φ_t is deterministic, we see that $P_k(u, \cdot)$ is the Dirac measure concentrated at $\psi^k(u)$, where ψ^k is the kth iteration of ψ. It follows that

$$\mathfrak{P}_k f(u) = f\big(\psi^k(u)\big), \quad \mathfrak{P}_k^*\mu = \mu\big(\psi_k^{-1}(\Gamma)\big).$$ □

Exercise 1.3.25 Using the group and cocycle properties, we write

$$\Theta_t\big(\Theta_s(\omega, u)\big) = \Theta_t(\omega_s\omega, \varphi_s^\omega u) = \big(\theta_t(\theta_s\omega), \varphi_t^{\theta_s\omega}(\varphi_s^\omega u)\big) = (\theta_{t+s}\omega, \varphi_{t+s}^\omega u) = \Theta_{t+s}(\omega, u).$$

The fact that Θ_0 is the identity operator is obvious. □

Exercise 1.3.29 By Corollary 1.3.22, the law of $\varphi_t u$ coincides with $\mathfrak{P}_t^*\mu$. Since μ is a stationary measure, we see that $\mathcal{D}(\varphi_t u) = \mu$ for all $t \in \mathcal{T}_+$. □

Exercise 2.1.1 Any (real-valued) function $u \in H^m_\sigma$ can be written as a Fourier series

$$u(x) = \sum_{k\in\mathbb{Z}^2} u_k e^{ikx}, \tag{5}$$

where $u_k \in \mathbb{C}^2$, $\langle u_k, k\rangle = 0$, and $\bar{u}_k = u_{-k}$. It straightforward to see that the functions defined by the truncated series $\sum_{|k|\le N} u_k e^{ikx}$ belong to $\mathcal{V}$ and approximate u in the H^m norm. □

Exercise 2.1.14 Let us note that if u is a solution of the Navier–Stokes system (2.19), then $\dot{u} = -\nu Lu - B(u) + \Pi f$ belongs to $L^2(0,T;V^*)$. Therefore, using (2.10) and (2.11), we obtain

$$\frac{1}{2}\big(|u(t)|_2^2 - |u_0|_2^2\big) = \int_0^t \langle -\nu Lu - B(u) + \Pi f, u\rangle\, ds = \int_0^t \big(-\nu|\nabla u(s)|_2^2 + \langle f, u\rangle\big)\, ds.$$

This coincides with the required relation (2.39). □

Exercise 2.1.17 The fact that ψ is unique up to an additive constant is obvious. Let us fix a function $u \in H$ and consider its Fourier series (5). It is easy to see that the coefficients u_k can be written as $u_k = c_k k^\perp$, where $k^\perp = (-k_2, k_1)$, $c_{-k} = -\bar{c}_k \in \mathbb{C}$, and

$$\|u\|_m^2 = \sum_{k\in\mathbb{Z}_0^2} \big(1+|k|^2\big)^m |k|^2 |c_k|^2.$$

It follows that the function

$$\psi(x) = -i \sum_{k\in\mathbb{Z}_0^2} c_k e^{ikx}$$

is real-valued, belongs to $H^1(\mathbb{T}^2)$, and satisfies the required properties. □

Exercise 2.1.23 Combining (2.39), (2.51), and the inequality $2|\langle f, u\rangle| \le \|f\|_1^2 + \|u\|_1^2$, we derive

$$\begin{aligned}|u(0)|_2^2 &= |u(t)|_2^2 + 2\nu \int_0^t |\nabla u|_2^2 ds - 2\int_0^t \langle f, u\rangle\, ds \\ &\le e^{-\alpha_1 t}|u(0)|_2^2 + \int_0^t \big(C_1\|u\|_1^2 + 2\,\|f\|_{-1}^2\big) ds.\end{aligned}$$

This estimate readily implies the required inequality. □

Exercise 2.1.26 Let $u \in \mathcal{H}$ be a solution of the Navier–Stokes system (2.19). Differentiating the function $\varphi(t) = t\,|L^{1/4}u(t)|_2^2$ in time and using some standard estimates, we get

$$\begin{aligned}\varphi'(t) &= |L^{1/4}u|_2^2 + 2t\langle L^{1/2}u, \dot{u}\rangle = |L^{1/4}u|_2^2 - 2\nu t\,|L^{3/4}u|_2^2 + 2t\,\langle L^{1/2}u, \Pi f - B(u)\rangle \\ &\le |L^{1/4}u|_2^2 - \frac{3\nu t}{2}\,|L^{3/4}u|_2^2 + C_1\nu^{-1}t\,|f|_2^2 + C_2 t\,|L^{3/4}u|_2 |L^{1/4}u|_2 \|u\|_1 \\ &\le |L^{1/4}u|_2^2 - \nu t\,|L^{3/4}u|_2^2 + C_1\nu^{-1}t\,|f|_2^2 + C_2\nu^{-1}t\,\|u\|_1^2 |L^{1/4}u|_2^2.\end{aligned}$$

It follows that

$$\partial_t\big(t\,|L^{1/4}u(t)|_2^2\big) + \nu t\,\|u\|_{\frac{3}{2}}^2 \le \|u\|_{\frac{1}{2}}^2 + C_3\nu^{-1}\|u\|_1^2\big(t\,|L^{1/4}u(t)|_2^2\big) + C_3\nu^{-1}t|f|_2^2.$$

A simple argument based on Gronwall's inequality and (2.24) now implies that

$$t\,\|u(t)\|_{\frac{1}{2}}^2 + \int_0^t s\,\|u(s)\|_{\frac{3}{2}}^2 ds \le C_4(\nu)\exp\Big(C_4(\nu)\int_0^t \|u\|_1^2 ds\Big)\Big(|u_0|_2^2 + \int_0^t |f(s)|_2^2 ds\Big). \tag{6}$$

Now let $u_1, u_2 \in \mathcal{H}$ be two solutions of (2.19). Their difference $u = u_1 - u_2$ satisfies Eq. (2.25). Multiplying it by $2tL^{1/2}u$ in H and carrying out some transformations, we derive

$$\partial_t\big(t\,|L^{1/4}u(t)|_2^2\big) + \nu t\,\|u\|_{\frac{3}{2}}^2 \le \|u\|_{\frac{1}{2}}^2 + C_1\nu^{-1}t\,|f_1 - f_2|_2^2 + C_5 t\,\|u\|_{\frac{3}{2}}^{4/3}|u|_2^{2/3}\big(|u_1|_2\|u_1\|_{\frac{3}{2}}^2 + |u_2|_2\|u_2\|_{\frac{3}{2}}^2\big)^{1/3}$$

whence it follows that

$$t\,\|u(t)\|_{\frac{1}{2}}^2 \le \int_0^t \big(\|u\|_{\frac{1}{2}}^2 + C_1\nu^{-1}s\,|f_1 - f_2|_2^2\big)\,ds + C_6(\nu)\int_0^t s|u|_2^2\big(|u_1|_2\|u_1\|_{\frac{3}{2}}^2 + |u_2|_2\|u_2\|_{\frac{3}{2}}^2\big)ds.$$

The required inequality can now be derived by the same argument as in the proof of Proposition 2.1.25. □

Exercise 2.1.27 To prove the existence of a solution, we assume without loss of generality that $g(0) = 0$ and seek a solution in the form $u = g + v$. Then v must be a solution of the problem

$$\dot v + \nu L v + B(v + g) = h, \quad v(0) = u_0. \tag{7}$$

The existence of a solution for this problem can be established using methods similar to those in Section 2.1.5.

To prove Lipschitz continuity, we assume again that $g_1(0) = g_2(0) = 0$. Representing the solutions in the form $u_i = g_i + v_i$, we see that v_i must be a solution of problem (7) with $h \equiv 0$ and $u_0 = u_i(0)$. It follows that the difference $v = v_1 - v_2$ satisfies the equations

$$\dot v + \nu L v + B(v + g, u_1) + B(u_2, v + g) = 0, \quad v(0) = u_1(0) - u_2(0),$$

where $g = g_1 - g_2$. Taking the scalar product in H of the first equation with $2v$ and repeating essentially the scheme used in the proof of assertion (i) of Proposition 2.1.25, we arrive at the required result. □

Exercise 2.2.3 Let us set $k_\varepsilon = [t_0/\varepsilon] + 1$. It follows from inequality (2.74) and the independence of η_k that

$$\mathbb{P}\Big\{\sup_{0\le t\le t_0} |\zeta_\varepsilon(t) - \tilde\zeta_\varepsilon(t)|_2 \le \delta\Big\} \ge \mathbb{P}\big\{\max\{|\eta_k|_2, 1 \le k \le k_\varepsilon\} \le \tfrac{\delta}{\sqrt{\varepsilon}}\big\}$$
$$\ge \prod_{k=1}^{k_\varepsilon}\Big(1 - \mathbb{P}\big\{|\eta_k|_2 > \tfrac{\delta}{\sqrt{\varepsilon}}\big\}\Big) \ge \big(1 - C_1(\delta)\varepsilon^{q/2}\big)^{k_\varepsilon}$$
$$\ge \exp\big(-C_2(\delta)k_\varepsilon\varepsilon^{q/2}\big).$$

Since $k_\varepsilon \sim 1/\varepsilon$ and $q > 2$, we see that the right-hand side of this inequality goes to 1 as $\varepsilon \to 0$. □

Exercise 2.2.4 Theorem 8.2 of [Bil99] implies that

$$\mathcal{D}(\tilde\zeta_\varepsilon^n) \to \mathcal{D}(\zeta^n) \quad \text{in } \mathcal{P}\big(C(0, t_0; H)\big) \text{ as } \varepsilon \to 0^+,$$

where the process ζ^n is defined in the proof of Theorem 2.2.2 and $\tilde\zeta_\varepsilon^n$ is a similar finite-dimensional approximation for $\tilde\zeta_\varepsilon$. Combining this convergence with inequality (2.73), its analogue for $\tilde\zeta_\varepsilon$, and Exercise 1.2.19, we arrive at the required result. □

Exercise 2.4.4 Let us denote by $\{\hat{e}_j\}$ an orthonormal basis in H composed of the eigenfunction of the Stokes operator L. For any integer $n \geq 1$, we denote by P_n the orthogonal projection in H to the vector space spanned by $\hat{e}_j$, $j = 1, \dots, n$. Let us consider the n-dimensional linear stochastic equation

$$\dot{z}_n + \nu L_n z_n = \partial_t P_n \zeta^n(t), \tag{8}$$

where $L_n = P_n L$ and $\zeta^n(t) = \sum_{j=1}^n b_j(t)\beta_j e_j$. In view of the well-known results on the existence and uniqueness of solutions for stochastic ODEs (e.g., see chapter 5 in [Øks03]), Eq. (8) has a unique solution z_n satisfying the initial condition $z_n(0) = 0$. It suffices to prove that there is a subsequence of $\{z_n\}$ that converges to $z(t)$ almost surely in the space $\mathcal{X}_T = C(0, T; H) \cap L^2(0, T; V)$ for any $T > 0$.

The proof of this fact is divided into two steps: we first derive some a priori estimates for z_n with the help of Itô's formula and Doob's moment inequality and then, using a similar technique, we show that the difference $z_n - z_m$ converges to zero in an appropriate space and that the limiting function satisfies (2.103).

Step 1: A priori estimates. We claim that

$$\mathbb{E}\biggl(\sup_{0\le t\le T} |z_n(t)|_2^2 + \int_0^T \|z_n(t)\|_1^2 dt\biggr) \le C(\mathfrak{B}, T)\,\nu^{-1}, \tag{9}$$

where the constant $C(\mathfrak{B}, T)$ does not depend on n. Indeed, by the finite-dimensional Itô formula (see chapter 4 in [Øks03]), we have

$$|z_n(t)|_2^2 = \int_0^t \Bigl(-2\nu\,|\nabla z_n(s)|_2^2 + \sum_{j=1}^n b_j^2 |P_n e_j|_2^2\Bigr)\,ds + 2\int_0^t \langle z_n, d\zeta^n\rangle. \tag{10}$$

Taking the mean value and using Friedrichs' inequality, we obtain

$$\sup_{0\le t\le T} \mathbb{E}\,|z_n(t)|_2^2 + 2\nu\,\mathbb{E}\int_0^T \|z_n\|_1^2 dt \le C_1(\mathfrak{B}, T) \quad \text{for all } n \geq 1. \tag{11}$$

Furthermore, in view of Doob's moment inequality, we have

$$\begin{aligned}\mathbb{E}\sup_{0\le t\le T}\biggl|\int_0^t \langle z_n, d\zeta^n\rangle\biggr|^2 &\le \mathbb{E}\int_0^T \sum_{j=1}^n b_j^2 (z_n(t), e_j)^2 ds\\ &\le C_2(\mathfrak{B})\,\mathbb{E}\int_0^T |z_n(t)|_2^2 dt \le C_3(\mathfrak{B}, T)\,\nu^{-1},\end{aligned}$$

where we used (11) in the last inequality. Combining this with (10) and (11), we arrive at (9).

Step 2: Convergence. We now prove that a subsequence of $\{z_n\}$ converges in $\mathcal{X}_T$ almost surely and that the limit function coincides with z. Let us fix some integers $m < n$. The difference $z_{mn} = z_n - z_m$ vanishes at $t = 0$ and satisfies the equation

$$\dot{z}_{mn} + \nu P_n L z_{mn} = \partial_t \zeta_{mn}(t), \tag{12}$$

where we used the relation $(P_n - P_m)L z_m = 0$ and set

$$\zeta_{mn}(t) = \sum_{j=1}^m b_j \beta_j(t)(P_n - P_m)e_j + \sum_{j=m+1}^n b_j\beta_j(t) P_n e_j.$$

Applying Itô's formula, we derive

$$|z_{mn}(t)|_2^2 = \int_0^t \bigl(-2\nu\,|\nabla z_{mn}(s)|_2^2 + F_{mn}\bigr)\,ds + 2\int_0^t \langle z_{mn}, d\zeta_{mn}\rangle, \tag{13}$$

where

$$F_{mn} = \sum_{j=1}^{m} b_j^2 |(P_n - P_m)e_j|_2^2 + \sum_{j=m+1}^{n} b_j^2 |P_n e_j|_2^2.$$

Taking the mean value in (13), we see that (2.107) remains true. The proof can now be completed by exactly the same argument as in the case of Proposition 2.4.2. □

Exercise 2.4.17 Let us set $\Lambda_N = \{j \in \Lambda : j \le N\}$. Then

$$\tilde{\zeta}_N(t,x) = \sum_{j \in \Lambda_N} \omega_t^{(j)} e_j(x), \quad t \in \mathbb{R},$$

is an H-valued process with continuous trajectories whose range is contained in the vector span of e_j, $j \in \Lambda_N$. Let us denote by $\mathcal{F}_t$, $t \in \mathbb{R}$, the natural filtration on Ω augmented[1] with respect to $(\mathcal{F}, \mathbb{P})$. The process $\tilde{\zeta}_N$ is the sum of finitely many Brownian motions and therefore is a zero-mean martingale with respect to $\mathcal{F}_t$. By Doob's moment inequality, for any $T > 0$ and $M < N$ we have

$$\mathbb{E}\Big(\sup_{|t| \le T} |\tilde{\zeta}_N(t) - \tilde{\zeta}_M(t)|_2^2\Big) \le 4\,\mathbb{E}\,|\tilde{\zeta}_N(T) - \tilde{\zeta}_M(T)|_2^2 = \sum_{j \in \Lambda_N \setminus \Lambda_M} b_j^2.$$

Thus, the series (2.149) converges in $L^2\big(\Omega, C(-T,T;H)\big)$ for any $T > 0$ and defines an H-valued process with continuous trajectories. Furthermore, since the laws of the finite-dimensional approximations for ζ and $\tilde{\zeta}|_{\mathbb{R}_+}$ coincide, this convergence implies that the same is true for the limiting processes. □

Exercise 2.4.18 The fact that $\boldsymbol{\theta}$ is a group is obvious. Now note that the restriction of θ_t to the space spanned by e_j with $j \in \Lambda_N$ is the usual shift on the canonical probability space of a finite-dimensional Brownian motion and therefore is measure preserving. The monotone class technique now implies the required result. □

Exercise 2.5.2 Let us set $\bar{\lambda}_t = t^{-1} \int_0^t P_s(u, \cdot)\, ds$. We claim that the family $\{\bar{\lambda}_t, t \ge 0\}$ is tight, so that the hypotheses of Theorem 2.5.1 are satisfied for the initial measure δ_u. Let us fix $\varepsilon > 0$ and choose $m \ge 1$ so large that $P_t(u, K_m^c) \le \varepsilon/2$ for $t \ge t_m$. On the other hand, the time continuity of trajectories implies that the mapping $t \mapsto P_t(u, \cdot)$ is continuous from $\mathbb{R}_+$ to the space $\mathcal{P}(X)$ endowed with the weak topology. Therefore the image of the interval $[0, t_m]$ is compact in $\mathcal{P}(X)$, whence it follows that there is a compact set $K_0 \subset X$ such that $P_t(u, K_0^c) \le \varepsilon/2$ for $0 \le t \le t_m$. Combining these two inequalities, we see that $\bar{\lambda}_t(K_0 \cup K_m) \ge 1 - \varepsilon$ for all $t \in \mathcal{T}_+$. □

Exercise 2.5.9 Let us establish the uniqueness of a stationary measure. To this end, it suffices to prove that if the right-hand side is sufficiently small, then the stochastic Navier–Stokes system satisfies (1.75). To this end, we repeat the argument used in the proof of inequality (2.54). Namely, if $u_0, v_0 \in H$ are two initial functions and u, v are the corresponding solutions, then their difference $w = u - v$ satisfies the equation (cf. (2.166))

$$\dot{w} + \nu L w + B(w,u) + B(v,w) = 0.$$

[1] In other words, we take the σ-algebra generated by the random variables $\omega_s^{(j)}$ with $s \le t$ and $j \in \Lambda$ and add zero-measure subsets of $\mathcal{F}$.

Taking the scalar product of this equation with $2w$ in H and carrying out some transformations, we derive

$$\partial_t |w|_2^2 + (\alpha_1 \nu - C_1 \nu^{-1} \|u\|_1^2)|w|_2^2 \le 0.$$

Applying Gronwall's inequality, we obtain

$$|w(t)|_2^2 \le |u_0 - v_0|_2^2 \exp\bigl(-\alpha_1 \nu t/2\bigr)\Psi(t, u_0), \quad t \ge 0,$$

where we set

$$\Psi(t, u_0) = \exp\left(-\frac{1}{2}\alpha_1 \nu t + C_1 \nu^{-1} \int_0^t \|u\|_1^2 ds\right).$$

It follows from (2.130) that if $|h|_2 + \mathfrak{B}$ is sufficiently small, then

$$\mathbb{P}\Bigl\{\sup_{t\ge 0} \Psi(t, u_0) \ge \rho\Bigr\} \le \exp(-\delta \ln \rho + \delta |u_0|_2^2),$$

where $\delta > 0$ is a small constant. We thus obtain inequality (1.75) with $\alpha(t) = e^{-\alpha_1 \nu t/2}$ and $\psi_{u_0, v_0} = \sup_{t\ge 0} \Psi(t, u_0)$. Further analysis based on inequalities (1.76) and (2.157) enables one to prove exponential convergence to the unique stationary measure in the dual-Lipschitz distance. □

Exercise 3.5.7 The existence and uniqueness of a solution whose almost every trajectory belongs to $C(\mathbb{R} \setminus \mathcal{T}_\omega; H) \cap L^2_{\mathrm{loc}}(\mathbb{R}_+; V)$ is obvious, since we can treat the equation pathwise (cf. the case of a random kick force considered in Section 2.3). Furthermore, the existence of left- and right-hand limits at t_k follows from the properties of the solution for a deterministic Navier–Stokes system. To prove the Markov property, let us denote by $\mathcal{F}_t$ the filtration generated by the random variables $\zeta(s)$ with $0 \le s \le t$. It is well known that the process $\zeta(t+s) - \zeta(t)$, $s \ge 0$, is independent of $\mathcal{F}_t$; e.g. see problem 3.2 of chapter 1 in [KS91] for the case of a Poisson process. The Markov property can now be established by an argument similar to that used in the proof of Proposition 1.3.21. □

Exercise 3.5.9 Let us denote by $\mathcal{F}_t$ the filtration associated with the problem in question, that is, the σ-algebra generated by the process

$$\int_0^t \eta(s)\, ds = \sum_{k=1}^{\infty} \eta_k(x) H(t - \tau_k),$$

where $H(t) = 1$ for $t \ge 0$ and $H(t) = 0$ for $t < 0$. Since the indicator function $\mathbb{I}_{[0,\tau_1]}(t)$ is $\mathcal{F}_t$-measurable, for any initial measure $\lambda \in \mathcal{P}(H)$ and any $s > 0$ we have

$$\begin{aligned}\mathbb{E}_\lambda \int_0^{\tau_1} f(u_{t+s})\, dt &= \mathbb{E}_\lambda \int_0^{\infty} \mathbb{I}_{[0,\tau_1]}(t) f(u_{t+s})\, dt \\ &= \mathbb{E}_\lambda \int_0^{\infty} \mathbb{I}_{[0,\tau_1]}(t)\, \mathbb{E}_\lambda\{f(u_{t+s}) \,|\, \mathcal{F}_t\}\, dt \\ &= \mathbb{E}_\lambda \int_0^{\infty} \mathbb{I}_{[0,\tau_1]}(t) (\mathfrak{P}_s f)(u_t)\, dt = \mathbb{E}_\lambda \int_0^{\tau_1} (\mathfrak{P}_s f)(u_t)\, dt.\end{aligned}$$

Therefore, setting $c = (\mathbb{E}\tau_1)^{-1}$ and using the definition of μ, for any $f \in C_b(H)$ we obtain

$$\begin{aligned}(\mathfrak{P}_s f, \mu) &= c\, \mathbb{E}_{\tilde{\mu}} \int_0^{\tau_1} (\mathfrak{P}_s f)(u_t)\, dt = c\, \mathbb{E}_{\tilde{\mu}} \int_s^{s+\tau_1} f(u_t)\, dt \\ &= (f, \mu) + c\, \mathbb{E}_{\tilde{\mu}} \int_{\tau_1}^{s+\tau_1} f(u_t)\, dt - c\, \mathbb{E}_{\tilde{\mu}} \int_0^{s} f(u_t)\, dt.\end{aligned}$$

It remains to note that, due to the strong Markov property, the second and third terms on the right-hand side coincide and, hence, $\mathfrak{P}_s^* \mu = \mu$ for any $s > 0$. □

Notation and conventions

$\mathbb{Z}$, $\mathbb{R}$, $\mathbb{C}$ denote the sets of integer, real, and complex numbers, respectively.

$\mathbb{R}^n$ stands for n-dimensional Euclidean space.

$\mathbb{Z}^n$ is the set of integer vectors $(s_1, \ldots, s_n)$ and $\mathbb{Z}_0^n$ is the set of non-zero vectors in $\mathbb{Z}^n$.

$\mathcal{T}$ stands for $\mathbb{Z}$ or $\mathbb{R}$, $\mathcal{T}_+ = \{t \in \mathcal{T} : t \geq 0\}$, and $\mathcal{T}_- = \{t \in \mathcal{T} : t \leq 0\}$.

If $Y \subset X$ is a subset, then Y^c stands for the complement of Y in X.

The infimum over an empty subset of $\mathbb{R}$ is $+\infty$. Given two real numbers a and b, we denote by $a \vee b$ and $a \wedge b$ their maximum and minimum, respectively.

We often deal with random forces decomposed in an orthonormal basis $\{e_j\}$ of the space $L^2(Q; \mathbb{R}^2)$. We denote by b_j the corresponding coefficients and always assume that

$$\mathfrak{B} := \sum_{j=1}^{\infty} b_j^2 < \infty.$$

When $\{e_j\}$ is a basis formed of the eigenfunctions of the Stokes operator, we set

$$\mathfrak{B}_k = \sum_{j=1} \alpha_j^k b_j^2,$$

where α_j stands for the eigenvalue associated with e_j. Note that $\mathfrak{B}_0 = \mathfrak{B}$. In the case of the standard torus $\mathbb{T}^2$, the eigenfunctions (as well as the eigenvalues and coefficients) are indexed by $s \in \mathbb{Z}_0^2$, and we obtain

$$\mathfrak{B}_k = \sum_{s \in \mathbb{Z}_0^2} |s|^{2k} b_s^2.$$

Abstract spaces and functions

For an arbitrary set Y, we denote by Id_Y the identity operator on Y (which takes an element $y \in Y$ to itself). If $\Gamma \subset Y$, then we write $\mathbb{I}_\Gamma$ for the indicator function of Γ (equal to 1 on Γ and to zero otherwise).

If Ω is an arbitrary set and $\mathcal{F}$ is a σ-algebra on Ω, then $(\Omega, \mathcal{F})$ is called a *measurable space*.

Let $(\Omega_i, \mathcal{F}_i)$, $i = 1, 2$, be measurable spaces and let $f : \Omega_1 \to \Omega_2$ be a mapping. We say that f is *measurable* if $f^{-1}(\Gamma) \in \mathcal{F}_1$ for any $\Gamma \in \mathcal{F}_2$. For any probability μ on $(\Omega_1, \mathcal{F}_1)$, we denote by $f_*(\mu)$ or $f_*\mu$ the image of μ under f, that is, a measure on $(\Omega_2, \mathcal{F}_2)$ defined by the relation $f_*(\mu)(\Gamma) = \mu(f^{-1}(\Gamma))$ for $\Gamma \in \mathcal{F}_2$.

X denotes either a Polish space (that is, a separable complete metric space) or a separable Banach space. We write dist_X and $\|\cdot\|_X$ for the corresponding metric and norm. All Polish spaces are endowed with the Borel σ-algebra $\mathcal{B} = \mathcal{B}(X)$ and are considered as measurable spaces. We denote by $L^\infty(X)$ the space of real-valued bounded measurable functions on X and by $\mathcal{P}(X)$ the set of probability measures on $(X, \mathcal{B}(X))$.

$C_b(X)$ denotes the space of bounded continuous functions $f : X \to \mathbb{R}$ endowed with the norm

$$\|f\|_\infty = \sup_{u \in X} |f(u)|.$$

$L_b(X)$ stands for the space of functions $f \in C_b(X)$ such that

$$\mathrm{Lip}(f) := \sup_{u_1, u_2 \in X} \frac{|f(u_1) - f(u_2)|}{\mathrm{dist}_X(u_1, u_2)} < \infty.$$

This is a Banach space with respect to the norm $\|f\|_L = \|f\|_\infty + \mathrm{Lip}(f)$.
$C_b^\gamma(X)$ denotes the space of functions $f \in C_b(X)$ such that

$$|f|_\gamma := \|f\|_\infty + \sup_{0 < \mathrm{dist}_X(u,v) \le 1} \frac{|f(u) - f(v)|}{\mathrm{dist}_X(u, v)^\gamma} < \infty.$$

The *canonical space* on an interval $J \subset \mathbb{R}$ is defined as the space $\Omega = \Omega_X$ of continuous functions $\omega : J \to X$ endowed with the metric of uniform convergence on bounded subintervals. We consider only two cases: $J = \mathbb{R}$ or $J = \mathbb{R}_+$. The *shift operator* $\theta_s : \Omega \to \Omega$ is given by $(\theta_s\omega)(t) = \omega(s + t)$, $t \in J$, where $s \in \mathbb{R}$ in the fist case and $s \ge 0$ in the second. The *canonical process* on Ω is defined by $x_t(\omega) = \omega(t)$ for $t \in J$.

Given a random process $\{\xi_t, t \ge 0\}$ valued in a Banach space, we denote by $\tau(B) \le \infty$ its *first hitting time* of a closed set B; that is, $\tau(B) = \inf\{t \ge 0 : \xi_t \in B\}$.

H stands for a separable Hilbert space with a scalar product $(\cdot, \cdot)_H$.

$(\Omega, \mathcal{F}, \mathbb{P})$ is a probability space. We often assume that it is complete, that is, if $A \in \mathcal{F}$ and $\mathbb{P}(A) = 0$, then any subset of A belongs to $\mathcal{F}$. If, in addition, $\{\mathcal{F}_t, t \ge 0\}$ is a filtration on $(\Omega, \mathcal{F}, \mathbb{P})$, then unless otherwise stated, we assume that $\mathcal{F}_t$ is right-continuous, and $\mathcal{F}_0$ contains all negligible sets of $\mathcal{F}$.

$B_X(u, r)$ denotes the closed ball in X of radius r centred at u. If X is a Banach space and $u = 0$, then we write $B_X(r)$.

If X is a Polish space, then $\boldsymbol{X}$ stands for the direct product $X \times X$ endowed with the natural metric, and we write $\boldsymbol{u} = (u, u')$ to denote elements of $\boldsymbol{X}$. Similarly, if $B \subset X$ is a subset, then $\boldsymbol{B}$ stands for $B \times B$. Finally, if $(\Omega, \mathcal{F})$ is a measurable space, then we denote by $(\boldsymbol{\Omega}, \boldsymbol{\mathcal{F}})$ the direct product of its two copies, that is, $\boldsymbol{\Omega} = \Omega \times \Omega$ and $\boldsymbol{\mathcal{F}} = \mathcal{F} \otimes \mathcal{F}$.

Functional spaces

Let Q be a domain in $\mathbb{R}^d$ or a manifold, let $J \subset \mathbb{R}$ be a closed interval, let $1 \le p \le \infty$, and let X be a Banach space. We use the following functional spaces.

$C_b^\infty(Q;\mathbb{R}^n)$ is the Fréchet space of infinitely smooth functions $f : Q \to \mathbb{R}^n$ that are bounded together with all their derivatives.

$L^p(Q;\mathbb{R}^n)$ is the space of measurable functions $f : Q \to \mathbb{R}^n$ such that

$$|f|_p := \left(\int_Q |f(x)|^p\right)^{1/p} < \infty.$$

In the case $p = \infty$, the above norm should be replaced by

$$|f|_\infty := \operatorname*{ess\,sup}_{x\in Q} |f(x)|.$$

$H^s(Q;\mathbb{R}^n)$ is the Sobolev space of order $s \in \mathbb{R}$ endowed with its standard norm $\|\cdot\|_s$.

$L^2_\sigma(\mathbb{T}^2;\mathbb{R}^2)$ denotes the space of functions $u \in L^2(\mathbb{T}^2;\mathbb{R}^2)$ such that $\operatorname{div} u = 0$ in $\mathbb{T}^2$ and H is the space of those $u \in L^2_\sigma(\mathbb{T}^2;\mathbb{R}^2)$ for which $\langle u\rangle := \int_{\mathbb{T}^2} u\,dx = 0$.

If Q is a bounded domain with smooth boundary, then $H = L^2_\sigma(Q;\mathbb{R}^2)$ stands for the space of functions $u \in L^2(Q;\mathbb{R}^2)$ such that $\operatorname{div} u = 0$ in Q and $\langle u, \boldsymbol{n}\rangle = 0$ on ∂Q, where $\boldsymbol{n}$ is the outward unit normal to ∂Q.

In the case of a torus, we set $V^k = H^k(\mathbb{T}^2;\mathbb{R}^2) \cap H$ and write $V = V^1$. In the case of a bounded domain, we define $V = H^1_0(Q;\mathbb{R}^2) \cap L^2_\sigma(Q;\mathbb{R}^2)$. In both cases, we denote by V^* the dual space of V with respect to the scalar product in L^2.

$L^p(J;X)$ denotes the space of Borel-measurable functions $u : J \to X$ for which $\|u\|_{L^p(J;X)} < \infty$; see page 4 for the definition of this norm.

$C(J;X)$ is the space of continuous functions on J endowed with the supremum norm $\|u\|_{C(J;X)}$ defined on page 4.

Measures and applications

A mapping $F : X \to Y$ between two Banach spaces is said to be *locally Lipschitz* if for any $R > 0$ there is a $C_R > 0$ such that

$$\|F(u_1) - F(u_2)\|_Y \le C_R\|u_1 - u_2\|_X \quad \text{for any } u_1, u_2 \in B_X(R).$$

$\mathcal{L}(X, Y)$ denotes the space of continuous linear applications from X to Y. In the case $X = Y$, we shall write $\mathcal{L}(X)$.

Given a measure μ on a Banach space X, we denote by $\mathfrak{m}_k(\mu)$ its kth moment:

$$\mathfrak{m}_k(\mu) = \int_X \|u\|_X^k \mu(du).$$

For any measurable space $(\Omega, \mathcal{F})$ and a point $a \in \Omega$, we denote by δ_a the *Dirac measure* concentrated at a, that is, $\delta_a(\Gamma) = 1$ if $\Gamma \in \mathcal{F}$ contains a and $\delta_a(\Gamma) = 0$ otherwise.

References

[Ada75] R. A. Adams, *Sobolev Spaces*, Academic Press, New York, 1975. Cited on p. 35.

[AKSS07] A. Agrachev, S. Kuksin, A. Sarychev, and A. Shirikyan, On finite-dimensional projections of distributions for solutions of randomly forced 2D Navier–Stokes equations, *Ann. Inst. H. Poincaré Probab. Statist.* **43** (2007), no. 4, 399–415. Cited on pp. 149, 172, 266.

[Ale43] A. D. Alexandroff, Additive set-functions in abstract spaces, *Rec. Math. [Mat. Sbornik] N.S.* **13(55)** (1943), 169–238. Cited on p. 274.

[App04] D. Applebaum, *Lévy Processes and Stochastic Calculus*, Cambridge University Press, Cambridge, 2004. Cited on p. 59.

[Arn98] L. Arnold, *Random Dynamical Systems*, Springer-Verlag, Berlin, 1998. Cited on pp. 35, 193, 210.

[AS05] A. A. Agrachev and A. V. Sarychev, Navier–Stokes equations: controllability by means of low modes forcing, *J. Math. Fluid Mech.* **7** (2005), no. 1, 108–152. Cited on pp. 149, 265.

[AS06] ______, Controllability of 2D Euler and Navier–Stokes equations by degenerate forcing, *Comm. Math. Phys.* **265** (2006), no. 3, 673–697. Cited on pp. 149, 265, 266.

[Bat69] G. K. Batchelor, Computation of the energy spectrum in homogeneous two-dimensional turbulence, *Phys. Fluids Suppl.* **11** (1969), 233–239. Cited on p. 243.

[Bat82] ______, *The Theory of Homogeneous Turbulence*, Cambridge University Press, Cambridge, 1982. Cited on pp. xi, 168, 209, 243.

[Bax89] P. Baxendale, Lyapunov exponents and relative entropy for a stochastic flow of diffeomorphisms, *Probab. Theory Related Fields* **81** (1989), no. 4, 521–554. Cited on p. 187.

[BBM00] J. Bourgain, H. Brezis, and P. Mironescu, Lifting in Sobolev spaces, *J. Anal. Math.* **80** (2000), 37–86. Cited on p. 292.

[BD95] F. Bethuel and F. Demengel, Extensions for Sobolev mappings between manifolds, *Calc. Var. Partial Diff. Equ.* **3** (1995), no. 4, 475–491. Cited on p. 292.

[BD07] V. Barbu and G. Da Prato, Existence and ergodicity for the two-dimensional stochastic magneto-hydrodynamics equations, *Appl. Math. Optim.* **56** (2007), no. 2, 145–168. Cited on p. 170.

[Ber00] D. Bernard, Influence of friction on the direct cascade of the 2d forced turbulence, *Europhys. Lett.* **50** (2000), 333–339. Cited on p. 244.

[BF09] Z. Brzeźniak and B. Ferrario, 2D Navier–Stokes equation in Besov spaces of negative order, *Nonlinear Anal.* **70** (2009), no. 11, 3902–3916. Cited on p. 100.

[Bil99] P. Billingsley, *Convergence of Probability Measures*, John Wiley & Sons, New York, 1999. Cited on pp. 59, 61, 300.

[BIN79] O. V. Besov, V. P. Il′in, and S. M. Nikol′skiĭ, *Integral Representations of Functions and Imbedding Theorems*. Vol. I, II, V. H. Winston & Sons, Washington, DC, 1979. Cited on p. 4.

[Bis81] J.-M. Bismut, Martingales, the Malliavin calculus and hypoellipticity under general Hörmander's conditions, *Z. Wahrsch. Verw. Gebiete* **56** (1981), no. 4, 469–505. Cited on p. 146.

[BK07] J. Bec and K. Khanin, Burgers turbulence, *Phys. Rep.* **447** (2007), no. 1–2, 1–66. Cited on pp. x, xiii, 110.

[BKL00] J. Bricmont, A. Kupiainen, and R. Lefevere, Probabilistic estimates for the two-dimensional stochastic Navier–Stokes equations, *J. Statist. Phys.* **100** (2000), no. 3–4, 743–756. Cited on pp. 92, 100.

[BKL01] ______, Ergodicity of the 2D Navier–Stokes equations with random forcing, *Comm. Math. Phys.* **224** (2001), no. 1, 65–81. Cited on pp. 146, 170.

[BKL02] ______, Exponential mixing of the 2D stochastic Navier–Stokes dynamics, *Comm. Math. Phys.* **230** (2002), no. 1, 87–132. Cited on pp. 147, 171.

[BMS11] W. Bołt, A. A. Majewski, and T. Szarek, *An invariance principle for the law of the iterated logarithm for some Markov chains*, Preprint (2011). Cited on p. 210.

[Bog98] V. I. Bogachev, *Gaussian Measures*, Mathematical Surveys and Monographs, vol. 62, American Mathematical Society, Providence, RI, 1998. Cited on p. 272.

[Bog07] ______, *Measure Theory*. Vol. I, II, Springer-Verlag, Berlin, 2007. Cited on p. 35.

[Bol82] E. Bolthausen, The Berry–Esseén theorem for strongly mixing Harris recurrent Markov chains, *Z. Wahrsch. Verw. Gebiete* **60** (1982), no. 3, 283–289. Cited on p. 181.

[Bor12] A. Boritchev, *Sharp estimates for turbulence in white-forced generalised Burgers equation*, Preprint (2012), arXiv:1201.5567. Cited on pp. x, xiii, 110.

[Bow75] R. Bowen, *Equilibrium States and the Ergodic Theory of Anosov Diffeomorphisms*, Lecture Notes in Mathematics, Vol. 470, Springer-Verlag, Berlin, 1975. Cited on p. 171.

[BPR06] P. Berti, L. Pratelli, and P. Rigo, Almost sure weak convergence of random probability measures, *Stochastics* **78** (2006), no. 2, 91–97. Cited on p. 274.

[Bri02] J. Bricmont, Ergodicity and mixing for stochastic partial differential equations, *Proceedings of the International Congress of Mathematicians, Vol.*

I (Beijing, 2002) (Beijing), Higher Education Press, 2002, pp. 567–585. Cited on p. 172.

[Bro71] B. M. Brown, Martingale central limit theorems, *Ann. Math. Statist.* **42** (1971), 59–66. Cited on p. 289.

[BT73] A. Bensoussan and R. Temam, Équations stochastiques du type Navier–Stokes, *J. Funct. Anal.* **13** (1973), 195–222. Cited on pp. 99, 257.

[BV92] A. V. Babin and M. I. Vishik, *Attractors of Evolution Equations*, North-Holland Publishing, Amsterdam, 1992. Cited on pp. 99, 232.

[BV11] F. Bouchet and A. Venaille, Statistical mechanics of two-dimensional and geophysical flows, *Phys. Rep.* (2011), to appear. Cited on p. 244.

[CC92] M. Capiński and N. Cutland, A simple proof of existence of weak and statistical solutions of Navier–Stokes equations, *Proc. Roy. Soc. London Ser. A* **436** (1992), no. 1896, 1–11. Cited on p. 257.

[CDF97] H. Crauel, A. Debussche, and F. Flandoli, Random attractors, *J. Dynam. Diff. Equ.* **9** (1997), no. 2, 307–341. Cited on p. 210.

[CF88] P. Constantin and C. Foiaş, *Navier–Stokes Equations*, University of Chicago Press, Chicago, IL, 1988. Cited on pp. 99, 145.

[CF94] H. Crauel and F. Flandoli, Attractors for random dynamical systems, *Probab. Theory Related Fields* **100** (1994), no. 3, 365–393. Cited on p. 210.

[CFNT89] P. Constantin, C. Foiaş, B. Nicolaenko, and R. Temam, *Integral Manifolds and Inertial Manifolds for Dissipative Partial Differential Equations*, Springer-Verlag, New York, 1989. Cited on p. 99.

[CG94] M. Capiński and D. Gątarek, Stochastic equations in Hilbert space with application to Navier–Stokes equations in any dimension, *J. Funct. Anal.* **126** (1994), no. 1, 26–35. Cited on p. 267.

[Che98] J.-Y. Chemin, *Perfect Incompressible Fluids*, Oxford Lecture Series in Mathematics and its Applications, Vol. 14, The Clarendon Press, Oxford University Press, New York, 1998. Cited on p. 224.

[CK08a] I. Chueshov and S. Kuksin, Random kick-forced 3D Navier–Stokes equations in a thin domain, *Arch. Rational Mech. Anal.* **188** (2008), 117–153. Cited on pp. 210, 247, 249, 250.

[CK08b] ———, Stochastic 3D Navier-Stokes equations in a thin domain and its α-approximation, *Phys. D* **237** (2008), no. 10–12, 1352–1367. Cited on p. 251.

[Cra91] H. Crauel, Markov measures for random dynamical systems, *Stoch. Stoch. Rep.* **37** (1991), no. 3, 153–173. Cited on p. 210.

[Cra01] ———, Random point attractors versus random set attractors, *J. London Math. Soc.* (2) **63** (2001), no. 2, 413–427. Cited on p. 187.

[Cra02] ———, *Random Probability Measures on Polish Spaces*, Taylor & Francis, London, 2002. Cited on p. 271.

[CV77] C. Castaing and M. Valadier, *Convex Analysis and Measurable Multifunctions*, Springer, Berlin, 1977. Cited on p. 271.

[CV02] V. V. Chepyzhov and M. I. Vishik, *Attractors for Equations of Mathematical Physics*, AMS Colloquium Publications, Vol. 49, AMS, Providence, RI, 2002. Cited on p. 210.

[DD02] G. Da Prato and A. Debussche, Two-dimensional Navier–Stokes equations driven by a space-time white noise, *J. Funct. Anal.* **196** (2002), no. 1, 180–210. Cited on p. 100.

[DD03] ———, Ergodicity for the 3D stochastic Navier–Stokes equations, *J. Math. Pures Appl.* **82** (2003), 877–947. Cited on p. 254.

[DDL+07] J. Dedecker, P. Doukhan, G. Lang, J. R. León, S. Louhichi, and C. Prieur, *Weak Dependence: With Examples and Applications*, Springer, New York, 2007. Cited on p. 210.

[DDT05] G. Da Prato, A. Debussche, and L. Tubaro, Coupling for some partial differential equations driven by white noise, *Stoch. Process. Appl.* **115** (2005), no. 8, 1384–1407. Cited on p. 170.

[Deb97] A. Debussche, On the finite dimensionality of random attractors, *Stoch. Anal. Appl.* **15** (1997), no. 4, 473–491. Cited on pp. 198, 210.

[Deb98] ———, Hausdorff dimension of a random invariant set, *J. Math. Pures Appl.* (9) **77** (1998), no. 10, 967–988. Cited on pp. 198, 210.

[Den04] D. Denisov, Personal communication (2004). Cited on p. 210.

[DG95] C. R. Doering and J. D. Gibbon, *Applied Analysis of the Navier–Stokes Equations*, Cambridge Texts in Applied Mathematics, Cambridge University Press, Cambridge, 1995. Cited on p. 51.

[DM78] C. Dellacherie and P.-A. Meyer, *Probabilities and Potential*, North-Holland, Amsterdam, 1978. Cited on p. 227.

[DM10] J. Dedecker and F. Merlevède, On the almost sure invariance principle for stationary sequences of Hilbert-valued random variables. In: I. Berkes, R. C. Bradley, H. Dehling, M. Peligrad and R. Tichy (eds), *Dependence in Probability, Analysis and Number Theory*, Kendrick Press, Heber City, UT, 2010, pp. 157–175. Cited on p. 210.

[DO05] A. Debussche and C. Odasso, Ergodicity for a weakly damped stochastic non-linear Schrödinger equation, *J. Evol. Equ.* **5** (2005), no. 3, 317–356. Cited on p. 172.

[DO06] ———, Markov solutions for the 3D stochastic Navier–Stokes equations with state dependent noise, *J. Evol. Equ.* **6** (2006), no. 2, 305–324. Cited on pp. 254, 259, 260.

[Dob68] R. L. Dobrušin, Description of a random field by means of conditional probabilities and conditions for its regularity, *Teor. Verojatnost. i Primenen* **13** (1968), 201–229. Cited on p. 35.

[Dob74] ———, Conditions for the absence of phase transitions in one-dimensional classical systems, *Math. USSR-Sb.* **22** (1974), no. 1, 28–48. Cited on pp. 18, 35.

[Doe38] W. Doeblin, Exposé de la théorie des chaînes simples constantes de Markov à un nombre fini d'états, *Rev. Math. Union Interbalkan* **2** (1938), 77–105. Cited on pp. 35, 101.

[Doe40] ———, Éléments d'une théorie générale des chaînes simples constantes de Markoff, *Ann. Sci. École Norm. Sup.* (3) **57** (1940), 61–111. Cited on pp. 35, 101.

[Doo48] J. L. Doob, Asymptotic properties of Markoff transition probabilities, *Trans. Amer. Math. Soc.* **63** (1948), 393–421. Cited on pp. 146, 170.

[DRRW06] G. Da Prato, M. Röckner, B. L. Rozovskii, and F.-Y. Wang, Strong solutions of stochastic generalized porous media equations: existence, uniqueness, and ergodicity, *Comm. Partial Diff. Equ.* **31** (2006), no. 1–3, 277–291. Cited on p. 170.

[Dud02] R. M. Dudley, *Real Analysis and Probability*, Cambridge University Press, Cambridge, 2002. Cited on pp. 6, 7, 13, 14, 15, 22, 30, 32, 35, 59, 84, 162, 223, 270, 273, 274, 287.

[DZ92] G. Da Prato and J. Zabczyk, *Stochastic Equations in Infinite Dimensions*, Cambridge University Press, Cambridge, 1992. Cited on pp. 70, 75, 278, 287.

[DZ96] ———, *Ergodicity for Infinite Dimensional Systems*, Cambridge University Press, Cambridge, 1996. Cited on pp. 110, 146, 286.

[EH01] J.-P. Eckmann and M. Hairer, Uniqueness of the invariant measure for a stochastic PDE driven by degenerate noise, *Comm. Math. Phys.* **219** (2001), no. 3, 523–565. Cited on p. 170.

[Elw92] K. D. Elworthy, Stochastic flows on Riemannian manifolds. In: M. Pinsky and V. Wihstutz (eds), *Diffusion Processes and Related Problems in Analysis, Vol. II* (Charlotte, NC, 1990), Birkhäuser, Boston, MA, 1992, pp. 37–72. Cited on p. 146.

[EMS01] W. E, J. C. Mattingly, and Ya. Sinai, Gibbsian dynamics and ergodicity for the stochastically forced Navier–Stokes equation, *Comm. Math. Phys.* **224** (2001), no. 1, 83–106. Cited on pp. 100, 171, 172.

[ES00] W. E and Ya. G. Sinaĭ, New results in mathematical and statistical hydrodynamics, *Russian Math. Surveys* **55** (2000), no. 4(334), 635–666. Cited on p. 172.

[Fel71] W. Feller, *An Introduction to Probability Theory and Its Applications*, Vol. II, John Wiley & Sons, New York, 1971. Cited on pp. 104, 110, 288.

[Fer97] B. Ferrario, Ergodic results for stochastic Navier-Stokes equation, *Stoch. Stoch. Rep.* **60** (1997), no. 3–4, 271–288. Cited on p. 170.

[Fer99] ———, Stochastic Navier–Stokes equations: analysis of the noise to have a unique invariant measure, *Ann. Mat. Pura Appl.* (4) **177** (1999), 331–347. Cited on pp. 146, 170.

[Fer03] ———, Uniqueness result for the 2D Navier–Stokes equation with additive noise, *Stoch. Stoch. Rep.* **75** (2003), no. 6, 435–442. Cited on p. 100.

[FG95] F. Flandoli and D. Gątarek, Martingale and stationary solutions for stochastic Navier–Stokes equations, *Probab. Theory Related Fields* **102** (1995), no. 3, 367–391. Cited on pp. 257, 267.

[Fla94] F. Flandoli, Dissipativity and invariant measures for stochastic Navier–Stokes equations, *NoDEA Nonlinear Diff. Equ. Appl.* **1** (1994), no. 4, 403–423. Cited on p. 100.

[Fla97] ———, Irreducibility of the 3-D stochastic Navier–Stokes equation, *J. Funct. Anal.* **149** (1997), no. 1, 160–177. Cited on p. 268.

[FM95] F. Flandoli and B. Maslowski, Ergodicity of the 2D Navier–Stokes equation under random perturbations, *Comm. Math. Phys.* **172** (1995), no. 1, 119–141. Cited on pp. 145, 146, 169, 170.

[FMRT01] C. Foias, O. Manley, R. Rosa, and R. Temam, *Navier–Stokes Equations and Turbulence*, Encyclopedia of Mathematics and its Applications, Vol. 83, Cambridge University Press, Cambridge, 2001. Cited on p. ix.

[FP67] C. Foiaş and G. Prodi, Sur le comportement global des solutions non-stationnaires des équations de Navier–Stokes en dimension 2, *Rend. Sem. Mat. Univ. Padova* **39** (1967), 1–34. Cited on p. 99.

[FR08] F. Flandoli and M. Romito, Markov selections for the 3D stochastic Navier–Stokes equations, *Probab. Theory Related Fields* **172** (2008), 407–458. Cited on pp. 257, 261, 262, 263.

[Fri95] U. Frisch, *Turbulence. The Legacy of A. N. Kolmogorov*, Cambridge University Press, Cambridge, 1995. Cited on pp. xi, 209.

[FT89] C. Foias and R. Temam, Gevrey class regularity for the solutions of the Navier–Stokes equations, *J. Funct. Anal.* **87** (1989), no. 2, 359–369. Cited on p. 51.

[Gal02] G. Gallavotti, *Foundations of Fluid Dynamics*, Springer-Verlag, Berlin, 2002. Cited on p. xi.

[GK03] I. Gyöngy and N. Krylov, On the splitting-up method and stochastic partial differential equations, *Ann. Probab.* **31** (2003), no. 2, 564–591. Cited on p. 100.

[GM05] B. Goldys and B. Maslowski, Exponential ergodicity for stochastic Burgers and 2D Navier–Stokes equations, *J. Funct. Anal.* **226** (2005), no. 1, 230–255. Cited on pp. 146, 170.

[Gor69] M. I. Gordin, The central limit theorem for stationary processes, *Dokl. Akad. Nauk SSSR* **188** (1969), 739–741. Cited on pp. 210, 290.

[GS80] Ĭ. Ī. Gīhman and A. V. Skorohod, *The Theory of Stochastic Processes*, Vol. I, Springer-Verlag, Berlin, 1980. Cited on p. 274.

[Hai02a] M. Hairer, Exponential mixing for a stochastic partial differential equation driven by degenerate noise, *Nonlinearity* **15** (2002), no. 2, 271–279. Cited on p. 170.

[Hai02b] ———, Exponential mixing properties of stochastic PDEs through asymptotic coupling, *Probab. Theory Related Fields* **124** (2002), no. 3, 345–380. Cited on pp. 154, 171.

[Hai05] ———, Coupling stochastic PDEs, *XIVth International Congress on Mathematical Physics*, World Scientific Publishing, Hackensack, NJ, 2005, pp. 281–289. Cited on p. 172.

[Har55] T. E. Harris, On chains of infinite order, *Pacific J. Math.* **5** (1955), 707–724. Cited on p. 35.

[Har64] P. Hartman, *Ordinary Differential Equations*, John Wiley & Sons, New York, 1964. Cited on p. 226.

[Has80] R. Z. Has′minskiĭ, *Stochastic Stability of Differential Equations*, Sijthoff & Noordhoff, Alphen aan den Rijn, 1980. Cited on p. 210.

[Hen81] D. Henry, *Geometric Theory of Semilinear Parabolic Equations*, Springer-Verlag, Berlin, 1981. Cited on p. 70.

[Hey73] C. C. Heyde, An iterated logarithm result for martingales and its application in estimation theory for autoregressive processes, *J. Appl. Probab.* **10** (1973), 146–157. Cited on p. 288.

[HH80] P. Hall and C. C. Heyde, *Martingale Limit Theory and Its Application*, Academic Press, New York, 1980. Cited on p. 181.

[HKL90] R. Hardt, D. Kinderlehrer, and F. H. Lin, The variety of configurations of static liquid crystals. In: *Variational Methods (Paris, 1988)*, Progr. Nonlinear Differential Equations Appl., Vol. 4, Birkhäuser Boston, Boston, MA, 1990, pp. 115–131. Cited on p. 292.

[HM06] M. Hairer and J. C. Mattingly, Ergodicity of the 2D Navier–Stokes equations with degenerate stochastic forcing, *Ann. Math.* (2) **164** (2006), no. 3, 993–1032. Cited on pp. 149, 172.

[HM08] ———, Spectral gaps in Wasserstein distances and the 2D stochastic Navier–Stokes equations, *Ann. Probab.* **36** (2008), no. 6, 2050–2091. Cited on pp. 149, 172, 210.

[HM11] ———, A theory of hypoellipticity and unique ergodicity for semilinear stochastic PDEs, *Electron. J. Probab.* **16** (2011), no. 23, 658–738. Cited on pp. 149, 172.

[JS87] J. Jacod and A. N. Shiryaev, *Limit Theorems for Stochastic Processes*, Springer-Verlag, Berlin, 1987. Cited on p. 210.

[Jud63] V. I. Judovič, Non-stationary flows of an ideal incompressible fluid, *Ž. Vyčisl. Mat. i Mat. Fiz.* **3** (1963), 1032–1066. Cited on p. 224.

[KA82] L. V. Kantorovich and G. P. Akilov, *Functional Analysis*, Pergamon Press, Oxford, 1982. Cited on pp. 16, 22.

[KB37] N. Kryloff and N. Bogoliouboff, La théorie générale de la mesure dans son application à l'étude des systèmes dynamiques de la mécanique non linéaire, *Ann. Math.* (2) **38** (1937), no. 1, 65–113. Cited on pp. 87, 100.

[KF75] A. N. Kolmogorov and S. V. Fomin, *Introductory Real Analysis*, Dover Publications, New York, 1975. Cited on p. 273.

[KH95] A. Katok and B. Hasselblatt, *Introduction to the Modern Theory of Dynamical Systems*, Cambridge University Press, Cambridge, 1995. Cited on p. 100.

[Kif86] Y. Kifer, *Ergodic Theory of Random Transformations*, Birkhäuser, Boston, MA, 1986. Cited on p. 35.

[KP05] S. Kuksin and O. Penrose, A family of balance relations for the two-dimensional Navier–Stokes equations with random forcing, *J. Statist. Phys.* **118** (2005), no. 3–4, 437–449. Cited on p. 244.

[KP08] S. Kuksin and A. Piatnitski, Khasminskii–Whitham averaging for randomly perturbed KdV equation, *J. Math. Pures Appl.* (9) **89** (2008), no. 4, 400–428. Cited on p. 239.

[KPS02] S. Kuksin, A. Piatnitski, and A. Shirikyan, A coupling approach to randomly forced nonlinear PDEs, II, *Comm. Math. Phys.* **230** (2002), no. 1, 81–85. Cited on p. 171.

[Kra67] R. H. Kraichnan, Inertial ranges in two-dimensional turbulence, *Phys. Fluids* **10** (1967), 1417–1423. Cited on p. 243.

[Kry73] N. V. Krylov, The selection of a Markov process from a Markov system of processes, and the construction of quasidiffusion processes, *Izv. Akad. Nauk SSSR Ser. Mat.* **37** (1973), 691–708. Cited on pp. 254, 256, 260.

[Kry74] ———, Some estimates for the density of the distribution of a stochastic integral, *Izv. Akad. Nauk SSSR Ser. Mat.* **38** (1974), 228–248. Cited on p. 284.

[Kry80] ———, *Controlled Diffusion Processes*, Applications of Mathematics, Vol. 14, Springer-Verlag, New York, 1980. Cited on p. 284.

[Kry86] ———, Estimates of the maximum of the solution of a parabolic equation and estimates of the distribution of a semimartingale, *Mat. Sb. (N.S.)* **130(172)** (1986), no. 2, 207–221, 284. Cited on p. 284.

[Kry87] ———, *Nonlinear Elliptic and Parabolic Equations of the Second Order*, D. Reidel Publishing, Dordrecht, 1987. Cited on p. 284.

[Kry96] ———, *Lectures on Elliptic and Parabolic Equations in Hölder Spaces*, American Mathematical Society, Providence, RI, 1996. Cited on p. 253.

[Kry02] ______, *Introduction to the Theory of Random Processes*, Graduate Studies in Mathematics, Vol. 43, American Mathematical Society, Providence, RI, 2002. Cited on pp. 75, 79, 221, 278, 287.

[KS91] I. Karatzas and S. E. Shreve, *Brownian Motion and Stochastic Calculus*, Springer-Verlag, New York, 1991. Cited on pp. 22, 25, 35, 253, 254, 282, 287, 303.

[KS00] S. Kuksin and A. Shirikyan, Stochastic dissipative PDEs and Gibbs measures, *Comm. Math. Phys.* **213** (2000), no. 2, 291–330. Cited on pp. 157, 158, 170, 171.

[KS01a] ______, A coupling approach to randomly forced nonlinear PDE's. *I, Comm. Math. Phys.* **221** (2001), no. 2, 351–366. Cited on p. 171.

[KS01b] ______, Ergodicity for the randomly forced 2D Navier–Stokes equations, *Math. Phys. Anal. Geom.* **4** (2001), no. 2, 147–195. Cited on pp. 67, 100, 170.

[KS02a] ______, Coupling approach to white-forced nonlinear PDEs, *J. Math. Pures Appl.* (9) **81** (2002), no. 6, 567–602. Cited on pp. 100, 147, 171.

[KS02b] ______, On dissipative systems perturbed by bounded random kick-forces, *Ergodic Theory Dynam. Systems* **22** (2002), no. 5, 1487–1495. Cited on p. 170.

[KS03] ______, Some limiting properties of randomly forced two-dimensional Navier–Stokes equations, *Proc. Roy. Soc. Edinburgh Sect. A* **133** (2003), no. 4, 875–891. Cited on pp. 92, 100, 210.

[KS04a] ______, On random attractors for systems of mixing type, *Funktsional. Anal. i Prilozhen.* **38** (2004), no. 1, 34–46, 95. Cited on p. 210.

[KS04b] ______, Randomly forced CGL equation: stationary measures and the inviscid limit, *J. Phys. A* **37** (2004), no. 12, 3805–3822. Cited on pp. 240, 244.

[Kuk02a] S. Kuksin, Ergodic theorems for 2D statistical hydrodynamics, *Rev. Math. Phys.* **14** (2002), no. 6, 585–600. Cited on pp. 172, 181, 210.

[Kuk02b] ______, On exponential convergence to a stationary measure for nonlinear PDEs perturbed by random kick-forces, and the turbulence limit, *Partial Differential Equations*, American Mathematical Society Translation Series 2, Vol. 206, American Mathematical Society, Providence, RI, 2002, pp. 161–176. Cited on p. 171.

[Kuk04] ______, The Eulerian limit for 2D statistical hydrodynamics, *J. Statist. Phys.* **115** (2004), no. 1–2, 469–492. Cited on p. 244.

[Kuk06a] ______, *Randomly Forced Nonlinear PDEs and Statistical Hydrodynamics in 2 Space Dimensions*, European Mathematical Society (EMS), Zürich, 2006. Cited on p. xvi.

[Kuk06b] ______, Remarks on the balance relations for the two-dimensional Navier–Stokes equation with random forcing, *J. Statist. Phys.* **122** (2006), no. 1, 101–114. Cited on pp. 216, 244.

[Kuk07] ______, Eulerian limit for 2D Navier–Stokes equation and damped/driven KdV equation as its model, *Tr. Mat. Inst. Steklova* **259** (2007), *Anal. i Osob. Ch.* 2, 134–142. Cited on p. 244.

[Kuk08] ______, On distribution of energy and vorticity for solutions of 2D Navier–Stokes equation with small viscosity, *Comm. Math. Phys.* **284** (2008), no. 2, 407–424. Cited on p. 244.

[Kuk10a] ———, Damped-driven KdV and effective equations for long-time behaviour of its solutions, *Geom. Funct. Anal.* **20** (2010), no. 6, 1431–1463. Cited on p. 239.

[Kuk10b] ———, Dissipative perturbations of KdV, In: *Proceedings of the 16th International Congress on Mathematical Physics* (Prague 2009), World Scientific, 2010, pp. 323–327. Cited on p. 244.

[Kuk12] ———, Weakly nonlinear stochastic CGL equations, *Ann. Inst. H. Poincaré Probab. Statist.* (2012), to appear. Cited on pp. 239, 240.

[Kup10] A. Kupiainen, *Ergodicity of two-dimensional turbulence*, Preprint (2010), arXiv:1005.0587. Cited on p. 149.

[Lad59] O. A. Ladyženskaja, Solution "in the large" of the nonstationary boundary value problem for the Navier–Stokes system with two space variables, *Comm. Pure Appl. Math.* **12** (1959), 427–433. Cited on p. 99.

[Lad63] O. A. Ladyzhenskaya, *The Mathematical Theory of Viscous Incompressible Flow*, Gordon and Breach, New York, 1963. Cited on p. 99.

[Le 87] Y. Le Jan, Équilibre statistique pour les produits de difféomorphismes aléatoires indépendants, *Ann. Inst. H. Poincaré Probab. Statist.* **23** (1987), no. 1, 111–120. Cited on p. 210.

[Led86] F. Ledrappier, Positivity of the exponent for stationary sequences of matrices. In: L. Arnold and V. Wihstutz (eds), *Lyapunov Exponents Proceedings* (Bremen, 1984), Springer, Berlin, 1986, pp. 56–73. Cited on p. 210.

[Ler34] J. Leray, Sur le mouvement d'un liquide visqueux emplissant l'espace, *Acta Math.* **63** (1934), no. 1, 193–248. Cited on p. 99.

[Lio69] J.-L. Lions, *Quelques Méthodes de Résolution des Problèmes aux Limites Non Linéaires*, Dunod, Paris, 1969. Cited on pp. 4, 35, 39, 99, 221.

[LM72] J.-L. Lions and E. Magenes, *Non-Homogeneous Boundary Value Problems and Applications*, Vol. I, Springer-Verlag, New York, 1972. Cited on pp. 35, 39, 221.

[LP59] J.-L. Lions and G. Prodi, Un théorème d'existence et unicité dans les équations de Navier–Stokes en dimension 2, *C. R. Acad. Sci. Paris* **248** (1959), 3519–3521. Cited on p. 99.

[LR06] J. A. Langa and J. C. Robinson, Fractal dimension of a random invariant set, *J. Math. Pures Appl.* (9) **85** (2006), no. 2, 269–294. Cited on pp. 198, 210.

[LS06] A. Lasota and T. Szarek, Lower bound technique in the theory of a stochastic differential equation, *J. Diff. Equ.* **231** (2006), no. 2, 513–533. Cited on p. 171.

[LY94] A. Lasota and J. A. Yorke, Lower bound technique for Markov operators and iterated function systems, *Random Comput. Dynam.* **2** (1994), no. 1, 41–77. Cited on p. 171.

[Mat99] J. C. Mattingly, Ergodicity of 2D Navier–Stokes equations with random forcing and large viscosity, *Comm. Math. Phys.* **206** (1999), no. 2, 273–288. Cited on pp. 100, 170.

[Mat02a] ———, The dissipative scale of the stochastic Navier–Stokes equation: regularization and analyticity, *J. Statist. Phys.* **108** (2002), no. 5-6, 1157–1179. Cited on p. 100.

[Mat02b] ———, Exponential convergence for the stochastically forced Navier–Stokes equations and other partially dissipative dynamics, *Comm. Math. Phys.* **230** (2002), no. 3, 421–462. Cited on pp. 20, 100, 147, 171.

[Mat03] ______, On recent progress for the stochastic Navier–Stokes equations, *Journées "Équations aux Dérivées Partielles"*, Univ. Nantes, Nantes, 2003, pp. Exp. No. XI, 52. Cited on p. 172.

[Maz85] V. G. Maz'ja, *Sobolev Spaces*, Springer-Verlag, Berlin, 1985. Cited on p. 35.

[McK69] H. P. McKean, *Stochastic Integrals*, Academic Press, New York, 1969. Cited on pp. 71, 79.

[Mey66] P.-A. Meyer, *Probability and Potentials*, Blaisdell Publishing Co. & Ginn and Co., Waltham, MA, 1966. Cited on p. 287.

[MP06] J. C. Mattingly and É. Pardoux, Malliavin calculus for the stochastic 2D Navier–Stokes equation, *Comm. Pure Appl. Math.* **59** (2006), no. 12, 1742–1790. Cited on pp. 149, 172.

[MT93] S. P. Meyn and R. L. Tweedie, *Markov Chains and Stochastic Stability*, Springer-Verlag, London, 1993. Cited on p. 210.

[MY02] N. Masmoudi and L.-S. Young, Ergodic theory of infinite dimensional systems with applications to dissipative parabolic PDEs, *Comm. Math. Phys.* **227** (2002), no. 3, 461–481. Cited on p. 171.

[Ner08] V. Nersesyan, Polynomial mixing for the complex Ginzburg–Landau equation perturbed by a random force at random times, *J. Evol. Equ.* **8** (2008), no. 1, 1–29. Cited on pp. 153, 172.

[Nov05] D. Novikov, *Hahn decomposition and Radon–Nikodym theorem with a parameter*, arXiv:math/0501215 (2005). Cited on p. 20.

[Oda07] C. Odasso, Exponential mixing for the 3D stochastic Navier–Stokes equations, *Comm. Math. Phys.* **270** (2007), no. 1, 109–139. Cited on pp. 254, 260.

[Oda08] ______, Exponential mixing for stochastic PDEs: the non-additive case, *Probab. Theory Related Fields* **140** (2008), no. 1–2, 41–82. Cited on pp. 20, 147, 154, 172.

[Øks03] B. Øksendal, *Stochastic Differential Equations*, Springer-Verlag, Berlin, 2003. Cited on pp. 75, 282, 285, 286, 301.

[Ons49] L. Onsager, Statistical hydrodynamics, *Nuovo Cimento* (9) **6** (1949), no. 2, 279–287. Cited on p. 242.

[Pit74] J. W. Pitman, Uniform rates of convergence for Markov chain transition probabilities, *Z. Wahrsch. Verw. Gebiete* **29** (1974), 193–227. Cited on p. 35.

[PSS89] N. I. Portenko, A. V. Skorokhod, and V. M. Shurenkov, *Markov Processes*, Current Problems in Mathematics. Fundamental Directions, Vol. 46 (Russian), Akad. Nauk SSSR Vsesoyuz. Inst. Nauchn. i Tekhn. Inform., Moscow, 1989, pp. 5–245. Cited on pp. 104, 110.

[Rev84] D. Revuz, *Markov Chains*, North-Holland, Amsterdam, 1984. Cited on p. 35.

[Rio00] E. Rio, *Théorie Asymptotique des Processus Aléatoires Faiblement Dépendants*, Springer-Verlag, Berlin, 2000. Cited on pp. 58, 210.

[Rom08] M. Romito, Analysis of equilibrium states of Markov solutions to the 3D Navier–Stokes equations driven by additive noise, *J. Statist. Phys.* **131** (2008), no. 3, 415–444. Cited on pp. 257, 262.

[Roz90] B. L. Rozovskii, *Stochastic Evolution Systems. Linear Theory and Applications to Non-Linear Filtering*, Kluwer, Dordrecht, 1990. Cited on p. 278.

[RS93] G. Raugel and G. R. Sell, Navier–Stokes equations on thin 3D domains, I. Global attractors and global regularity of solutions, *J. Amer. Math. Soc.* **6** (1993), no. 3, 503–568. Cited on p. 246.

[Rue68] D. Ruelle, Statistical mechanics of a one-dimensional lattice gas, *Comm. Math. Phys.* **9** (1968), 267–278. Cited on p. 171.

[Rut96] M. A. Rutgers, X-I Wu, and W. I. Goldburg. The onset of 2-D grid generated turbulence in flowing soap films, *Phys. Fluids* **8** (1996), no. 9. Cited on p. xv.

[RY99] D. Revuz and M. Yor, *Continuous Martingales and Brownian Motion*, Springer-Verlag, Berlin, 1999. Cited on pp. 25, 31, 287.

[Sei97] J. Seidler, Ergodic behaviour of stochastic parabolic equations, *Czechoslovak Math. J.* **47(122)** (1997), no. 2, 277–316. Cited on p. 146.

[Shi02] A. Shirikyan, Analyticity of solutions of randomly perturbed two-dimensional Navier–Stokes equations, *Russian Math. Surveys* **57** (2002), no. 4, 785–799. Cited on pp. 82, 92, 100.

[Shi04] ———, Exponential mixing for 2D Navier–Stokes equations perturbed by an unbounded noise, *J. Math. Fluid Mech.* **6** (2004), no. 2, 169–193. Cited on pp. 100, 172.

[Shi05a] ———, Ergodicity for a class of Markov processes and applications to randomly forced PDE's, I, *Russian J. Math. Phys.* **12** (2005), no. 1, 81–96. Cited on p. 172.

[Shi05b] ———, Some mathematical problems of statistical hydrodynamics. In: *XIVth International Congress on Mathematical Physics*, World Scientific Publishing, Hackensack, NJ, 2005, pp. 304–311. Cited on p. 172.

[Shi06a] ———, Approximate controllability of three-dimensional Navier–Stokes equations, *Comm. Math. Phys.* **266** (2006), no. 1, 123–151. Cited on p. 267.

[Shi06b] ———, Ergodicity for a class of Markov processes and applications to randomly forced PDE's, II, *Discrete Contin. Dyn. Syst. Ser. B* **6** (2006), no. 4, 911–926 (electronic). Cited on p. 172.

[Shi06c] ———, Law of large numbers and central limit theorem for randomly forced PDE's, *Probab. Theory Related Fields* **134** (2006), no. 2, 215–247. Cited on pp. 181, 210.

[Shi07a] ———, Contrôlabilité exacte en projections pour les équations de Navier–Stokes tridimensionnelles, *Ann. Inst. H. Poincaré Anal. Non Linéaire* **24** (2007), no. 4, 521–537. Cited on p. 267.

[Shi07b] ———, Qualitative properties of stationary measures for three-dimensional Navier–Stokes equations, *J. Funct. Anal.* **249** (2007), 284–306. Cited on pp. 261, 267.

[Shi08] ———, Exponential mixing for randomly forced partial differential equations: method of coupling, *Instability in Models Connected with Fluid Flows, II*, Int. Math. Ser. (N.Y.), vol. 7, Springer, New York, 2008, pp. 155–188. Cited on pp. 147, 155, 172.

[Shi11a] ———, *Control and mixing for 2D Navier–Stokes equations with space-time localised noise*, Preprint (2011), arXiv:1110.0596. Cited on p. 172.

[Shi11b] ———, Local times for solutions of the complex Ginzburg–Landau equation and the inviscid limit, *J. Math. Anal. Appl.* **384** (2011), 130–137. Cited on pp. 240, 244.

[Sin91] Ya. G. Sinaĭ, Two results concerning asymptotic behavior of solutions of the Burgers equation with force, *J. Statist. Phys.* **64** (1991), no. 1–2, 1–12. Cited on p. 170.

[Soh01] H. Sohr, *The Navier–Stokes Equations*, Birkhäuser, Basel, 2001. Cited on pp. 99, 261.

[Ste70] E. M. Stein, *Singular Integrals and Differentiability Properties of Functions*, Princeton University Press, Princeton, NJ, 1970. Cited on pp. 4, 117, 217.

[Ste94] L. Stettner, Remarks on ergodic conditions for Markov processes on Polish spaces, *Bull. Polish Acad. Sci. Math.* **42** (1994), no. 2, 103–114. Cited on p. 146.

[Str93] D. Stroock, *Probability. An Analytic Viewpoint*, Cambridge University Press, Cambridge, 1993. Cited on pp. 28, 138.

[SV79] D. Stroock and S. R. S. Varadhan, *Multidimensional Diffusion Processes*, Springer, Berlin, 1979. Cited on pp. 254, 255, 256, 257.

[Sza97] T. Szarek, Markov operators acting on Polish spaces, *Ann. Polon. Math.* **67** (1997), no. 3, 247–257. Cited on p. 171.

[Tay97] M. E. Taylor, *Partial Differential Equations*, Vols. I–III, Springer-Verlag, New York, 1996-97. Cited on pp. 4, 35, 38, 51, 225.

[Tem68] R. Temam, Une méthode d'approximation de la solution des équations de Navier–Stokes, *Bull. Soc. Math. France* **96** (1968), 115–152. Cited on p. 99.

[Tem79] ———, *Navier–Stokes Equations*, North-Holland, Amsterdam, 1979. Cited on pp. 38, 43, 47, 99.

[Tem88] ———, *Infinite-Dimensional Dynamical Systems in Mechanics and Physics*, Springer-Verlag, New York, 1988. Cited on p. 99.

[TZ96] R. Temam and M. Ziane, Navier–Stokes equations in three-dimensional thin domains with various boundary conditions, *Adv. Diff. Equ.* **1** (1996), no. 4, 499–546. Cited on p. 246.

[VF88] M. I. Vishik and A. V. Fursikov, *Mathematical Problems in Statistical Hydromechanics*, Kluwer, Dordrecht, 1988. Cited on pp. x, 77, 99, 100, 131, 257, 267.

[Vio75] M. Viot, Équations aux dérivées partielles stochastiques: formulation faible. In: *Séminaire sur les Équations aux Dérivées Partielles* (1974–1975), III, Exp. No. 1, Collège de France, Paris, 1975, p. 16. Cited on p. 99.

[Vio76] ———, *Solutions faibles d'équations aux dérivées partielles stochastiques non linéaires*, Thèses de Doctorat (1976), Paris–VI. Cited on p. 99.

[VKF79] M. I. Vishik, A. I. Komech, and A. V. Fursikov, Some mathematical problems of statistical hydromechanics, *Uspekhi Mat. Nauk* **34** (1979), no. 5(209), 135–210. Cited on p. 99.

[Yos95] K. Yosida, *Functional Analysis*, Springer-Verlag, Berlin, 1995. Cited on pp. 4, 35.

Index